MOLECULAR
MEDICINE

MOLECULAR MEDICINE

An Introductory Text

Third Edition

Ronald J Trent

PhD, BSc(Med), MB BS (Sydney), DPhil (Oxon), FRACP, FRCPA
Professor of Molecular Genetics, Faculty of Medicine, University of Sydney
and
Head, Department of Molecular & Clinical Genetics, Royal Prince Alfred Hospital, NSW 2050,
Australia

ELSEVIER
ACADEMIC
PRESS

Amsterdam • Boston • Heidelberg • London
New York • Oxford • Paris • San Diego
San Francisco • Singapore • Sydney • Tokyo

Acquisition Editor: Tari Paschall
Project Manager: Carl M. Soaers
Associate Editor: Karen Dempsy

Elsevier Academic Press
30 Corporate Drive, Suite 400, Burlington, MA 01803, USA
525 B Street, Suite 1900, San Diego, California 92101-4495, USA
84 Theobald's Road, London WC1X 8RR, UK

This book is printed on acid-free paper. ∞

Library of Congress Cataloging-in-Publication Data
Trent, R. J.
   Molecular medicine / Ronald J. Trent.
      p. ; cm.
   Includes bibliographical references and index.
   ISBN-13: 978-0-12-699057-7 ISBN-10: 0-12-699057-3 (alk. paper)
   1. Medical genetics.   2. Molecular biology.
   [DNLM: 1. Genetics, Medical.   2. Molecular Biology.]   I. Title.
   RB155.T728 2005
   616'.042—dc22

                                            2004028087

British Library Cataloguing in Publication Data
A catalogue record for this book is available from the British Library
ISBN-13: 978-0-12-699057-7
ISBN-10: 0-12-699057-3

For all information on all Elsevier Academic Press publications visit our Web site
at www.books.elsevier.com

Printed in the United States of America
06 07 08 09 10 9 8 7 6 5 4 3 2

CONTENTS

PREFACE

The landmark event since the second edition of *Molecular Medicine—An Introductory Text* has been the completion of the Human Genome Project, which is already living up to the promise that it will provide the framework for important new medical discoveries in the twenty-first century.

It is worthwhile making one observation about the writing of the third edition, and this relates to the enormous amount of information available on any subject because of the Internet. This is mentioned to emphasise the impact that bioinformatics and the Internet will have on health professionals; i.e., information overload coupled with the rapid changes occurring in molecular medicine will be a formidable challenge for continuing professional education.

The format for the third edition is changed from the previous ones. I hope that this reflects a more contemporary view of DNA in medicine—a field with no discipline boundaries, with far-reaching applications in many areas of practice.

I thank Carol Yeung, who has worked with me since the first edition and provides the talented artistic skills, as well as a lot of patience. I thank Sr Regis Dunne RSM for her helpful suggestions about Chapter 10. I am fortunate to have dedicated and professional scientists working with me, and they have all contributed towards this edition.

I dedicate *Molecular Medicine—An Introductory Text* to my sister Lynette for her courage and inspiration during 2003, and the two new arrivals—Charlotte and Timothy—who have, with great enthusiasm, given life a new dimension.

Ronald J Trent
Camperdown, December 2004

1

HISTORY OF MOLECULAR MEDICINE

THE FOUNDATIONS: 1869–1980s

INTRODUCTION

The term **molecular medicine** is used to describe the role that knowledge of DNA is having on medical practice. Because of the broad implications of DNA in medicine, molecular medicine cannot be considered a traditional discipline since DNA crosses natural boundaries such as the species, as well as artificial ones exemplified by the medical disciplines.

Other terms that overlap with molecular medicine include **molecular biology** (the use of DNA-based knowledge in research), **genetic engineering** or **recombinant DNA (rDNA) technology** (the use of DNA-based knowledge for new products in industry or research). The common thread in these names is how an understanding of DNA and the ability to manipulate it *in vivo* and *in vitro* have greatly advanced the options that are available in clinical practice, research and industry. The impact made by molecular medicine is seen from the choices made by the prestigious journal *Science* for molecules or achievements of the year—an annual event since 1989 that identifies what development that year has led to a major contribution in advancing science and providing a benefit to society (Table 1.1).

DNA

In 1869, a Swiss physician named F Miescher isolated an acidic material from cell nuclei which he called nuclein. From this was derived the term nucleic acid. The next discovery came some time later in 1944, when O Avery and colleagues showed that the genetic information in the bacterium *Pneumococcus* was found within its DNA. Six years later, E Chargaff demonstrated that there were equal numbers of the nucleotide bases adenine and thymine as well as the bases guanine and cytosine in DNA. This finding, as well as the X-ray crystallographic studies by R Franklin and M Wilkins, enabled J Watson and F Crick to propose the double-stranded structure of DNA in 1953, an event described as the beginning of molecular biology. Subsequently, it was shown that the complementary strands that made up the DNA helix separated

1

Table 1.1 Scientific breakthroughs involving molecular medicine

Year	Considered by the journal *Science* to be the outstanding achievement during the year
1989	Polymerase chain reaction (PCR)
1993	*TP53* (p53) gene
1994	DNA repair enzyme
1996	AIDS research: New hope in HIV disease (chemokines)
1997	Cloning: The lamb that roared
1999	Stem cells—capturing the promise of youth
2000	Genomics comes of age
2002	Small RNAs make a big splash
2003	SARS: A pandemic prevented

during replication. In 1956, a new enzyme was discovered by A Kornberg. This enzyme, called DNA polymerase, enabled small segments of double-stranded DNA to be synthesised (see Chapter 2).

Other discoveries during the 1960s included the finding of mRNA (messenger RNA), which provided the link between the nucleus and the site of protein synthesis in the cytoplasm, and the identification of autonomously replicating, extrachromosomal DNA elements called plasmids. These elements were shown to carry genes such as those coding for antibiotic resistance in bacteria. Plasmids would later be used extensively by the genetic engineers (or molecular biologists)—terms used to describe individuals who manipulated DNA. A landmark in this decade was the definition of the full genetic code, which showed that each amino acid was encoded in DNA by a nucleotide triplet (see Table 2.1). In 1961, M Lyon proposed that one of the two X chromosomes in female mammals was normally inactivated. The process of X-inactivation enabled males and females to have equivalent DNA content despite differing numbers of X chromosomes. In 1966, V McKusick published *Mendelian Inheritance in Man*, a catalogue of genetic disorders in humans. This became a forerunner to the many databases that would subsequently be created to store and transfer information on DNA from humans and many other species (see Bioinformatics in Chapter 5 and the Human Genome Project later in this chapter).

TECHNOLOGICAL DEVELOPMENTS

The dogma that DNA → RNA → protein moved in only one direction was proven to be incorrect when H Temin and D Baltimore showed, in 1970, that reverse transcriptase, an enzyme found in the RNA retroviruses, allowed RNA to be copied back into DNA; i.e., RNA → DNA. This enzyme would later provide the genetic engineer with a means to produce DNA copies (known as complementary DNA or cDNA) from RNA templates. Reverse transcriptase also explained how some viruses could integrate their own genetic information into the host's genome (see Chapters 2, 6, 8 and Appendix).

Enzymes called restriction endonucleases were isolated from bacteria by H Smith, D Nathans, W Arber and colleagues during the late 1960s and early 1970s. Restriction endonucleases were shown to digest DNA at specific sites determined by the underlying nucleotide base sequences. A method now existed to produce DNA fragments of known sizes (see Appendix). At about this time an enzyme called DNA ligase was described. It allowed DNA fragments to be joined. The first recombinant DNA molecules comprising segments that had been stitched together were produced in 1972. P Berg was later awarded a Nobel Prize for his contribution to the construction of recombinant DNA molecules. This was one of many Nobel Prizes that resulted from work that had or would have a direct impact on the development of molecular medicine (Table 1.2). In the same year, S Cohen and colleagues showed that DNA could be inserted into plasmids, which were then able to be reintroduced back into bacteria. Replication of the bacteria containing the foreign DNA enabled unlimited amounts of a single fragment to be produced; i.e., DNA could be cloned. The first eukaryotic gene to be cloned was the rabbit β globin gene in 1976 by T Maniatis and colleagues (see Appendix for further discussion on DNA cloning).

The development of DNA probes followed from a 1960 observation that the two strands of the DNA double helix could be separated and then re-annealed. Probes comprised small segments of DNA labelled with a radioactive marker such as ^{32}P. DNA probes were able to identify specific regions in DNA through their annealing (the technical term for this is hybridisation) to complementary nucleotide sequences. The specificity of the hybridisation reaction relied on the predictability of base pairing; i.e., the nucleotide base adenine (usually abbreviated to A) would always anneal to the base thymine (T); guanine (G) would anneal to cytosine (C). Thus, because of nucleotide base pairing, a single-stranded DNA probe would hybridise in solution to a predetermined segment of single-stranded DNA (see Chapter 2).

Solution hybridisation gave way in 1975 to hybridisation on solid support membranes when DNA digested with restriction endonucleases could be transferred to these membranes by Southern blotting, a process named after its discoverer, E Southern. The ability of radiolabelled DNA probes to identify specific restriction endonuclease fragments enabled DNA maps to be constructed (see Appendix for details). This was the fore-

Table 1.2 Molecular medicine and some Nobel Prize winners (1958–2004)[a]

Year	Recipients	Subject
1958	G W Beadle, E L Tatum and J Lederberg	Regulation and genes, and genetic recombination in bacteria
1959	S Ochoa, A Kornberg	*In vitro* synthesis of nucleic acids
1962	J D Watson, F Crick, M H F Wilkins	Structure of DNA
1965	F Jacob, A L Woff, J Monod	Genetic control enzyme and virus synthesis
1968	R W Holley, H B Khorana, M W Nirenberg	Interpretation of the genetic code
1975	D Baltimore, H Temin and R Dulbecco	Reverse transcriptase and oncogenic viruses
1978	W Arber, D Nathans, H D Smith	Restriction endonucleases
1980[a]	P Berg and W Gilbert, F Sanger	Creation of first recombinant DNA molecule and DNA sequencing
1989	J M Bishop, H E Varmus	Oncogenes
1989[a]	S Altman, T R Cech	RNA ribozymes
1993	R Roberts, P Sharp	Gene splicing
1993[a]	K Mullis and M Smith	Polymerase chain reaction (PCR) and site directed mutagenesis
1995	E Lewis, C Nüsslein-Volhard, E Wieschaus	Genetic mechanisms in early embryonic development
2001	L H Hartwell, R T Hunt, P M Nurse	Key regulators of the cell cycle
2002	S Brenner, J E Sulston, H R Horvitz	Genetic regulation of organ development and programmed cell death
2004	R Axel, L B Buck	Sense of smell including the genes involved in the olfactory system

[a] Nobel Prize in Chemistry; all others Nobel Prize in Physiology or Medicine.

runner to DNA mutation analysis discussed further in Chapter 2.

In the 1970s, the impressive developments in molecular medicine were matched by growing concern in both the public and scientific communities that genetic engineering was a "perversion of nature" and a potential source of much harm. In 1975, a conference was convened at Asilomar in California USA to discuss these issues. Subsequently, regulatory and funding bodies issued guidelines for the conduct of recombinant DNA work. These guidelines dealt with the types of experiments allowable and the necessity to use both vectors (e.g., plasmids) and hosts (e.g., bacteria such as *Escherichia coli* which carried the vectors) that were safe and could be contained within laboratories certified to undertake recombinant DNA work. Guidelines began to be relaxed during the late 1970s and early 1980s when it became apparent that recombinant DNA technology was safe when carried out responsibly. However, government and private funding bodies insisted that a form of monitoring be maintained which has continued to this day (see Chapter 10).

GENE STRUCTURE AND FUNCTION

The anatomy of the gene became better defined with descriptions in 1975 and 1977 of methods to sequence individual nucleotide bases in DNA. The significance of DNA sequencing was acknowledged with a Nobel Prize to F Sanger and W Gilbert. In 1977, an unexpected observation revealed that eukaryotic genes were discontinuous; i.e., coding regions were split by intervening segments of DNA. To distinguish these two components in the gene, the terms exons and introns were first used by W Gilbert in 1978. Splicing, the mechanism by which genes were able to remove introns to allow the appropriate exons to join, was described by R Roberts and P Sharp, for which they were awarded a Nobel Prize in 1993. In the 1970s and 1980s, three scientists—E Lewis, C Nüsslein-Volhard and E Wieschaus—laid the foundations for what would be exciting revelations about genes and their roles in the development of the fruit-fly *Drosophila melanogaster*. The same genes were also found in animals including humans, illustrating the importance of evolutionary conservation of key genes across the species. Characterisation of these would subsequently show the molecular basis for normal and abnormal development in vertebrates. For their pioneering work, the three were awarded a Nobel Prize in 1995.

Variations in the lengths of DNA segments between individuals (called DNA polymorphisms) were reported in the mid-1970s, although their full potentials were not realised until the early 1980s when D Botstein and colleagues described how it might be possible to use DNA polymorphisms as markers to construct a map of the human genome. Subsequently, DNA polymorphisms would form a component to many studies involving DNA (see Chapters 2, 3, 4, 9). These markers also became useful in comparing different populations or species to identify evolutionary affinities and origins. In 1987, R Cann and colleagues proposed, on the basis of mitochondrial DNA polymorphisms and sequence data, that *homo sapiens* evolved from a common African female ancestor. Although evolutionary data based on DNA markers have evoked a number of controversies, many of which remain unresolved, the contribution of DNA polymorphisms in comparative studies has been very significant.

COMMERCIALISATION

The ability to take DNA *in vitro* and from it produce a protein became an important step in the commercialisation of molecular medicine. The latter half of the 1970s saw the development of the biotechnology industry based on this type of DNA manipulation. The first genetic engineering company called Genentech Inc. was formed in 1976 in California. Human insulin became the first genetically engineered drug to be marketed in 1982 following the successful *in vitro* synthesis of recombinant human growth hormone a few years earlier (see Chapter 6).

During the 1980s, transgenic mice were produced by microinjecting foreign DNA into the pronucleus of fertilised oocytes. Injected DNA became integrated into the mouse's own genome, and the transgenic animal expressed both the endogenous mouse genes and the foreign gene. A "supermouse" was made when a rat growth hormone gene was microinjected into a mouse pronucleus. In 1988, the first US patent was issued for a genetically altered animal. In the years to follow, particularly the 1990s, the debate about patents and genes would provoke considerable controversy. The first constraints on DNA clinical testing because of patenting issues were reported in the early 2000s (discussed further in Chapter 10).

MODERN ERA: 1980s–2000s

GENE DISCOVERY

Until the mid-1980s, conventional approaches to understanding genetic disease relied entirely on the identification and then characterisation of an abnormal protein. This could be taken one step further with molecular medicine, since it became possible to use the protein to clone the relevant gene. From the cloned gene more information could then be obtained about the underlying genetic disorder. This was called **functional cloning** and is illustrated by reference to the thalassaemia syndromes (Box 1.1). However, the identification of an abnormal protein was not always easy or indeed possible. For example, the genetic disorder Huntington's disease was first described by G Huntington in 1872, and more than 100 years later no abnormal protein had been found.

In the late 1980s, an alternative approach to the study of genetic disease became available through positional cloning. This method bypassed the protein and enabled direct isolation of genes on the basis of their chromosomal location and certain characteristics that identified a segment of DNA as "gene-like." Identifica-

tion of the mutant gene as well as knowledge of the genetic disorder could then be inferred from the DNA sequence. The strategy was initially called reverse genetics. Subsequently, the name was changed to the more appropriate one of **positional cloning**. The first success stories involving positional cloning for human genes came in 1986 with the isolation of the gene for chronic granulomatous disease by S Orkin and colleagues, and in 1987 with the Duchenne muscular dystrophy gene by L Kunkel and colleagues. Successes were slow to follow initially, but by the mid-1990s, it became difficult to keep up with the output from positional cloning (Table 1.3).

A variation of positional cloning enabled genes to be identified on the basis that they were "candidates" for genetic disorders rather than their positions on a chromosome. In other words, prior knowledge of the gene's function suggested it was worthwhile looking further at this candidate as there was a strong likelihood that it would be involved in the genetic disorder. This provided a shortcut to gene discovery but required information about likely genes that might be involved (Table 1.4 and Chapter 3).

POLYMERASE CHAIN REACTION (PCR)

In 1985, work by K Mullis, R Saiki and colleagues at the Cetus Corporation, California, made it possible to target segments of DNA with oligonucleotide primers and then amplify these segments with the polymerase chain reaction, or PCR as it is now best known. PCR soon became a routine procedure in the molecular biology laboratory. In a short period of time, this technology has had a profound and immediate effect in both diagnostic and research areas. Reports of DNA patterns obtained from single cells by PCR started to appear. Even the dead were

Table 1.3 Examples of some important genes found by positional cloning

Cystic fibrosis

Neurofibromatosis types 1, 2

Testis determining factor

Fragile X mental retardation

Familial adenomatous polyposis

Myotonic dystrophy

Huntington's disease

DNA repair defects (ataxia telangiectasia, a rare form of colon cancer)

Bloom's syndrome

Breast cancer genes in familial cases

Haemochromatosis

Box 1.1   Historical example: Functional cloning to identify the globin genes.

During the 1950s and 1960s, extensive analyses were undertaken to characterise the globin proteins. These studies showed that there were at least two different proteins, and in 1959, V Ingram and A Stretton proposed that the thalassaemias could be subdivided into α and β types corresponding to the α globin and β globin proteins. Following the discovery of reverse transcriptase, it became possible to take immature red blood cells from patients with homozygous β thalassaemia and isolate from them α globin mRNA, which could then be converted to complementary DNA (cDNA) (see Figure 2.2). In this way DNA probes were made, and they were then used to find the α globin genes. DNA probes specific for the β globin genes were isolated next. This was assisted by the cloning of the rabbit β globin gene since it had considerable homology (similarity) to its human equivalent. In contrast to positional cloning, which would develop later, functional cloning was difficult and limiting because it required knowledge of proteins. This was possible with the globins. However, for most other diseases, particularly those with no known protein abnormalities, the genes could not be found with a functional cloning approach. With the availability of restriction endonuclease enzymes, it became possible to construct DNA maps for the globin genes. DNA maps for the α thalassaemias were quite abnormal, indicating that the underlying gene defects involved a loss of DNA, i.e., deletions. On the other hand, the maps for the β thalassaemias were normal. Therefore, the molecular (DNA) abnormalities in these were either point mutations (changes involving a single nucleotide base) or very small deletions.

Table 1.4 Genes that have been cloned and characterised by the candidate gene approach[a]

Genetic disorder	Underlying abnormality	Candidate gene(s) and relevance
Retinitis pigmentosa	Retina	Rhodopsin, the gene for human rhodopsin (visual) pigment was isolated first. Subsequently, retinitis pigmentosa was localised to the same chromosome. Thus, rhodopsin became an obvious candidate gene for retinitis pigmentosa.
Familial hypertrophic cardiomyopathy (FHC)	Heart muscle	The genes for β myosin heavy chain, α tropomyosin and troponin T were initially selected as candidates because they are basic components of the muscle sarcomere. Subsequently, FHC is shown to involve eight more genes of the muscle sarcomere (see Table 3.1).
Marfan's syndrome	Connective tissue	A good candidate for this disease was fibrillin (the fibrillin protein is a constituent of elastin-associated myofibrils), and this was subsequently confirmed to be correct.
Alzheimer's disease (AD)	Early onset dementia	Amyloid protein was isolated in plaques from patients with AD. Hence, the *APP* (amyloid precursor protein) gene became a likely candidate for AD.
Long QT syndrome	Cardiac arrhythmias	A gene named *HERG* (predicted to have potassium channel-like activity) was situated in chromosomal proximity to the long QT syndrome locus determined from linkage analysis.

[a] A candidate or likely gene is used to narrow the field when searching DNA clones isolated on the basis of their chromosomal location by positional cloning.

not allowed to rest as it soon became possible to study DNA patterns from ancient Egyptian mummies, old bones and preserved material of human origin. The extraordinary contributions made by PCR in medicine, industry, forensics and research were recognised by the award of a Nobel Prize to K Mullis in 1993.

The availability of automation meant that DNA amplification had unlimited potential for mutation analysis in genetic disorders, as well as the identification of diagnostically significant DNA sequences from infectious agents. A patent was obtained to cover the use of PCR and illustrated the growing importance of commercialisation in recombinant DNA technology (see Chapters 2, 10 and Appendix).

GENES AND CANCER

During the 1980s–1990s, the application of molecular biology techniques to medicine or molecular medicine proved to be critical for our understanding of cancer. The first breakthrough actually came in 1910, when P Rous implicated viruses in the aetiology of cancer by showing that a filterable agent (virus) was capable of inducing cancers in chickens. For this work, he was awarded the Nobel Prize in 1966. However, the next big discovery did not come until the early 1980s, when a segment of DNA from a bladder cancer cell line was cloned and shown to have the capacity to induce cancerous transformation in other cells. The cause of the neoplastic change in both the above examples was soon demonstrated to be dominantly acting cancer genes called **oncogenes**. These oncogenes have assumed increasing importance in our understanding of how cancers arise and progress. For their work on oncogenes during the early-1980s, J Bishop and H Varmus were awarded a Nobel Prize in 1989.

More recently, the identification of cellular sequences that normally repress or control cellular growth led to the discovery of **tumour suppressor genes**. Loss or mutation of tumour suppressor gene DNA through genetic and/or acquired events was shown to be associated with unregulated cellular proliferation and hence neoplasm. Just as occurred with positional cloning, the list of known oncogenes and tumour suppressor genes has expanded rapidly. Information from both classes of genes has provided insight into the mechanisms that can lead to cancer but, just as importantly, the way the normal cell functions (see Chapters 3, 4). This focus on genes controlling cellular division both physiologically and in cancer led to a Nobel Prize in 2001 to L Hartwell, R Hunt and P Nurse for their work on key regulators of the cell cycle.

DNA changes found in cancers subsequently provided evidence for A Knudson's two-hit hypothesis that was proposed to explain the development of certain tumours. Thus, earlier epidemiological data on which Knudson devised his model in 1971 were now confirmed.

MUTATION ANALYSIS

Nucleic acid mutation analysis (i.e., identifying defects in DNA or RNA) was not considered a legitimate activity by some until the 1990s when, through the efforts of many, particularly R Cotton, it assumed a higher profile. This followed from the increasingly larger and more complex genes that were being isolated by positional cloning, as well as the realisation that DNA diagnosis could provide useful information for the clinical management of patients with genetic disorders or infectious diseases. The profile for mutation analysis increased further in the mid-1990s as genes for important cancers (e.g., bowel, breast) began to be discovered.

The gold standard in terms of DNA mutation analysis is sequencing because it allows the mutation to be defined. However, the size of genes involved, as well as the very broad range of mutations observed in genetic disorders, made this impractical as a routine means to detect mutations. Although automated DNA sequencing became more accessible in the 1990s, it remained expensive. PCR quickly made a major impact on mutation analysis, but the very focused nature of PCR (i.e., a specific region of the genome is targeted and then amplified) was its weakness when it came to mutations that could be found anywhere in a very large gene; e.g., there are more than 1300 mutations in the large 250 kb cystic fibrosis gene (1 kb is 1 kilobase or 1000 base pairs [bp]). Hence, the 1990s became a decade for designing new methods to allow large DNA fragments to be **scanned** for potential changes in their DNA sequence. Methods such as SSCP (single stranded conformation polymorphism), DGGE (denaturing gradient gel electrophoresis) and CCM (chemical cleavage of mismatch) were reported. However, none was ideal in all situations. Today, dHPLC provides a semi-automated and more sophisticated approach to scanning DNA for potential sites of mutations (see Chapter 2, Appendix).

Another aspect of DNA mutation analysis to achieve prominence in the 1990s was the potential for commercialisation through the availability of DNA kits. In the first instance, these kits were provided to facilitate the implementation of DNA technology in less experienced diagnostic laboratories. Ultimately, it was predicted that the family practitioner or local health centres would utilise DNA kits to identify specific genetic parameters. Perhaps over-the-counter kits would become available. The implications of this development in molecular medicine will need careful consideration. It is interesting to note that the community response to these potential developments was less vocal than on previous occasions, e.g., when genetic engineering was first started or the use of gene therapy was considered. However, important questions will need to be asked; e.g., what DNA testing is necessary, and what will cause more harm than good? How will the costs be met? The advantages and disadvantages of DNA mutation analysis will require clear forward thinking by both health professionals and the community (see Chapter 10).

GENETIC THERAPIES

In 1987, the first recombinant DNA vaccine against the hepatitis B virus was produced by inserting a segment of viral DNA into a yeast expression vector. Successes with the expression of DNA *in vitro* (e.g., the production of recombinant DNA-derived drugs) and *in vivo* (e.g., transgenic mice) highlighted the potential for transferring DNA directly into cells for therapeutic benefits (i.e., gene therapy). However, concern about manipulation of the human genome through a gene therapy approach was followed by a moratorium and extensive public debate until guidelines for the conduct of this type of recombinant DNA work became firmly established. Permission for the first human gene therapy trial to proceed was obtained only in 1990. A child with the potentially fatal adenosine deaminase (ADA) immunodeficiency disorder was given somatic gene therapy using her own lymphocytes that had been genetically engineered by retroviral insertion of a normal adenosine deaminase gene. Lymphocytes with the normal ADA gene were then reinfused into the individual. Initial clinical and laboratory responses were promising although efficacy could never be proven as these children had also been treated with a newly released drug. It took more than a decade before gene therapy was shown to be effective. Again, the model was an immunodeficiency (SCID X1), but in this situation alternative forms of treatment were not always available. Fourteen children were eventually treated and shown to have responded very well to the introduced gene. Unfortunately, 2 of the 14 children developed acute leukaemia as a direct result of the gene therapy (see Chapter 6 for further discussion of gene therapy).

During the late-1980s to the mid-1990s, alternative approaches to gene therapy that avoided the use of retroviruses began to be discussed as potential ways to transfer DNA into cells or interfere with foreign RNA or DNA (e.g., virus) in cells. New viral and non-viral vectors were developed, but each had its own particular drawback. A novel way to inhibit unwanted gene expression was identified following observations made in 1981 that RNA from the protozoan *Tetrahymena thermophila* had enzyme-like activity (called ribozyme). The discoverers, S Altman and T Cech, were subsequently awarded a Nobel Prize for this work. Ribozymes provided a new approach to gene therapy because they made use of naturally occurring RNA sequences that could cleave RNA targets at specific sites. Diseases for which ribozymes had particular appeal included cancer (to inhibit the mRNA from oncogenes) and AIDS (to inhibit mRNA from the human immunodeficiency virus).

By the mid-1990s, an unexpected change in the indications for gene therapy became apparent. Until this time, gene therapy was considered to be an experimental approach to treat rare genetic disorders for which there was no known treatment. However, the number of proposals involving cancer and HIV infection soon surpassed those for genetic disorders, showing that gene therapy had a much broader potential. Methodologies that enabled genes to be targeted to their normal genetic locus through a process called homologous recombination were also described. This was attractive since it would enable therapeutic manipulations to be conducted with greater accuracy (see Chapter 6). However, unless the targeted cells were in the germline (and this was prohibited in gene therapy), homologous recombination was not an efficient mechanism. Similarly, the major limitation with gene therapy today continues to be the inefficient means, both physical and viral vectors, by which DNA is inserted into a cell.

The cloning of cattle by nuclear transplantation in 1987 highlighted ethical and social issues that could arise from irresponsible use of recombinant DNA technology in humans. Legislative prohibitions relating to certain types of human embryo experimentation and human gene therapy involving the germline were enacted in many countries. However, few were prepared for the next development that came in 1997 when I Wilmut and colleagues successfully cloned Dolly, the sheep. This technologic tour de force seemed unachievable because it required the reprogramming of adult cells. However, using the procedure of somatic cell nuclear transfer, cells isolated from the udder of an adult female sheep were inserted into an enucleated ovum. From this, Dolly the sheep was derived, an identical copy of her adult mother (see Chapter 6). Subsequently, other animals were cloned in this way, leading to the proposal by some that humans would be cloned next.

In the midst of the cloning debate, the finding that human stem cells (both adult and embryonic) could redifferentiate into tissues such as nerve and other cells identified a potentially novel approach to treating a range of problems from degenerative conditions such as Parkinson's disease to spinal cord injuries resulting from trauma. Like cloning, stem cells became the focus for emotive discussions with scientists, politicians and laypersons providing their personal views, which at times became difficult to separate from scientific facts (see Chapters 6, 10 for more discussion on stem cells).

DNA POLYMORPHISMS

The number and types of DNA polymorphisms first described in the 1970s to early 1980s rapidly expanded. The inherent variability in DNA polymorphisms led to the concept of DNA fingerprinting in 1985 when A Jeffreys and colleagues described how more complex DNA polymorphisms (minisatellites) were able to produce unique DNA profiles of individuals. DNA testing for minisatellites has subsequently had an increasingly important role to play in forensic practice. The courts of law became involved in molecular medicine when the potential for identification of individuals on the basis of their minisatellite DNA patterns was realised. In 1987, DNA fingerprints were allowed as evidence in the first court case. Just as was mentioned earlier when describing the use of DNA polymorphisms in studies to identify the origins and affinities of peoples and species, DNA fingerprinting has also produced its share of controversies, which continue to be debated in the courts (see Chapters 9, 10).

Another type of DNA polymorphism, known as a microsatellite, was shown in the 1980s to be dispersed throughout the human genome and so very useful when it came to studies involving positional cloning (see Chapters 2–4). In the late 1990s, it became apparent from data coming out of the Human Genome Project that the genomes of individuals had many single base changes. These polymorphisms were called SNPs (single nucleotide polymorphisms—pronounced *snips*) and had the potential, when measured collectively, to produce a unique profile of an individual. The number of SNPs in the human genome was estimated to be more than 3×10^6, i.e., 1 SNP approximately every 1 kb of DNA sequence. By definition, a SNP is present when the rare allele in a single base change reaches a frequency of 1% or more in a population (otherwise, by convention, the single base change is referred to as a mutation). SNPs had the potential to provide considerable information because the numbers available for assay were considerably greater than would be possible with other polymorphisms. This meant that rare or complex traits could now be studied more effectively (see Chapter 4).

WORLD WIDE WEB

The Internet, which is described in more detail in Chapter 5 and Box 5.2, comprises the cables and computers linking many computer networks. It is the most important development in communication since the telephone and television. Many services run on the Internet, one of which is the World Wide Web (WWW). The WWW was first developed in 1989 and then made available on the Internet in 1991. Its original purpose was to facilitate communication in the physics community within an institution in Switzerland and across the world. Today, the WWW is an indispensable component of our lives, from scientific, medical and social perspectives. Its inventor, T Berners-Lee, developed: (1) a new computer language called HTML, (2) a protocol known as HTTP that allowed the transfer of web pages written in HTML between web browsers and servers, and (3) URL addresses that provided unique identifiers for files accessible over the Internet. With the WWW, users need only a web browser, and they have access to a wealth of information. For his invention, T Berners-Lee was awarded the inaugural Millennium Technology Prize in April 2004.

RNAi

A final historical highlight in molecular medicine to end the twentieth century was the discovery of another function for RNA (see Tables 1.1, 2.2). This was called RNA interference (RNAi) by A Fire and colleagues in 1998 using as their model the worm *Caenorhabditis elegans*. RNAi enabled long double-stranded (ds) RNA to be converted into short interfering RNA (siRNA) that cleaved homologous RNA species. Apart from a physiologic role likely to be as a protective mechanism against invasion of the genome by parasites such as viruses, the siRNA molecules had the ability to provide another level of gene regulation. There was also potential for siRNA to be used in gene therapy (see Chapters 2, 6). It has been predicted that the long-term impact of RNAi will rival the discovery of PCR.

HUMAN GENOME PROJECT: 1990–2000

THE START

The Human Genome Project (HGP) represents a landmark scientific feat. It is comparable to the moon landing in terms of the technologic challenges that had to be overcome and the many benefits both direct and indirect now emerging. It also demonstrated how scientists throughout the world could work together to bring about a dream that many considered was impossible. The consequences of the HGP will influence medical practice and the conduct of medical research for many years to come. Apart from the obvious gains in knowledge about genetics, advances in bioinformatics, biotechnology and clinical care will ensure the financial costs of the HGP are repaid many times over. Potential negative outcomes from the HGP reflect the ethical, social and legal implications that might emerge from inappropriate use of genetic information. However, these issues are being openly debated, and there is optimism that they will be avoided and so not detract from the HGP (see Chapter 10).

The United States Department of Energy (DOE) was a key player in proposing the HGP in 1987. The DOE had a long-term research interest in DNA because of its work with nuclear weapons, and the only way to understand fully the mutagenic effects of irradiation from these weapons was to characterise individual differences in DNA, i.e., sequence the human genome. However, in the mid-1980s, DNA sequencing was technically difficult; hence, only a few selected genes within the genome had been sequenced. Most of the 50 000–100 000 genes in the human genome (the number of genes considered to be present at the time—this number has now been reduced to about 30 000) had not been discovered, and the great majority of the 3×10^9 base pairs making up the human haploid genome did not contain gene sequences. This was (unfortunately) called junk DNA and would not normally be the target for DNA sequencing. Therefore, vast tracks of DNA remained unexplored, and the technology to sequence such large areas was not available. No group or groups of researchers were big enough to take on the mammoth task being proposed.

Despite what would appear to be insurmountable obstacles, scientists overall felt that the HGP was feasible, and in 1988 the US Congress funded both the DOE and the NIH (National Institutes of Health) to explore further the potential of an HGP. However, not all scientists were unanimous in their enthusiasm, and there was considerable misapprehension that the work involved was not research in its purest sense, but a monumental exercise in data gathering. The potential costs involved were also a major worry, particularly if funds from more traditional research activities were diverted to the HGP.

The HGP started in late 1990 with a planned completion by 2005 and a US$3 billion budget. Politically, the HGP promised both health and wealth outcomes. Health would come from medical benefits, and wealth would be gained from technological developments leading to economic growth and job creation. D Smith, then Director of the DOE's Human Genome Program, described the HGP as "developing an infrastructure for future research." In reply to the potential for shrinking research funds because moneys were going to the HGP, he made the prescient comment that following the HGP "individual investigators would do things that they would never be able to do otherwise."

Three additional points should be mentioned about the HGP. (1) The first is that the term "human" is a misnomer since it was also planned to characterise the genome of model organisms including mouse, fruit fly, various microorganisms, a worm, a plant and a fish. The model organism work, called "comparative genomics," was considered necessary for a complete understanding of the human genome, since the same genes are found in all organisms, and having model organisms would facilitate our understanding of gene function (Figure 1.1). (2) A non-laboratory component was also added. This was to consider the ethical, legal and social implications (abbreviated to ELSI) of the HGP. Three percent of the total budget has been set aside to research issues such as

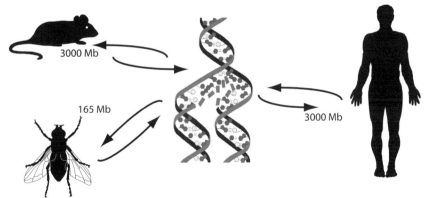

Fig. 1.1 Comparative genomics.
Gene discovery. The genomes of various organisms compared to humans are very similar. Genes with an important function will be conserved during evolution; i.e., the same gene will be identifiable in different organisms. Therefore, because of its relatively small size, determining the DNA sequence of the fruit fly *Drosophila melanogaster* (its genome size is about 165 mega bases [Mb] compared to the 3000 Mb of the human) is a more achievable goal. Knowledge of the fruit fly's genes has direct relevance to humans because once a gene is identified in the fruit fly, its homologue (equivalent) in the human can be sought in the DNA databases that have DNA sequences derived from the Human Genome Project. **Gene function.** The determination of function in a newly discovered human gene can be a formidable challenge. Looking at natural or induced mutants in the fruit fly is relatively simple and does not involve ethical dilemmas present in human research. From these mutants some knowledge is gained about gene function. Another strategy to determine function is to identify the same gene in the mouse and then knock it out by genetic manipulation. Knockout mice have become potent means by which to determine human gene function (see Chapter 5 for more discussion about model organisms).

Table 1.5 Components of the Human Genome Project

Goal	Purpose
1	Mapping and sequencing of the human genome—ultimately to determine the sequence of the ~3 billion bases that make up the human genome.
2	Mapping and sequencing the genomes of model organisms. They included Bacteria: *Escherichia coli, Bacillus subtilis,* two species of mycobacteria; Yeast: *Saccharomyces cerevisiae;* Simple plant: *Arabidopsis thaliana;* Nematode: *Caenorhabditis elegans;* The fruit fly: *Drosophila melanogaster;* and as an example of a mammal: the mouse.
3	Identifying the 30 000 or so genes making up the human genome (initially, the number of human genes was thought to be closer to 50 000–100 000).
4	Developing software and database designs to support large-scale collections of data, their storage, distribution and access. Developing tools for analysing large data sets. This goal would require very sophisticated bioinformatics capability. So, a by-product of the HGP has been the rapid development of bioinformatics, a discipline involving computational skills.
5	Creating training posts particularly in interdisciplinary sciences related to genome research, providing training courses (some of this work was subsequently taken on by HUGO, the—Human Genome Organization).
6	Transferring technologies to the private sector. For both technology development and training to be effective, private industry needed to be involved; hence, an early goal of the HGP was to enhance interchange of information between public and private enterprises. Developments coming from HGP were meant to be rapidly disseminated to users, an ideal that would come into conflict with the more entrepreneurial commercial sector.
7	Developing a flexible distribution system so that results and developments were quickly transferred to potential users and the community.
8	The final goal was directed to addressing the ethical, legal and social implications arising from the HGP. Included in this would be issues related to privacy, confidentiality, stigmatisation, discrimination, equity and education of the public and health professionals.

privacy and confidentiality (e.g., who will have access to an individual's genetic makeup), stigmatisation or discrimination (e.g., what might insurance companies or employers do with the information generated from the HGP, what adverse effects might result from knowledge of the genome's sequence and so the potential to predict health or disease in an individual). Educating the public and professionals about the HGP was also an important aim. (3) Although finding new genes (gene discovery) was not an early goal of the HGP, this was added soon after the HGP started.

The HGP had a number of goals or programs that are summarised in Table 1.5. The first involved the construction of comprehensive genetic and physical maps of the human genome (see Chapter 2 and Appendix for more details of genetic and physical maps). From these maps, which were tedious and time consuming to make, genes could be found, and segments of DNA were able to be sequenced. The distance between markers on a genetic map is defined as a centimorgan (cM) with 1 cM equal to ~1 Mb (megabase or 1×10^6 base pairs). An initial aim of the HGP was to produce a genetic map to cover the entire genome with DNA markers that were 1 cM apart. Each of the DNA markers generated would require a unique identifier, and for this, the concept of sequence tagged sites (STSs) was proposed. This meant that sequencing of DNA markers would be required. Each

marker would then be identified by the part of its sequence that was unique.

From genetic maps it was possible to construct physical maps so that the distance between DNA markers could be determined in absolute terms (i.e., bp, kb or Mb). This was a mammoth task since it became necessary to characterise entire regions of the genome on the basis of overlapping DNA clones that would ultimately need to be sequenced. This strategy, which was followed by the publicly funded HGP effort, contrasted with the approach subsequently adopted by the commercial company Celera, which is discussed further below. To accomplish the above, new DNA sequencing technologies were required, and apart from being more efficient, these technologies had to be cheaper. One goal of the HGP was to reduce the cost for each base pair sequenced to less than 50 cents. The development of robotics was another essential requirement.

The work of constructing genetic and physical maps was undertaken by many different laboratories across the world, particularly in the USA, UK (funded by the Wellcome Trust) and France (funded by the Muscular Dystrophy Association). Those involved will remember with affection the single chromosome workshops (i.e., workshops dedicated to one particular chromosome) that attracted scientists from all corners of the globe. At these workshops there would be focused and earnest discussions relating to the relative positions of different DNA markers in a particular segment of the genome. A few years on, these activities seem trivial when it is now possible to sequence any region of the genome with comparative ease, and from this pinpoint the location of any DNA base.

A separate goal of the HGP focused on the mapping and sequencing of model organisms. This goal was undertaken for two reasons. First, model organisms would provide less complex genomes to facilitate technology development. Second, the models would enable comparative studies to be made between the human genome and those from non-human sources. Information generated from these comparisons would be essential for an understanding of the evolutionary processes, the way genes are regulated and the aetiology of some genetic disorders (for examples, see the discussions on imprinting in Chapters 4, 7).

Another HGP goal involved bioinformatics. This goal was essential to develop computer-based methods to store the vast amount of data that would be generated by the HGP, i.e., genome maps and DNA sequences. A considerable amount of software development would also be needed to allow the various databases to be analysed, and from this identify the sites of genes and what these genes did, i.e., function of genes. The key elements in bioinformatics were computer networks and databases. The Internet and local servers provided the former, and the development of resource centres and databases acted as a focus into which information could be fed and processed.

Programs were also set in motion to train individuals who would have a sound knowledge of genome research methodologies. Skills resulting from this would be not only in the area of molecular biology but would include computer science, physics, chemistry, engineering and mathematics. Interdisciplinary approaches to training and skill acquisition were to be encouraged. Technology transfer and outreach were considered important goals. To expand the pool of researchers and resources, funding and interactions with private industry were considered essential. Five-year plans were developed.

TWO FIVE-YEAR PLANS

Years 1 to 5 of the HGP (1991–1995) could be described as a time of enthusiasm and steady achievements. During this period, individuals in many laboratories throughout the world carefully constructed maps of the genome and then identified, by DNA sequencing, each base in the segment they were allotted. Although the DOE was the key leading player, it was soon partnered by the NIH in view of the latter's support of biomedical research and its vast network of scientists. The NIH, through its National Human Genome Research Institute led by F Collins, sub-

sequently became the leading public-sector player in the HGP. Another influential body was HUGO (Human Genome Organization). HUGO's role was to coordinate international efforts as well as to facilitate education and rapid exchange of information.

The second five years of the HGP (1996–2000) were more turbulent. By 1998, the impressive developments in technology, particularly automation, had meant that the timing for specific goals needed to be moved forward. A new estimate for the complete sequence of the human genome was now given as 2003. The first success stories coming from the HGP involved the completed sequencing of genomes from model organisms. In particular, the reporting by the commercial company Celera in 1995 that the *Haemophilus influenzae* genome had been sequenced was a major achievement. This was followed soon by the sequence for *Mycoplasma genitalium*, and in 1996, the first eukaryote genome to be sequenced was that of *Saccharomyces cerevisiae*. With these successes, the momentum for the human work increased, since it became evident that genomes could be completely sequenced, and the information coming from this work had both scientific and medical significance.

Towards the end of the second five years, the influence and contribution of the commercial sector grew. This became a source of tension as the HGP's ethos required that genomic information and DNA sequencing results were to be communicated freely and without delay. To some extent, this was at odds with the protection of intellectual property through patenting. A high-profile example of commercialisation came with the Celera company, a privately funded organisation sponsored by Applied Biosystems. Celera (its motto was "Speed counts" and the name was derived from the Latin for "quick") took on the might of the NIH and the world, when its leader C Venter publicly boasted that it had the resources (e.g., more than 200 of the most modern automated DNA sequencers, backed by super-computers second only to those in the US military for bioinformatics) to finish the first draft of the human sequence before the NIH or other countries, and at a much reduced cost of US$2 million. To do this, Celera adopted a different strategy to sequence whole genomes of human and many model organisms.

The Celera strategy was a "shot-gun" approach in that it did not require ordered genetic or physical maps but blasted the entire human genome into small fragments. Each fragment was individually sequenced, and then computer software matched the fragments based on overlapping sequences. In effect, a giant jigsaw puzzle of DNA sequences was created, and the computers were used to place the right parts together. At the time there was much speculation about which of the two different approaches was superior. This question will never be resolved although it is interesting to observe that in sequencing the zebrafish by both strategies, the

Table 1.6 Initial interesting facts to emerge from the Human Genome Project

Observation	Comments or explanation
Number of human genes	About 30 000, which is considerably less than originally anticipated.
Number of genes in model organisms	Surprisingly, humans and mice have a similar number of genes. Humans have only twice the number of genes as the fruit fly or worm.
Proteome	The difference between humans and other species may be explicable by the more complex protein repertoire (proteome) in the former. This suggests that genes can encode for different proteins, and other mechanisms are likely to exist to explain the two to three times increase in the human proteome complexity.
Similarity between humans	At the DNA level humans are 99.9% similar.
Similarity with other organisms	Humans share their genes with model organisms (e.g., about 50% of their genes with nematodes, 20% with yeast).
Proportion of coding to "junk" DNA	Only 1–2% of bases in DNA encode proteins. Approximately 50% of the genome consists of highly repetitive DNA sequences with no known function. The remainder of the genome is non-coding, non-repetitive DNA, also without an apparent function.
Most common type of human sequence	This is the SNP (single nucleotide polymorphism), which involves a single base change occurring at least 1 in each 1 kb of DNA. Total number of SNPs exceeds 3×10^6.
Most mutations occur in males	There is a 2 : 1 excess of mutations in the male versus female germline suggesting either inefficient repair in the male or a consequence of the greater number of cell divisions in male meioses.
Bacterial DNA in human genome	It is likely that hundreds of human genes have originated from the horizontal transfer of DNA from bacteria at some point in the vertebrate lineage. Genes have also evolved from transposable elements.

Cambridge investigators at the Sanger Institute have observed that the approach adopted by Celera is faster, but the quality is not as good as that from the traditional sequencing strategy (see Chapter 5).

THE COMPLETION

The Celera challenge had many positive effects; for example, it focused the cumbersome and slow-moving multicampus, multinational publicly funded HGP. On the negative side, it highlighted that big corporations and money can get there, but at a cost—the availability and access to future databases or knowledge would not necessarily be free. In June of 2000, US President Bill Clinton, flanked by Dr Francis Collins (NIH) and Dr Craig Venter (Celera), announced simultaneously with the UK Prime Minister Tony Blair that the first draft of the human DNA sequence was now complete with contributions from both the public and private sectors. In February 2001, *Science and Nature* published the first draft of the complete sequence of the human genome. Although the HGP had officially ended in mid-2000 (five years earlier than its anticipated completion), the sequence produced

was only the first draft, and considerable work remained to ensure that DNA sequencing errors and ambiguities were removed. In April 2003, 50 years after Watson and Crick had described the structure of DNA, the NIH announced the completion of a high-quality comprehensive sequence of the human genome. The challenge to sequence the human genome had been successfully accomplished (Figure 1.2).

A number of interesting facts have emerged from the initial analysis of the human genome sequence. These facts, summarised in Table 1.6, will provide the focus for much work into the future. A particular challenge will be the understanding of why humans are so similar at the gene level (99.9%), yet they are clearly very heterogeneous in terms of their phenotypes. Another puzzling observation is the smaller than expected number of human genes. This, as well as the fact that only about 1–2% of the human genome codes for protein, cannot explain the human's complexity at the proteome level; i.e., about 30 000 human genes are associated with hundreds of thousands of proteins. Other explanations are needed, and they are likely to involve RNA as well as the "junk" DNA.

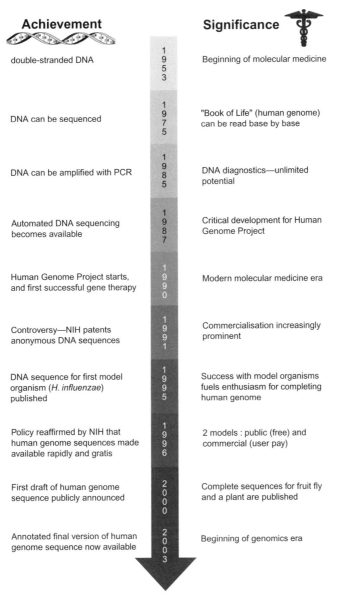

Fig. 1.2 Landmarks associated with the Human Genome Project.

FUTURE

The term "post-genome" is used by many to describe the genome-based activities that will emerge with the completion of the HGP. However, this seems an inappropriate description as the HGP has provided the tools to enter the genomics era, not leave it! Today, there is less talk about genetics and more on genomics. With the term **genetics**, the focus was on single genes and their associated disorders. **Genomics** looks at the big picture and encompasses the entire complement of the genetic material, with some also including how this genetic material interacts with the environment (Figure 1.3, see Chapter 5 for further discussion of genomics).

At the end of the HGP, the DNA sequence from the approximately 30 000 human genes was deposited in various databases. What is now left is the monumental task of working out where the genes are in this sequence and their function; i.e., the DNA sequence needs to be annotated, and for this, better bioinformatics is needed (see Chapter 4 for further discussion of DNA annotation). Hence, the post-genome era has also been called functional genomics, and included in this is proteomics—the technology and strategies required to determine the func-

tion of proteins. How this will be accomplished remains to be determined, but new technologies will be needed. The use of microarrays (a method by which the expression of many thousands of genes can be identified very rapidly with microchips) is an early promising strategy in functional genomics (see Chapter 5 for further discussion on genomics, proteomics and bioinformatics). The role of bioinformatics will also be critical. It is likely that the traditional wet-laboratory approach to molecular research will give way to a complete *in silico* (i.e., computer) positional cloning strategy for future gene discovery and functional analysis.

Novel approaches to interrogate the human genome are already starting to appear. An example of this is the International Hap Map (haplotype map) project. This project involves laboratories in many countries attempting to reduce the size of the human genome by defining specific regions that are inherited as a block. In this way, instead of needing a large number of DNA polymorphisms (i.e., SNPs) to study the genome, it will become possible to rationalise the SNPs to the few that define a particular block. Apart from the concept of a haplotype block, the other assumption is that SNPs that are co-located will be inherited together; i.e., they are in linkage disequilibrium. It has been suggested that a hap map could reduce the need to use 10 million SNPs to investigate a complex trait to a more manageable (and affordable) number like 500 000 key SNPs.

The old concept of molecular medicine as the study of DNA → DNA → RNA → protein in a single gene or genetic abnormality has evolved to the study of many genes, RNA species and proteins in a particular cell. A new term has also crept into the molecular medicine vocabulary—it is the **phenome**. The phenome follows on the "all" theme as the total phenotypic characteristics of an organism reflecting the interaction of the complete genome with the environment (Figure 1.4).

To articulate the future of genome research, the NIH (through the National Human Genome Research Institute directed by F Collins) organised a series of meetings and forums over a two-year period involving scientists and the public. From these, three major themes emerged: (1) Genomics to Biology, (2) Genomics to Health and (3) Genomics to Society. For each of the themes, a series of grand challenges were identified as a way forward in the genomics era (Table 1.7).

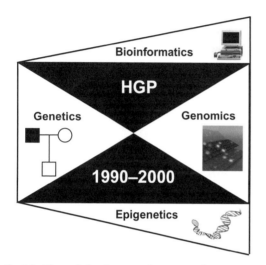

Fig. 1.3 The evolution from genetics to genomics.
The molecular **genetics** era allowed a better understanding of mutations in single genes and their effects on individuals and families. With the completion of the **Human Genome Project**, the mass of data now available (as well as new technologies developed) has moved molecular genetics into **genomics**, with the potential for technologies such as microarrays to analyse hundreds to thousands of genes (the genome) and mRNAs (the transcriptome) simultaneously. In parallel with the development of genomics has been the increasing requirement for **bioinformatics** in storing, handling and analysing large data sets. Another interesting development has been epigenetics (perhaps more appropriately called epigenomics), which is the study of how the expression of DNA can be influenced without changing the actual DNA base sequence. Presently, **epigenetics** remains an unknown quantity in terms of its likely impact on gene function, although many interesting claims are being made; they will be discussed further in Chapters 2, 4 and 7.

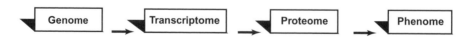

Fig. 1.4 The spectrum of molecular medicine.
Molecular genetics with its focus on single genes has evolved into genomics (large number of genes) including the transcriptome (all mRNA species in a cell at a given time) and the proteome (all proteins in a cell at any given time). Ultimately, this knowledge must lead to the phenome—the total phenotypic characteristics of an organism, which includes the interactions of the genome with the environment.

Table 1.7 Challenges and visions for future genomics research (Collins *et al* 2003)

Major themes	Challenges
Genomics to Biology	1 Identify the structural and functional components in the human genome. 2 Determine the organisation of genetic networks and protein pathways and the way they contribute to phenotypes. 3 Understand the heritable variation in the human genome. 4 Understand evolutionary variation across species and their mechanisms. 5 Develop policies to facilitate the widespread use of genome information in both research and clinical settings.
Genomics to Health	1 Develop strategies for identifying the genetic contributions to disease and drug response. 2 Develop strategies for identifying gene variants that contribute to good health and disease resistance. 3 Develop genome-based approaches to prediction of disease susceptibility, drug response, early detection of disease and molecular taxonomy of disease states. 4 Use new understanding of genes and pathways to develop novel therapeutic approaches to disease. 5 Investigate how genetic risk information is conveyed in clinical settings, how the information influences health strategies and behaviours and how these affect health outcomes and costs. 6 Develop genome-based tools that improve the health of all.
Genomics to Society	1 Develop policy options for uses of genomics in medical and non-medical settings. 2 Understand the relationships between genomics, race and ethnicity, and the consequences of uncovering these relationships. 3 Understand the consequences of uncovering the genomic contributions to human traits and behaviours. 4 Assess how to define the ethical boundaries for uses of genomics.

FURTHER READING

The Foundations: 1869–1980s

King RC, Stansfield WD. A Dictionary of Genetics. 6th edition. 2002. Oxford University Press, Oxford UK.

Watson JD. The Double Helix. 1968. Touchstone, New York (*a historical overview of the events leading to the discovery of the double helix*).

http://www.nobel.se/medicine/laureates/ (*Site listing all Nobel Prize winners as well as giving a brief summary of their achievements*).

Modern Era: 1980s–2000s

Nature Genetics supplement vol 33 March 2003 (*provides a retrospective from 1992–2002 relating to 11 different areas of genetics. Available free online*).

Stevenson M. Therapeutic potential of RNA interference. New England Journal of Medicine 2004; 351:1772–1777 (*useful introduction to RNAi*).

http://www.technologyawards.org/ (*Internet address for the Millennium Technology Prize awarded to T Berners-Lee for developing the World Wide Web*).

Human Genome Project: 1990–2000

Green ED, Chakravarti A. The human genome sequence expedition: views from the "base camp." Genome Research 2001; 11:645–651 (*describes key early findings from the human genome sequence*).

Nature 15 February 2001 (vol 409, pp 860–921) and Science 16 February 2001 (vol 291, pp 1304–1351) (*contain the two papers describing the first draft of the human genome sequence*).

Venter JC, Levy S, Stockwell T, Remington K, Halpern A. Massive parallelism, randomness and genomic advances. Nature Genetics 2003; suppl 33:219–227 (*provides a view of the HGP from the industry perspective including a comparison of the two different strategies adopted*).

http://www.genome.gov/10001772 (*historical overview of the HGP from the National Human Genome Research Institute*).

http://www.genome.gov/Pages/Hyperion/educationkit/index.html (*Internet resource for Human Genome Project and many topics related to genetics and molecular medicine*).

Future

Collins FS, Green ED, Guttmacher AE, Guyer MS. A vision for the future of genomics research. Nature 2003; 422:835–847 (*this article provides a blueprint for future genomics directions*).

http://www.genome.gov/10001688 (*Internet reference to the International Hap Map [haplotype map] project, which is attempting to reduce the size of the human genome by dividing it into blocks*).

DNA, RNA, GENES AND CHROMOSOMES

DNA

FIFTY YEARS OLD

"We wish to suggest a structure for the salt of deoxyribose nucleic acid (D.N.A.). This structure has novel features which are of considerable biological interest." This is the somewhat modest introduction to the April 1953 letter to *Nature* by James Watson and Francis Crick when they first described the structure of DNA.

DNA is deoxyribonucleic acid, a double-stranded macromolecule (double helix) containing the organism's genetic information (genetic code). Its chemical structure allows the inheritable characteristics to be passed on to the next generation of cells. These properties of DNA form the basis for molecular medicine.

Double Helix

DNA comprises two polynucleotide strands twisted around each other in the form of a double helix (Figure 2.1). Each strand has a sugar phosphate backbone linked from the 5' and 3' carbon atoms of deoxyribose. At the end of a strand is either a 5' phosphate group (5' end) or a 3' phosphate group (3' end). One strand of DNA (sense-strand) contains the genetic information in a 5' to 3' direction in the form of the four nucleotide bases adenine (A), thymine (T), guanine (G) and cytosine (C). Its partner strand (antisense strand) has the complementary sequence; i.e., A pairs with T; G with C and vice versa. This is called Watson and Crick base pairing. For example, the

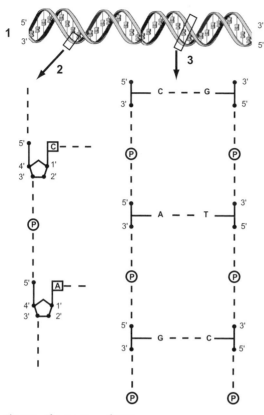

Fig. 2.1 The structure of DNA.
(1) A schematic drawing of the DNA double helix. Two complementary strands (i.e., the base adenine always pairs with thymine; cytosine with guanine) run in opposite directions: The sense strand 5′→3′ and the antisense 3′→5′. **(2)** An expanded view of a single strand showing the three basic components: the bases (C—cytosine, A—adenine; not shown are T—thymine and G—guanine). Nitrogenous bases are of two types: purines (A, G) and pyrimidines (T, C). The deoxyribose sugar with the position of its 5 carbons identified as 1′ to 5′. Finally, the phosphodiester (P) linkage between the deoxyribose sugars. **(3)** An expanded view of the two strands. The two strands are held together by hydrogen bonds between the bases (two hydrogen bonds between A/T and three between G/C). The direction for transcription is 5′→3′.

double-stranded DNA sequence for the polypeptide described in the next paragraph is

5′-ATG GGT TCT GTT GCT GCT TGG TAA-3′
    = sense strand
3′-TAC CCA AGA CAA CGA CGA ACC ATT-5′
    = antisense strand

In biological terms, the double-stranded DNA structure is essential for replication to ensure that each dividing cell receives an identical DNA copy.

Table 2.1 The genetic code
Nucleotides code in sets of three, or triplets, for individual amino acids. The triplets or codons are shown as they appear in DNA (T = thymine, C = cytosine, A = adenine and G = guanine). In mRNA, T is replaced by U (uracil). The code is degenerate, i.e., there can be more than one codon per amino acid. The genetic code is read from left to right, i.e., TTT = phe (phenylalanine); TCT = ser (serine); TAT = tyr (tyrosine)[a].

First nucleotide [5′]	Second nucleotide				Third nucleotide [3′]
	T	C	A	G	
T	Phe	Ser	Tyr	Cys	T
T	Phe	Ser	Tyr	Cys	C
T	Leu	Ser	STOP	STOP	A
T	Leu	Ser	STOP	Trp	G
C	Leu	Pro	His	Arg	T
C	Leu	Pro	His	Arg	C
C	Leu	Pro	Gln	Arg	A
C	Leu	Pro	Gln	Arg	G
A	Ile	Thr	Asn	Ser	T
A	Ile	Thr	Asn	Ser	C
A	Ile	Thr	Lys	Arg	A
A	Met	Thr	Lys	Arg	G
G	Val	Ala	Asp	Gly	T
G	Val	Ala	Asp	Gly	C
G	Val	Ala	Glu	Gly	A
G	Val	Ala	Glu	Gly	G

[a] Amino acid abbreviations are Cys = cysteine; Trp = tryptophan; Leu = leucine; Pro = proline; His = histidine; Gln = glutamine; Arg = arginine; Ile = isoleucine; Met = methionine; Thr = threonine; Asn = asparagine; Lys = lysine; Val = valine; Ala = alanine; Asp = aspartic acid; Glu = glutamic acid; Gly = glycine.

Genetic Code

The genetic code in DNA is represented by nucleotide triplets called **codons** (Table 2.1). This means that the signal for individual polypeptides is coded by different triplet combinations. For example, the codons for a polypeptide such as glycine-serine-valine-alanine-alanine-tryptophan will read GGT TCT GTT GCT GCT TGG. The positions indicating where a polypeptide starts and where it ends are also defined by triplet codons. For example, ATG is found at the start, and the end or stop codons are TAA or TAG or TGA. Point mutations (single changes in the nucleotide bases) in any of the above codons can lead to genetic disease because a new amino acid will form (see DNA Mutations and Polymorphisms below). Deletions or insertions affecting the codons will produce a frameshift, which is discussed further below. The genetic code needs to be read from the sense strand. Hence, transcription to give the appropriate mRNA sequence is taken from the antisense strand so that the single-stranded mRNA will have the sense sequence (antisense RNA is discussed further in Chapter 6).

DNA has a number of properties that can be exploited in the laboratory. In terms of genetic diseases, DNA in all cells of an organism is identical in its sequence. Therefore, obtaining a tissue specimen for DNA studies is relatively simple since a small volume of blood will suffice. Ten ml of blood yields approximately $250\,\mu g$ of DNA. Isolation of DNA is straightforward. Nuclei are first separated from cellular debris by enzymes and detergents. DNA is then separated from protein by chemicals or physical means. Apart from blood, convenient sources of DNA used in routine genetic diagnosis include the exfoliated cells from mouth swabs or washes, or hair roots (see the Appendix for details on DNA preparation).

DNA PROBES AND PRIMERS

The double-stranded structure of DNA is used by the molecular biologist to make DNA probes, or DNA primers for DNA amplification that will be described later. Probes comprise single-stranded fragments of DNA. This allows them to bind to the complementary DNA

sequences in another single-stranded DNA fragment. For example, if the single-stranded target has the sequence 5'-GGTTACTACGT-3' the single-stranded DNA probe will be 3'-CCAATGATGCA-5'. The specificity of a probe resides in its nucleotide sequence. Since double-stranded DNA is held together by hydrogen bonds, it is relatively easy to make both DNA probe and target DNA single-stranded; e.g., heating breaks hydrogen bonds. On cooling, the complementary DNA strands will reanneal, i.e., reform into base-paired double strands. Reannealing will occur between the following combinations: DNA probe + DNA probe; target DNA + target DNA and DNA probe + target DNA. If the DNA probe is labelled with a fluorescent marker, then the DNA probe + target DNA hybrids can be detected by a laser. DNA probes are of three types: cDNA, genomic and oligonucleotide. The latter will be described in more detail under Polymerase Chain Reaction and DNA Mutations and Polymorphisms.

RNA

STRUCTURE AND FUNCTION

RNA differs from DNA in three ways: (1) The nucleotide base thymine is replaced with uracil; (2) the backbone is ribose rather than the 2'-deoxyribose of DNA; and (3) RNA is usually single stranded. Except for some viruses, RNA does not contain the cell's genetic material. RNA is a lot less robust than DNA, and so isolation techniques require the addition of chemicals to ensure that any RNase (also written RNAase) enzymes present are inactivated to avoid its degradation (see the Appendix for further details).

Compared to DNA, RNA shows tissue-specificity. Thus, the relevant mRNA can be isolated only from a tissue that is transcriptionally active in terms of the target protein. For example, reticulocytes, the red blood cell precursors, would contain predominantly erythroid-specific mRNAs (i.e., mRNAs for α and β globin genes). The reticulocyte would be inappropriate as a source of neuronal-specific mRNA. The tissue-specificity requirement limited use of mRNA until fairly recently. Now it has been observed that mRNA production in cells that are easy to access, such as peripheral blood lymphocytes, can be leaky; i.e., there is transcription of mRNA species that are not directly relevant to the lymphocytes' function. These "ectopic" or "illegitimate" mRNAs, as they are called, are found in minute amounts, but the amplification potential of the polymerase chain reaction (PCR) can be utilised to isolate rare species or minute amounts of either DNA or RNA. In terms of the latter, this means that the tissue specificity constraint becomes less of an issue.

For recombinant DNA technology, mRNA has one important advantage over DNA; i.e., it contains only the

essential genetic data found in exons without the superfluous information associated with introns (i.e., mRNA is much smaller than its corresponding DNA). cDNA refers to complementary (or sometimes called copy) DNA. The usual progression in the cell of DNA to RNA to protein can be perturbed *in vitro* and *in vivo* with the enzyme reverse transcriptase. Now it is possible to take an mRNA template and produce from this a second strand that is the complement of the mRNA. The double-stranded structure formed from this is called cDNA. Unlike the starting or native DNA (known as genomic DNA), the cDNA does not have introns but contains only coding (exon) sequences (Figure 2.2).

RNA's key contribution in transcription and translation is well known, and will be discussed further under Gene Structure to follow. During the 1980s, a new role for RNA was described when the catalytic RNA species called ribozymes were discovered (see Chapter 1 and Table 2.2). Today, the focus on RNA is the non-coding RNA molecules shown to be abundant in the genome. For example, 98% of the transcriptional output in humans is non-coding RNA species, including the various types of small RNAs that will be discussed later. The function of non-coding RNA remains to be determined, but there is mounting evidence that these molecules might explain an unexpected finding from the Human Genome Project; i.e., humans and other species are similar at the genome level, yet their proteomes (i.e., the protein pools) are quite different (see Table 1.6).

It is proposed that one way non-coding RNA species can influence gene output is through networking to allow various loci to be interconnected including gene-to-gene communication. Another mechanism to explain the

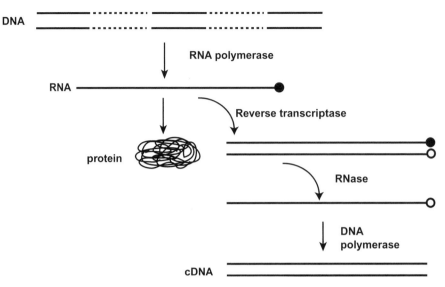

Fig. 2.2 Making cDNA with RT-PCR.
Double-stranded DNA (exon represented as a solid line; introns as broken lines) is transcribed into RNA (solid line with ●). In the normal course of events, the RNA is then translated into protein. However, reverse transcriptase allows a copy (cDNA) of the RNA to be made (solid line with ○). Once this event occurs, the RNA component of the cDNA is removed with an enzyme such as RNase. A DNA polymerase enzyme will then allow the second DNA strand of the cDNA to be formed. From the initial DNA template, a synthetic double-stranded segment containing only exon(s) has now been made. The type of PCR approach described above is usually abbreviated to RT-PCR (reverse transcriptase-PCR).

Table 2.2 Functions of RNA

Type of RNA	Function
mRNA (messenger RNA)	RNA that functions as the intermediary between DNA in the nucleus and protein production in the cytoplasm
tRNA (transfer RNA)	RNA that transfers an amino acid to a growing polypeptide chain during translation
rRNA (ribosomal RNA)	RNA that is a component of ribosomes
ribozyme	RNA that functions as an enzyme by catalysing chemical reactions in the cell and *in vitro*
RNAi (RNA interference)	RNA's regulatory role in terms of gene expression as well as protecting the genome from invasion by molecular parasites such as viruses

human proteome is alternative splicing, and this will be discussed below. As the sequencing and annotation of the human chromosomes continue, it will become possible to identify the source and numbers of the various non-coding RNA species and so their role (see Chromosomes 13 and 19 below). Already a web site has been developed to enable the increasing number of micro RNA (miRNA) species to be catalogued. To date, more than 1300 miRNA entries are to be found on http://www.sanger.ac.uk/Software/Rfam/mirna/.

RNA INTERFERENCE

Apart from the catalytic activity demonstrated by RNA in the form of ribozymes (see Chapters 1, 6), a more recent development in molecular medicine has been the finding that RNA is also involved in regulating gene activity. One mechanism for this is called RNA mediated gene silencing (post-transcriptional gene silencing) or RNA mediated interference (RNAi). This highly coordinated sequence-specific activity involves RNA in post-transcriptional gene regulation. RNAi works through a number of RNA species including (1) siRNA (small interfering RNA)—small, double-stranded (ds) RNA species that degrade mRNA; (2) miRNA (micro RNA)—small, double-stranded RNA species that interfere with translation by imperfect base pairing with mRNA. Both siRNA and miRNA share common intermediaries including DICER and RISC (Figure 2.3). Small amounts of dsRNA have been shown to silence a vast excess of target mRNA. Evidence is also mounting that dsRNA may regulate gene function through an epigenetic DNA change such as methylation. siRNA would also function in defending the genome from invasion by molecular parasites such as viruses.

More recently, this activity of RNA has been reproduced by chemically synthesising small 21–23 nucleotide dsRNA and then showing that they can act through siRNA to inhibit specific gene expression in mammalian cells. This observation opens up the possibility that siRNA has potential as an alternative form of gene therapy. In the research laboratory, the siRNAs could be used instead of the more complex and costly knockout mouse to determine gene function (see Chapters 5, 6).

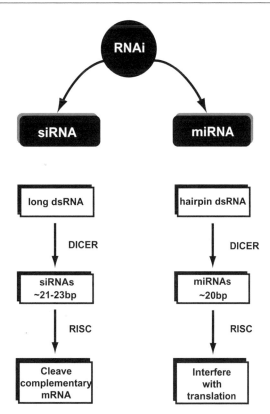

Fig. 2.3 RNA interference (RNAi).
siRNA: Long double-stranded (ds) RNA is digested by an RNase enzyme called DICER. This produces a number of small dsRNA species about 21–23 bp (bp = base pair) in size. The latter are small interfering RNAs (siRNAs), and they next interact with RISC, a protein complex with endonuclease activity. This allows the antisense strand of the siRNA to bind to the complementary sequence in mRNA, leading to its degradation. **miRNA:** A parallel process follows with the less understood miRNAs (micro RNAs). These small (about 20 bp), non-coding, double-stranded RNA species are derived from hairpin precursor RNAs (hairpins are formed by RNA folding on itself). The miRNAs do not have complementarity to mRNA species, and so they do not cleave mRNA like siRNA, but appear to regulate gene activity via translation through non-specific binding to the 3′ untranslated ends of genes. Note: There is some confusion in the literature with nomenclature. The term "miRNA" is sometimes used as a generic description of all small dsRNA species including siRNA.

GENE STRUCTURE

EXONS, INTRONS

Eukaryotic genes (eukaryotes are organisms ranging from yeast to humans with the characteristic feature of nucleated cells) are usually discontinuous; i.e., they have coding regions called exons broken up by non-coding regions called introns. Another name for intron is intervening sequence or IVS (Figure 2.4). During the process of transcription, the entire gene sequence is copied, and then the introns are removed by a process known as splicing. This produces the mature mRNA, which is smaller than the genomic structure since it comprises only genetic information derived from exons. Hence, cDNA, which is made from mRNA, is the equivalent of DNA but in an abbreviated form in respect to the gene's genomic structure.

At the exon-intron boundaries, signals indicate where splicing should occur. For example, introns always begin with a GT and end with an AG. At the end of the gene, the 3′ end in respect to the DNA double helix (Figure 2.1), there is (1) a stop codon (TAA, TGA or TAG); and (2) a run of adenine nucleotides (called poly A tail) that stabilise mRNA. At the beginning of the gene (5′ end), an ATG codon indicates where the protein starts, and further 5′ are transcription initiation signals called promoters.

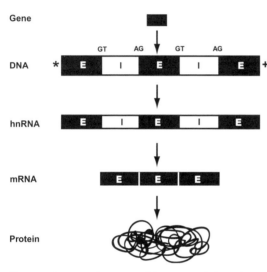

Fig. 2.4   Anatomy of a gene, and how a protein is made.
Gene: A gene is a segment of DNA that contains genetic information. As indicated in Chapter 1, only about 1–2% of our genome is made up of genes. **DNA:** DNA, which comprises a gene, is made up of a number of components. The beginning (left end or 5′ end) of the gene contains regulatory sequences (*), and the tail end (right end or 3′ end) has a poly A tail (+). The gene itself comprises coding regions called exons (black) separated by non-coding regions called introns (clear). Introns are also known as intervening sequences (IVS). The border between introns and exons is demarcated by splicing signals. At one intron-exon boundary, the splicing signal is a dinucleotide GT (called the donor junction). The intron-exon boundary on the other side of the intron is an AG dinucleotide (acceptor junction). In addition to the GT and AG that are constant at intron-exon boundaries, additional nucleotide signals help to define when a gene should splice. Although introns have no known function, they may contribute to genomic complexity through alternative splicing (see text), and regulatory regions are found within introns. **hnRNA:** Precursor RNA formed initially during transcription copies the entire gene sequence (exons and introns). **mRNA:** The introns are next spliced out, leaving the mature messenger RNA (mRNA) with only the exons (i.e., the code for the protein). **Protein:** The last step that will make the appropriate protein is called translation.

The latter are important because they control the amount of mRNA that is produced and play a role in the gene's tissue specificity (remember that all tissues have the same complement of genes). Mutations in any of the above key regions in the gene can lead to genetic disease.

Gene expression is controlled by the promoter regions, as well as more distant regulatory sequences known as enhancers. Promoters work because they bind proteins known as transcription factors. Increasing access of transcription factors to the promoters will activate genes, whereas hiding or mutating the promoter regions will down-regulate the gene's function. Therefore, a major influence on gene expression occurs through folding of the chromatin (the complex of DNA and histone proteins in which the genetic material is packaged inside the cells of eukaryotes). Chromatin structure is dynamic, and changes can be inherited by the next generation, independent of the DNA sequence.

ALTERNATIVE SPLICING

An important finding from the Human Genome Project mentioned earlier is that genes occupy only about 1–2% of the human genome sequence (see Table 1.6). The rest of the DNA used to be called "junk" DNA, although as described earlier, a component of the "junk" includes the non-coding small RNA species, and they are likely to play an important role in the diversity and regulation of the human genome. Despite the gene pool being relatively small, most genes have many exons, and their introns are very large (see the cystic fibrosis gene in Figure 3.10). The simple gene structure shown in Figure 2.4 comprising three exons and two introns is provided for illustrative purposes but is unusual as most genes are much larger. Therefore, another way in which protein complexity can be increased is through alternative splicing; i.e., a gene with five exons depicted as 1i2i3i4i5 (number = exon; i = intron) can produce a protein encoded by this genetic information (i.e., 12345). On the other hand, alternative splicing could leave out exons, which would be reflected in different proteins being produced: (1) Protein 1345 (exon 2 is left out), (2) Protein 145 (exons 2, 3 are missed), (3) Proteins 234, 15 and so on. The proteins produced from alternative splicing share some structure in common but otherwise would be different. Therefore, they are called isomorphs.

Alternative splicing helps to explain how the protein repertoire could be increased without changing the numbers of genes. How important it is and how it compares with the role played by small non-coding RNAs are not known. It is said that up to 50% of genes demonstrate alternative splicing and so have the potential to produce isomorphs. The example given above in which a gene made up of 5 exons can produce different isomorphs is likely to be conservative since there is now increasing evidence that introns and non-gene segments may expand the options for alternative splicing because they contain cryptic splice signals. As shown in Figure 2.4, signals at intron-exon boundaries, such as GT and AG dinucleotides, indicate where splicing should occur. If these signals are changed (or new ones are created) through mutations, then it is possible for the cell to mis-interpret the signals and splice in wrong regions. If an incorrect splice occurs, a genetic disease is possible. On the other hand, an alternative splice created in this way could produce different isomorphs.

EPIGENETIC CHANGES

The conventional view is that human genetic diseases result from changes in the sequence of DNA, i.e., muta-tions. However, this might be an oversimplification since it ignores the fact that identical twins have essentially the same DNA content and share a lot of the environment, yet they can develop different genetic defects. Further-more, all cells in an individual have the identical DNA profile, yet the expression of genes is tightly regulated depending on tissue needs. A mechanism that might explain the above anomalies is epigenetics, i.e., herita-ble changes in the pattern of gene expression mediated by mechanisms other than changes in the primary DNA sequence of a gene. One epigenetic way to influence gene expression is via DNA methylation.

The methylated form of the base cytosine is sometimes called the fifth nucleotide base, i.e., A, T, G, C and methyl C. An enzyme, DNA methyltransferase, found in many species adds a methyl group to some cytosines at the C5 position in DNA. For some time now, it has been known that genes with methylated cytosine do not express; i.e., they are turned off. Regulation of gene expression without altering the nucleotide sequence is known as an epigenetic change. Epigenetic processes are considered to be essential to the development and differentiation of the organism, although how they arise and how they are controlled remain a matter of debate. Even the basic observation that methylated cytosine means the gene is no longer functioning is still not fully understood; i.e., is the methylation a primary event, or does it occur once the gene is no longer expressing?

Methylation has been implicated in the development of cancer as well as imprinting, an unusual form of genetic inheritance. Hence, epigenetics is a focus of much research to determine how factors apart from the genetic sequence can alter gene expression (see Chap-ters 4, 7). It is worthwhile emphasising that epigenetic changes can be inherited, and the methylation state has been shown to influence histones and so chromatin folding and gene expression. Methylated histones are usually associated with a transcriptionally inactive state, whereas acetylated histones are usually found in tran-scriptionally active chromatin. It was also mentioned earlier that dsRNA may also play a role in gene regula-tion through methylation. An interesting observation has been made about animals that have been cloned by somatic cell nuclear transfer (SCNT), for example, Dolly, the sheep (see Chapter 6 and Figure 6.9). A common feature in these animals is a low frequency of develop-ment to term and a high incidence of malformations. To explain this, it has been proposed that cloning by SCNT is limited by the inability to reprogram the epigenetic status of the transferred cell nucleus or the recipient ovum cytoplasm. Some interest is now being shown in the possibility that epigenetic changes may explain why babies born by assisted reproductive technologies may be smaller than normal, and perhaps have a higher risk for certain genetic abnormalities (discussed further in Chapter 7).

POLYMERASE CHAIN REACTION (PCR)

DNA AMPLIFICATION

The ability to target a segment of DNA (also RNA) and then produce multiple amounts of that segment is known as DNA amplification. The usual method to achieve this is by the polymerase chain reaction (abbreviated to PCR). Although details about molecular technologies including PCR are found in the Appendix, PCR is also described separately in this chapter because it is a core technique likely to be involved in most molecular medicine appli-cations. It is one of the very few technologies that should be understood by health professionals.

PCR utilises a DNA extension enzyme (DNA poly-merase) that can add nucleotide bases once a template is provided (Figure 2.5). There are three basic steps in PCR:

(1) Denaturation of double-stranded DNA into its single-stranded form. (2) Annealing of oligonucleotide primers to both ends of a target sequence. The oligonucleotides are a type of DNA probe; i.e., they are constructed so that they are complementary to target DNA, but unlike DNA probes, primers are more likely to be used in a technique such as PCR than for detecting DNA mutations. This com-plementarity, which extends over a distance of about 20 bases, is enough to give PCR its specificity; i.e., it will not bind to other regions of the genome. Oligonucleotide primers are readily available commercially, and so pro-vided there is knowledge of the DNA sequence, setting up a PCR-based test is relatively straightforward. (3) Addition of the four nucleotide bases and a polymerase. *Taq* poly-merase is used since it is relatively heat resistant and so

the denaturation step can be incorporated into the overall cycle without interfering with the polymerase activity. One PCR cycle comprises steps 1–3 described. After a cycle, each of the single-stranded DNA target segments has become double-stranded through the polymerase's activities. The cycle is then repeated, and each time a new target segment of DNA is synthesised. Theoretically, the number of templates produced equals 2^n; i.e., after 20 cycles of amplification there should be somewhere near 1×10^6 templates.

Amplified DNA products can be visualised in a number of ways. The simplest makes use of elec-

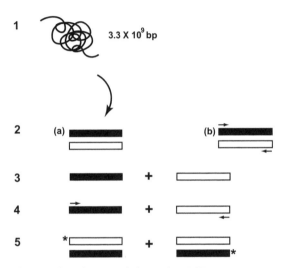

1 3.3 X 10^9 bp

2 (a) (b)

3 +

4 +

5 * +

Fig. 2.5 The polymerase chain reaction (PCR).
This technique allows amplification of a targeted DNA sequence by using a DNA thermostable extension enzyme (polymerase) to make new copies of the sequence. Oligonucleotide primers give PCR its specificity. **(1)** In this example a small sequence of DNA (say 600 bp in size) from the total 3.3×10^9 bp human genome is required. **(2a)** The sequence of interest is depicted, and in **(b)** the PCR primers are designed to flank the ends of this sequence. The primers ($\rightarrow$,$\leftarrow$) are in fact single-stranded DNA sequences complementary to the ends of the targeted sequence, which is double-stranded DNA (dark and light blocks). **(3)** Double-stranded DNA is made single-stranded by heating to about 94°C. **(4)** The DNA is allowed to cool to about 55°C, which allows the primers to stick to the single-stranded DNA at either end. **(5)** *Taq* DNA polymerase (a thermostable DNA polymerase) and a mixture of the four nucleotide bases are added and the temperature elevated to about 72°C, which allows the *Taq* polymerase to work. The combination of primers, nucleotide bases and the polymerase will lead to a copying of the single-stranded segment from the primer. The new copied fragments of DNA are indicated *. The final product is double-stranded DNA, which comes from the region defined by the primers. At this stage of the PCR, an initial DNA template has been duplicated. Steps 3–5 are repeated to produce (in theory if the process is 100% efficient) 2^n times the amount of template DNA (where n = number of cycles); e.g., 20 cycles should amplify the original segment about 1×10^6 times.

trophoresis and staining of DNA with ethidium bromide, which intercalates between the nucleotide bases and fluoresces under ultraviolet light (Figure 2.6). Radiolabelled nucleotides can also be incorporated into the PCR steps and the amplified products visualised by autoradiography. Color-based assays are now more commonly used with PCR. They depend on oligonucleotide primers that have been modified to allow them to be labelled with chemicals such as biotin or fluorescein.

A feature of PCR is its exquisite sensitivity so that even DNA from a single cell can be amplified. The ability of PCR to amplify small numbers of target molecules has been put to use in detecting illegitimate transcription. As described earlier, mRNA is tissue-specific except for some leakiness that occurs in cells such as the lymphocyte. Thus, mRNA specific for muscle tissue in disorders like Duchenne muscular dystrophy and the hereditary cardiomyopathies has been characterised by amplification of mRNA from lymphocytes. PCR is rapid and automated with a 30-cycle procedure taking approximately 0.5–3 hours to complete. New modifications or innovations are constantly being described.

The uses for PCR are many. It is a core technique for genetic DNA testing. PCR enables a segment of DNA to be screened for polymorphisms or point mutations. In the infectious diseases, PCR can detect DNA or RNA from microorganisms that are present in too few numbers to be visualised, or whose growth characteristics mean there will be a delay in diagnosis. The sensitivity of PCR is useful in forensic pathology where tissue left at the scene of the crime is small in amount or degraded. Research applications utilising PCR are extensive. Amplification has enabled cloning, direct sequencing and the

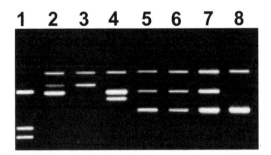

1 2 3 4 5 6 7 8

Fig. 2.6 Ethidium bromide intercalates into dsDNA, and its presence can be detected with UV light.
This photograph is of a gel in which DNA has been electrophoresed after PCR. Track 1 is the DNA size marker, and tracks 2–8 represent different DNA samples. The band patterns are complex because this is a multiplex PCR looking for various types of deletions in the α globin complex (and so producing α thalassaemia). To distinguish the patterns, the gel is immersed in ethidium bromide. The excess stain is washed off, and then the presence of DNA is detected by using ultraviolet light.

creation of mutations in DNA segments. Investigation of DNA/protein interactions by a method called footprinting is possible. RNA may also be studied along the same lines as described for DNA. An additional step, which incorporates the enzyme reverse transcriptase, enables RNA to be first converted to cDNA, from which amplification is then able to proceed.

VARIANTS OF PCR

The utility of PCR is limited only by the ingenuity of the scientist. Variations of this technique include: **Multiplex PCR** refers to the mixing of various primer combinations so that simultaneous amplifications are occurring. This approach is valuable when testing for multiple DNA mutations simultaneously (see Figure 2.6). **Nested PCR** describes the use of two sets of primers, the second of which lies within the first set of primers. This approach can increase the sensitivity and specificity of PCR. *In situ* **PCR** allows genes, or more commonly mRNA, to be identified in tissue sections. **RT-PCR** (reverse transcriptase-PCR) is used to amplify RNA. **Long PCR** allows large segments of DNA to be amplified. Conventional PCR products are usually relatively small fragments measuring in the hundreds of base pairs to 5 kb (kb—kilobase or 1000 base pairs). With long PCR, amplified DNA up to 40 kb in size is possible.

An important development since the second edition of *Molecular Medicine* has been **quantitative PCR** (Q-PCR). Until real-time PCR became available, quantification of end-point DNA (or mRNA) products was imprecise because individual amplification reactions had variable efficiency, and once PCR had progressed to a certain extent (plateau phase), the amplification stopped. The variability in PCR reflected the product concentration, limiting substrates in the reaction mixture and the presence of PCR inhibitors. However, that has now changed because real-time PCR allows the amplification to be monitored as it progresses. In one method, a dye is released with each amplification cycle, and measurement of this dye makes it possible to monitor the PCR in real time to ensure that quantitation of DNA occurs in the exponential phase. A graph plotting dye versus number of PCR cycles is drawn, and the quantitation is based on the number of PCR cycles required to reach a designated threshold. This is called the cycle threshold (Ct). Ct values are directly proportional to the amount of starting template, and so mRNA expression levels or DNA copy number can be calculated (Figure 2.7).

ERRORS WITH PCR

Like any laboratory technique, PCR can produce the wrong result. Because of PCR's exquisite sensitivity, contamination from another DNA source is always a potential problem. Contaminating DNA can come from other samples, or even the operator. The commonest source for contamination is amplified products from previous tests. For the genetic disorders, contamination is not usually a problem if the laboratory maintains a high standard; i.e., it is avoidable. Contamination becomes more of an issue with infectious disease DNA testing (see Chapter 8).

The sequence fidelity of amplified products is an additional consideration when assessing the usefulness of DNA amplification since *in vitro* DNA synthesis is an error-prone process. The error rate associated with *Taq* DNA polymerase activity is relatively low and has been estimated to be about 0.25%, i.e., one misincorporation per 400 bases over 30 cycles. The more recent commercially produced *Taq* polymerases have much lower reading errors, and there are a number on the market with different properties depending on the type of test being undertaken.

False negative or false positive results with PCR comprise another source of error. The purity of the DNA can be critical. Usually, this is not an issue if DNA has been prepared from a large, clean source, e.g., a blood sample. However, if DNA is isolated from a small or inadequate specimen, there is always the potential that PCR will not work optimally because the DNA is contaminated. In this scenario, "allele drop-out" might occur; i.e., one of the two alleles does not amplify efficiently, if at all. This produces an error through a misinterpretation of the PCR products that appear to be present. Another source of error occurs if there is a DNA polymorphism in the DNA primer binding site. This has the potential to interfere with the binding of the primer to the target DNA, and so it will not amplify, producing the same error described when allele drop-out occurs (Figure 2.8).

The health professional ordering a PCR-based test must not be overawed because it involves DNA but should always remember that mistakes (including clerical ones) can occur. The potential for error in genetic DNA testing is an important issue since patients, including the fetus, are likely to have no clinical features of the disease to guide the health professional. This is a real concern in the predictive tests (see DNA Mutations and Polymorphisms below). In this circumstance, an incorrect result may not be discovered for many years, and by then a number of clinical and personal decisions will have been taken. Remember also that a DNA test provides information not only about the patient, but has potential implications for others in the family who share that DNA (see Chapter 10 for more discussion on duty of care). An error in terms of the patient is also an error for other family members. Because of this, it is a wise practice that all important DNA tests, particularly the predictive ones, should be repeated, or tested in duplicate blood samples to reduce the potential for clerical mistakes (probably the most common error in the pathology setting), and errors associated with the assay itself.

Amplification Plot

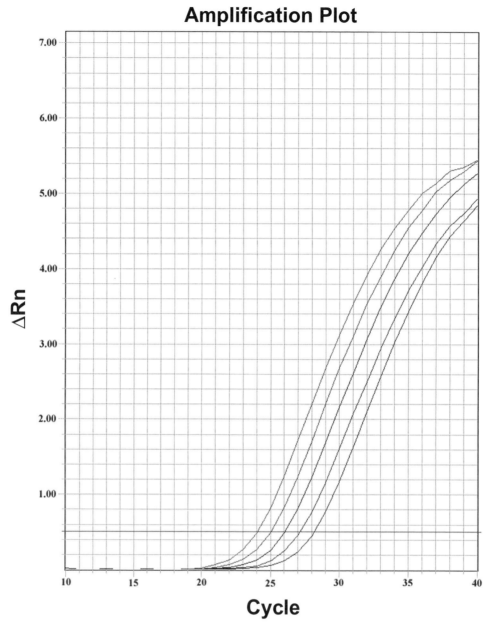

Fig. 2.7   Real-time PCR to quantitate DNA (or mRNA).
As PCR progresses, dye that has been incorporated into the reaction is released, allowing real-time monitoring of the DNA amplification. PCR products can now be measured during the exponential phase. The plot illustrated shows on the R_n axis the fluorescence signal plotted against the PCR cycle number. A threshold for the dye signal is determined. The threshold is calculated as ×10 the standard deviation of the average signal of the baseline dye signal. The threshold in this experiment is the horizontal line just above the Cycle axis. The threshold becomes the point to compare all PCR products. Five samples are illustrated. They represent 50% serial dilutions of mRNA for cardiac actin. The number of cycles required to reach the threshold is called the cycle threshold (Ct). In the experiment shown, the left reaction Ct is 24 cycles (the next sample has a Ct of 25 cycles and so on). The Ct values of the two samples are used to calculate the relative amount of the templates; i.e., 25 − 24 = 1 cycle difference or a Ct of 1. Because of the exponential nature of PCR, this means a two-fold difference; i.e., the two samples with Ct 24 and Ct 25 are confirmed to be 50% dilutions. *Photo courtesy of Dr Bing Yu, Department of Molecular & Clinical Genetics, Royal Prince Alfred Hospital.*

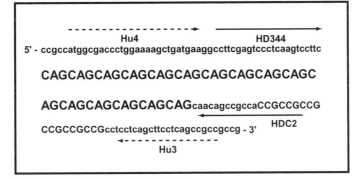

Fig. 2.8 PCR and primer binding as a source of error.
Depicted is the DNA sequence for the beginning of the Huntington's disease gene (see Chapter 3, Table 3.4). The sequence is read from left to right starting with ccg . . . in the top line. Large uppercase letters identify the $(CAG)_n$ triplet associated with the development of Huntington's disease, and small uppercase letters at the end of line 3 show the adjacent $(CCG)_n$ repeat, which is non-pathogenic. In determining whether a patient has Huntington's disease, the size of the $(CAG)_n$ is measured. If there are 40 or more repeats, the patient has the disease. To measure the repeats, two PCR primers were designed that flank the repeats (HD344 and HDC2). These primers can give a false result if there is a polymorphism along the primer binding site; i.e., one allele may not be amplified. Let us assume that the non-amplifying allele has 41 repeats. The remaining (second) allele will amplify because it does not have the polymorphism to interfere with primer binding. Let us assume that the second allele has 20 repeats. Because there is only one allele, the result could be falsely interpreted as being homozygous 20 and 20 (usually written 20 / 20). To avoid this error, a second set of primers (Hu4 and Hu3) is used for a confirmatory PCR. In this case, the PCR measures both the $(CAG)_n$ and the $(CCG)_n$ but this does not matter because it will show quickly that the patient's DNA could not possibly be homozygous 20 / 20. The second set of primers is designed so that they do not overlap the first. References to the above primer sets are: Clinical Genetics 1999; 55:198–202 and Human Molecular Genetics 1993; 2:637.

CHROMOSOMES

STRUCTURE

Chromosomes are thread-like elements in the cell nucleus. Each chromosome has a constriction called the centromere, which divides chromosomes into short (p for petite) and long (q) arms. The tip of each chromosome is the telomere. The latter is important for sealing the end of the chromosome and maintaining stability and integrity. The telomere comprises mainly tandem DNA repeats, and the size of the telomere is maintained by an enzyme known as telomerase. Shortening of the telomere may be involved in cell death and the ageing process (see Chapter 4 for further discussion of ageing). Each somatic cell contains two sets of chromosomes inherited from the parents. Humans have 22 sets of autosomes and 2 sex chromosomes, i.e., a total of 46 chromosomes.

Chromosomes comprise both DNA and histone protein. This combination is called the chromatin. In the nucleus, chromosomes are packed tightly, which allows a large amount of DNA to be located within a small space. Packing also plays a role in gene regulation as discussed earlier. When stained, chromosomes demonstrate light and dark bands. The light bands identify euchro-matin, which contains actively expressing genes. The dark bands are the heterochromatin, which is largely composed of repetitive nonexpressing DNA.

The study of chromosomes is called cytogenetics. This discipline has relied heavily on a number of standard techniques such as cell culture, chromosomal isolation, staining and recognition patterns to identify normal and abnormal karyotypes (karyotype—the number, size and morphology of chromosomes in a cell, individual or species). More recent developments have enabled auto-mated analysis of karyotypes and DNA-based chromo-some analytic tools.

KARYOTYPE

A karyotype describes an individual's chromosomal con-stitution. It was only in 1956 that the human diploid chromosome number was shown to be 46. During the 1970s, methods were developed to distinguish bands within individual chromosomes. Each of the 44 human autosome chromosomes and the X or Y sex chromosomes can now be counted and characterised by banding tech-niques. The most common, called G-banding, involves trypsin treatment of chromosomes followed by staining with Giemsa. G-banding produces a pattern of light and dark staining bands for each chromosome (Figure 2.9). The banding patterns, the size of the chromosome and the position of the centromere enable the accurate identification of each individual chromosome.

The short and long arms of a chromosome are divided into regions that are marked by specific landmarks. Regions comprise one or more bands. Regions and bands are numbered from the centromere to the telomere along each arm (Figure 2.10). Therefore, each band will have four descriptive components. For example, the cystic fibrosis locus on chromosome 7q31 defines a band involving chromosome 7, on the long arm at region 3 and band 1. Additional information is available by higher

resolution banding techniques that enable sub-bands to be identified. In the case of the cystic fibrosis locus, this becomes 7q31.3, where the .3 defines the sub-band (see Figure 2.10).

FLUORESCENCE IN SITU HYBRIDISATION

Even greater resolution than was possible by chromosomal banding became available through the development in the early 1980s of in situ hybridisation. This technique combined conventional cytogenetics with DNA probes (molecular genetics technology). Radiolabelled DNA probes were initially used to define the chromosomal localisation of single copy DNA sequences. The emergence of non-isotopic in situ hybridisation, particularly

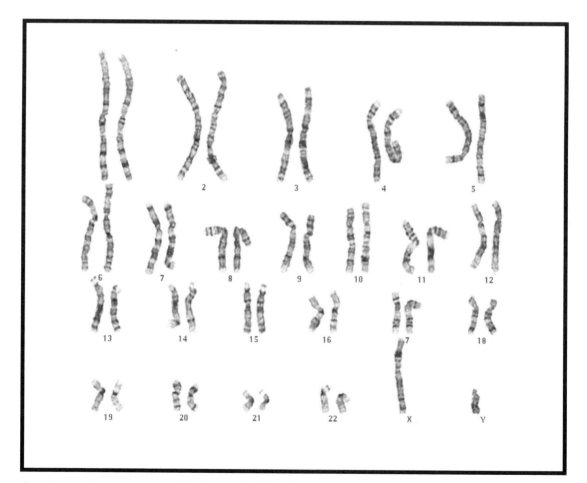

Fig. 2.9 A normal human karyotype (46,XY) illustrating G-banding.
A normal individual has 44 autosomal (non-sex) and 2 sex chromosomes. They can be counted and characterised by staining techniques that produce bands. Note the light and dark bands on the chromosomes. *Karyotype provided by Pauline Dalzell, Molecular & Cytogenetics Unit, South Eastern Area Laboratory Services, Sydney.*

CHROMOSOME 7

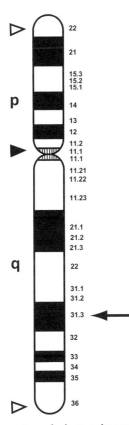

Fig. 2.10 Banding patterns for human chromosome 7.
The individual bands are designated by numbers. The short and long arms are shown by p and q, respectively; the centromere, by a filled triangle; and the telomeres, by open triangles. An arrow marks position q31.3.

fluorescence in situ hybridisation (abbreviated to FISH), has greatly enhanced the utility of this technique. FISH works on the same basis as a DNA probe; i.e., a single-stranded DNA probe can be made to anneal to its complementary single-stranded sequence in the genome. In the case of FISH, the genome is not total DNA used for DNA mutation testing, but chromosomes on a metaphase spread. Resting (interphase) chromosomes can also be studied with FISH (Figure 2.11).

The potential to use a number of DNA probes each labelled with a different fluorochrome in the same procedure means that separate loci can be identified, comparisons can be made and relationships to the centromere and telomeres established. Probes can be purchased allowing different colours to be assigned to the chromosomes, thereby identifying them more easily by their unique colour. With FISH, genes can be localised on chromosomes, or chromosomes and chromosomal rearrangements can be identified.

Chromosomal abnormalities include (1) Numerical, e.g., aneuploidies (monosomy, trisomy), polyploidies (triploidy, tetraploidy); (2) Structural, e.g., translocations, deletions, inversions, isochromosomes; (3) Cell line mixtures, e.g., mosaicism, chimaerism. Although the great majority of these abnormalities are detectable by conventional cytogenetic approaches, some cannot and they provide an important use of FISH. The advantage of FISH is that DNA probes enable specific chromosomes or segments of chromosomes to be accurately identified. Apart from detecting chromosomal changes associated with genetic disorders, particularly in relation to reproductive genetics, FISH has been very useful in characterising somatic cell chromosomal rearrangements that occur in cancers and haematological malignancies such as leukaemia (see Chapter 4). FISH has now emerged as the bridge between conventional cytogenetics and molecular DNA genetic testing, and has proven to be a rapid, efficient and highly sensitive molecular cytogenetic technique.

CHROMOSOME 21

An important genetic disorder on chromosome 21 is Down's syndrome, which occurs in approximately 1 in 650 to 1 in 1000 live births (Box 2.1). Chromosomal abnormalities in Down's syndrome include (1) Free trisomy ~95%; (2) Translocations ~5%; and (3) About 2–4% of cases of free trisomy 21 also have mosaicism for a trisomy and normal cell lines. Most trisomy cases involve an additional maternal chromosome 21 that has arisen by non-disjunction (see also Uniparental Disomy in Chapter 4 for other examples of non-disjunction). In the trisomy cases, causation and recurrence risks relate to maternal age. Because of its relatively small size and the association with Down's syndrome, chromosome 21 was a priority for sequencing in the Human Genome Project. The chromosome's long arm was completely sequenced in 2000, and a number of genes are now being studied as likely candidates. In addition to explaining the mental retardation and the characteristic phenotype of Down's syndrome, molecular characterisation is expected to provide additional information on the mechanism for non-disjunction, an important cause of autosomal trisomies.

Two other genes identified on chromosome 21 are (1) *SOD1*—superoxide dismutase 1, which is involved in a proportion of familial autosomal dominant amyotrophic lateral sclerosis (motor neuron disease); (2) *APP*—amyloid precursor protein, which is associated with some rare forms of autosomal dominant Alzheimer's disease discussed further in Chapter 4.

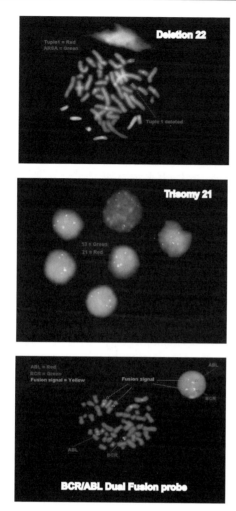

Fig. 2.11 FISH.
Top: The metaphase FISH of a deleted chromosome 22 shows a chromosome spread. Chromosome 22 is identified using two DNA probes (coloured red and green). One of the two chromosomes 22 is abnormal; i.e., there is only the red probe because the area of chromosome 22 detected by the green probe has been deleted. **Middle:** This photo shows an interphase FISH, which has several advantages over metaphase FISH. It can score a larger number of cells, thus increasing the chance of detecting a chromosomal rearrangement, particularly in low-level mosaic states. Interphase FISH is more rapid, which is desirable in prenatal diagnosis. This particular picture illustrates trisomy 21 (Down's syndrome) because it has an additional signal with the red (chromosome 21) DNA probe. **Bottom:** This composite photo shows a metaphase and interphase FISH for the *BCR-ABL* fusion gene found in chronic myeloid leukaemia. The red and green DNA probes for the *ABL* and *BCR* genes will show up as a yellow signal when both are present in the translocated rearrangement (read more about chronic myeloid leukaemia and the *BCR-ABL* rearrangement in Chapter 4). *FISH provided by Dr Michael Buckley, Molecular & Cytogenetics Unit, South Eastern Area Laboratory Services, Sydney.* (see colour insert)

Box 2.1   Chromosomal disorder: Down's syndrome (Trisomy 21).

Down's syndrome is the most frequent genetic cause of mental retardation. It results from trisomy 21 described in 1959. Chromosome 21, one of the smallest human chromosomes, has been completely sequenced. Despite this, the pathogenesis of Down's syndrome remains elusive although the region of involvement in Down's syndrome has been narrowed to about 5 Mb (Mb = megabase or 1×10^6 base pairs) called the Down's syndrome critical region (DSCR). In this region, four genes (*DSCR1*, *DSCR2*, *DSCR3* and *DSCR4*) are implicated in this disorder. *DSCR1* is particularly interesting because (1) It is over-expressed in the Down's syndrome fetal brain, and (2) It has been possible to mutate a gene known as *nebula* in *Drosophila melanogaster*. This gene is the ortholog of *DSCR1*, and flies that overexpress or demonstrate a loss of function of *nebula* have severe learning defects. *DSCR1* forms part of the calcineurin inhibitor family called calcipressin, thereby providing a focus for further study in Down's syndrome. Chromosome 21 has about 400 genes, and, interestingly, there seems to be about 16 genes predicted to be involved in mitochondrial energy generation and reactive oxygen species metabolism. This is relevant since there is evidence linking Down's syndrome to mitochondrial dysfunction. There is also a cluster of genes involved in folate metabolism, which is another intriguing association because folate supplementation in early pregnancy reduces the risk for neural tube defects. This is discussed further in Chapter 7 (Roizen and Patterson 2003; Chang *et al* 2003).

CHROMOSOMES 13 AND 19

The most recent of the human chromosomes to be completely sequenced in 2004 were chromosomes 13 and 19. They have shown some interesting contrasts. Chromosome 19 is about 55 Mb in size, and codes for nearly 1500 genes. As such, it is the most gene rich of all chromosomes; i.e., it has about twice the average number of genes expected for its size. Chromosome 19 also shows that large areas are strongly conserved, based on comparisons with DNA sequences in other species. Of particular interest is the observation that some conserved areas include non-coding DNA. This refutes further the concept of "junk" DNA and provides additional proof that the non-coding regions of the genome are likely to play an important role in gene function and regulation.

Compared to chromosome 19, chromosome 13 is larger at 96 Mb, but has only about 633 genes. Surprisingly, within chromosome 13 there is a large segment described as a desert with very few genes (only 47 genes in about 40 Mb of DNA). Approximately 100 of the genes found in chromosome 13 are thought to involve non-coding RNAs. Two of these (miR-15 and miR-16) are located within a region that is frequently deleted in chronic lymphocytic leukaemia. Both these genes are deleted or down-regulated in about 70% of cases involving this form of leukaemia, which is the most common found in adults from the Western world.

DNA MUTATIONS AND POLYMORPHISMS

DNA MUTATIONS

An important and practical use of molecular medicine is in DNA diagnosis. For this, mutations are sought in DNA using a range of techniques. The word **heterogeneity** will frequently appear when describing DNA mutations since, with very few exceptions, the number and types of DNA mutations that can inactivate a gene are extensive, and range from single base changes to complex chromosomal rearrangements. Even within each particular group, many different mutations are associated with the same clinical disorder. For example, more than 1300 mutations produce cystic fibrosis. A recent survey of mutations showed that single base changes were the most common, occurring in about 60% of genetic disorders, with deletions the next most common at 22% (Table 2.3).

Because of the heterogeneity of mutations, there has been some confusion with nomenclature. Therefore, attempts have been made to develop a uniform but informative system (Table 2.4). For example, ΔF508 means deletion (Δ) of the amino acid phenylalanine at position 508 (the commonest mutation in cystic fibrosis), and Cys282Tyr means replacement of cysteine by tyrosine at position 282 (the commonest mutation in hereditary haemochromatosis). While this degree of detail might seem esoteric to the practising clinician, it is worth noting that patients (and families) with genetic disease often know a lot about their disorder and regularly access the Internet to learn about new developments. Hence, a health professional who does not understand what a mutation means is disadvantaged very early on in the consultation if the patient knows.

Despite the heterogeneity of DNA mutations, PCR remains useful provided there is DNA sequence information so that primers can be designed to bind to either side of the abnormality (Figure 2.12). Sometimes it is difficult to be sure that a DNA change is a mutation or is simply a neutral change (such as a DNA polymorphism). A number of criteria are used in this circumstance to distinguish whether the change in the DNA is likely to lead to a genetic defect (hence, it is a mutation) or is simply a DNA polymorphism (Table 2.5).

DNA POLYMORPHISMS

Most changes in the DNA sequence do not lead to disease. These changes are called polymorphisms. By convention, a polymorphism is a difference in DNA sequence between individuals that occurs in >1% of the

Table 2.3 Relative frequency of DNA mutations (data from the Human Gene Mutation Database quoted in Botstein and Risch 2003)

Type of mutation[a]	Percentage of total
Deletion	21.8
Insertion/duplication	6.8
Complex rearrangement	1.8
DNA repeat abnormality	0.1
Missense/nonsense	58.9
Splicing	9.8
Regulatory	0.8

[a] Total number of mutations counted were 27 027.

Table 2.4 Nomenclature for DNA mutations (Antonarakis *et al* 1998)

Abbreviation	Explanation of mutation
R97X	X = stop codon; hence, this mutation indicates that the amino acid arginine (R) at position 97 has been replaced by a stop codon.
Y97S	The amino acid Y (tyrosine) at codon 97 is replaced by the amino acid S (serine). Another way to write this is Tyr97Ser.
T97del	The amino acid threonine at position 97 has been deleted. However, the common mutation in cystic fibrosis is written ΔF508, which means the amino acid phenylalanine is deleted at position 508, so there is some discrepancy in the way some mutations are written.
1997delT	This indicates that there is a deletion of the nucleotide T (thymine) at nucleotide position 1997.
1997–1998insT	This describes an insertion of one nucleotide T between nucleotides 1997 and 1998.
IVS1,110 G>C	This indicates a G to C change in the first intervening sequence (IVS1) at nucleotide position 110 (this type of mutation might affect splicing).

Table 2.5 Distinguishing disease-causing DNA mutations from neutral DNA polymorphisms—most dilemmas will involve single base changes (Cotton and Scriver 1998)

Finding in DNA	Explanation
Single base change	Either a disease-causing mutation or a neutral (non-pathogenic) change called a DNA polymorphism.
Single base change with a second mutation in cis	Finding a single base change in a gene does not automatically mean that the change is the cause of the disease. Another explanation is a second and causative mutation in cis, i.e., located nearby at the same locus as the first "mutation" that was detected.
Single base change leading to an alteration in the amino acid	Changing one amino acid into another (non-synonymous change) may or may not produce disease. It is more likely to be disease causing if the original amino acid was conserved and/or the change in amino acid was significant, e.g., producing a change in charge.
Single base change in the third nucleotide of a codon (see Table 2.1)	Because the genetic code is redundant, the third nucleotide change may not alter the amino acid (synonymous change), and so there is no overall effect. However, the new third nucleotide may create a new splice site or interfere with mRNA stability, and so synonymous changes can still be pathogenic.
Single base changes leading to a nonsense codon (TAA, TGA or TAG) or a frameshift.	These are usually pathogenic.

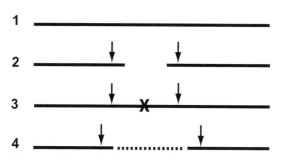

Fig. 2.12 PCR can identify a range of DNA mutations.
(1) A normal stretch of DNA sequence. PCR can be used to amplify any segment of this DNA provided the sequence is known to allow appropriate primers to be designed. (2) A deletion in DNA is shown. PCR will detect this if primers are designed on either side of the deletion (shown by ↓). If the deletion is small, both normal and deleted fragments will be detected with the same primers. If the deletion is very large, primers depicted will detect only the deleted fragment. (3) X indicates a single base change in the DNA. The primers shown by the ↓ on either side of the X will allow that region to be amplified by PCR. Then the single base change can be detected using various DNA methods (described in the Appendix). (4) The . . . represents a DNA rearrangement. DNA primers designed on either side of this rearrangement will detect it.

population. Since only about 1–2% of the human genome contains sequence for genes (see Table 1.6), the great majority of polymorphisms will not directly affect gene function although a polymorphism falling within a regulatory region in the genome might have functional significance. A polymorphism that alters an amino acid in the gene is better described as a mutation although there will continue to be overlap (and some confusion)

between what is a mutation and what constitutes a polymorphism.

DNA polymorphisms are used for many purposes in molecular medicine—from forensic DNA typing (Chapter 9) to DNA linkage analysis, a technique that allows diseases to be traced through families. There are a number of different DNA polymorphisms: RFLP, VNTR, SSR, SNP (Figure 2.13, Table 2.6 and additional discussion in Chapter 9). In clinical medicine, the two relevant polymorphisms are the microsatellites (also called simple sequence repeats or SSRs) and single nucleotide polymorphisms (SNPs). SSR-type polymorphisms involve 2–4 base repeats such as $(AC)_n$ or $(GAA)_n$ where n can be any number. Up to the present, microsatellites have been the workhorse in molecular medicine. Now, considerable interest is being shown in SNPs. They are found at least 1 in each 1000 bp (bp—base pair). Because they are more common than microsatellites, SNPs are increasingly being used in gene discovery strategies for complex human traits. Pharmaceutical companies are also starting to target SNPs as a way by which individual responsiveness to drugs can be predicted. This would then allow patients in clinical trials to be stratified better into likely responders and likely non-responders, which should improve the evaluation of such trials (see Pharmacogenetics in Chapter 5 for more discussion).

DIRECT DETECTION

The detection of mutations in DNA has relevance to many fields in medicine. They include Genetics, to identify genetic disorders; Oncology, to look for mutations in cancer-producing genes; and Microbiology, for identifying infectious agents. Wherever possible, DNA tests

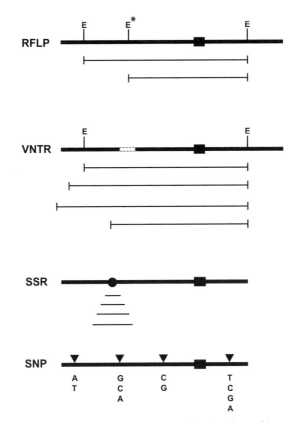

Fig. 2.13 Four types of DNA polymorphisms: RFLP, restriction fragment length polymorphisms; VNTR, variable number of tandem repeats; SSR, simple sequence repeats, i.e., microsatellites; and SNPs, single nucleotide polymorphisms.
DNA polymorphisms are produced by changes in the nucleotide sequence. They result from (1) variations in the fragment length pattern produced after digesting DNA with restriction enzymes, or (2) variations in the size of a DNA fragment after PCR amplification. **RFLP:** A segment of DNA is digested with a restriction enzyme "E." This segment can be identified in Southern analysis by using a DNA probe that will hybridise against the segment marked (■), or PCR can be used to amplify this specific region. RFLPs are caused by point mutations affecting a single restriction enzyme recognition site (E*), which will either be absent or present. If absent, enzyme E will digest the DNA at the two outside E sites; if present, enzyme E will digest DNA at E* as well as the two outer E sites, although because of the place where the probe is located, only the fragment generated from the E* and the E site on the right will be detected. Therefore, RFLPs are biallelic; i.e., they give two options (large or small) depending on whether the polymorphic restriction fragment site (*) is absent or present, respectively. **VNTR:** In contrast, the multiallelic VNTR has the potential to be more polymorphic (and so more informative) since the changes in the E-specific restriction fragment are brought about by the insertion of a variable number of repeat units at the polymorphic site (hatched area). Thus, the number of polymorphic DNA fragments generated is potentially much greater (e.g., the four different sized fragments illustrated). Because of their greater intrinsic variability, VNTRs are usually more informative in polymorphism studies since there is a greater chance that heterozygous patterns will be detected at any one locus. Examples of VNTRs detected by restriction enzyme digests are the minisatellites. **SSR:** In contrast to RFLPs or VNTRs, which can be identified by Southern analysis and PCR, SSRs are much smaller in size and so are detectable only if PCR is used. SSRs are the microsatellites, and they are polymorphic because of repeats in simple sequences (●) such as a $(CA)_n$ where n can be any number. Amplification of DNA containing an SSR will produce fragments of variable size. **SNPs:** These single base changes are different to the RFLPs because they do not necessarily result in a change in the restriction enzyme digestion pattern. They are single base changes that can be biallelic, or there can be more than two alleles for any SNP. Because these are found frequently throughout the genome (four are depicted here in relationship to ■, which could represent a gene), they have the potential to be powerful DNA polymorphisms. At present, their individual detection is technically difficult and expensive, and so they are not being used to their capacity.

Table 2.6   DNA polymorphisms

Class of polymorphism	Description
Macrosatellite	Small units of DNA are repeated in tandem thousands of times—hence, VNTRs (variable number of tandem repeats). This large polymorphism is found mostly in centromeres and telomeres.
Minisatellite	Repeat units are larger than macrosatellites, but there are fewer. Also an example of a VNTR.
Microsatellite	These involve small repeats (e.g., 2–4 bases in size); hence, they are called SSR (simple sequence repeat). Microsatellites are very useful in linkage analysis and form the basis of unstable triplet repeats in some neurologic disorders.
Single nucleotide polymorphism (SNP)	Single base changes throughout the genome occur at least 1 in every 1000 base pairs. The Human Genome Project has greatly facilitated their discovery, and there is expectation that SNPs will allow the complex genetic disorders to be characterised.

involving direct detection for a mutation are preferred. However, other strategies are available if the direct test is not applicable, for example, scanning or linkage analysis that will be described below.

The gold standard for mutation detection is DNA sequencing. This is relatively time consuming and expensive, so it is usually required only for the initial identification and characterisation of a specific mutation. Thereafter, more rapid means are devised to detect the same mutation in other specimens. Just as DNA mutations are heterogeneous, so also could this term be used to describe the varied number of ways in which DNA mutations can be investigated. A selection of some approaches is given in Table 2.7. More specific details on methodology are provided in the Appendix.

Whichever mutation detection test is used, oligonucleotide DNA probes have widespread application in the diagnosis of genetic disorders because PCR allows the target DNA to be amplified, and with amplified DNA the relatively small oligonucleotide probe can hybridise efficiently to its target. These probes can be labelled with chemicals, enzymes or fluorochromes, thereby allowing changes in DNA to be detected on the basis of hybridisation or ligation. As indicated earlier, oligonucleotides are readily available commercially and can be designed in-house to ensure that they bind to a specific segment of target DNA.

DNA SCANNING

Because of the heterogeneity associated with DNA mutations, it is found that some mutations recur (i.e., they are present in many families), whereas others are family specific. The latter are sometimes called private mutations. Some mutations localise to particular regions of a gene known as hot spots, whereas others occur randomly throughout the gene. With current available technology, it is not possible to look for all mutations. In the future, this may change with technologies such as microarrays, described in Chapter 5. Therefore, DNA mutation proto-

cols usually involve the testing for common recurring mutations. The less common or family-specific ones may or may not be sought, depending on the laboratory's interest and available resources. If the uncommon mutations are to be found, an additional DNA technique is usually required to try and identify what region of the gene is likely to have this mutation. This is called DNA scanning (some refer to it as DNA screening, but this can be confused with another form of DNA screening described below).

In the past few years, the ability to scan amplified DNA products to identify which one is likely to contain a mutation has improved the detection rate for mutations. For example, with Duchenne muscular dystrophy or familial hypertrophic cardiomyopathy, there are a number of underlying mutations, and to complicate this further, the genes are too large (Duchenne muscular dystrophy) or there are too many (familial hypertrophic cardiomyopathy) to make DNA sequencing or any other standard DNA test a practical diagnostic approach. One solution involves the amplification of DNA in a number of adjacent segments. For example, the DNA might be 5 kb in size, but it is amplified in 10 segments, each of which is ~0.5 kb in size. The 10 segments are next scanned to see which might have a mutation in it. The technique now most effective for scanning is called dHPLC (denaturing High Performance Liquid Chromatography; see the Appendix for details). dHPLC will identify a segment with an altered nucleotide sequence because this will be reflected in a changed mobility for that DNA fragment. Once a likely fragment is found, the mutation must be confirmed by DNA sequencing. In terms of the example just given, the advantage of the dHPLC step is it allows the sequencing of a single fragment of 0.5 kb rather than the entire 5 kb.

LINKAGE ANALYSIS

An alternative approach to detecting a mutation in a gene with many mutations (for example, familial hypertrophic

Table 2.7 Strategies for identifying mutations in DNA[a]

Type of approach	Description	Applications
Direct sizing of a PCR product	A deletion of two or more bases (or insertions) can be detected by sizing a PCR fragment (see Figure A.9).	The ΔF508 deletion involving 3 bp is seen on electrophoresis by measuring the size of the PCR fragment generated.
RFLP (restriction fragment length polymorphism)	DNA is digested with restriction enzymes, and the presence of a single base change can be detected (see Figure A.1).	Restriction enzymes are less frequently used to detect DNA changes but remain useful approaches.
ASO (allele specific oligonucleotide)	A single-stranded labelled probe is used to hydridise against single-stranded target DNA, looking usually for single nucleotide changes (see Figure A.10).	ASOs are used to identify a wide range of DNA mutations as well as polymorphisms.
OLA (oligonucleotide ligation assay)	Two oligonucleotide probes are designed to hybridise adjacent to each other on the target sequence. Once adjacent, the two probes can be joined by DNA ligase (Figure A.11).	Useful for a range of mutations including insertions, deletions and single base changes. Can be multiplexed.
ARMS (amplification refractory mutation system)	Oligonucleotide primers are designed to amplify preferentially one of the two alleles (Figure A.12).	Useful for a range of mutations and can be used in multiplex PCR.
PTT (protein truncation test)	cDNA or genomic DNA in the region of the mutation is made into mRNA, and then the mRNA is translated into a protein (Figure A.13).	Used where there are nonsense or frameshift mutations leading to a truncated gene product.

[a] The above technologies are described in more detail in the Appendix.

cardiomyopathy; see Chapter 3) or investigating diseases that do not have common or recurring mutations (for example, haemophilia B) is linkage analysis. This is an indirect strategy since it does not specifically identify a defect in DNA but works on the basis of finding co-segregation between DNA polymorphisms and the disease phenotype being investigated in members of a family. Once established, the DNA polymorphism then becomes an indirect marker of disease for that family. The same DNA polymorphism has no particular relevance for another family with the same disease unless it can be established once again that the polymorphism and the clinical defect are linked.

The advantage of using a DNA polymorphism in a linkage study is that it provides an indirect marker for a locus or gene. A minimum requirement for a DNA linkage study is a family in which can be identified a key individual, i.e., someone who is unequivocally normal or affected. The key person is essential to allow assignment of the polymorphic markers to the wild-type (normal) or mutant alleles. DNA from each of the relevant family members is studied using DNA polymorphisms such as the microsatellites, and then linkage between the disease phenotype and a polymorphic marker is attempted (Figure 2.14).

In some circumstances, DNA polymorphisms are not informative; i.e., family members are homozygous for the polymorphic alleles. Therefore, the normal and mutant alleles are unable to be distinguished. In this case, additional DNA polymorphisms are sought until

informative ones are found. Studies based on a polymorphic linkage analysis strategy can be very useful, but they also have a number of potential problems. They include (1) The requirement for a family study that may not always be possible; (2) Non-paternity can produce erroneous linkage patterns; and (3) Recombination can lead to error.

Recombination is a function of the distance between a polymorphic marker and the gene of interest. Although grossly oversimplified, a **physical** distance of 1 Mb in DNA is roughly equivalent to a **genetic** distance of 1 cM (cM—centimorgan). 1 cM indicates a 1% recombination potential (i.e., in 100 meioses there will be one recombination event between the DNA polymorphism and the target DNA of interest). The use of intragenic polymorphisms (for example, SNPs located within the introns or exons of genes, or microsatellites found within introns) or polymorphisms located in the immediate 5′ or 3′ region of genes reduces the risk of recombination considerably.

Another trick when using DNA polymorphisms is to group a number together across a specific region into what is called a **haplotype**. In other words, a single DNA polymorphism may not be informative, but when it is used in conjunction with other polymorphisms, its value increases. In addition to increasing the informativeness of polymorphisms, haplotypes help to identify recombination events (Figure 2.15).

The problems described above preclude DNA linkage analysis as the primary method to detect mutations in

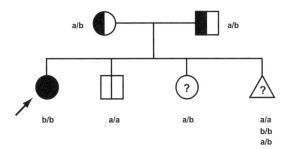

Fig. 2.14   DNA linkage study in β thalassaemia.
Understanding how DNA polymorphisms are used to follow a
disease within a family (called linkage analysis) is a difficult
concept to grasp. Essentially, a polymorphism is a DNA
marker that is used as a sign post for a chromosomal location
or to identify a gene. In the case of the β globin gene, each
individual has two genes (one on each chromosome 11),
and so two polymorphic markers should be detectable. To
undertake linkage analysis, the first step involves identifying
family-specific DNA polymorphic markers that will distinguish
the two β globin genes. The polymorphisms are not mutations
but simply DNA sequence changes that allow the two genes
to be marked. Once the polymorphisms are identified,
they are traced in a family and compared to the clinical
phenotypes. In the pedigree given, the two parents are
β thalassaemia carriers. Their carrier status is easily
determined by blood counts and special haematology tests for
thalassaemia. They have a female child who has homozygous
β thalassaemia (β thalassaemia major, indicated by →), and
they also have a normal male. The thalassaemia status for a
third (female) child is unknown (?), although a conventional
blood count would sort this out quickly. The mother is also
pregnant, and the fetus (indicated by a triangle) has an
unknown thalassaemia status. For the purpose of linkage
analysis, let us assume that the underlying β globin gene
mutations cannot be identified in this family. Therefore,
linkage analysis is the next approach to use. The two
polymorphisms that distinguish the two β globin genes in this
family are defined by the letters "a" and "b." Both the parents
are carriers and have the a/b polymorphic markers. This
information alone is not enough for diagnosis. The key
individual for this is the homozygous-affected child who is
b/b. This shows that the polymorphic marker "b" identifies the
mutant β thalassaemia gene **in this family**. Therefore, it can be
assumed that the marker "a" defines the normal gene, and this
is confirmed because the normal child is a/a. The child with
the unknown status is a/b, and so she must be a carrier (which
would have been more appropriately determined by a blood
count than a DNA test). Of more relevance, the fetus can have
three combinations, and they will predict the genetic status,
i.e., a/a (= normal), b/b (= homozygous-affected) and a/b (=
carrier).

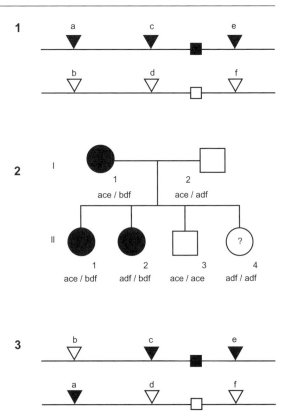

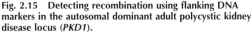

**Fig. 2.15   Detecting recombination using flanking DNA
markers in the autosomal dominant adult polycystic kidney
disease locus (*PKD1*).**
It is assumed that in this family the *PKD1* defect involves the
chromosome 16p locus. **(1)** The three polymorphic markers
and their alleles for the *PKD1* locus are a or b; c or d;
e or f. The filled box is the normal gene, and its associated
polymorphisms are the filled triangles; the open box is the
mutant gene, and its associated polymorphisms are the open
triangles. **(2)** The pedigree illustrates the segregation patterns
for the above three polymorphisms. I-1 (female) has *PKD1*.
Two of her children (II-1, II-2) are affected, and so they allow
the mutant-specific haplotype to be identified as bdf/since this
is what the three have in common. The one male offspring (III-
3) has not inherited the maternal bdf/haplotype, which is
consistent with his normal phenotype at age 50 years. The
remaining female sibling (II-4) is a problem. The adf/adf
genotype in her does not fit. Non-paternity can be excluded
since it is the maternal haplotype that is the problem. This is
an example of recombination that has occurred somewhere
between the a/b and the c/d loci (shown in panel **3**). The
mutant-specific haplotype has now become adf/rather than
bdc/. Therefore, II-4 has actually inherited the *PKD1* mutation.
Note: If only the one set of polymorphisms had been used in
this linkage study (i.e., a and b), the recombination event
would not have been detected, and II-4 would have been
incorrectly diagnosed as being normal.

genetic disorders. However, this approach is useful for research and, in the clinical context, for cases in which a mutation cannot be found or the gene is too large to make scanning a practical option. See the Appendix for further discussion of linkage analysis, and Figure A.16 for a worked example of linkage analysis in cystic fibrosis.

DNA TESTS

CLASSES OF DNA TESTS

A feature of genetic DNA tests that contrasts with all other pathology tests is their applicability to different clinical scenarios from a relatively simple (diagnostic) test to confirm a clinical diagnosis to the more complex situation in which predictions are being made about an individual's future risk for developing a genetic disease (Table 2.8). DNA tests are continually evolving, so examples given also contain reference to pharmacogenetic DNA tests that are not routinely in use, although they promise much for the future. DNA screening tests are currently the topic of much debate since many promote this approach to community genetics, whereas others are less convinced.

In some cases, the same DNA test falls into multiple classes; e.g., the finding of the common mutation Cys282Tyr for the haemochromatosis diagnostic DNA test described in Table 2.8 will confirm the clinical suspicion. A positive result will also automatically raise the risk for family members, particularly the siblings. DNA testing of the family members at risk would then constitute a different class of test, i.e., DNA predictive test (unless the latter were also showing clinical or other laboratory signs to suggest that they were developing haemochromatosis).

DNA tests are often perceived to be unduly complex, and many health professionals lack the confidence to utilise these tests optimally, or alternatively do not appreciate their limitations and order the tests inappropriately, or provide the wrong information about the test's significance. However, knowledge of the human genome will impact across many disciplines in clinical practice, and so it is essential that health professionals develop confidence and skills to deal with DNA tests that are relevant to their area of work. This will become particularly important when the interest in single gene disorders like cystic fibrosis expands into the complex but more common multifactorial disorders such as diabetes, dementia and heart disease. The multifactorial disorders involve interactions between genes and the environment, and it is likely that many genes contribute to the same phenotype, complicating further the interpretation of these tests (see Chapter 4).

Some DNA tests deal with sensitive issues and require careful consideration before they are ordered. Examples of these would be predictive tests, since these are the ones likely to harm patients (and families) if tested for inappropriately, or the results are interpreted incorrectly. This class of DNA tests has the potential to stigmatise or discriminate against individuals, to impact on privacy, and so should be restricted. Health professionals ordering these tests must have the knowledge and resources to provide the appropriate counselling and support.

Another class of DNA test that requires some thought before it is ordered is the screening test (Table 2.9). Indeed, in some cases, these tests need to be formally reviewed by an appropriate institutional ethics committee because a research question is often a component of the testing strategy. Like predictive tests, DNA screening essentially involves the testing of asymptomatic individuals and, in some cases, the testing of individuals without any prior risk. This class of tests has the potential to discriminate or stigmatise individuals or even populations (see Chapter 10 for further discussion about these issues).

VALIDITY OF DNA TESTS

As indicated above, DNA tests are not equal because they can be undertaken for a range of indications. DNA tests are also not equal because they are associated with variable sensitivity, specificity and positive and negative predictive values (Table 2.10). The ideal DNA diagnostic test should be sensitive and specific. For predicting the likelihood of disease on the basis of the test result, the parameters of positive predictive value and negative predictive value become important. Although well established in practice, the four parameters in Table 2.10 reflect population-based values, and so for the individual, they are less meaningful.

As well as the conventional parameters used in assessing the value of pathology tests, another important variable in the DNA test is the penetrance of the underlying genetic disorder. Penetrance is an important concept to grasp since DNA testing now allows an accurate estimate of this to be made particularly in autosomal dominant disorders. For example, if 20 out of 100 individuals with a known DNA mutation show the corresponding clinical picture, the penetrance would be 20%. To increase the accuracy further, the penetrance would be estimated at a particular age group, because with time more at-risk individuals are likely to manifest a disease.

The ideal DNA test scenario would include (1) a highly sensitive and specific test, (2) a genetic disease

Table 2.8 Classes of DNA tests

Class and description	Examples
Diagnostic: DNA test to confirm a clinical suspicion that the patient has an established disorder. The DNA test in this circumstance is comparable to other laboratory tests (e.g., measurement of haemoglobin), although a positive result has implications for family members.	Haemochromatosis: A case suspected on clinical grounds can be confirmed by a liver biopsy, a procedure associated with morbidity and mortality. The alternative is a DNA test to look for the common haemochromatosis mutation (C282Y).
Prenatal: DNA test to detect a genetic disorder in the fetus or embryo.	(i) Prenatal diagnosis: DNA test in an at-risk fetus. Sources of fetal DNA include desquamated cells in the amniotic fluid (obtained during amniocentesis) or chorionic villi (chorion villus sampling or CVS). (ii) Preimplantation genetic diagnosis (PGD): A few cells obtained from the developing pre-embryo allow genetic diagnosis *in vitro* (i.e., IVF). This allows early detection of an abnormality and avoids termination of pregnancy. PGD is still not robust enough to be routinely available.
Predictive (presymptomatic): DNA test to predict the development of a genetic disorder in advance of any signs or symptoms.	(i) Neurologic adult onset disorders: Huntington's disease (HD) develops around the fourth decade of life. The DNA predictive test is used to identify whether an at-risk individual has inherited the mutant or normal gene from an affected parent. (ii) Cancers: Individuals with a risk of familial breast cancer can have DNA testing of the *BRCA1* and *BRCA2* genes to look for a mutation that might guide them in future treatment decisions, as well as identify risks in other family members. However, the penetrance with *BRCA1* and *BRCA2* is variable (36–85% lifetime risk). Therefore, deciding when to test and what the result means is not straightforward.
Screening: DNA test to look at asymptomatic individuals (or populations) to determine who are carriers or have a genetic predisposition.	(i) Reproduction screening: DNA tests are undertaken to identify couples at risk for a genetic defect in the fetus. Presently, most reproduction screening is undertaken because there is a family history or the couple has a risk factor, e.g., advanced maternal age. (ii) Newborn screening: Newborns are screened for a number of genetic disorders that are treatable or preventable provided they are diagnosed early (see Chapter 7). (iii) Community screening: This involves the DNA testing of populations with no prior risk as whole communities or communities selected because of a general increased risk. An example of whole community testing would be DNA testing all pregnant women for cystic fibrosis. A selected community screening program is exemplified by testing for Tay Sachs disease in Ashkenazi Jews.
Pharmacogenetic: A form of DNA screening involving drug metabolism pathways to predict a patient's response to treatment. Unlike screening mentioned above, this type of test is directed to the individual. Like many forms of screening, this class of DNA test is still in the evaluative phase.	(i) Malignant hyperthermia: Individuals exposed to anaesthetic agents can develop life-threatening complications if they have a genetic abnormality involving the ryanodine receptor genes (*RYR1*). (ii) Oral anticoagulation with warfarin: Side effects of anticoagulation treatment include life-threatening bleeding episodes. Two genetic variants of the important warfarin metabolizing enzyme (hepatic microsomal enzyme CYP2C9) are associated with decreased enzymatic activity.
Somatic cell: This class differs from the above because somatic rather than germ cell mutations are being sought. This is a type of DNA diagnostic test, but there are no implications for family members.	(i) Lymphoproliferative disorders: DNA testing for the leukaemias and the lymphomas by PCR to detect specific chromosomal translocations that interrupt gene function. For example, the Ph chromosome found in chronic myeloid leukaemia leads to a fusion gene detectable by PCR. (ii) Minimal residual disease: Following chemotherapy for haematologic malignancies, the risk of recurrence is to some extent dependent on tumour tissue remaining behind (minimal residual disease). A sensitive way in which this can be detected is PCR.

with high penetrance, and (3) a disease that is treatable. Examples to illustrate the potential value of a DNA test include Huntington's disease, which has a penetrance of 100%. Almost all who have Huntington's disease will have a mutation involving an expansion of a $(CAG)_n$ triplet; i.e., the test is very sensitive. The rare exceptions have clinical features of the disease but normal $(CAG)_n$

repeats. These are phenocopies. In some cases a second Huntington's disease locus (HDL2) has been shown to involve expansion in a triplet repeat for the *junctophilin* 3 gene. No one with triplet repeats <26 will develop Huntington's disease; i.e., the test is highly specific. Complexities arise with intermediate level of repeats (between 27–39; see Table 3.4), but the cut-off at 40 or more

Table 2.9 Population screening strategies[a]

Screen	Explanation
Family screening	Family members at increased risk for a genetic disorder are targeted. DNA testing can be very useful in this circumstance if a mutation is known to be present in a family. DNA polymorphic markers also allow a linkage approach. For example, cystic fibrosis screening in parents and siblings can be implemented if a newborn child is shown to be heterozygous for the ΔF508 mutation. A variant of family screening is cascade screening. Here, grandparents of the newborn child described above would also be tested to determine from which side of the family the cystic fibrosis mutation has been inherited. Once this information is known, family members on that side are identified as being at risk and screened.
Population screening	The testing of a subset or an entire population group without an *a priori* risk usually looking for a common disorder, e.g., cervical cancer screening. To be effective, a large percentage of women must participate, and the test needs to be sufficiently sensitive and specific to make the program cost-effective. A contemporary DNA population screening dilemma is haemochromatosis.
Newborn screening	This accepted approach tests for reversible or treatable genetic or congenital disorders in newborns who are not at any specific risk (discussed further in Chapter 7).
Workplace screening	Screening here can have two aims: (1) Identifying those at risk for industrially related complications; e.g., the combination of α1-antitrypsin deficiency in a worker and a dusty workplace is more likely to lead to chronic lung disease. (2) Detecting DNA damage related to the workplace could provide objective data in terms of cause and effect. Screening at the workplace has potential, but a criticism (apart from the concern about discrimination) is that emphasis is placed on exclusion of the at-risk worker rather than making the workplace a safer environment.

[a] The term "screening" has many different meanings. To government, it suggests population testing and a large financial commitment. To the family, it means at-risk members are being tested. It has been suggested that the latter should not be called screening because of its focused and limited nature. However, "screening" will be used as a general term and its exact meaning defined as required.

Table 2.10   Characteristics of the DNA test to be considered for screening purposes (see also Table 8.3 for an infectious disease example)

Feature	Explanation	Criteria for a good test
Sensitivity	The proportion of those with disease who test positive	A sensitive test will have few false negatives. A highly sensitive test will be positive in nearly all patients with the disease but may also be positive in many patients without disease. To be useful clinically, a test should have high sensitivity and high specificity.
Specificity	The proportion of those without disease who test negative	A highly specific test should have few false positives.
Negative predictive value	The proportion of those tested with a negative result who do not have the disease	For predicting the likelihood of disease on the basis of the test result rather than the converse, the positive and negative predictive values are used.
Positive predictive value	The proportion of those tested with a positive result who have the disease	The positive predictive value falls as the prevalence of a disease falls, so tests for rare conditions will have more false positive results than true positive results.

repeats represents a definite value as does 26 or less. Overall, the Huntington's disease DNA test is useful, despite the fact that there is no known treatment.

Another example is MEN2 (multiple endocrine neoplasia type 2), which requires mutation testing of the *RET* oncogene. The risk (penetrance) of developing thyroid cancer with MEN2 is high (90%), and prophylactic thyroidectomy prevents the development of cancer particularly when it is undertaken at a young age. Therefore, the test is useful despite the fact that there are many mutations to be sought in the large *RET* gene. Testing for *BRCA1* and *BRCA2* DNA mutations is a different

story since the penetrance is considerably less than 100%, and there is no guarantee that the mutation will be found as the genes are large. Therapeutic options like prophylactic bilateral mastectomy may provide a better outlook, but this is balanced by the radical nature of this treatment and the deficiencies in the DNA test (see Chapter 4 for additional discussion on the breast cancer genes).

Haemochromatosis is a topical example illustrating the equivocal value of DNA testing in a community screening program. There is little doubt that screening for haemochromatosis is useful because there is a very simple therapeutic option available (venesection), and there is the potential, although it is yet to be proven, that this will prevent the serious iron overload-related complications of liver cirrhosis and hepatocellular carcinoma. On the other hand, the DNA test provides most information when individuals of northwestern European ethnic background are tested, and a homozygous Cys282Tyr mutation is detected. In addition, there is no guarantee that an individual with this mutation will always develop haemochromatosis since environmental and other genetic factors influence progression and severity of this disorder (i.e., the penetrance is low). In ideal circumstances (minimal costs, educated community and health professionals, adequate counselling and support services), it would be possible to screen populations, particularly males who are most at risk, and then follow carefully those with the Cys282Tyr defect with serial iron measurements. As soon as the iron started to rise, preventative venesection would be implemented. However, the ideal circumstances are difficult to achieve, and more research is needed to understand better the low penetrance in this disorder. On balance, therefore, community screening for haemochromatosis would not seem

justified (see Chapter 3 and Figure 3.13 for more discussion of haemochromatosis DNA testing).

A final but important comment about DNA tests: The finding of a DNA mutation does not necessarily mean that the individual has a disease. In many cases, this simply indicates that the individual has a genetic predisposition which may or may not progress to disease. The situation described could be called a **predisease**. This can be a difficult concept to grasp for health professionals, patients and the community, and it becomes even more so when it involves a risk estimate that is not 100%. However, it will increasingly become more relevant as DNA testing expands the options for predictive medicine. In some genetic disorders, the defect alone is sufficient to produce the full clinical picture (phenotype); for example, Huntington's disease will always occur if the (CAG) triplet repeat is 40 or more. In contrast, DNA testing for factor V Leiden is undertaken to detect a genetic predisposition to thromboembolism. The finding of this mutation in an asymptomatic person does not necessarily mean that the individual will develop thromboembolic disease, only that the lifetime risk is now increased to about 12–30%.

FUTURE

The concept of junk DNA will forever be put to rest when the function of about 98% of the human genome can be determined. It is likely that this class of DNA will explain why humans seem to have a comparable genome to other organisms, yet the human phenotype is more complex.

RNA, which has had a relatively low profile in molecular medicine, is starting to provide interesting insights into a molecule with a range of activities that extend beyond the traditional transcription and translation. RNA is now implicated in gene regulation and perhaps even the explanation for how the relatively simple human genome can produce a complex proteome. There is more to learn about the underestimated RNA.

The availability of newer methods to amplify DNA or the removal of the contamination problem with the present techniques will see greater use of DNA amplification across many areas including clinical practice, research and industry. Add to this nanotechnology leading to miniaturisation of analytic platforms, and it is reasonable to predict that DNA amplification will become so user-friendly that testing at the bedside or consulting office will become commonplace (nanotechnology is discussed further in Chapter 5).

Technology developments resulting from the Human Genome Project have been very impressive, particularly the ability to sequence large segments of DNA. Technol-

ogy improvements will continue, and it will become routine to test for the 1300 plus mutations causing a disorder such as cystic fibrosis.

Another challenge will come when population screening becomes more achievable with the availability of genomic-based technologies. Although there are criteria identifying a DNA test that is suitable for population screening (Table 2.11), these guidelines can be bypassed by pressure from industry, researchers or even misplaced popular demand. It will be important to identify appro-

Table 2.11 Some criteria that a disease should meet prior to its being considered for population screening (from Grody 2003)

Relatively common
Relatively serious
Manageable number of predominant mutations
High penetrance
Defined and consistent natural history
Effective preventative or surveillance interventions
Mutation detection relatively inexpensive
Screening test acceptable to population
Infrastructure in place for pre- and post-test counselling

priate processes that can be used to ensure that screening tests are used to benefit the community.

The genetics era has passed, and we are now talking about genomics. However, the epigenetic era is still to come, and this will identify interesting new ways in which genes can be controlled. This knowledge will lead to novel therapies.

FURTHER READING

DNA
http://www.ehgonline.net/ (online reference to the Encyclopedia of the Human Genome published by Nature Publishing Group; expensive but comprehensive).

RNA
Calin GA et al. Frequent deletions and down-regulation of micro-RNA genes miR15 and miR16 at 13q14 in chronic lymphocytic leukemia. Proceedings of the National Academy of Sciences USA 2002; 99:15524–15529 (proposes an interesting role for miRNA species in cancer).

Tijsterman M, Ketting RF, Plasterk RHA. The genetics of RNA silencing. Annual Reviews in Genetics 2002; 36:489–519 (a comprehensive review of the subject; Figure 2 helpful for understanding).

Gene Structure
Falsenfeld G, Groudine M. Controlling the double helix. Nature 2003; 421:448–453 (a sophisticated summary of how gene expression is regulated).

Mattick JS. Non-coding RNAs: the architects of eukaryotic complexity. EMBO reports 2001; 2:986–991 (provides a novel perspective on how non-coding [junk] DNA may play a vital role in explaining eukaryote proteome complexity).

Polymerase chain reaction
Ginzinger DG. Gene quantification using real time quantitative PCR: an emerging technology hits the mainstream. Experimental Hematology 2002; 30:503–512 (provides useful description of Q-PCR and how real-time PCR works for this).

Mullis K. The unusual origin of the polymerase chain reaction. Scientific American 1990; 262:56–65 (easy-to-read account describing the history as well as the basis for PCR).

Chromosomes
Chang KT, Shi Y-J, Min K-T. The Drosophila homolog of Down's syndrome critical region 1 gene regulates learning: implications for mental retardation. Proceedings of the National Academy of Sciences USA 2003; 100:15794–15799 (shows how model organisms provide invaluable information about complex human diseases).

Dunham A et al. The DNA sequence and analysis of human chromosome 13. Nature 2004; 428:522–528 (describes the sequencing of human chromosome 13).

Grimwood J et al. The DNA sequence and biology of human chromosome 19. Nature 2004; 428:529–535 (describes the sequencing of human chromosome 19).

Mueller RF, Young ID. Emery's Elements of Medical Genetics. 10th edition. 1998. Churchill Livingstone, Edinburgh.

Roizen NJ, Patterson D. Down's syndrome. Lancet 2003; 361:1281–1289 (provides a summary of Down's syndrome, particularly clinical issues).

Xu J, Chen Z. Advances in molecular cytogenetics for the evaluation of mental retardation. American Journal of Medical Genetics 2003; 117C:15–24 (in-depth overview of a number of molecular cytogenetic techniques useful for studying mental retardation).

DNA Mutations and Polymorphisms
Antonarakis SE and the Nomenclature Working Group. Recommendations for a nomenclature system for human gene mutations. Human Mutation 1998; 11:1–3 (provides a number of examples of mutations and how they should be described).

Botstein D, Risch N. Discovering genotypes underlying human phenotypes: past successes for mendelian disease, future approaches for complex disease. Nature Genetics 2003; 33:228–237 (comprehensive overview describing how gene discovery for simple mendelian traits has advanced knowledge of genetics but the complex genetic diseases will be more challenging).

Bryant-Greenwood P. Molecular diagnostics in obstetrics and gynecology. Clinical Obstetrics and Gynecology. 2002; 45:605–621 (although the title suggests that it is written for a particular discipline, it is in fact a very general and easy-to-read overview of technologies as well as the types of DNA mutations that can be identified).

Cotton RGH, Edkins E, Forrest S (eds). Mutation Detection. A Practical Approach. 1998. IRL Press, Oxford (comprehensive but technical overview of the various methods involved in DNA mutation testing).

Cotton RGH, Scriver CR. Proof of "disease causing" mutation. Human Mutation 1998; 12:1–3 (criteria are given on how to define a polymorphism, and when a change in DNA can be called a mutation).

Guttmacher AE, Collins FS. Genomic medicine—a primer. New England Journal of Medicine 2003; 347:1512–1520 (provides an excellent overview of DNA, genetic disorders and mutations).

DNA Tests
Burke W. Genetic testing. New England Journal of Medicine 2002; 347:1867–1875 (basic and informative introduction to DNA testing).

Grody WW. Molecular genetic risk screening. Annual Reviews Medicine 2003; 54:473–490 (provides a good overview of the types of DNA tests, focusing particularly on the screening group).

Higashi MK et al. Association between CYP2C9 genetic variants and anticoagulation-related outcomes during warfarin therapy. Journal of the American Medical Association 2002; 287:1690–1698 (an interesting example of how pharmacogenetic DNA testing might prove clinically useful).

Manolio T. Novel risk markers and clinical practice. New England Journal of Medicine 2003; 349:1587–1589 (defines and shows how to measure sensitivity, specificity, etc).

http://www.gendia.net/ (useful online link to provide a perspective on the range of DNA testing options that are available).

MENDELIAN GENETIC TRAITS

INTRODUCTION

PATTERNS OF GENETIC INHERITANCE

The standard approach to classifying genetic diseases is based on their mendelian inheritance, i.e., autosomal dominant, autosomal recessive, and X-linked. An additional dimension can now be added to reflect the molecular basis of genetic disease. A molecular classification of genetic diseases would include defects that involve the following:

- Single genes
- Many genes (polygenic)
- Genes and the environment (multifactorial)
- Chromosomal abnormalities
- Somatic cells

Genetics traditionally refers to conditions that are inherited; i.e., the underlying mutation is found in somatic and germ cells. However, genetics has become more complex with the knowledge that DNA changes are found in a number of sporadic (non-inherited) disorders. Acquired mutations in DNA affecting only the somatic (non-germ) cells are well described in a range of cancers, both solid and haematologic. Since DNA is involved, they have a genetic component to their aetiology

although the abnormality is not inheritable. Thus, genetics can now be thought of in terms of inherited genetics and somatic (acquired or non-inherited) genetics.

Mendelian Inheritance in Man is a compendium of human genes and genetic disorders that evolved into an encyclopaedia of gene loci. The first edition was published in 1966 with a total of 1487 entries. The November 2003 version has 29698 entries (Figure 3.1). In 1987, this compendium became available online as OMIM (*Online Mendelian Inheritance in Man*—http://www.ncbi.nlm.nih.gov/entrez/query.fcgi?db=OMIM). Entries in OMIM relate to autosomal disorders in 94% of cases, sex linked in 6% and mitochondrial DNA in 0.4%. The majority of entries involve single genes, which are the topic of this chapter, because the mendelian inheritance is easily identifiable. More complex genetic disorders, both inherited and acquired, will be discussed in Chapter 4.

Key words used in this chapter include the following: Different forms of a gene at a locus are called **alleles**. The **haplotype** refers to a set of closely linked DNA markers at one locus that are inherited as a unit. The **genotype** is the genetic (DNA) make-up of an organism. In the present context, genotype would also refer to the genetic

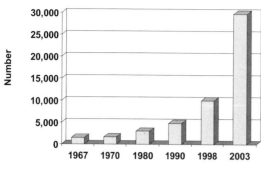

Entries in OMIM

Fig. 3.1 Entries in McKusick's *Mendelian Inheritance in Man* and its online version known as OMIM.
The increasing number of entries reflects the intensive work being undertaken in the area of genome mapping. Access to this compendium is now obtained through the Internet (http://www.ncbi.nlm.nih.gov/entrez/query.fcgi?db=OMIM).

constitution of alleles at a specific locus, i.e., the two haplotypes. The **phenotype** reflects the recognisable characteristics determined by the genotype and its interaction with the environment. An individual is **homozygous** if both alleles at a locus are identical, and **heterozygous** if the alleles are different. **Autosomal** inheritance involves traits that are encoded for by the 22 pairs of human autosomes. **X-linked** inheritance refers to genes located on the X chromosome. The products of both normal (wild-type) alleles at a particular locus need to be non-functional in a **recessive** disorder, e.g., cystic fibrosis. On the other hand, a **dominant** disorder results if only one of the two wild-type alleles is mutated, e.g., familial hypertrophic cardiomyopathy or Huntington's disease.

GENE DISCOVERY

The finding of new genes or gene discovery has been greatly facilitated by a molecular strategy known as positional cloning, which involves the isolation and cloning of a gene on the basis of its chromosomal position rather than its functional properties. For positional cloning it is necessary to have (1) chromosomal location, (2) DNA polymorphisms, (3) genetic and physical maps, and (4) confirmation that the gene is the right one. Variations of positional cloning involve the identification of candidate genes or *in silico* positional cloning (see also the Appendix and Figure A.4 for further discussion of positional cloning).

Chromosomal Location

The first step in positional cloning is to obtain, if possible, a clue to the likely locus or chromosome involved.

This information usually comes from case reports or observations in which chromosomal rearrangements have been noted to occur in association with the clinical picture. Alternatively, whole genome scans are undertaken with a large number of DNA polymorphisms known as microsatellites. These studies require large families to identify likely chromosomal loci. Another approach, if a disease locus has not been identified, is to look with DNA markers derived from "candidate genes." For example, a good candidate gene for a heart muscle disorder such as familial hypertrophic cardiomyopathy would be cardiac β myosin heavy chain, a component of muscle. The gene for this protein is found on chromosome 14q12. The candidate gene has identified a locus to start the positional cloning.

DNA Polymorphisms

Microsatellites (also called SSRs—simple sequence repeats) comprise 2–4 base pair repeats, for example, $(CA)_n$, that are easily identified by a technique such as PCR (see Chapters 1, 2, 9 for further discussion of polymorphisms). The SSRs that are dispersed throughout the genome allow linkage studies to be undertaken either in relation to specific locations, or once sufficient SSRs had been found, it becomes possible to consider whole genome linkage studies.

Genetic and Physical Maps

The genetic and physical maps allow the chromosomal locus to be narrowed down to the point that individual genes are able to be identified within that locus. The genetic map is made by looking at DNA polymorphisms within affected families. The closer the polymorphism is to the gene, the fewer will be the recombinations (breaking and rejoining of the DNA) that are observed. Eventually, a polymorphism associated with the gene itself will produce no recombination events. In contrast, a physical map is based on actual measurements (e.g., kb or Mb) and allows the area of interest to be narrowed even further.

Gene Confirmation

Genes of interest are sought within the area identified by positional cloning. Which is the right gene? To answer this question, the most likely gene, usually in the form of cDNA, is sequenced and the results entered via the Internet into the various DNA and protein databases available. Computer programs then enable searches to be made that compare sequences in the databases with the recently discovered gene. Three outcomes of such a search are possible: (1) A perfect match is found; i.e., the gene has already been described—this is bad luck because it means the gene was found by others! (2) No match is found. This means the gene is novel, but there is no clue as to what it might do. Again, it is bad luck because a lot of work will be needed to determine function. (3) Some homology (i.e., similarity) is found to another entry in the database. This is the best result since the gene is still novel, and a clue to its function can come from the gene in the database with which it shares some DNA sequence.

Candidate Genes and In Silico Positional Cloning

An alternative approach is simply to sequence what is thought to be the right (candidate) gene in a number of affected patients, and if a mutation can be found, it is confirmed to be the one involved in the disorder being studied. The vast amount of information generated from the Human Genome Project now allows a shortcut in positional cloning called *in silico* **cloning**. Through linkage analysis (or in the case of more complex genetic disorders, linkage analysis is often replaced by an association study—see Chapter 4 and the Appendix), a likely location is found. The databases are then searched to identify what genes are there, and these genes are then studied to look for mutations in affected individuals. The *in silico* step avoids the tedious and very time-consuming construction of physical and genetic maps. For more information on *in silico* positional cloning, particularly in the more complex genetic disorders, see Breast Cancer in Chapter 4.

AUTOSOMAL DOMINANT DISORDERS

OVERVIEW

The characteristic feature in a pedigree with autosomal dominant inheritance is a **vertical** mode of transmission. This appearance comes from the fact that the disorder can appear in every generation of the pedigree. Both males and females are affected, and their offspring are at 50% risk (Figure 3.2). A number of additional features need to be considered when dealing with autosomal dominant disorders. The following become important in counselling.

Sporadic cases occur, and they become increasingly more common as the mutation in question interferes with fertility. This may represent a secondary effect of the disorder or because death occurs before reproductive age is reached. For example, mutations in unrelated families with X-linked Duchenne muscular dystrophy are usually of independent origins because affected individuals are unlikely to survive to a reproductive age. At the other end of the spectrum, Huntington's disease does not have a direct effect on reproduction. Thus, sporadic cases of Huntington's disease are rare, and even some of them can now be shown to have inherited an unstable premutation from one parent. The meaning of premutation is discussed below.

Penetrance is an important concept to understand. It is an all-or-nothing phenomenon that describes the clinical expression of a mutant gene in terms of its presence or absence at a stated age. Thus, an individual carrying a mutant gene may not express the clinical phenotype; i.e., the condition is non-penetrant. In the autosomal dominant disorders penetrance can be determined a number of ways: (1) From family studies it is possible to identify the number of obligatory carriers for a mutant allele. If 7 out of 10 show the clinical phenotype, the disorder is 70% penetrant. That is, there is a 70% probability that an individual carrying a mutant gene at a certain

age will display the clinical phenotype. (2) If the underlying DNA mutation is known, the penetrance calculation is based on those who have the DNA mutation and manifest the disease at a particular age. Penetrance will be discussed further under Predictive Medicine—Huntington's disease and Gene Discovery—Familial hypertrophic cardiomyopathy. Apart from spontaneous mutations and death before onset of symptoms, penetrance is an additional mechanism accounting for affected offspring having an apparently normal parent.

Expressivity refers to the severity of the phenotype. There are genes that can produce apparently unrelated

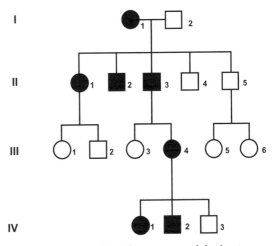

Fig. 3.2 Idealised pedigree for an autosomal dominant disorder.
Affected individuals are indicated by ● and ■. Note the vertical disease pattern (compare with Figure 3.9) with the disease apparent in every generation and affecting males and females.

effects on the phenotype or act through involvement of multiple organ systems. This is called **pleiotropy**. Such genes often show variable expressivity. An example of this is Marfan's syndrome, which has autosomal dominant inheritance and involves connective tissues in the skeletal system, the eye or the heart. Individuals with Marfan's syndrome have any combination of manifestations, which can also be present in different degrees of severity. Such variability can occur even within families in which it is presumed the same mutant allele is present. To date, the underlying basis for expressivity has not been defined, but it is thought to represent either gene/environment or gene/gene interactions. Somatic instability, a recently described mechanism, may provide another explanation (see Chapter 4).

Severity of an autosomal disorder can also be influenced by the sex of the transmitting parent (see Huntington's Disease) or the sex of the affected person. An example of the latter is otosclerosis, a cause of deafness in adults due to overgrowth of bone in the ear. The female-to-male ratio in otosclerosis is approximately 1.8 to 1. The reason for this is unknown. One hypothesis suggests that affected males may have a selective disadvantage compared to females, and this selection is having its effect prenatally.

GENE DISCOVERY—FAMILIAL HYPERTROPHIC CARDIOMYOPATHY

Familial hypertrophic cardiomyopathy (FHC—also abbreviated to HCM) is an autosomal dominant disorder involving heart muscle. About one third of cases have no family history, and they are thought to have arisen from spontaneous mutations. Familial hypertrophic cardiomyopathy was first described in 1958. The characteristic findings are ventricular hypertrophy and myocyte disarray. Apart from cardiac failure, the major complication is sudden cardiac-related death. Familial hypertrophic cardiomyopathy is one of the most common monogenic cardiac disorders, and the most frequent cause of sudden death from cardiac disease in children and adolescents. Its frequency in the general population has been estimated at 1 in 500. Diagnosis is made by a combination of clinical examination, ECG and echocardiography. However, this approach will not detect all affected individuals, particularly the young. Although familial hypertrophic cardiomyopathy is an autosomal dominant disorder, males are more likely to present with familial hypertrophic cardiomyopathy or have the sudden death complication.

Positional Cloning

No abnormal cytogenetic studies were reported in familial hypertrophic cardiomyopathy, and so there was no clue to a likely location for this genetic heart disorder. Thus, positional cloning to find the familial hypertrophic cardiomyopathy gene would either require a whole genome scan, which is a huge undertaking, or a shorter route via candidate genes could be tried. Candidate genes in familial hypertrophic cardiomyopathy would be those that encoded for muscle proteins expressed in the heart such as actin and myosin. Linkage studies could then be undertaken using DNA probes derived from these genes' sequences. In 1990, linkage was established between the β isoform of the cardiac β myosin heavy chain gene on chromosome 14 and familial hypertrophic cardiomyopathy. The gene was confirmed to be the right one when point mutations in various exons were found in affected individuals.

The connection between the β myosin heavy chain gene and familial hypertrophic cardiomyopathy was only the start of the molecular story, since there were families who did not show linkage to the chromosome 14 locus or mutations in the β myosin heavy chain gene. Therefore, familial hypertrophic cardiomyopathy was heterogeneous at the DNA level, and other candidate genes were sought. Today, at least 11 different sarcomere genes are implicated in familial hypertrophic cardiomyopathy, and more than 200 different mutations have been described (Table 3.1). Genes involved in the usual familial hypertrophic cardiomyopathy phenotype are associated with the muscle sarcomere. Other, rare forms of familial hypertrophic cardiomyopathy involve non-sarcomeric genes including mitochondrial DNA. The phenotypes in these cases can be atypical.

The sarcomere has seven major proteins and several minor ones organised into thin and thick filaments. Muscle contraction and force generation result from the relative sliding between these filaments. Although the sarcomere is the key element in muscle contraction, the unit itself is susceptible to a number of influences such as ion channels, calcium and adrenergic receptors, to name a few. Muscle contraction is a complex process, but positional cloning allowed the genes in familial hypertrophic cardiomyopathy to be discovered, and because of this familial hypertrophic cardiomyopathy is now described as a disease of the sarcomere.

Although positional cloning provided greater insight into the cause of familial hypertrophic cardiomyopathy, it has been disappointing because the considerable heterogeneity (number of genes and mutations) has meant that DNA testing to detect those at risk, or confirm those with suspicious cardiologic findings, is not particularly helpful with current technology. Other unanswered questions in familial hypertrophic cardiomyopathy include Why are males more likely to be affected, and why can the severity be different even within the same family when it is clearly a monogenic disorder? The environ-

Table 3.1 Mutations in FHC involving genes of the muscle sarcomere (see mutation database on http://www.angis.org.au/Databases/Heart)[a]

Gene (symbol)	Frequency	Type of mutations
Cardiac β Myosin heavy chain (*MYH7*)	46%	93 missense, 4 deletions, 1 deletion/insertion
Myosin binding protein C (*MYBPC3*)	24%	38 missense, 7 deletions, 3 insertions/duplications, 1 insertion/deletion
Cardiac troponin T (*TNNT2*)	9%	19 missense, 1 deletion
Cardiac troponin I (*TNNI3*)	7%	12 missense, 2 deletions
Regulatory myosin light chain (*MYL2*)	5%	11 missense
αTropomyosin (*TPM1*)	3%	7 missense
Cardiac α Actin (*ACTC*)	2%	5 missense
Essential myosin light chain (*MYL3*)	Rare	3 missense
α Myosin heavy chain (*MYH6*)	Rare	1 missense, 1 rearrangement
Titin (*TTN*)	Rare	1 missense
Cardiac troponin C (*TNNC1*)	Rare	1 missense

[a] The 11 genes above encode muscle sarcomere proteins. Recently, two non-sarcomere genes associated with familial hypertrophic cardiomyopathy have been described in this disorder.

Table 3.2 Genes associated with the long QT syndrome[a]

Type	Gene	Comments
Autosomal dominant	KVLQT1, minK, HERG, MiRP1	Potassium channel genes
	SCN5A	Sodium channel gene
Autosomal recessive	KVLQT1, minK	Potassium channel genes
Familial ventricular tachycardia	Ryanodine receptor	Calcium channel gene

[a] From Nabel (2003), Marian and Roberts (2001).

ment is known to play a role in the pathogenesis of familial hypertrophic cardiomyopathy, but it is also considered that other genetic factors known as modifying genes may be important (see Future for more discussion).

The molecular cardiology era started with familial hypertrophic cardiomyopathy. Other forms of heart muscle disease (particularly dilated cardiomyopathy) have now been extensively investigated at the DNA level, and a number of genes have been identified. Genes that predispose to cardiac arrhythmias have also been found. One condition, known as the long QT syndrome, has been studied, and it has been shown at the molecular level that, in the autosomal dominant form of this disorder, mutations in at least five genes involving either potassium or sodium channels are responsible (Table 3.2). In the future, knowledge of molecular defects will make it possible to devise treatment that is individualised to the patient's particular mutation.

PREDICTIVE MEDICINE—HUNTINGTON'S DISEASE

Huntington's disease is a neurodegenerative disorder with autosomal dominant inheritance. Offspring of affected individuals have a 50% risk of inheriting the disease, which can present in various ways including a progressive movement disorder (typically chorea), psychological disturbance and dementia. Disease onset is usually between 35–45 years of age, and there is complete penetrance by the age of 80. Although Huntington's disease was described in 1872 by Dr George Huntington, a general practitioner, it took more than 100 years for the first major advance in this disorder. Until the breakthrough came in 1983, Huntington's disease could not be definitively diagnosed early on in its course. Those at risk had to wait until their mid adult life to see whether they had inherited the abnormal gene, by which time reproduction and other life decisions had been made.

Positional cloning for Huntington's disease proved to be particularly difficult. A cytogenetic location had not been identified, and in the early 1980s whole genome-based DNA scans were not available. Candidate genes for Huntington's disease could not be identified. Therefore, a trial-and-error approach was required to find DNA polymorphisms linked to the clinical phenotype. Success with this blind approach would not have been possible without the large pedigrees identified in Venezuela. In 1983, a DNA marker located on chromosome 4p16.3 was found to co-segregate with Huntington's disease. This showed that the disease gene was located on chromosome 4. From 1983, different genetic and physical mapping strategies were used in many laboratories to find the relevant gene. This approach was successful in 1993 when a gene called *IT15* (IT—interesting transcript) was isolated. The related protein (and now the gene) is called "huntingtin." Although finding the right gene took 10 years, the molecular approaches enabled DNA testing to be started once a locus on chromosome 4 was identified.

DNA Predictive (Presymptomatic) Testing

Unlike conventional pathology tests, the use of DNA allows predictions to be made because it is possible to look for mutations in DNA long before any signs or symptoms develop. Hence, predictive testing has become a unique feature of molecular medicine. For convenience, the terms "predictive" and "presymptomatic" DNA testing will be used interchangeably (see glossary for further discussion of these two terms).

By using DNA polymorphisms linked to the Huntington's disease locus, it became possible from 1983 to undertake predictive testing within the confines of a family unit, i.e., a linkage study (see Figure 2.14). Individuals with a family history of Huntington's disease now had an opportunity to alter their *a priori* risks by DNA studies. Once the Huntington's disease gene was found, DNA predictive testing became easier, since direct mutation detection, rather than a family linkage study, was possible. Predictive testing programs were able to expand because family studies were no longer essential. DNA testing for the gene mutation became a new option to assist physicians in the differential diagnosis of a neurological disorder, e.g., gait disturbances or dementia. This type of DNA testing (called diagnostic testing) is different from a predictive test because the patient has established signs or symptoms of the disorder (see Chapter 2, Table 2.8).

Two important issues emerged from the Huntington's disease predictive testing programs. First, key individuals could be lost through death (including suicide), and this prevented a number of families from having access to predictive testing through linkage analysis. The concept of a **DNA bank**, which will be discussed further in Chapter 5, assumed increasing importance in this circumstance.

A second consideration related to the comprehensive clinical, counselling and support facilities that were necessary in a predictive testing program. They had major resource implications. It should also be noted that in some instances DNA tests placed further stress on individuals and/or their families because they were able to show who would get Huntington's disease and who was spared. The potential ethical/social issues resulting from DNA testing are discussed in Chapter 10. Prenatal detection also became possible (see Chapter 7).

Molecular Pathogenesis

The molecular defect in Huntington's disease involves a novel mechanism shown in 1991 to result from expansions of triplet nucleotide repeats. The first example of this was the fragile X syndrome (triplet repeat is CCG), followed by myotonic dystrophy (CTG triplet repeat) and then spinal and bulbar muscular atrophy (CAG triplet repeat). In Huntington's disease, it was shown that there

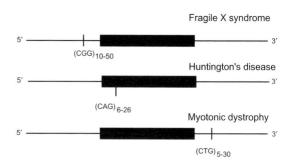

Fig. 3.3 DNA triplet repeats and disease pathogenesis.
The fragile X syndrome (CGG) repeat is in the 5' flanking region of the gene. Normally, there are about 10–50 repeats. Expansion beyond 200 repeats is associated with methylation (silencing) of the *FMR1* gene. For Huntington's disease, the $(CAG)_n$ triplet repeat (the normal number of repeats ranges from 6–26) is located within the gene's first exon. Therefore, adding more polyglutamines to this protein (called huntingtin) will interfere with its structure or function. Studies in humans and mouse models suggest that huntingtin has its deleterious effect through a gain of function. For myotonic dystrophy, the (CAG) repeat is located in the 3' flanking region, and normally there are about 5–30 repeats. Mildly affected patients have 50–80 repeats, whereas severely affected individuals have 2000 or more repeats. How expansion in the number of repeats located at the 3' non-coding region affects function of the myotonic dystrophy gene (*DMPK*) is not known. Possible explanations include the altering of the chromatin conformation following triplet repeat expansion, and an effect on the expression of genes located near *DMPK*. In all three examples, the repeat numbers between the normal values and those required to interfere with gene function represent examples of premutations.

was a DNA triplet involving $(CAG)_n$ in the first exon. The normal number of repeats is 6–26 (Figure 3.3). Expansions greater than 39 repeats are associated with the development of disease. Statistically, it was also shown that the greater the number of repeats, the earlier the onset of the disorder. Another observation related to instability in the repeat numbers with the possibility that repeats could expand when transmitted through sperm (and could also contract in numbers when transmitted through the ovum). These observations explained the occasional presentation of Huntington's disease in children or young adults (the CAG repeat is very high) and the reason cases of juvenile Huntington's disease invariably inherited the mutant gene from their fathers.

Triplet repeat expansion as a mechanism for developing genetic disease has been a recent finding, and it has explained a number of unusual observations such as the failure to identify a traditional inheritance pattern in some disorders and **anticipation**, i.e., the earlier onset and more severe phenotype as the mutant gene passed through succeeding generations (Figure 3.4). There are now a number of disorders that have an associated triplet repeat as the underlying aetiology (Table 3.3). Although

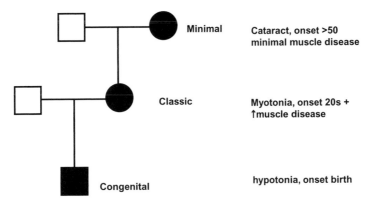

Fig. 3.4 Anticipation.
Myotonic dystrophy is an autosomal dominant, multisystem disorder that is the most common form of adult muscular dystrophy. A feature of this disease is its variable expressivity, including a very severe congenital form. Molecular characterisation has now explained the phenomenon of anticipation seen in myotonic dystrophy. The diagram illustrates the increasing severity and earlier onset of symptoms expected in anticipation. A corresponding expansion in the myotonic dystrophy $(CTG)_n$ triplet as it is passed through the female germline would parallel the clinical changes.

Table 3.3 Neurologic diseases caused by expansions of triplet (and other) repeats

Class	Disorder[a]	Repeat	Mode inheritance	Comments[b]
Gain of function	HD	CAG	Autosomal Dominant	In exon 1. Demonstrates anticipation.
	DRPLA	CAG	Autosomal Dominant	Exonic repeat. Demonstrates anticipation.
	SCA1,2,3,7,17[c]	CAG	Autosomal Dominant	5' UTR or exonic repeat. Demonstrates anticipation.
	SBMA	CAG	X linked	In Exon 1.
Loss of function	FMR	CGG	X linked	Located 5' to gene, expansion of triplet repeats leads to gene methylation (inactivation) Demonstrates anticipation.
	FA	GAA	Autosomal Dominant	Intron 1 (most involve expansion of both alleles). Demonstrates anticipation.
Non-coding repeat (RNA defect)	DM1	CTG	Autosomal Dominant	Repeat located in 3' UTR in *DMPK* gene. Demonstrates anticipation (Figure 3.4).
	DM2	CCTG	Autosomal Dominant	Intron 1 repeat in *ZNF9* gene.
	SCA8,10,12	CTG, ATTCT, CAG	Autosomal Dominant	Repeats located in 3' UTR, intron 9 or 5' UTR, respectively.
	HDL2	CTG	Autosomal Dominant	Alternatively spliced transcripts but suggestion that in two forms the CTG may be in non-coding region of the gene.

[a] HD—Huntington's disease; DRPLA—dentatorubral and pallidoluysian atrophy; SCA—spinocerebellar ataxia; SBMA—spinal and bulbar muscular atrophy (Kennedy's syndrome); FMR—familial mental retardation; FA—Friedreich's ataxia; DM—myotonic dystrophy; HDL—Huntington's disease like. SCA2 also called Machado-Joseph disease. [b] UTR—untranslated region (5'—front of gene; 3'—end of gene). [c] A limited number of SCA disorders are listed. Others not described include SCA5, SCA6 (CAG repeat), SCA16, SCA19, SCA21.

the gene for Huntington's disease was found more than a decade ago, it is still not understood how the expansion of the triplet repeat leads to genetic disease, although it is generally believed that in the case of Huntington's disease, the CAG (glutamine) expansion leads to a **gain of function**. That is, the product of the mutant allele results in the gain of a new abnormal protein with

the potential to inhibit the normal gene or normal protein. Evidence for this includes the autosomal dominant nature of the disease and the finding of similar phenotypes in the rare cases that are homozygous for an expanded triplet repeat.

There is also some evidence that triplet repeats, for example, the CGG in the fragile X syndrome, produce

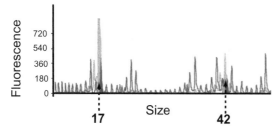

Fig. 3.5 Measurement of the Huntington's disease triplet repeat.
DNA on either side of the Huntington-disease specific triplet repeat, i.e., $(CAG)_n$, is amplified (review Figure 2.8, which shows the triplet repeats and the location where the DNA primers for PCR are placed). Fluorescent-labelled primers allow two alleles to be detected by capillary electrophoresis. There are two alleles because each chromosome 4 has a Huntington's disease gene. Capillary electrophoresis enables automated and accurate measurement of the two alleles. Normal "n" for the $(CAG)_n$ is 26 or fewer repeats and the Huntington's disease range is 40 or more repeats. In between is the grey zone (see Table 3.4 for further explanation). In the example given, the two alleles are identified by an arrow. The other bands in the profile represent size markers (they are distinguished from the CAG repeats by their different shading in this diagram, but in actual fact are better distinguished because they are labelled with different colours). The alleles in this example are 17 and 42. The 42 repeats indicate that the person has Huntington's disease or will develop it at some future date.

Table 3.4 Interpretation of $(CAG)_n$ repeat expansion in Huntington's disease[a]

Number of triplet repeats	Interpretation
26 or fewer	Normal phenotype.
27–35	Normal phenotype, but there is a risk that offspring will develop Huntington's disease.
36–39	This is associated with the Huntington's disease phenotype, but there is the potential for reduced severity; some individuals with these repeat numbers might not develop Huntington's disease. For this repeat range, there is the chance that offspring will develop Huntington's disease.
40 or more	The predicted phenotype is Huntington's disease.

[a] Interpretation of the results based on guidelines from American College of Medical Genetics/American Society of Human Genetics (American Journal of Human Genetics 1998; 62:1243–1247).

disease through a **loss of function**. In other words, when the expanded CGG in fragile X reaches a certain threshold, it triggers methylation of the gene, which turns it off. A third mechanism to explain triplet repeat expansion is through interference with RNA function, which may explain the effects of triplet repeats located outside the gene. In SCA8 there is some evidence that the CTG triplet located in the 3′ UTR is transcribed but not translated and acts as a gene regulator (see Table 3.3).

Although sizing of the triplet repeat made DNA testing more accurate and available to a larger group of at-risk individuals, some results remained equivocal. For example, an expanded repeat of 40 or more had a 100% probability of leading to Huntington's disease, and similarly a repeat 26 or fewer indicated a normal allele (Figure 3.5). However, repeat expansions in the 36–39 range are described as "intermediate" and require more careful consideration, as do repeats of 27–35 (Table 3.4). The significance of intermediate repeats remains topical. They are considered to be **premutations** that, when expanded, lead to Huntington's disease in future generations. The concept of a premutation has helped to explain cases of apparently sporadic Huntington's disease because there was no family history. In these circumstances, parents who were able to be tested invariably demonstrated that one of them, usually the

father, had a triplet repeat size in the intermediate range, i.e., a premutation (Box 3.1).

CANCER PATHOGENESIS—FAMILIAL ADENOMATOUS POLYPOSIS

Colorectal cancer is one of the most common cancers in Western countries with a lifetime risk about 5–6%. Five percent of these cancers have a strong familial risk that is inherited as a Mendelian autosomal dominant trait, for example, familial adenomatous polyposis, and hereditary non polyposis colorectal cancer (HNPCC). An additional 20% of colorectal cancers have a less well-defined genetic factor involved. Apart from the genetic variants present, the feature that distinguishes colorectal cancer from other frequently occurring malignancies is the distinct precancerous state associated with the adenomatous polyp. This makes colon cancer a unique model to study the evolution of a solid tumour because progression can be followed from the premalignant (polyp) stage to the locally advanced and then invasive (metastatic) cancer.

Familial adenomatous polyposis (FAP) is a rare form of colon cancer (~0.5% of all cases) and is inherited as an autosomal dominant disorder with close to 100% penetrance although there is considerable variation in the phenotypic expression of this disease. It is characterised by hundreds to thousands of polyps in the colon, with the risk of cancer closely related to the number of polyps present. Because of this high risk, treatment involves prophylactic removal of the colon.

A clue that the gene for familial adenomatous polyposis would be found on the long arm of chromosome 5 came from the chance observation of a deletion involving this chromosome and the finding of familial adenomatous polyposis in the same family. Positional cloning

1 in 2 (50%) risk to offspring of an affected parent. The at-risk children will need to be followed carefully with colonoscopy because, at some time in the future, prophylactic colectomy will need to be considered before the pre-malignant polyps become cancerous. DNA predictive testing will also exclude half the at-risk children from further follow-up (because they do not have the affected parent's mutation), thereby avoiding an unpleasant procedure such as colonoscopy.

Up to 25% of familial adenomatous polyposis patients are now considered to have spontaneously developed an *APC* mutation. In these circumstances, the risks for siblings will be the same as the general population, unless the spontaneous appearance is actually due to germinal mosaicism (see Chapter 4). Hence, knowledge of the parent's status (clinical or molecular) is necessary to confirm the true genetic inheritance of familial adenomatous polyposis. Children of a parent with a spontaneous mutation still have a 50% risk of inheriting the mutant allele.

Genotype/Phenotype Correlations

Familial adenomatous polyposis provides an example of how knowledge at the gene level (the genotype) can predict the clinical picture (the phenotype). Although the penetrance for cancer is 100% in this disorder, there are a number of associated conditions in familial adenomatous polyposis. They include (1) Attenuated familial adenomatous polyposis, which is characterised by a smaller number of adenomatous polyps although the risk for cancer is still increased. (2) Congenital hypertrophy of the retinal pigment epithelium in about 60% of families. This is not a premalignant state and does not affect vision but is useful in detecting at-risk individuals before polyps develop. (3) Desmoid tumours which contribute to morbidity and mortality. (4) A number of other tumours.

The risk of developing associated complications of familial adenomatous polyposis is to some extent determined by the position of the *APC* mutation (Table 3.5). As well as having predictive value, the knowledge of the underlying DNA mutation may assist in planning colectomy. For example, to delay the development of desmoid tumours, surgery is postponed as long as possible in the case of mutations in *APC* codons 1403–1578 since this region is associated with desmoid development, and surgery is an additional risk factor for this complication. Mutations in codons 1250–1464, particularly codon 1309, are associated with severe familial adenomatous polyposis, i.e., many polyps with involvement of the rectum likely. In this circumstance, a more radical colectomy may be considered so that the rectum is also removed. Generally, mutations in the central portion of the *APC* gene are associated with a severe phenotype and extracolonic manifestations, while mutations at either end of the gene lead to a milder disease.

was started at the chromosome 5q locus, and this led to the identification of the familial adenomatous polyposis gene (called *APC*—<u>a</u>denomatous <u>p</u>olyposis <u>c</u>oli). The *APC* gene extends over 8.5 kb and has 21 exons. Exon 15 is responsible for >75% of the coding sequence. It has two hot spots for mutations at codons 1061 and 1309 although all codons between 200 and 1600 are sites for mutations. The *APC* gene is associated with both germline and somatic cell mutations. Germline mutations are found in most familial adenomatous polyposis patients. About 95% of mutations in *APC* involve nonsense changes or frameshift mutations, leading to the production of a truncated protein (see the Appendix for description of the protein truncation test).

Since only one gene causes familial adenomatous polyposis, and the penetrance is very high, predictive DNA testing is worthwhile. There is also justification for testing children (in contrast to the Huntington's disease example) because now the information from the predictive test can be put to practical use; i.e., there will be a

Table 3.5 Genotype/phenotype correlations in the *APC* gene (Fearnhead *et al* 2001)

Phenotype	Genotype
Severe disease	Usually, there are mutations between codons 1286–1513 (mutation cluster region—see Figure 3.6) and particularly at codon 1309.
Attenuated disease	Mutations are usually in the 5′ or 3′ ends of the gene, or in the alternatively spliced region of exon 9.
Congenital hypertrophy of the retinal pigment epithelium	*APC* mutations are found between codons 457–1444 (mutation cluster region as well as a portion 5′ to it).
Desmoid disease	*APC* mutations usually in codons 1403–1578 (end of mutation cluster region and a portion 3′ to it).
Non-FAP patients with multiple adenomas or carcinoma developing at an early age	*APC* missense germline mutations (rather than protein-truncating ones) such as I1307K, E1317Q.

Genetic Model of Cancer

In the early 1970s, epidemiological studies of both retinoblastoma and Wilms' tumour led A Knudson to propose his two-hit model of tumourigenesis. Knudson's hypothesis required, in either the sporadic or genetic forms of retinoblastoma, the tumour cells to acquire two separate genetic changes in DNA before a tumour developed. The first or predisposing event could be inherited either through the germline (familial retinoblastoma), or it could arise *de novo* in somatic cells (sporadic retinoblastoma). The second event occurred in somatic cells. Thus, in sporadic retinoblastoma both events arose in the retinal (somatic) cells. In familial retinoblastoma the individual had already inherited one mutant gene and required only a second hit affecting the remaining normal gene in the somatic cells. The frequency of somatic mutations was sufficiently high that those who had inherited the germline mutation were likely to develop one or more tumours. On the other hand, sporadic forms of the tumour required two separate somatic events. The second hit must occur in the same cell lineage that has experienced the first or predisposing hit. The probability of this is relatively rare, and so sporadic forms of the tumour occur later in life and have the additional features of being unifocal and unilateral.

Like the retinoblastoma gene, *APC* is a tumour suppressor gene (see Chapter 4 for more discussion of tumour suppressor genes) and to some extent follows the two-hit model, with mutations being found in both the germline and somatic cells in familial adenomatous polyposis. Somatic mutations in *APC* are found in the majority of colorectal adenomas and sporadic carcinomas, and inactivation of both alleles is also common. The majority of somatic mutations in *APC* occur within a small region of the gene (between codons 1286 and 1513), and this area has been called the mutation cluster region (MCR). Another mechanism by which the *APC* gene can be inactivated is through hypermethylation of its 5′ promoter region. Because of the large number of mutations found in *APC*, methylation has been considered a relatively minor component in pathogenesis. However, some recent findings in colon cancer would suggest that perhaps this may not be the case, and the role of epigenetic changes in the causation of cancer may be more significant (see Epigenetics in Chapter 4).

APC encodes a large multi-domain protein that includes N-terminal, an oligomerisation domain, an armadillo region, a number of 15 and 20 amino acid repeats in the centre and towards the C-terminal there is a basic domain, an EB1 binding site and a DLG binding site (Figure 3.6). The different domains in *APC* allow various interactions with proteins. The oligomerisation domain is involved with wild-type protein-forming dimers with both wild-type and truncated (mutated) protein. This can explain the dominant-negative effect of the *APC* mutation; i.e., the mutant protein has the potential to inactivate the remaining normal protein. The armadillo domain binds to proteins associated with the wnt signalling pathway. The central repeat regions play a key role in *APC* function through binding of β catenin, thereby promoting its degradation. A mutation in *APC* or activation by the wnt pathway is associated with an accumulation of β catenin, which stimulates transcription of a wide variety of genes, and so tumours develop. It should be noted that about half of non-*APC* related colorectal cancers also have an accumulation of β catenin through mutations in other pathways, showing that the β catenin step is a crucial one in carcinogenesis. The final domains in *APC* are the C-terminal binding sites. They are implicated in microtubule binding and cell cycle activities necessary for chromosomal stability.

Apart from the germline and acquired somatic mutations in the *APC* gene, a number of other acquired DNA changes are observed when comparing precancerous adenomatous polyps to established malignant tissue:

- Both proto-oncogenes and tumour suppressor genes develop mutations.
- Multiple mutations are seen: The numbers present correlate approximately with the stage of evolution; i.e., there are fewer changes in adenomatous polyps compared to carcinoma.
- The *APC* gene defect occurs early and persists. Other changes develop concurrently with the evolution of

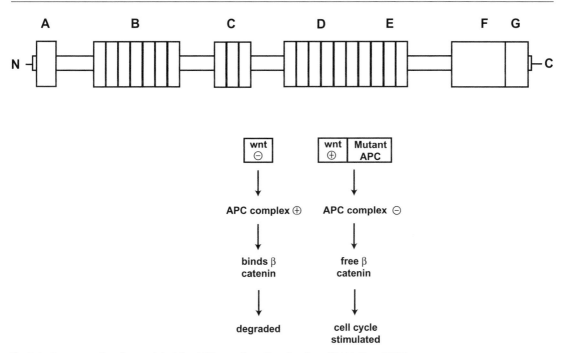

Fig. 3.6 Structure—function model of the *APC* gene (from Fearnhead *et al* 2001; Yang 2002).
Schematic representation of the functional regions in the *APC* gene. The seven important regions are indicated A–G and include A—oligomerisation, B—Armadillo repeat, C—β catenin binding, D—β catenin down regulation, E—axin/conductin binding, F—EB1 binding and G—DLG binding. The various functional regions are described in the text. A key segment of *APC* that produces numerous adenomas in knockout transgenic mice is the mutation cluster region (MCR) (more than 60% of the mutations in *APC* are found within this region). The MCR overlaps regions D and E involved in β catenin down regulation and binding of axin and conductin proteins. When the wnt signalling pathway is shut off (–), *APC* complexes with axin/conductin and glycogen synthase kinase 3β. This complex binds β catenin. Bound β catenin undergoes proteolysis. In contrast, when the wnt signalling pathway is turned on (or there is a mutation in *APC*), the *APC* complex described above is disassociated, and so it cannot bind β catenin. This leads to stimulation of the cell cycle via myc + cyclin D1 and retinoblastoma (see Figure 4.13 and the section in Chapter 4 on the cell cycle for more discussion about myc, cyclin D1 and retinoblastoma and their effect on the cell cycle).

the tumour. No specific combination of mutations leads to a predictable phenotype.

A number of hypotheses have been advanced to explain the clinical and genetic findings in familial adenomatous polyposis. One suggestion for the (precancerous) adenomatous polyp formation involves a dose effect from the *APC* gene enabling a single mutation to initiate the growth of a polyp. This, as well as a number of additional genetic changes, enables the slow but steady progression to carcinoma. Important components allowing tumour progression include genetic instability and clonal growth advantage, e.g., loss of cell cycle control by the tumour clone. Invasion and metastasis require further genetic changes (Figure 3.7). The net effect is multiple well-recognised mutations in key regions, superimposed on which are other effects (genetic or environmental). Clinical heterogeneity (e.g., invasiveness, the number of adenomatous polyps present) reflects the various genetic

combinations or types of mutations present. Since similar changes in oncogene and tumour suppressor gene loci are seen in other malignancies, there must be common pathways in tumourigenesis. An example of this was given earlier with β catenin in colorectal, as well as other, tumours including skin, stomach and pancreas.

DNA REPAIR—HEREDITARY NON POLYPOSIS COLORECTAL CANCER

Hereditary non polyposis colorectal cancer (HNPCC) is the other rare form of colorectal cancer with an autosomal dominant mode of inheritance. It is more common than familial adenomatous polyposis and has a lifetime risk for colorectal cancer of about 80%. In women with HNPCC, there is also a 50–60% risk for endometrial cancer. A number of other cancers are associated with HNPCC. Like familial adenomatous polyposis, and in

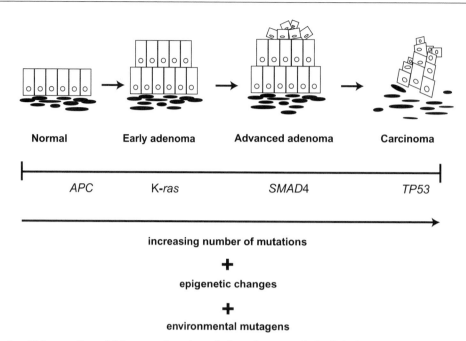

| Normal | Early adenoma | Advanced adenoma | Carcinoma |

| APC | K-*ras* | SMAD4 | TP53 |

increasing number of mutations

+

epigenetic changes

+

environmental mutagens

Fig. 3.7   A multistep genetic model for tumourigenesis producing colon cancer in familial adenomatous polyposis.
An initial insult affecting the colonic tissue can involve any number of genes. The example given here is *APC*—adenomatous polyposis coli. This is inherited in familial adenomatous polyposis but may be acquired in sporadic colon cancer. This initiates the tumour pathway by the development of the early adenoma, and the genomic instability leads to other mutations in genes. The colonic epithelium with these accumulating mutations develops a growth advantage over normal tissue. Additional mutations involving the proto-oncogene K-*ras* and tumour suppressor genes such as *SMAD4* contribute to the adenoma progressing to the development of carcinoma. One of the late genetic changes involves the tumour suppressor gene *TP53*. During the above stepwise progression, epigenetic factors such as hypomethylation of DNA predispose to further genomic instability. Throughout this process the environment (e.g., mutagens in food) can contribute to the DNA damage. The development of cancer from the first mutated cell relies on an accumulation of genetic defects until the appropriate combination of oncogenes and tumour suppressor genes and DNA damage is present.

contrast to sporadic colorectal cancer, onset of this disorder is early (40 years of age versus 60 years for sporadic cancers). In HNPCC, accelerated carcinogenesis also occurs; i.e., a tiny adenoma may progress to carcinoma in a much shorter time frame than would be found in sporadic cancer. Hence, closer surveillance is required in HNPCC, and this includes annual colonoscopy.

In contrast to familial adenomatous polyposis, making a diagnosis of HNPCC is not easy. Hence, an understanding and management of this condition have been very challenging. International criteria have been established to define the entity, but even so there are difficulties with a firm diagnosis (see DNA Mismatch Repair below). An important observation in HNPCC was the finding of microsatellite instability in these tumours; i.e., simple tandem repeats known as microsatellites (see Chapter 2) were shown to be disrupted (i.e., the PCR microsatellite pattern in tumour tissue compared to normal tissue had different sized alleles; Figure 3.8). This gave the first clue that HNPCC might be caused by DNA repair genes.

DNA Repair

Unlike RNA, proteins and other cellular components that are replaced regularly, DNA does not undergo a regular turnover. Furthermore, DNA is exposed to many damaging agents: (1) Endogenous: products of metabolism including oxidation, methylation and errors related to replication. (2) Exogenous: ultraviolet and ionising radiation, chemicals and mutagens. Therefore, in response to damage, a DNA repair system is required. The importance of this system is confirmed by finding that many of its genes are conserved throughout evolution; i.e., the identical genes are found in bacteria, yeast and humans. A number of pathways are involved in DNA repair. The three of particular relevance to molecular medicine are (1) mismatch repair, (2) nucleotide excision repair and (3) base excision repair (Table 3.6). The response to DNA damage is broader than straight-out repair and includes cellular ageing (via arrest of the cell cycle) and cellular death (via apoptosis). However, mutations in the DNA repair mechanisms are associated with the development

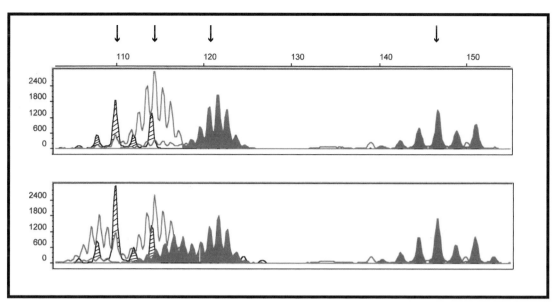

Fig. 3.8 Microsatellite instability detectable by DNA testing.
The ↓ indicate the position of four DNA markers that are used to detect microsatellite instability (from left to right they are D5S346 (stippled—black), BAT26 (open—green), BAT25 (filled—blue) and D17S250 (filled—green). The upper series shows the normal patterns for these markers; i.e., there is no microsatellite instability. The lower series comes from a tumour that has microsatellite instability. This is seen in BAT26 and BAT25 since the profile now shows that at these two loci a number of additional fragments are generated following PCR; i.e., there has been slippage reflecting the microsatellite instability. The lower series comes from a tumour that has microsatellite instability. This is seen in BAT26 and BAT25 since the profile now shows that at these two loci a number of additional fragments are generated following PCR; i.e., there has been slippage reflecting the microsatellite instability.
Electrophoresis pattern courtesy of Le Huong, Department of Molecular & Clinical Genetics, Royal Prince Alfred Hospital, Sydney.
(see colour insert)

Table 3.6 DNA repair mechanisms

Mechanism	Explanation	Genes[a]	Associated diseases
Mismatch repair	Removes nucleotides that have been misincorporated as DNA is being copied. Acts on single base mismatches as well as small displaced loops a few bases in size, which occur in repetitive regions (e.g., microsatellites).	*MLH1, MSH2, MSH6 PMS2*	HNPCC (see text)
Nucleotide excision repair	This system is predominantly involved in removing bulky helix-distorting lesions from DNA. In this mechanism the damaged site is excised in a ~30bp segment.	>12 genes (proteins) are involved	Xeroderma pigmentosa (predisposition to skin cancer) (OMIM 278700)[b], Cockayne syndrome (developmental disorder) (OMIM 216400)
Base excision repair	DNA damage secondary to exogenous or endogenous causes leads to the damaged base being excised and replaced with the correct one.	*MYH, OGG1, MTH1*	Non-FAP multiple adenomas

[a] For more information about genes, mutations and HNPCC, see the HNPCC web site at http://www.insight-group.org. [b] More information about these disorders can be found in OMIM (*Online Mendelian Inheritance in Man*—http://www.ncbi.nlm.nih.gov/entrez/query.fcgi?db=OMIM).

of cancer and a number of other serious genetic disorders confirming the pre-eminent role DNA repair plays in cellular function and normal development.

DNA Mismatch Repair

The DNA mismatch repair genes associated with HNPCC are *hMSH2* and *hMLH1* (a third but less common gene is *hMSH6*). As for familial adenomatous polyposis, those who have a germline mutation in the above genes develop cancer only when a second hit occurs and inactivates the remaining normal allele. Apart from cancer developing, the occurrence of the second hit can be seen in the finding of microsatellite instability mentioned earlier. More than 90% of HNPCC patients have microsatellite instability. Interestingly, microsatellite

instability is also present in about 15% of sporadic colorectal cancer. The molecular mechanism in the latter group is thought to be inactivation of the mismatch repair genes by epigenetic changes such as hypermethylation. Therefore, the finding of microsatellite instability does not necessarily indicate HNPCC. Indeed, since sporadic colorectal cancer is so much more common than HNPCC, microsatellite instability is more likely to be found in sporadic cancer unless it is present within the context of a family with features of HNPCC.

For the same reasons mentioned earlier in Cancer Pathogenesis—Familial adenomatous polyposis, DNA predictive testing in HNPCC is valuable. However, DNA testing in HNPCC is complex because (1) Determining what is likely to be a case of HNPCC is difficult. (2) At least three genes are involved. (3) A proportion of the mutations are missense changes, so it can be more difficult to be sure that they are pathogenic. (4) Some mutations may be large deletions that could be missed by a technique such as PCR unless there was prior knowledge of the deletion. Therefore, it is helpful to have some assistance in determining whether DNA testing is going to be fruitful since a lot of work will be involved. Microsatellite instability is used as a guide. For example, a family with the clinical features of HNPCC and a positive test for

DNA microsatellite instability would be worthwhile studying for DNA mismatch repair mutations. An alternative to the DNA test for microsatellite instability is to look for protein products of the three mismatch repair genes by immunohistochemical staining.

Base Excision Repair

The next development after familial adenomatous polyposis and HNPCC has been the finding that patients with an autosomal recessive form of colonic adenomatous polyps and carcinoma had somatic mutations consistent with oxidative-induced DNA damage. The number of polyps in this variant are significantly reduced compared to the autosomal dominant form. When the *MYH* gene (involved in base excision repair) was studied, the affected patients were shown to have both *MYH* alleles mutated in the germline (consistent with an autosomal recessive disorder). This syndrome has now implicated the base excision repair pathway in the development of adenomas and colorectal cancer. Earlier, attenuated familial adenomatous polyposis was described, and it was associated with a certain site of mutations in the *APC* gene. A second mechanism involving DNA repair has now been shown to give a similar phenotype.

AUTOSOMAL RECESSIVE DISORDERS

OVERVIEW

The appearance of an autosomal recessive disorder in a pedigree gives rise to a **horizontal** rather than **vertical** pattern. This occurs because affected individuals tend to be limited to a single sibship, and the disease is not usually found in multiple generations (Figure 3.9). Males and females are affected with equal probability. Consanguinity can be demonstrated in some affected families. The usual mating pattern leading to an autosomal recessive disorder involves two heterozygous individuals who are clinically normal. From this union, there is a one in four (25%) chance that each offspring will be homozygous-normal or homozygous-affected for that trait or mutation. There is a two in four (50%) chance that offspring will themselves be carriers (heterozygotes) for the trait or mutation. The same risks apply to each pregnancy.

The inheritance patterns described may not be apparent, particularly in communities where the numbers of offspring are few. In these instances, the genetic trait or mutation can appear to be sporadic in occurrence. Therefore, the finding of a negative family history in the autosomal recessive disorders should not be ignored, since the genetic defect can still be transmitted to the next generation, particularly if the mutant gene occurs at a high frequency in that population, e.g., cystic fibrosis.

DISEASE TO SYNDROME—CYSTIC FIBROSIS

Traditionally, cystic fibrosis was a disorder of children. Today, with knowledge of its molecular changes, cystic fibrosis is better described as a syndrome with far-reaching effects involving children and adults. Cystic fibrosis is the most common autosomal recessive disorder in Caucasians (Box 3.2). It affects approximately one in 2000 to 2500 live births with a carrier rate in northern Europeans of 1 in 20 to 1 in 25. The high incidence of cystic fibrosis remains unexplained although there is increasing evidence that the carrier state for cystic fibrosis provides an evolutionary selective advantage. Evidence for this comes from experimental data showing the heterozygote for cystic fibrosis is less susceptible to bacterial toxin-mediated diarrhea caused by cholera and typhoid. For example, transgenic mice homozygous for a cystic fibrosis mutation did not secrete fluid when challenged with cholera toxin. Heterozygous mice secreted 50% of the normal fluid and chloride ion in response to the cholera toxin. Whether the selective advantage reflects a relative inability to lose chloride in response to infection because of the cystic fibrosis carrier state (and so less risk for dehydration and its consequences) or a more sophisticated explanation is not clear. An example

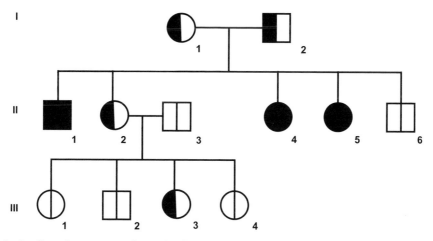

Fig. 3.9 Idealised pedigree for an autosomal recessive disorder.
Females are represented by ○; males, by □. Carriers of the genetic trait are indicated as half-filled circles or squares; affected as
● or ■. Note the horizontal distribution for the affected. The carriers are called heterozygotes, and the affected are sometimes
called homozygous affected (contrasting with normals who are also called homozygous normal).

Box 3.2 Clinical features of cystic fibrosis.

Cystic fibrosis is one of the most common lethal genetic disorders. It involves many organ systems, with the common feature being an abnormality in exocrine gland function manifesting as raised sweat chloride, recurrent respiratory infections leading to bronchiectasis, and malabsorption related to pancreatic disease. The first comprehensive description of this disorder was given in 1938. Approximately 1 in 2000 newborns of northern European descent is affected although considerable differences exist within ethnic groups; e.g., the incidence in some Hawaiians is 1 in 90 000. Complications associated with cystic fibrosis include (1) Respiratory—chronic bacterial infections, particularly *Pseudomonas aeruginosa*. (2) Gastrointestinal tract—10% of newborn infants with cystic fibrosis present with obstruction of the ileum (meconium ileus). More than 85% of children show evidence of malabsorption due to exocrine pan-

creatic insufficiency, which requires dietary regulation and supplementation with vitamins and pancreatic enzymes. Rare gastrointestinal tract complications include biliary cirrhosis. (3) Infertility affecting 85% of males and, to a lesser extent, females. The long-term outlook for cystic fibrosis has improved dramatically over the years. Previously, it was a fatal disease within the first 12 months for most affected babies. Today, many patients reach adult life, with some having children of their own. This dramatic improvement in survival reflects a higher standard of living, improved nutrition and the availability of specialised clinics to allow a multidisciplinary approach for follow-up and treatment. Support groups have ensured that families are aware of recent developments, which include expanded diagnostic options. Gene therapy is also being evaluated as a form of treatment.

---

of the latter is the apparent requirement for the normal cystic fibrosis gene for entry of *S. typhi* into epithelial cells. Diminished activity of this gene would therefore be protective.

After asthma, cystic fibrosis is the commonest cause of chronic respiratory distress in childhood and is responsible for the majority of deaths from respiratory disease in this age group. Clinical features relate to the thick tenacious secretions that can manifest with intestinal obstruction in the newborn (called meconium ileus), pancreatic insufficiency and chronic respiratory infections in childhood. Although first described as a clinical entity in

the 1930s, the pathogenesis of cystic fibrosis remained elusive despite clinical, electrophysiological and other conventional approaches used to study this disorder. The only clue to the underlying defect related to the elevated chloride in sweat. In the mid-1980s, chloride ion conductance across the apical membranes of respiratory epithelial cells or sweat ducts was shown to be decreased. Whether this represented an abnormal chloride channel or aberrant control of a normal channel was unknown. This remained the state of knowledge until 1989, when the cystic fibrosis gene was isolated by positional cloning. This represented a *tour de force* at the time (Box 3.3).

Box 3.3 Positional cloning in cystic fibrosis.

Initial attempts at chromosome localisation in cystic fibrosis were unsuccessful. This delayed isolation of the gene since a trial-and-error approach was required to determine which DNA polymorphic markers would co-segregate with cystic fibrosis. In 1985, linkage of cystic fibrosis to DNA markers on chromosome 7q31 was shown. Subsequently, a huge amount of work was conducted to narrow the distance and identify likely candidate genes within this region. The search narrowed to ~0.5 Mb of DNA, which contained a number of genes. Clues which suggested that one was the cystic fibrosis gene included (1) Conservation of DNA sequence across a number of species; i.e., the gene carried out an important function; (2) mRNA was present in tissues connected with cystic fibrosis, i.e., lung, pancreas, intestine, liver and sweat glands. The *CFTR* gene was found in 1989. Its genomic structure (i.e., the exons and introns) extended over 250 kb of DNA (see Figure 3.10). The mRNA transcript was 6.5 kb in size. The protein encoded by *CFTR* had considerable similarity to a family of membrane-associated ATP-dependent transporter proteins involved in the active transport of substances across membranes. Common structural findings in the above transporter proteins include hydrophobic transmembrane domains and nucleotide-binding folds for ATP attachment from which the energy for transport is obtained. *CFTR* has these features, and a highly charged central R domain (R for regulatory) involved in phosphorylation by protein kinase. It was subsequently proven that *CFTR* codes for a chloride ion channel. Activation of this channel can occur by cyclic AMP or following phosphorylation by protein kinase. The latter involves the R domain and may work through a conformational change that allows the passive flow of chloride ions. Whether *CFTR* has other functions remains to be determined. One recent suggestion is that the *CFTR* gene acts as a receptor for *Ps. aeruginosa*. This allows internalisation into the epithelial cell and so clearance of these organisms commonly found in cystic fibrosis. Impaired *CFTR* function would diminish the ability to clear *Ps. aeruginosa*.

Table 3.7 Classes of *CFTR* mutations based on function

Class	*CFTR* abnormality and phenotype	Mutation examples
I	Absent protein because of defective production; therefore, severe phenotype.	Nonsense and frameshift mutations, e.g., G542X, W1282X, R553X, 621+1 G→T
II	Absent protein because of defective processing or transport. Variable phenotype. Homozygous ΔF508 severe disease.	ΔF508
III	Defective activation and regulation of *CFTR* at plasma membrane leading to normal amount but non-functional protein. Variable phenotype.	G551D
IV	Decreased protein conductance leading to normal amounts but impaired function. Milder phenotype.	R117H
V	Reduced amounts of functioning protein. Milder phenotype.	3849 + 10 kb C → T

amino acid (F = phenylalanine) at position 508. This mutation is known as ΔF508. When the gene was discovered, it was predicted that few mutations would be found because it appeared that relatively few haplotypes were associated with this disorder. However, that prediction was incorrect, and although ΔF508 is the most common defect, there are now more than 1300 other mutations (Table 3.8).

DNA Testing

The sweat test is considered to be the gold standard for detecting cystic fibrosis. A raised sweat chloride concentration greater than 60 mmol/l after two days of life indicates cystic fibrosis. However, disadvantages of the sweat test are its inability to detect heterozygotes (carriers), and it is not suitable for prenatal diagnosis. Once DNA polymorphisms became available, it was possible to utilise a linkage study approach to undertake prenatal diagnosis and carrier testing provided the appropriate family structure was available (further discussion and an example are found in the Appendix). Since 1989, direct identification of the ΔF508 and other *CFTR* mutations by PCR has become the method of choice for laboratory testing. DNA testing options include prenatal diagnosis, newborn screening, carrier testing and diagnostic testing to distinguish those disorders that resemble cystic fibrosis but have atypical features.

The ΔF508 mutation involves about 68% of the cystic fibrosis chromosomes in northern Europeans, with a

The cystic fibrosis gene is called *CFTR* (cystic fibrosis transmembrane conductance regulator). It is very large, with 24 exons initially described, although this was soon revised to 27 and the 3 additional exons are numbered 6b, 14b and 17b (Figure 3.10). The protein is a 1480 amino acid glycosylated membrane glycoprotein that functions as a cyclic AMP regulated chloride channel. *CFTR* is involved in the movement of ions and water predominantly in respiratory, gastrointestinal, hepatobiliary and reproductive systems. The most common mutation in cystic fibrosis involves a deletion (Δ) of a single

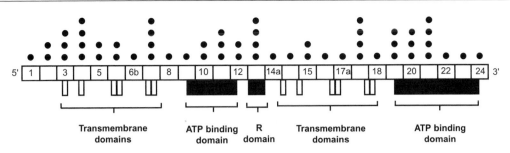

Fig. 3.10   CFTR gene.
The diagram, which is not drawn to scale, shows the 27 exons of the *CFTR* gene. There are 24 exons numbered, but the total is 27 since after the exons were first described and numbered, introns were detected dividing exon 6 (so that there are now exons 6a and 6b), exon 14 (exons 14a and 14b) and exon 17 (exons 17a and 17b). Above each exon is an estimate of the number of mutations affecting that exon as well as its corresponding intron, e.g., exon 3 and intron 3. Each • represents about 20 mutations. Below the exons are depicted the major functional domains, i.e., transmembrane domains, ATP-binding domains and an R (regulatory) domain. The ΔF508 defect is found on exon 10 and so interferes with ATP-binding. Mutations in *CFTR* can now be classified on the basis of their function (Table 3.7).

Table 3.8   Frequencies of the CFTR mutations in 17 853 cystic fibrosis patients in the USA (McKone et al 2003)

CFTR mutation	Allele frequency (%)
ΔF508	69.4
Unknown	15.7
G542X	2.3
G551D	2.2
ΔI507	1.6
W1282X	1.4
N1303K	1.2
R553X	0.9
621+1 G → T	0.8
R117H	0.7
3849+10 kb C → T	0.7
1717-1 G → A	0.5

Table 3.9   Geographical distribution of the ΔF508 mutation of cystic fibrosis (Cystic Fibrosis Mutation Database 2003— http://www.genet.sickkids.on.ca/cftr/)
More than 1300 mutations produce cystic fibrosis. Mutations other than ΔF508 occur less frequently (see Table 3.8).

Continent[a]	Percentage ΔF508 detection rate
Northern Europe	68
Southern Europe	43
North America	66
Middle East	33
Africa (Algeria, Tunisia)	29
Australia	69
All continents	66

[a] Ethnic heterogeneity within continents is a further consideration in assessing the local frequency of the ΔF508 mutation.

lower frequency in other regions, e.g., 43% in southern Europe, 33% in the Middle East (Table 3.9). The multiplicity of mutations makes detection of all cystic fibrosis defects an unrealistic goal with present technology. Therefore, DNA-based diagnostic tests incorporate ΔF508 plus a limited number of other mutations (e.g., 12–25) selected on the basis of their prevalence in each population. This enables about 70–80% of the cystic fibrosis mutations to be detected. DNA mutation testing has high specificity; i.e., finding two relevant mutations indicates cystic fibrosis with 100% certainty. However, the DNA test has low sensitivity because so many mutations are involved. Tissues that are suitable for PCR include blood, hair follicles, chorionic villus, blood spots such as those taken from neonatal heel pricks (Guthrie spots), cells shed in amniotic fluid (amniocytes) and buccal cells obtained from mouth washes. DNA linkage analysis is another option that can be helpful in the prenatal diagnosis scenario when the underlying DNA mutation cannot be found (Figures 3.11 and A.16). By using intragenic DNA polymorphisms, the risk of recombination with the *CFTR* gene is not a significant source of error in the indirect (linkage) approach.

The potential of PCR to be automated and so screen for the ΔF508 defect on a widespread basis has produced

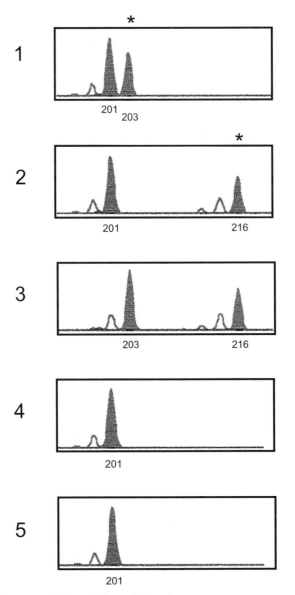

Fig. 3.11 Linkage testing during a prenatal diagnosis for cystic fibrosis.
Because mutations in the *CFTR* gene were not detectable in a couple with a cystic fibrosis child, prenatal testing during the current pregnancy was possible only by linkage analysis. Using an intragenic (IVS8) microsatellite marker, the family could be studied to identify which allele co-segregated with the cystic fibrosis defect in the affected child. **(1)** and **(2)** The parents, each of whom has two polymorphic markers detectable, i.e., 201 and 203 bp; 201 and 216 bp. **(3)** Homozygous-affected child. **(4)** and **(5)** Duplicate samples from a chorion villus biopsy. From **(3)** it is apparent that the 203 and 216 alleles co-segregate with cystic fibrosis because this is DNA from the affected child, and so in the parents these alleles (indicated by *) identify the cystic fibrosis chromosomes. In **(4)** and **(5)**, the fetus has inherited only the 201 allele from each parent; i.e., the fetus is homozygous for this marker. This shows that the fetus is normal. From the results in **(1)**, it is evident that the IVS8 marker is a dinucleotide microsatellite because the two alleles are 201 and 203 bp in size; i.e., they differ by two nucleotides consistent with a marker such as $(CA)_n$. See also Figure A.16 for another, more complex example of linkage with cystic fibrosis.

a controversy, i.e., whether there should be population-based cystic fibrosis screening. A recent recommendation in the USA is that all pregnant women should be offered cystic fibrosis DNA testing even if there is no family history of this disorder. However, this recommendation is not followed in Europe and most other countries. Some studies are under way looking at the implication of DNA testing in schoolchildren. The protagonists for screening point out the importance that carrier status knowledge would have on decisions about reproduction. Those against population screening note that the DNA test has low sensitivity, and indicate that benefits and risks of population screening are uncertain unless 90–95% of carriers can be detected, which is presently not an economical or realistic goal with the *CFTR* gene, particularly in multicultural communities.

Pathogenesis

As well as being many in number, mutations in the *CFTR* gene are, not surprisingly, heterogeneous in their effect. Some mutations have a significant detrimental effect on the function of the *CFTR*, whereas others are associated with mild deficiencies in function. The potential to predict the phenotype, particularly clinical severity based on the genotype, then becomes possible (see Table 3.7, Figure 3.10). A retrospective cohort study involving 17 853 cystic fibrosis patients in the USA has allowed clinical outcomes to be predicted on the basis of the underlying DNA mutation (genotype). The study showed that homozygous ΔF508 results in a severe disorder with comparable mortality rates in compound heterozygotes for ΔF508 and a range of mutations including G551D, G542, R553X, 621 + 1 G → T and 1717—1 G → A. On the other hand, compound heterozygotes for ΔF508 and mutations ΔI507 or R117H or 3849 + 10kb C →T had a significantly lower standardised mortality rate (McKone *et al* 2003).

Considerable effort has gone into the correlation of genotypes (DNA defects) with phenotypes (clinical features) because of the genetic and clinical heterogeneity found in cystic fibrosis. For example, normal pancreatic function is often present in mild forms of the disorder. Molecular defects that are associated with pancreatic sufficiency can now be identified. They usually involve missense codon changes. In these circumstances, prognosis for cystic fibrosis is considerably improved. The molecular defects in those with pancreatic insufficiency (i.e., cystic fibrosis is of the severe type) are more likely to involve the ΔF508 deletion, which is located within the first ATP binding site (see Figure 3.10). Similarly, meconium ileus is frequently found in the newborn with the ΔF508 deletion. Mutations associated with severe phenotypes produce premature stop codons, frameshifts or splicing defects. However, the story is not that simple since exceptions are found.

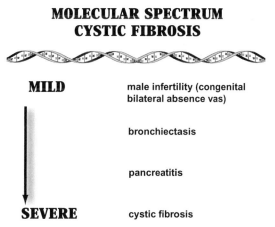

MOLECULAR SPECTRUM CYSTIC FIBROSIS

MILD
↓
SEVERE

male infertility (congenital bilateral absence vas)

bronchiectasis

pancreatitis

cystic fibrosis

Fig. 3.12 Spectrum of cystic fibrosis.
Mutations in the *CFTR* gene are usually associated with cystic fibrosis. However, molecular understanding of cystic fibrosis has demonstrated a more complex picture. This shows cystic fibrosis to represent a broad spectrum from male infertility (due to bilateral absence of the vas deferens) to bronchiectasis or pancreatitis and then cystic fibrosis. Even the latter has mild (pancreatic sufficient) and more severe (pancreatic insufficient) forms. Not surprisingly, the molecular spectrum is based on different DNA mutations with the milder ones leading to subtle phenotypes and the more severe ones associated with the complete phenotype.

Disease Spectrum

Another development following on from the discovery of the *CFTR* gene is that cystic fibrosis can no longer be considered a single disorder, but a spectrum of diseases with cystic fibrosis the most severe form. A number of other disorders have been shown to be caused by mild forms of *CFTR* mutations. Included here is male infertility due to congenital bilateral absence of the vas deferens, idiopathic bronchiectasis and idiopathic pancreatitis (Figure 3.12). In these three scenarios, the clue that there was a link to cystic fibrosis came from large population studies which identified in these diseases a greater frequency of *CFTR* mutations than would be expected in the normal population. Subsequently, in congenital bilateral absence of the vas deferens, it was shown that in about 50% of these individuals, usually one *CFTR* mutation is detected. In some reports, the R117H defect occurs more often. Hence, it was proposed that the mild phenotype reflects the genotype, i.e., a double heterozygote for a known *CFTR* mutation described in Table 3.8 and a second (unknown but presumably very mild) defect. One interesting new mutation was found in male infertility caused by cystic fibrosis. This is a polymorphism within intron 8 of the *CFTR* gene, which has three alleles: 5T, 7T and 9T. Those with congenital bilateral absence of the vas deferens were more likely to have a

Table 3.10 Causes of iron overload

Type	Classification	Comments
Genetic	Autosomal recessive and autosomal dominant forms described	Subtypes include (i) common type is *HFE* related, (ii) non-*HFE* including juvenile forms and autosomal dominant form.
Acquired	Haematologic disease	(i) thalassaemias, (ii) sideroblastic anaemias (iii) chronic haemolytic anaemia
	Dietary, parenteral	
	Liver disease	Alcohol, fatty liver disease, porphyria cutanea tarda
	Miscellaneous	African iron overload, other rare conditions.

Table 3.11 Milestones in the hereditary haemochromatosis story

Year	Name	Finding
1865	A Trousseau	Described the triad of diabetes, cirrhosis, pigmentation
1886	FD von Recklinghausen	Pigment confirmed to be iron—coined the term "hemochromatose"
1935	JH Sheldon	First suggestion that haemochromatosis is a genetic disease (otherwise considered to be acquired)
1977	M Simon	Linkage to HLA A3 and B7 shown, confirming genetic association
1996	JN Feder, working in biotechnology company	Positional cloning finds the gene *HFE* and the two common mutations C282Y and H63D

Box 3.4 Novel molecular mechanism for disease: 5T *CFTR* polymorphism.

In intron 8 of the *CFTR* gene (between exons 8 and 9), there is a run of T (thymidine) nucleotide bases. Three types are found: 5Ts, 7Ts and 9Ts. There is now good evidence that the 5T polymorphism (but not 7T or 9T) predisposes to alternative splicing so that exon 9 is excluded from the mRNA. In congenital bilateral absence of the vas deferens, there is a good correlation between the 5T polymorphism that is co-inherited with another *CFTR* mutation and the development of this type of infertility in the male. Recently, there is some evidence that 5T in association with the mutation M470V and another associated polymorphism (TG12) may be more commonly found in bronchiectasis. However, more needs to be done to understand what is occurring here because 5T alone appears to cause no particular problem. If it is a real mutation, it must have low penetrance.

---

standard *CFTR* mutation plus the 5T variant of this polymorphism (Box 3.4).

The story in pancreatitis and bronchiectasis is not as advanced as that for congenital bilateral absence of the vas deferens. All that can be said in these two conditions is that, on a population basis, there are a greater number of *CFTR* mutations. However, no mutations are specific for pancreatitis or bronchiectasis. DNA testing in these circumstances is often requested but rarely beneficial because little information will be provided about individuals (in contrast to populations) since at best it can be expected that one *CFTR* mutation will be found. Therefore, it will be difficult to decide whether this is simply an asymptomatic carrier or an individual who is a compound heterozygote with the second mutation unde-

tectable. Again, like the congenital bilateral absence of the vas scenario, the second mutation in these disorders will be very mild, and so unlikely to be included in the usual profile of *CFTR* mutations that are generally the more severe ones (since invariably these mutations are found in patients with cystic fibrosis).

Another condition associated with cystic fibrosis is meconium ileus in the fetus or newborn. The former is more often being detected because of increasing ultra-sound monitoring of pregnancies. Meconium ileus in the fetus is said to be associated with the ΔF508 defect, although the evidence for this association is not very strong. Another tentative link with cystic fibrosis is nasal polyps, although like meconium ileus, this needs confirmation.

PREDISEASE—HEREDITARY HAEMOCHROMATOSIS

Iron overload can be acquired or genetic (Table 3.10). Hereditary haemochromatosis is an autosomal recessive genetic disorder of iron metabolism affecting about 1 in 300 individuals of northern European origin, and having a carrier frequency of 1 in 8 to 1 in 10. The highest incidences are found in populations with a Celtic background (Ireland, Wales and other regions in the world where there has been migration from Ireland). Clinical features range from non-specific symptoms such as lethargy or arthralgia to more florid but less common presentations such as diabetes mellitus, liver disease and generalised pigmentation. Life-threatening complications include cardiomyopathy and hepatocellular carcinoma.

Some milestones in the hereditary haemochromatosis story are listed in Table 3.11. Again, as has been illustrated previously, developing some understanding of hereditary haemochromatosis took many years, but it was not until the gene was isolated by positional cloning in

1996 that real progress was possible. The common genetic form of hereditary haemochromatosis is now known to be caused by the gene *HFE*, which codes for a protein that has some features of the HLA class I molecules. Hence, the *HFE* gene was originally named *HLA-H*, but this was soon changed when it became apparent that the gene was not strictly part of the HLA complex. The *HFE* gene codes for a protein that binds $\beta2$ microglobulin (like other MHC Class 1 molecules) and interacts with transferrin receptor 1.

Three mutations have been reported in hereditary haemochromatosis. They are C282Y (also written Cys282Tyr), i.e., at amino acid position 282, a cysteine is replaced with a tyrosine; H63D (His63Asp), i.e., histidine is replaced by aspartic acid at position 63; and S65C (Ser65Cys), i.e., serine is replaced by cysteine at position 65. The C282Y defect alters the ability of the *HFE* protein to bind to $\beta2$ microglobulin, which is essential for its subsequent interaction with the transferrin receptor 1. This reduces the latter's affinity for transferrin and produces a reduction in the amount of *HFE* protein expressed on the cell surface. Neither of the two remaining mutations has an effect on transferrin receptor 1, and so their modes of action are different from C282Y but little more is known. Approximately 85% of patients with clinical haemochromatosis of northern European origin will be homozygous for the C282Y mutation. The only other known genetic risk factor is a double heterozygote for H63D/C282Y or S65C/C282Y, although with this combination, the iron loading is a lot less.

HFE-related hereditary haemochromatosis occurs mostly in those of northern European ethnic background. There is growing evidence based on DNA studies that the C282Y mutation is likely to have arisen spontaneously once or twice at most and was then spread throughout the world. It has been proposed recently that the migratory patterns of the Vikings would explain the distribution of C282Y in northern Europe. Just as was discussed with cystic fibrosis and later on thalassaemia, the common carrier frequency for this mutation implies the possibility of some type of evolutionary selective advantage. For hereditary haemochromatosis, it has been proposed that women who were carriers had a reproductive advantage since they would be less likely to be iron deficient (a common problem in women particularly if there is malnourishment). This does not explain the carrier frequency in males although an evolutionary advantage would come from having some resistance to iron deficiency, and this might also strengthen immunity to infections.

Other forms of hereditary haemochromatosis must exist, particularly in southern Europeans, where C282Y is less common. These forms are broadly called non-*HFE* haemochromatosis, but relevant genes are now starting to be identified. They include (1) Ferroportin disease—autosomal dominant inheritance involving mutations in the gene *SCL40A1* and generally associated with a milder phenotype. Unlike the *HFE* mutation, ferroportin disease is geographically more widespread, including Asian populations. (2) Transferrin receptor 2 gene mutations. Autosomal recessive with a lower frequency than ferroportin, and like *HFE*, it interferes with the uptake of transferrin bound iron, but the affinity of the transferrin receptor 2 is considerably less than that for transferrin receptor 1. (3) Two genes are implicated in the juvenile forms—hepcidin (*HAMP*) and hemojuvelin (*HFE2*).

Predisease

The concept of predisease was raised earlier in relation to predictive DNA testing; i.e., it is possible with a DNA test to look for a mutation and so predict well into the future that an individual is at higher risk for developing a genetic disorder. The Huntington's disease model showed predictive testing that was very accurate and able to determine many years in advance that Huntington's disease would develop. In these circumstances, the DNA test converts a patient who is "at risk" into an individual who has a "predisease," i.e., a definite risk because the mutation is there, but the patient is well and asymptomatic. Another complexity with the predisease concept is that the risk for developing disease is variable from a 100% certainty with conditions such as Huntington's disease to a 36–85% lifetime risk with breast cancer caused by mutations in *BRCA1* and *BRCA2* genes (see Chapter 4). Haemochromatosis is another example of predisease.

An individual with hereditary haemochromatosis has the genetic predisposition, but there are environmental factors, and perhaps other genetic contributors, that will determine if there will be progression to clinical haemochromatosis. An important environmental factor is sex; i.e., the male-to-female ratio for haemochromatosis is as high as 3 : 1 although it is an autosomal disorder. However, loss of blood through menstruation for women is protective, and so women prior to menopause have a much lower risk. Therefore, the distinction between **hereditary haemochromatosis** and **clinical haemochromatosis** is important. Although a person may be labelled as hereditary haemochromatosis, there is considerable uncertainty about the long-term risk. Some studies have indicated that between 24–58% of those homozygous for C282Y will not express the clinical features of this disorder (for example, a raised ferritin or a transferrin saturation greater than 45%). Another study has suggested that <1% of C282Y homozygotes will develop clinical haemochromatosis (Beutler *et al* 2002). On the other hand, uncertainty is not an issue if the patient presents with clinical features of haemochromatosis, and the DNA test is undertaken to confirm the diagnosis. Previously, this confirmation was possible only with an invasive procedure such as a liver biopsy. Today, the main reason for a liver biopsy is to determine whether there is liver damage.

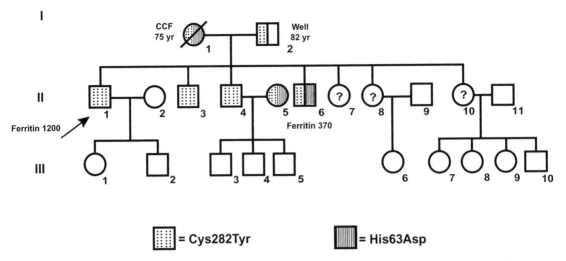

= Cys282Tyr ||| = His63Asp

Fig. 3.13 *HFE* **family.**
The propositus (→) is a 50-year-old male of northwestern European origin, who was investigated because he had clinical features of haemochromatosis, including a raised serum ferritin of 1200 μg/L (normal in males 25–300). A DNA test (this would be a DNA **diagnostic** test) confirmed the haemochromatosis because it showed he was homozygous for the common Cys282Tyr mutation. In this family, it would be important to recommend that others get tested because they are at risk, particularly the male siblings. This type of DNA test, in contrast to the above, would be a **predictive** DNA test as the people being tested are asymptomatic. As can be seen, two brothers (II-3, II-4) are also homozygous for Cys282Tyr, and so they would be called genetic haemochromatosis cases because they are at risk and should have regular follow-up with ferritin levels. In addition II-4's partner is a double heterozygote for Cys282Tyr and His63Asp. This would mean that all children (III-3, III-4, III-5) will also be homozygous Cys282Tyr or double heterozygotes. It is interesting that a third brother (II-6) is a double heterozygote. This particular individual had his ferritin checked, and the slightly raised level of 370 μg/L is consistent with the genotype that is associated with mild disease. DNA testing can provide more information about this family. The mother of the propositus (I-1) is deceased, but she must be at least a heterozygote for Cys282Tyr (because she has sons who are homozygotes). Her living spouse, aged 82, is well and healthy. As expected, after DNA testing, he is shown to be a carrier for the same mutation, but he does not have the His63Asp defect. Therefore, the source of this defect must be the deceased mother; i.e., she is a double heterozygote like her son II-6. This fact is relevant because it might explain why she was very healthy until the age of 75 years, when she developed intractable congestive cardiac failure, which was unresponsive to conventional treatment. A likely explanation for this cardiac failure is an underlying haemochromatosis that took many years to develop because the iron loading was relatively mild.

DNA Screening

As indicated earlier, the implications for the same haemochromatosis DNA will vary depending on the clinical scenario. The pedigree in Figure 3.13 is an example of a family that was ascertained through a male who had the clinical features of this disorder. He had DNA testing to confirm the diagnosis, and with his Anglo-Saxon ethnic background, it is not surprising that he was homozygous for C282Y. However, because he shares his genes with family members, it is necessary in this situation to recommend testing, at least, the first degree family members (i.e., siblings and children) since these individuals are also at risk. In this particular case, the parents must be obligatory carriers at least. Family members showing *HFE* mutations would need to be followed with a biochemical marker of iron excess such as ferritin. As the family pedigree demonstrates, a number of brothers are at risk (i.e., two are homozygous for C282Y and so they have genetic-type haemochromatosis, and one is

a double heterozygote for C282Y and H63D; hence, his risk is less, which is consistent with his mildly elevated ferritin level).

Although screening for the hereditary haemochromatosis DNA mutation is most valuable when it is known to be present in a family, the question about population screening is often raised. The main reason is that the treatment for clinical haemochromatosis is simple and cheap; i.e., regular venesection is likely to prevent liver cirrhosis from developing, and one hopes, the consequent complication of hepatocellular carcinoma. This venesection can be undertaken simply by being a regular blood donor as blood transfusion services will usually accept blood from a patient with haemochromatosis provided the other requirements are met and infection screens are negative.

Screening for haemochromatosis can be **phenotypic** (biochemical markers such as ferritins or transferrin saturation) or **genotypic** (i.e., DNA mutations). Which, if any, approach is better remains problematic. It is difficult

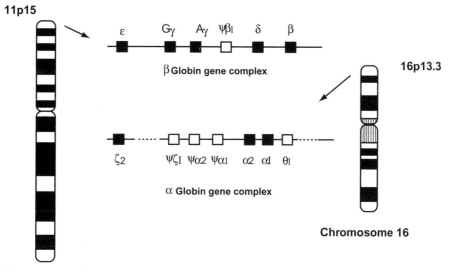

Fig. 3.14 The globin gene complexes on chromosomes 11 and 16.
Functional genes are indicated as ■ and non-functioning genes (called pseudogenes) as □. On the short arm of chromosome 11 at band position 15 is found the β globin gene complex. One gene is active during embryonic life (ε), two are fetal specific (Gγ, Aγ), and two are expressed in adult life (δ, β). The α globin complex is on the short arm of chromosome 16 at band 13.3. A lot more genes are situated in this complex, but many are non-functional. The embryonic/fetal gene is ζ_2 and the two adult genes are $\alpha2$ and $\alpha1$. The evolution of the globin gene clusters from a common ancestral gene is seen by the similarity in structure and sequence, which the above genes share even though they are on different chromosomes. The . . . on the α globin complex marks the position of a DNA polymorphism.

to screen biochemically, and when the ferritin is raised, some damage may already have been caused. The raised ferritin does not distinguish genetic from non-genetic causes. On the other hand, the DNA test is more expensive, and the progression from hereditary haemochromatosis to clinical haemochromatosis is an unresolved issue. In a multicultural community, the C282Y test can be less helpful. Because of these uncertainties, no universal screening programs are under way, but the debate will continue for some time.

GENE INTERACTIONS—THALASSAEMIA SYNDROMES

Haemoglobinopathies are inherited disorders of globin. They are classified into the **thalassaemia syndromes** (e.g., α thalassaemia, β thalassaemia) and the **variant haemoglobins** (e.g., sickle cell haemoglobin [HbS]). It is estimated that about 7% of the world population are carriers of the haemoglobinopathies. The thalassaemias were one of the first human disorders to be characterised at the molecular level. As such, they provide a useful starting point in understanding the molecular basis for human genetic disorders.

Haemoglobin, the pigment in red blood cells, comprises iron and a protein called globin. Four polypeptide

chains make up globin, including two α globin chains and two β globin chains. A genetic defect that impairs globin synthesis produces thalassaemia, and this manifests as anaemia with red blood cells that are small and pale. The clinical picture in the thalassaemia syndromes is diverse and ranges from an asymptomatic disorder that is detected fortuitously to a lifelong blood transfusion-dependent anaemia or an anaemia that is fatal *in utero* or soon after birth. Following cloning of the human α and β globin genes in the late 1970s, it became apparent that the globins represented a gene family with a larger number of genes than initially considered. Cell fusion studies localised a cluster of such genes on chromosome 16—α globin cluster—and a second cluster on chromosome 11—β globin cluster (Figure 3.14). DNA studies also confirmed that the globin genes had evolved by duplication from a single ancestral gene. Another feature of the globin genes was their developmental regulation with embryonic, fetal and adult genes being identified within each cluster. During development, there was a change in the haemoglobin profile, with the complete switch to adult globins occurring about six months after birth (Table 3.12).

The first accurate clinical description of thalassaemia was given by T Cooley in 1925. Cooley's anaemia, as it became known, was shown to be genetic in origin during the late 1930s and early 1940s when relatives of severely affected individuals were observed to have similar but milder changes in their red blood cells. The word "thalassaemia" was coined in 1936 and comes from the Greek $\theta\alpha\lambda\alpha\sigma\sigma\alpha$, which means "the sea." The name arose since it was initially considered (erroneously) that thalassaemia was a disease which affected those who lived near the Mediterranean Sea. Today, populations at risk come from many parts of the world, including the Mediterranean, southern China, South East Asia, India, Middle East, Africa and other regions. The high frequency of thalassaemia and sickle haemoglobin carriers reflects the protection from malaria provided by these disorders. Molecular epidemiology studies confirmed the close relationship between thalassaemia and malaria, with the mechanism being an evolutionary survival advantage for red blood cells that carried the thalassaemia trait; i.e., pale, small red blood cells provided a poor environment for the growth of malarial parasites. A similar positive selective advantage against malaria was also shown with the variant haemoglobin known as sickle cell defect, a mutation affecting the β globin chain.

Molecular Pathology

During the 1960s, the biochemical defect in the thalassaemias was identified as an **imbalance** in the number of α and β globin chains with the normal α/β ratio being 1. Once this ratio becomes imbalanced, the red blood cell precursors are prematurely destroyed in the bone marrow. Failure to produce α globin gave rise to α thalassaemia, which is fatal in its most severe form. Failure to produce any β globin (β thalassaemia) was associated with a life-long blood transfusion-dependent anaemia. Carriers of either thalassaemia defect were clinically asymptomatic although their blood counts ranged from normal to mildly abnormal. Despite the very elegant biochemical studies, an unexplained feature of the thalassaemias remained, i.e., the considerable variation in phenotypes involving both the clinical and laboratory pictures.

Heterozygotes for α thalassaemia and β thalassaemia demonstrate variable clinical and laboratory manifestations ranging from asymptomatic with normal blood pictures to mild anaemia with microcytic hypochromic red blood cells. Based on the molecular pathology, the α thalassaemias are divided into α^+ and α^0, determined by whether one of the two or none of the two α globin genes at the one locus is deleted; i.e., $\alpha\alpha/\alpha\alpha$ is the normal complement of α globin genes (two on each chromosome). A loss of one (i.e., $-\alpha/\alpha\alpha$) is heterozygous α^+ thalassaemia. A loss of both genes in the one chromosome (i.e., $-/\alpha\alpha$) is heterozygous α^0 thalassaemia. Various combinations of α^+ and α^0 can occur. Because there are only two β globin genes (one on each chromosome), the permutations are fewer. However, β globin gene mutations are divided into β^+ and β^0 on the basis of whether there is some (+) or nil (0) β globin production. Hence, the phenotype can be variable, just like α thalassaemia. A clinical and molecular classification for the thalassaemias is given in Table 3.13.

Table 3.12 Globin gene switching in development
Different genes are functional during specific periods of development, resulting in sequential changes in the globin subunits of haemoglobin.

Functional component	Yolk sac	Fetal liver	Adult marrow
α Globin genes	α and ζ	α	α
β Globin genes	ε	γ	β
Haemoglobins	$\zeta_2\varepsilon_2$ and $\alpha_2\varepsilon_2$	$\alpha_2\gamma_2$	$\alpha_2\beta_2$

Table 3.13 Clinical and molecular classification of the thalassaemias

Type	Mutations	Clinical consequences	Molecular defects
β	β^+ and β^0	Heterozygous (carriers) have microcytic hypochromic red blood cells. β^+ forms (some β globin gene function) less severe than β^0 forms (no β globin gene function) Homozygous affected—severe anaemia requiring life-long treatment	Predominantly point mutations
α	$-\alpha$ (α^+) $-/$ (α^0)	$-\alpha/\alpha\alpha$ (heterozygous α^+ thalassaemia; asymptomatic) $-\alpha/-\alpha$ (homozygous α^+ thalassaemia—as for β thalassaemia carrier) $-/\alpha\alpha$ (heterozygous α^0 thalassaemia—as for β thalassaemia carrier) $-\alpha/-$ (double heterozygote. HbH disease. Variable phenotype from asymptomatic to severe blood transfusion dependent) $-/-$ (Hb Barts Hydrops Fetalis—fatal *in utero* or soon after baby born)	Predominantly deletions
HPFH	Variable but leading to continued expression of the fetal (γ) globin genes	Asymptomatic because missing β globin is replaced by the persisting γ globin gene expression as HbF	Mixture of point mutations and deletions

Homozygotes for β globin gene mutations giving rise to β thalassaemia or sickle cell anaemia usually develop a severe clinical disorder. On the other hand, homozygotes for a rare thalassaemia called deletional hereditary persistence of fetal haemoglobin (HPFH) or $(\delta\beta)^0$ thalassaemia have no β globin gene activity (just like β thalassaemia) because both β globin genes are deleted. However, they are clinically normal or very mildly affected compared to homozygous β thalassaemia. This occurs because there is greater γ globin gene activity and so increased amounts of fetal haemoglobin (HbF) in the blood (in normal adults, HbF comprises <1% of the total haemoglobin). In effect, this replaces the mutant β globin chains, and so the α/β ratio reverts back to 1 because β globin is being replaced by γ globin.

It is known that individuals who have the ability to produce an excess of HbF for whatever reason will have milder forms of β thalassaemia and sickle cell disease. Although the molecular basis of the high HbF thalassaemias is well understood, it has been difficult to induce HbF production artificially. Mutations in β thalassaemia become apparent once the switch is complete as only then do the β globin chains become fully operational. Therefore, a long sought after but elusive goal has been to manipulate the globin genes to prevent switching. If this were possible, the β thalassaemias and HbS disorders would no longer be clinical problems. Evidence from DNA linkage analysis suggests that other gene loci not on chromosome 11 (where the β globin gene complex is found) are also involved in the regulation of γ globin gene expression. Further work is required to define the multiple molecular mechanisms producing the hereditary persistence of fetal haemoglobin (HPFH) disorders.

Another way to reduce the severity of β thalassaemia is through co-inheritance of α thalassaemia. Surprisingly, the loss of both α and β globin genes does not lead to a more severe disorder but a milder one because the α/β ratio remains close to unity, and so the red blood cells do not become prematurely destroyed in the bone marrow. Co-inheritance of α thalassaemia has also been shown to play a role in explaining some mild forms of sickle cell disease. Phenotypes in the thalassaemia syndromes can sometimes be difficult to understand. In these circumstances it is essential to draw a pedigree and study all family members since it is possible that within families there is more than one type of thalassaemia being inherited, and the various gene-gene interactions as described above are very confusing if only the patient is studied.

The thalassaemia syndromes are the best understood, and arguably, the most studied of the genetic disorders at the molecular level. However, despite a wealth of knowledge about genes and gene-gene interactions, what is known about genotype-phenotype correlations often does not help the couple who are both carriers and planning a family. The question on their mind is: What risks will their future children face? Gene-gene interactions as described above are difficult to identify in individual cases. The environment plays a role, particularly in sickle cell disease, and it is likely that other modifying genes (comparable to the situation in familial hypertrophic cardiomyopathy) influence severity in this condition (modifying genes are discussed under Future).

X-LINKED DISORDERS

OVERVIEW

X-linked disorders result from mutated genes on the X chromosome. Males, who have only one X chromosome (i.e., they are **hemizygous**), will fully express an X-linked disorder. On the other hand, females, who have two X chromosomes, will be carriers of the defect in the majority of cases, and so they are usually asymptomatic. Although females have two X chromosomes to the male's one, products from this chromosome are quantitatively similar in both sexes because one of the two X chromosomes in females is inactivated.

Lyonisation (named after M Lyon) describes the random X-inactivation of an X chromosome that occurs during embryonic development. Because of the early onset and randomness of the process, female carriers of X-linked disorders can demonstrate variable amounts of the gene product (i.e., a protein), which will depend on

the proportion of normal-to-mutant X chromosomes that remain functional. The majority of the X chromosome is inactivated although some segments escape this process (see Figure 7.3). The molecular basis for X inactivation is unknown. Methylation may play some role (methylation is discussed further in Chapter 4).

The shape of a pedigree illustrating X-linked inheritance is shown in Figure 3.15. The X-linked pedigree has an oblique character through involvement of uncles and nephews related to the female consultand. The usual mating pattern involves a heterozygous female carrier and a normal male. Each son has a 50% risk of being affected by inheriting the mutant maternal allele. Similarly, each daughter has a 50% chance of inheriting the mutant gene from her mother but will remain unaffected since she has her father's normal X chromosome. Male-to-male transmission is not seen but may appear to occur if the trait is sufficiently common that by chance the

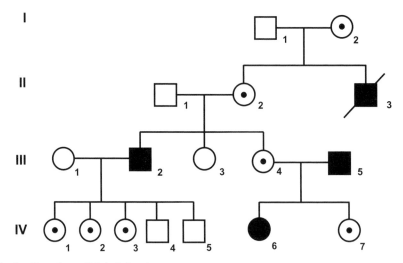

Fig. 3.15   Idealised pedigree for an X-linked disorder.
Females are represented by ○ and males by □. Female offspring of affected males are obligatory carriers. IV-6 is an affected female since both her parents carry the X-linked disorder. Affected males are shown by ■; carrier females by a dot within the circle; individual II-3 is deceased. Because the disease is X-linked, a male cannot transmit the disorder to his sons (to whom he contributes a Y chromosome). Construction of pedigrees are key steps in the understanding of genetic diseases and the way they are transmitted (see also Figure 4.5, which reinforces the value of a pedigree in the situation of imprinting).

mother also carries the mutant gene. An example of this would be glucose-6-phosphate dehydrogenase deficiency in those of black African origin. Approximately 10–20% of blacks in the United States are carriers or hemizygous for this defect.

Females can be symptomatic carriers or develop X chromosome-related disorders in a number of ways. (1) A disproportionate number of normal X chromosomes are inactivated. This can be a chance event or following a translocation between an X chromosome and an autosome. In the latter scenario, X inactivation appears to be non-random since the normal X chromosome is preferentially inactivated. This may represent a selective process as cells with the normal X inactivated are least imbalanced and so will have a survival advantage. (2) Hemizygosity in the female, e.g., Turner's syndrome or 45,X. (3) Inheritance from both parents of a frequently occurring X-related gene, e.g., glucose-6-phosphate dehydrogenase deficiency. (4) The recently defined heritable unstable DNA repeats, which were Discussed earlier under Predictive Medicine—Huntington's Disease.

Just as for autosomal dominant conditions, the frequency of spontaneous mutations in the X-linked disorders needs to be considered, particularly when counselling females who are potential carriers. Haemophilia does not interfere with the reproductive capacity of the affected individual. In contrast, Duchenne muscular dystrophy is usually fatal in the second decade of life. Therefore, spontaneous mutations occurring in the latter disorder would be greater in numbers, and correspond-

ingly, the proportion of females who are carriers will be less.

CARRIER TESTING—HAEMOPHILIA

Coagulation factors involved in haemostasis function as a cascade; i.e., the first activates a second, which then activates a third and so on. In mammals, five proteases (factor VII, factor IX, factor X, protein C and prothrombin) interact with five cofactors (tissue factor, factor V, factor VII, protein S and thrombomodulin) to generate fibrin. Deficiencies in these proteins lead to bleeding. Abnormalities in two of the above factors (factor IX and factor VIII) are well recognised because haemophilia results (Table 3.14). Factor VIII and factor IX circulate as inactive precursors that become activated by a haemostatic challenge. Factor IX's serine protease activity has an absolute requirement for factor VIII. Activation of these two products in the presence of calcium and phospholipid forms the tenase complex that activates factor X and sets off the final steps of coagulation leading to the deposition of fibrin. Because of the interacting effects of factors VIII and IX, it is not surprising that the clinical features of haemophilia A (factor VIII deficiency) and haemophilia B (factor IX deficiency) are identical. The factor VIII and factor IX genes are found on the X chromosome; hence, only males get haemophilia, while females are carriers, unless they have inherited a haemophilia mutation from both their father and mother. Rare examples of symptomatic female haemophilia car-

Table 3.14 Clinical, laboratory and molecular features of the haemophilias

Property	Haemophilia A (factor VIII deficiency)	Haemophilia B (factor IX deficiency)
Frequency	1 in 10 000 males, all ethnic groups	1 in 60 000 males, all ethnic groups
Defect	Clotting cofactors VIII and IX, produced in the liver	
Clinical	Prolonged bleeding spontaneously or after minor trauma into joints, muscles, subcutaneous tissues and organs. About half have a severe disorder (factor VIII level <1%); others are moderately severe (factor VIII 2–5%) or mild (factor VIII 5–30%).	
Genetics	X linked; 10–30% spontaneous mutations	
Gene	Large = 26 exons over 186 kb of genomic DNA	Smaller, 8 exons over 34 kb
Chromosome location	Distal to Xq28	Xq27

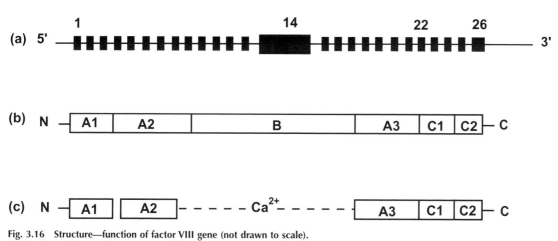

Fig. 3.16 Structure—function of factor VIII gene (not drawn to scale).
(a) The factor VIII gene has 26 exons (the largest are exons 14 and exon 26). **(b)** The inactive factor VIII precursor protein has a number of functional domains. The three A domains are homologous to factor V and caeruloplasmin (the copper-binding protein). The two C domains are important for binding to von Willebrand factor as well as thrombin and activated factor X. The B domain is predominately needed for intracellular processing, and it plays no direct role in haemostasis. **(c)** Active factor VIII is a heterotrimer of the domains shown (dimeric N-terminal heavy chain and monomeric C-terminal light chain) held together by calcium.

riers are also described. The underlying mechanism is considered to be non-random X inactivation, although this may be an oversimplification.

Knowledge of the factor VIII's molecular structure was greatly expanded in 1984 when the gene was cloned. From the gene structure, it was possible to predict potential protein domains with their related functions (Figure 3.16). The factor VIII protein has 2332 amino acids and circulates as an inactive co-factor. Factor VIII is activated by proteolysis induced by thrombin, with the net effect being removal of the B domain. The active factor VIII comprises both A1 and A2 complexes joined by calcium to A3, C1 and C2 domains. To maintain its stability, active factor VIII in plasma is bound to another protein (von Willebrand factor). The factor IX gene is much smaller

and simpler than factor VIII although the protein produced has characteristic functional domains.

DNA Mutations

The haemophilias provide a useful model to illustrate the range of DNA mutations seen in genetic disorders. They include single base changes, deletions, insertions and rearrangements (Table 3.15). Probably the most interesting of the mutations is the factor VIII rearrangement, which has also been called the "flip tip." It comes about because, within intron 22 of the factor VIII gene, there are two mini genes (F8A and F8B). F8A comprises one small exon (<2 kb) and is transcribed in the opposite direction to factor VIII (i.e., towards exon 22). F8B com-

Table 3.15 Range of DNA mutations in haemophilia
(See http://archive.uwcm.ac.uk/uwcm/mg/hgmd0.html for links to factor VIII and factor IX databases describing the various mutations.)

Class of mutation	Examples	Phenotype
Point mutations	(i) Missense—produces a change in amino acid that can be similar or dissimilar and may or may not involve an important functional region. (ii) Nonsense—a premature stop codon will lead to a truncated protein. (iii) Splicing—single base change that either creates a new splice site or destroys an existing one.	(i) Phenotype variable from mild to severe depending on the site and the changed amino acid. (ii) Invariably, a severe phenotype because the protein has been shortened. (iii) Variable from mild (because the splicing defect has a variable effect on mRNA processing) to severe (now the splicing defect leads to complete loss of mRNA processing).
Deletions	These can be heterogeneous involving a part of the entire gene. Considerably more rare than point mutations.	Severe phenotype usually because a portion of the gene (protein) is missing.
Insertions	Variable size like deletions. Relatively rare.	Severe phenotype because likely to lead to frameshifts and so a significant effect on the protein structure.
Rearrangements	See text for full description. Found in about 30–40% of severe haemophilia A.	Severe phenotype because gene function and protein structure disrupted.

prises one exon, which then merges with exons 23–26 of the factor VIII gene, and so it transcribes in the same direction as factor VIII. The functions of these mini genes are unknown. Two additional F8A sequences are located a considerable distance from the factor VIII gene towards the end of the long arm of the X chromosome. Because the three F8A regions are the same, they provide a hot spot for homologous recombination. In fact, the recombination occurs predominantly in males because the single X chromosome predisposes to an intrachromosomal recombination event (Figure 3.17). Therefore, in families with the haemophilia A flip tip mutation (it is found in about 30–40% with severe factor VIII deficiency), the mutation will have arisen in an unaffected male relative, or in the case of haemophilia with no family history, the mutation will have arisen in the maternal grandfather.

As indicated above, there are well-described functional domains within the factor VIII and IX genes. Therefore, mutations will have variable effects and so differing severities depending on the domains involved. They include impaired secretion of the co-factor, interference with binding of factor VIII to factor IX or von Willibrand factor, and a range of missense changes interfering with cleavage to produce the active co-factor. Although factor VIII deficiency is transmitted as an X-linked trait, it can rarely appear to be dominant in transmission if the primary defect involves von Willebrand factor (the gene is located on chromosome 12) and the defect interferes with the binding of von Willebrand factor to factor VIII. Another interesting manifestation of haemophilia involves a mutation in the 5' region of the factor IX gene (called haemophilia B Leyden). In this example, the haemophilia B is a severe disorder during childhood but then improves after puberty. This unexpected behaviour was explained through an understanding of the molecular pathology showing that the mutation interfered with protein binding at the 5' promoter site. This produced a down regulation of gene function. However, overlapping this binding site is also an androgen-responsive element that does not come into play until after puberty when the level of androgens increases. At this time, the effect of the androgens on the promoter override the effect of the mutation, and the haemophilia B spontaneously improves.

Carrier Detection

Carrier detection is usually undertaken to determine whether a female is a carrier and so is at risk of having an affected male offspring. Just as was described for haemochromatosis, two approaches are possible in screening: (1) Phenotypic-based assay, i.e., measuring the level of a protein associated with causing the disease; (2) Genotypic-based assay, i.e., looking for mutations in DNA and from this determining the carrier status.

Phenotypic Assays

Protein levels for factors VIII and IX demonstrate a wide normal range in blood. Because of random X-inactivation, the levels of factors VIII and IX can vary considerably in females who are carriers of haemophilia. This scatter makes an accurate assessment of carrier status difficult if the subject being tested demonstrates a normal or borderline result for the coagulant protein (Figure 3.18). The level may reduce the individual's *a priori* risk but does not provide definitive proof of her carrier status. In addition to X-inactivation, physiological fluctuations are seen with the coagulation factors, e.g.,

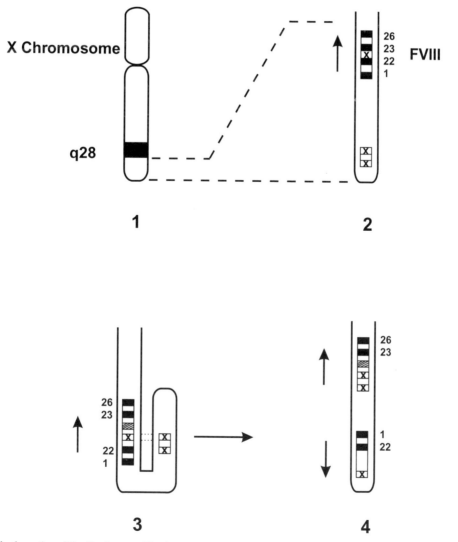

Fig. 3.17 The formation of the flip tip recombination mutation in factor VIII.
(1) The region of the X chromosome below band q28, which contains the factor VIII gene, is magnified. **(2)** Only relevant exons (1, 22, 23, 26) in the gene are shown as ■. The x indicates the location within intron 22 of an inverted DNA repeat known as F8A. DNA homologous to F8A and located more telomeric is also displayed (x x). The vertical ↑ indicates the direction that the factor VIII gene is transcribed. **(3)** This shows an intrachromosomal crossing-over event between the two F8A homologous regions (. . .). The additional (banded) section in intron 22 above F8A is a second gene that has been found in intron 22. **(4)** The final result from the cross-over is a factor VIII gene that has been flipped around (inverted) and is now in two sections—exons 1 to 22 and one F8A segment transcribe in a telomeric direction; two F8A segments and exons 23–26 transcribe towards the centromere. This gross structural rearrangement has a major effect on factor VIII production, resulting in a severe form of haemophilia A. The cross-over depicted has occurred between the intron 22 F8A and the more distal of the F8A homologous regions. Proximal cross-overs can also occur. The flip tip mutation is detectable by PCR.

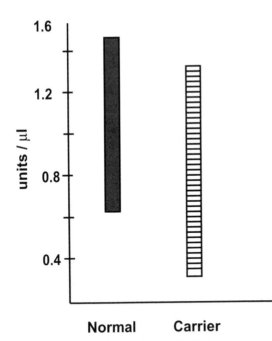

Fig. 3.18   Random X inactivation and its effect on gene dosage reflected by protein levels.
Levels of the factor VIII protein have a broad normal range in blood. In haemophilia carriers they can also vary considerably because of random inactivation of an X chromosome. This makes accurate assessment of carrier status difficult. As shown here for the factor VIII protein in blood, considerable overlap exists between normal females and obligatory female carriers. Better discrimination can be obtained by measuring ratios (e.g., factor VIII coagulant/factor VIII antigen) although there is also overlap with ratios. Thus, a woman who wants to know her carrier status will get an unequivocal result only if her factor VIII level is very low. Any other result is meaningless.

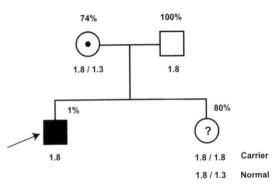

Fig. 3.19   Simple DNA linkage analysis for an X-linked disorder.
The child with severe haemophilia B ($\rightarrow$) has a factor IX level of 1%. The mother is an obligatory carrier because she has an uncle who is affected (not shown in the pedigree). Factor IX coagulant levels are given as percentages in the pedigree (normal is >50%). Factor IX levels for the mother and her daughter are within the normal range (74% and 80%, respectively), but as indicated previously, this does not exclude the carrier state because of random X-inactivation in females. Therefore, a DNA study is undertaken to determine the carrier status of the daughter. From the DNA polymorphism patterns, it is evident that the haemophilia B defect co-segregates with the 1.8 kb DNA polymorphism since this is the marker present in the haemophiliac son. The haemophiliac boy has only one polymorphic DNA marker (1.8) compared to his female relatives; i.e., he is hemizygous since he does not inherit an X chromosome from his father. Therefore, the daughter's carrier status can be determined on the basis of which DNA polymorphism she inherits from her mother; i.e., if the daughter is homozygous for the 1.8 kb marker (she will always inherit one 1.8 kb marker from her father), she is a carrier. If the daughter has both 1.8 and 1.3 kb markers, then the latter must have come from her mother; i.e., the daughter is not a carrier since the 1.3 kb polymorphism is a marker for the normal maternal X chromosome.

pregnancy (or taking the oral contraceptive) at which times the baseline levels for coagulation factors can increase. Finally, there is the not infrequent problem of assessing whether an affected relative is an example of a spontaneous mutation rather than the transmission of a haemophilia defect within a family. This occurs when there is only one affected male in the family (see factor VIII inversion below, which shows how a molecular marker will help in this dilemma).

Genotypic Assays

Testing for DNA mutations has advantages over protein assays: (1) Access to DNA is unlimited, whereas an abnormal protein may not be easy to obtain; and (2) Unlike protein, DNA is not affected by physiological fluctuations. The former is not a problem in haemophilia because a blood sample is adequate. The latter is an important consideration for the reasons mentioned previously. An indirect DNA linkage approach for diagnosis has been used in haemophilia since (1) The majority of defects are point mutations; (2) The genes are large; and (3) There are many mutations (see Table 3.15). A number of DNA polymorphisms have been described, which are located within (intragenic) and in close proximity to (extragenic) the factor VIII and factor IX genes. These polymorphisms allow DNA diagnosis (prenatal or carrier) to be made in ~70–80% of families (Figure 3.19). Intragenic polymorphisms have the advantage that recombination is unlikely to occur since the markers are located within the gene. Despite the many different mutations described for haemophilia A and B, the number of polymorphisms in these genes are relatively small. Hence, options for linkage analysis are limited.

The disadvantages inherent in DNA linkage testing must also be considered. They include (1) Key family

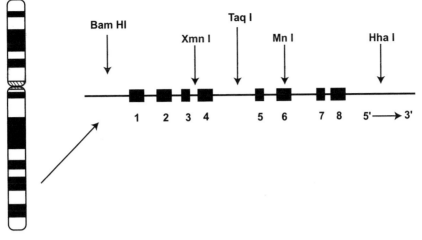

Fig. 3.20 DNA polymorphisms in the factor IX gene.
The eight exons of the factor IX gene are shown. Five polymorphic restriction enzyme sites giving restriction fragment length polymorphisms (RFLPs) are indicated by ↓. Three occur within the gene (intragranic), and two (*Bam*HI and *Hha*I) are extragenic. Some of these polymorphisms are inherited in a preferential association known as linkage disequilibrium (e.g., *Xmn*I and *Mnl*I) and are therefore less useful in DNA testing.

members are required to allow phase of the polymorphism to be determined, i.e., identify which polymorphic marker in that family is co-inherited with the disease phenotype. Key family members may be deceased or unavailable. (2) It is difficult to determine whether mutations are spontaneous events if there is no family history of haemophilia. (3) Germline mosaicism, in which an individual has two or more cell lines of different chromosomal content derived from the same fertilised ovum, cannot be excluded. This is discussed further in Chapter 4. (4) An additional problem with the DNA linkage approach, particularly in haemophilia B, is the effect that linkage disequilibrium (preferential association of linked markers) can have on the informativeness of polymorphisms. For example, five biallelic DNA polymorphisms are associated with the factor IX gene (Figure 3.20). Some of these polymorphisms are inherited in a preferential association; i.e., the *Xmn*I and *Mnl*I polymorphisms are in linkage disequilibrium, which means that results obtained with either are similar since one allele of the polymorphism is nearly always inherited with the same allele of the other. Therefore, not all five polymorphisms will necessarily be informative. This is a particular problem with the factor IX gene locus in Chinese and Asian Indian populations. (5) Non-paternity and its effect on DNA polymorphisms is not an issue if male offspring are studied because the father does not contribute his X chromosome to males. However, the source of the paternal X chromosome is important if a female is being assessed for carrier status.

Direct Mutation Detection

Direct detection of mutations would overcome many of the problems described with linkage analysis. In 1993, it was reported that a factor VIII inversion (flip tip) was frequently found in patients with severe haemophilia A. The inversion produced by an intrachromosomal recombination event flipped a part of the gene. This mutation is now detectable by a variation of PCR known as long-PCR because a large fragment of DNA is amplified.

Today, direct sequencing is increasingly being used to detect mutations in haemophilia genes. This approach would be particularly applicable for the factor IX gene, which is small (8 exons). Because there is only one X chromosome, DNA sequencing is easier since one is not looking for a heterozygous change. Two strategies are used: (1) Either the genomic DNA is sequenced in segments including 5′ UTR, exons, exon-intron boundaries and 3′ UTR regions; or (2) mRNA is prepared from peripheral blood lymphocytes, and the mRNA is converted to cDNA. If cDNA is used, the size of the gene is smaller than the genomic equivalent because now only the exons are represented. The cDNA is sequenced in a number of segments. As indicated above, mutations in the factor VIII and factor IX genes are quite heterogeneous. Therefore, interpreting the significance of a mutation that has not been described previously can be difficult unless it has obvious pathogenicity, for example, a nonsense mutation (see Table 3.15).

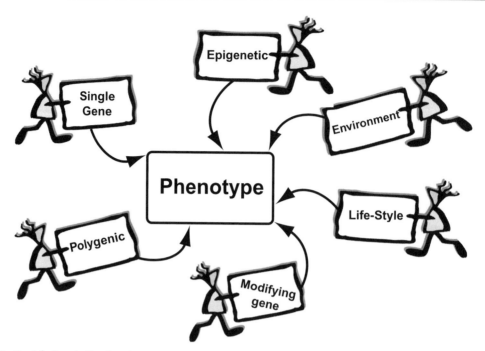

Fig. 3.21 Contributions to the phenotype.
Even when dealing with classic Mendelian traits such as cystic fibrosis, the phenotype can be subtly influenced by a number of factors, both genetic and non-genetic. Mutations in single genes—for example, the cystic fibrosis *CFTR* gene—can produce a well-recognised genetic disease. Mutations in a number of genes (polygenic)—for example, the α and β globin genes—produce a variety of thalassaemia syndromes, and interactions between these genes can influence the disease severity. A third genetic effect can occur from modifying genes. These comprise multiple gene contributions, but each effect is small (i.e., on their own, modifying genes do not cause disease). Epigenetic effects influence gene function without changing the DNA sequence and will be discussed in more detail in Chapter 4. Non-genetic effects include the environment, e.g., exposure to irradiation from the sun, food additives and so on. Life-style effects include alcohol intake, smoking and so on.

FUTURE

This chapter is titled Mendelian Genetic Traits because it is possible to draw a pedigree and follow a disease through the family; i.e., classic Mendelian inheritance can be demonstrated. However, the advances in molecular medicine are showing that even these straightforward disorders are becoming more complex. In particular, there is considerable interest in finding other genes that might influence the phenotype. These genetic factors are called **modifying genes**. Figure 3.21 highlights the various components that can influence a particular phenotype. The relative importance of each component will vary. For example, in single gene disorders (cystic fibrosis) or polygenic ones (α and β thalassaemia), the causative genes have the predominant effect. However, even in these examples, the environment can influence the final phenotype. Can other genes have an effect, albeit a subtle one? The β thalassaemias confirm that this is so because a loss of β globin is easily replaced with γ

globin chains (both are β-like chains). Therefore, even though a patient might be homozygous for a β^0 thalassaemia (and so producing no β globin; i.e., a severe phenotype is likely), the actual phenotype might be milder than expected if the missing β globin is replaced by γ chains. This type of modifying gene effect is easy to understand as the γ chains are simply helping to correct the globin chain imbalance.

A different type of modifying gene is being observed in the autosomal dominant disorder familial hypertrophic cardiomyopathy. Although familial hypertrophic cardiomyopathy is a disease of the muscle sarcomere, it is evident that even within families when individuals share the same mutation, the phenotypes can be quite variable, including the risk for sudden cardiac-associated death. One study has shown in identical twins with familial hypertrophic cardiomyopathy that a sedentary life in one of the twins was associated with milder disease compared to the other who was more active physically. Nevertheless, there is growing evidence that, apart from the environment, there are modifying genes that influence the severity of familial hypertrophic cardiomyopathy. These genes are not able to cause the disease, but their effects can impact on the degree of hypertrophy, or perhaps the complication of sudden death.

Using crude association studies (more about these in Chapter 4 and the Appendix) and selecting likely candidates genes, it has been shown that a number of genetic factors apart from the primary mutation in the sarcomere gene are likely to contribute to the final phenotype. Strong candidates would be genes of the renin-angiotensin-aldosterone pathway because this pathway plays a key role in cardiovascular haemostasis. Not surprisingly, a number of associations have come up positive, although there is still some dispute about the overall validity of these results. The most controversial but still interesting association occurs with the ACE (angiotensin I converting enzyme) gene. ACE has two distinct polymorphic forms due to a 287 bp Alu repeat in intron 16. Therefore, some individuals have this repeat in both their gene copies (i.e., genotype I,I where I = insertion), others have this repeat missing in both gene copies (i.e., genotype D,D where D = deletion), and the remainder are a mix of I and D (i.e., genotype I,D). It has also been shown that the plasma level of ACE in a DD subject is higher than a II subject with the ID individual somewhere in between. When the distribution of the I and D polymorphisms in ACE is compared between mildly affected familial hypertrophic cardiomyopathy patients and those with severe left ventricular hypertrophy, it is seen that there are more individuals with D in the latter group. Hence, the D allele is considered to be associated with a poorer outlook in terms of hypertrophy. Other modifying genes have been implicated in influencing severity in familial hypertrophic cardiomyopathy, including angiotensin II receptor 1, endothelin 1 and tumour necrosis factor α. At this stage, more work is needed to confirm these findings.

Modifying genes will be an interesting development to follow in molecular medicine. When fully understood and characterised, they will allow a more complete understanding of pathogenesis, and from this, a more accurate prognosis. These genes will also be new targets for therapies. Here, the aim will not be to cure the disease, but to modulate the modifying genes to improve clinical well-being. Although they play only a small part in the phenotype, there are likely to be a number of theses genes, and so their cumulative effects will be important. They are difficult to identify and characterise at present because they represent a more complex mode of genetic inheritance.

In the context of a family that is thought to have HNPCC, **microsatellite instability** testing can be useful for deciding whether mutation testing of the DNA repair genes is indicated. There is now some evidence that microsatellite instability might also provide a guide to treatment options in colorectal cancer. For example, the drug fluorouracil is used as adjuvant treatment following surgery. Improved survival seems to be predictable by the microsatellite instability profile; i.e., tumours that do not have microsatellite instability are reported to respond better to fluorouracil than tumours with high-frequency microsatellite instability. If confirmed, this is an important finding because it will avoid the use of a potentially toxic agent such as fluorouracil in patients who are unlikely to respond to this drug.

FURTHER READING

Autosomal Dominant Disorders

ACMG/ASHG Statement: Laboratory Guidelines for Huntington Disease Genetic Testing. American Journal of Human Genetics 1998; 62:1243–1247 (*Huntington's disease triplet repeat interpretation*).

Almqvist EW, Elterman DS, MacLeod PM, Hayden MR. High incidence rate and absent family histories in one quarter of patients newly diagnosed with Huntington disease in British Columbia. Clinical Genetics 2001; 60:198–205 (*illustrates various clinical scenarios that become more meaningful with knowledge from DNA testing*).

Fearnhead NS, Britton MP, Bodmer WF. The ABC of *APC*. Human Molecular Genetics 2001; 10:721–733 (*provides comprehensive overview of the molecular pathology of familial adenomatous polyposis*).

Jarman PR, Wood NW. Genetics of movement disorders and ataxia. Journal of Neurology, Neurosurgery and Psychiatry 2002; 73 (Suppl II):ii22–ii26 (*brief review of triplet expansions in a group of neurologic disorders characterised by gait disturbances*).

Lynch HT, de la Chapelle A. Hereditary colorectal cancer. New England Journal of Medicine 2003; 348:919–932 (*useful overview of clinical and molecular changes in genetic forms of colorectal cancer*).

Marian AJ, Roberts R. The molecular genetic basis for hypertrophic cardiomyopathy. Journal of Molecular and Cellular Cardiology 2001; 33:655–670 (*overview of molecular genetics*).

Nabel EG. Cardiovascular disease. New England Journal of Medicine 2003; 349:60–72 (*one of the articles in the NEJM series on Genomic Medicine. Provides a summary of familial hypertrophic cardiomyopathy and a number of other cardiovascular conditions*). See also Marian and Roberts above.

Ranum LPW, Day JW. Dominantly inherited, non-coding microsatellite expansion disorders. Current Opinion in Genetics and Development 2002; 12:266–271 (*provides an insight into three different mechanisms causing repeat expansion-related disease*).

Sieber OM et al. Multiple colorectal adenomas, classic adenomatous polyposis and germline mutations in MYH. New England Journal of Medicine 2003; 348:791–799 (*description of the base excision DNA repair defect in colon cancer*).

Yang VW. *APC* as a checkpoint gene: the beginning or the end? Gastroenterology 2002; 123:935–939 (*editorial providing a nice overview of the APC gene structure and function*).

http://www.angis.org.au/Databases/Heart (*web-based DNA mutation database for familial hypertrophic cardiomyopathy, illustrating the genes involved and the range of mutations possible*).

http://www.ncbi.nlm.nih.gov/entrez/query.fcgi?db=OMIM (*OMIM—Online Mendelian Inheritance in Man—an online catalogue that describes in great detail all examples of the triplet repeat diseases*).

Autosomal Recessive Disorders

Beutler E, Felitti VJ, Koziol JA, Ho NJ, Gelbart T. Penetrance of 845G→A (C282Y) hereditary haemochromatosis mutation in the USA. Lancet 2002; 359:211–218 (*provides interesting data about the clinical significance of the C282Y mutation*).

Brennan AL, Geddes DM. Cystic fibrosis. Current Opinion in Infectious Diseases 2002; 15:175–182 (*a good clinical and molecular overview of cystic fibrosis, with emphasis on infectious aspects*).

Doull IJM. Recent advances in cystic fibrosis. Archives of Diseases in Childhood 2001; 85:62–66 (*provides a broad overview of clinical and molecular aspects of cystic fibrosis*).

McKone EF, Emerson SS, Edwards KL, Aitken ML. Effect of genotype on phenotype and mortality in cystic fibrosis: a retrospective cohort study. Lancet 2003; 361:1671–1676 (*useful facts about mutations and clinical issues in cystic fibrosis*).

National Institutes of Health Consensus Development Conference Statement on Genetic Testing for Cystic Fibrosis. Genetic Testing for Cystic Fibrosis. Archives of Internal Medicine 1999; 159:1529–1539 (*provides a balanced perspective of population screening for cystic fibrosis*).

Pietrangelo A. Hereditary hemochromatosis—a new look at an old disease. New England Journal of Medicine 2004; 350:2383–2397 (*concise overview of the genetic and clinical aspects of hereditary haemochromatosis including non-HFE genes*).

Weatherall DJ. Phenotype-genotype relationships in monogenic disease: lessons from the thalassaemias. Nature Reviews Genetics 2001; 2:245–255 (*provides a useful assessment of how knowledge of genes often opens up more questions about how the genes interact with the environment*).

http://www.ncbi.nlm.nih.gov/entrez/query.fcgi?db=OMIM (*OMIM—Online Mendelian Inheritance in Man—OMIM numbers 219700 and 602421 provide extensive information on cystic fibrosis and the CFTR gene*).

http://www.genet.sickkids.on.ca/cftr/ (*address for the Cystic Fibrosis Mutation Database, which provides interesting data on mutations and other topics*).

X-linked Disorders

Bowen DJ. Haemophilia A and haemophilia B: molecular insights. Molecular Pathology 2002; 55:1–18 (*overview of the molecular basis for haemophilia including aspects of DNA testing*).

http://archive.uwcm.ac.uk/uwcm/mg/hgmd0.html (*address for Human Gene Mutation Database in Cardiff—this provides links to a large number of locus-specific databases including factor VIII and factor IX*).

http://europium.csc.mrc.ac.uk (*mutation database for factors VIII and VII*).

Future

Marian AJ. Modifier genes for hypertrophic cardiomyopathy. Current Opinion in Cardiology 2002; 17:242–252 (*comprehensive discussion of modifier genes and their effects*).

Peltomaki P. Role of DNA mismatch repair defects in the pathogenesis of human cancer. Journal of Clinical Oncology 2003; 21:1174–1179 (*discusses hereditary non polyposis colon cancer and sporadic colorectal cancer including the relationship between DNA repair and chemotherapy response*).

http://www.acmg.net Site for the American College of Medical Genetics (*provides information about genetic diseases. In particular, "Standards and Guidelines for Clinical Genetics Laboratories" provides examples of some of the diseases discussed in this chapter and discusses relevant laboratory aspects*).

COMPLEX GENETIC TRAITS

INTRODUCTION

In contrast to Chapter 3, which illustrates a number of genetic disorders with an obvious Mendelian inheritance pattern on family study, the disease models in Chapter 4 are considerably more complex to understand at the molecular level. This is due to (1) There is no apparent traditional (Mendelian) mode of genetic inheritance. (2) The influence or effect of the genetic component is difficult to quantitate. (3) It is likely that many genes are involved in pathogenesis, with each gene contributing a relatively small effect. (4) The phenotypes are difficult to define since multiple genes as well as the environment are involved, and each can influence the phenotype. (5) Few of the underlying genes have been discovered for the common sporadic forms of the complex disorders. (6) Novel (including epigenetic) changes in DNA may explain the genetic effect. Figure 4.1 summarises some of the basic differences between the genetic disorders in Chapters 3 and 4.

GENE DISCOVERY—COMPLEX TRAITS

The strategy of positional cloning (and the recent addition to this of *in silico* positional cloning) has been a powerful molecular tool to discover new genes in the straightforward Mendelian diseases. In the past decade, this has enabled more than 1000 genes that caused human genetic diseases to be detected. However, this strategy has not been successful with the complex genetic disorders because one key determinant for positional cloning is an accurate phenotype that is then matched to a genetic marker (usually a DNA polymorphism such as a microsatellite). Positional cloning starts with linkage analysis although this step can be shortened by selecting an appropriate candidate gene and finding a disease-causing mutation in it (see the Appendix and Figure A.4). In the complex genetic traits, neither the phenotype nor the mode of genetic inheritance is easy to follow or deter-

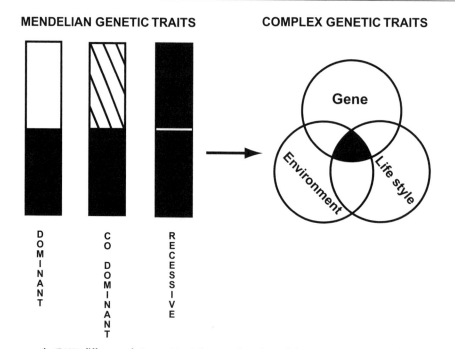

MENDELIAN GENETIC TRAITS

D
O
M
I
N
A
N
T

C
O
D
O
M
I
N
A
N
T

R
E
C
E
S
S
I
V
E

COMPLEX GENETIC TRAITS

Gene

Environment

Life style

Fig. 4.1 Some genetic (DNA) differences between Mendelian genetic traits and the more common complex genetic traits.
Mendelian genetic traits: In Chapter 3, all genetic diseases described have a well-defined Mendelian mode of inheritance, and the gene effect is substantial; i.e., an autosomal dominant disorder leads to development of disease when one of the two inherited genes is mutated (□ normal gene; ■ mutant gene). For an autosomal recessive disorder, both copies of the gene are mutated. Carriers with only one mutated gene are not usually detectable by phenotypic-based assays. In between is the recessive condition (for example, thalassaemia) in which carriers with only one mutated gene can still be detected by phenotypic assays (hatched box), for example, a blood count; hence, this group of recessive disorders is also called co-dominant. **Complex genetic traits:** There is a genetic (DNA) component to disease, but its contribution is more difficult to measure because additional effects include the environment and lifestyles. These are also called multifactorial disorders. Although the three basic components to complex traits are known, their relative contributions remain to be determined and will vary with each disease. In some cases, the genetic aspect is trivial, whereas in others it may be more substantive. Compared to the mendelian traits, in complex genetic traits, any individual gene has a less marked effect on the phenotype.

mine. Therefore, new approaches have been attempted for gene discovery. The current strategy is called an **association** study, which can be a case control study or a family-based association study. This strategy involves selecting a cohort of affected patients and comparing their genetic DNA markers with a comparable control population. Genetic differences detected are then tested to confirm that they relate to the underlying phenotype.

Components of an Association (Case Control) Study

(1) Large numbers of subjects are required to reduce the chances of false associations, which are considerably greater than the traditional linkage approaches. These large numbers (in the thousands in conditions such as diabetes) are also needed because the phenotype is more difficult to define since it can be affected by the environment.

(2) DNA markers (polymorphisms) are needed to compare their distributions between a patient group and a control group. Because of the Human Genome Project, a new type of DNA polymorphism is now available. The SNP (single nucleotide polymorphism) is defined as a single base polymorphism with the rare allele having a frequency of at least 1% in the population. SNPs are estimated to occur approximately 1 in every 600–1000 bp; i.e., there are well over 3 million SNPs in the human genome, and some suggest that the actual number is closer to 10 million SNPs.

(3) Shortcuts are possible if candidate genes (i.e., genes likely to be involved) can be identified. This means that the number of SNPs can be reduced because only those closely located to the candidate gene are required.

(4) The complexity of an association study requires sophisticated bioinformatics and statistical ap-

proaches to compare genetic data between the two cohorts being tested. Many such analytic programs are available, but to date none has shown itself to be superior to others. Therefore, association studies continue, but results are often equivocal and will continue to be an issue until better analytic software can be written.

There are many problems with association studies, including (1) The necessity to test very large numbers of individuals because the phenotypic effect of the genetic determinant is small. (2) The finding of an appropriate matched control population to avoid stratification errors, i.e., differences between cases and controls due to ancestry. (3) The use of sufficient numbers of DNA polymorphisms at a suitable density across the genome. (4) The requirement for many thousands of subjects and many thousands of SNPs produces a time-consuming and

costly study (a conservative cost for detecting an SNP might be close to 50c, i.e., it would cost $0.5 M if 1000 subjects are typed with 1000 SNPs). (5) Better bioinformatics and statistical tools to analyse the data are still needed. (6) Finally, proof that the genetic change has functional significance is even more difficult than for the traditional Mendelian traits because the genetic effect is small, and it interacts with the environment.

Of the limitations described, only (3) is being addressed through the increasing availability of SNPs. Costs remain limiting, but this may also be resolved because it is now evident that chunks of the human genome are inherited as "blocks." Therefore, within a block, it would not be necessary to assay all available SNPs since they will be inherited in linkage disequilibrium; i.e., they are inherited together as a unit. An important spin-off of the Human Genome Project is a new initiative to develop a "haplotype map—HAPMAP"

Table 4.1 Gene discovery strategies for simple and complex genetic disorders

Strategy	Simple, mendelian genetic disorders	Complex genetic traits and disorders
Finding a genetic locus to start the gene discovery pathway	(1) Chance event, e.g., chromosomal abnormality present and (2) linkage analysis using large family and DNA polymorphisms such as microsatellites, which are dispersed throughout the region in question.	Linkage analysis is possible but more difficult because now looking for a QTL[a] like effect and so: (1) Inheritance pattern in a family study is difficult to follow, and (2) Phenotype is subtle and may also be influenced by environment. Alternatively, association studies are used to get some idea where to look in the genome. The rationale is that the distribution of DNA polymorphisms is compared between affected and controls. Any differences detected might indirectly implicate a gene which is located in or around that polymorphism.
How to identify what is a likely (candidate) gene	Can be used without knowing a genetic locus, but this is still looking for a needle in a haystack. Best if complemented by a known chromosomal locus, and from this site, a smaller number of candidate genes are able to be identified.	Without a chromosomal location, a direct candidate gene approach may be required but is even more of a challenge than for a single gene effect.
Confirming that the gene, once it is found, is relevant to the disease being studied	Relatively easy because when dealing with mendelian conditions, a mutation is usually found that clearly impairs the gene's function.	This is particularly difficult because it is unlikely that QTL effects are all or nothing (as would be found with mutations). Perturbations in gene function with complex traits are more likely to involve subtle changes in gene expression.
Proving that the gene and changes in the gene have functional significance	Again, relatively easy. A knockout mouse model will often reproduce the phenotype, thereby providing final proof for the gene's identity and function.	Difficult because the subtle effects expressed by the gene may not be apparent without an environmental interaction (which is often unknown). The human phenotype may not be easily detectable in an animal model, making it difficult to confirm that a gene change has contributed to the complex phenotype.
Future directions	Because the genetic sequence for all human genes is now in databases, an increasingly popular approach involves *in silico* positional cloning; i.e., computer programs are used to identify where a DNA sequence is likely to represent a gene, and computer programs are used to simulate the gene's function.	A similar approach is likely in the complex disorders although the computer program will need to be more sophisticated (to detect and predict from more subtle phenotypic changes).

[a]QTLs (quantitative trait loci) represent polygenic traits that can be measured in some quantitative manner. They are usually the result of multiple genes having small cumulative effects, and the gene-environment interactions also contribute to the phenotype.

of the human genome. This will allow a more rational and cost-effective strategy to select the minimum number of SNPs needed for an association study. Table 4.1 compares gene discovery strategies for simple (Mendelian) traits and the more complex (multifactorial) genetic disorders. Chapter 5 provides more information on association and linkage studies, with the emphasis on the bioinformatics components and needs.

MULTIFACTORIAL DISORDERS

DEFINITION

The term **polygenic** can have a number of meanings, for example, genetic effects resulting from the interaction of multiple genes. A trait in the population such as intelligence is frequently used to illustrate polygenic inheritance. However, the environment (i.e., non-genetic effect) plays a substantial role in the development of intelligence, and this is not acknowledged by the term "polygenic." Therefore, these types of traits are more appropriately called **multifactorial**. Polygenic is best reserved for genetic diseases that result from mutations in a number of genes. These mutations produce a similar phenotype, and they can also interact to modify the phenotype. The best example would be the thalassaemia syndromes (see Chapter 3). In these disorders, the predominant abnormal phenotypes result from mutations in the α or β globin genes, and they produce similar effects, i.e., small, pale red blood cells with a reduced life span. Moreover, co-inheritance of mutations involving both these genes can further alter the phenotype.

Figure 3.21 illustrates the aetiological complexity associated with multifactorial traits. The phenotype starts off as a genetic predisposition or abnormality based on a mutation or mutations in single or multiple genes. Other genetic contributions can come from the effects of modifying genes, or epigenetic changes in DNA. The environment also comes into play and may trigger the development of disease or influence its progression. The environmental effect is variable from a very strong influence, to a minimal contribution to the phenotype. Finally, lifestyles comprise another variable that works through the environment. Lifestyles are considered separately because they are easier to manipulate than the physical environment. At the molecular level, multifactorial traits have been extensively studied, but there are always limiting factors to success. For example, in the case of hypertension, one important barrier has been the definition of the phenotype; i.e., where is the line drawn between normal and elevated blood pressures?

Another concept that should be understood in the multifactorial genetic disorders is the quantitative trait locus (or generally called QTL). Polygenic traits measurable in some quantitative manner can be mapped to what are called QTLs. Thus, complex characteristics—for example, height (normal trait) or hypertension (abnormal trait)—are contributed to by a number of genes usually found on different chromosomes. In addition, the environment also plays a role in these traits. As the various examples for multifactorial genetic disorders are discussed in this chapter, the reader will see that a genetic component to the disease contributes a certain percent; for example, with schizophrenia, the genetic effect is estimated to be about 70–80%. However, no single gene can be responsible because schizophrenia does not behave like a typical Mendelian trait with a mode of inheritance that is easily discernable.

Common Sporadic Human Disorders

Study of the single gene disorders illustrated in Chapter 3 has provided significant insight into their pathogenesis. Nevertheless, a number of these conditions are relatively rare when compared to the important public health problems of today, which are the multifactorial disorders. There is now considerable interest in the latter because of the potential to identify the underlying genes based on the information provided from the Human Genome Project. Although considerably more complex genetically, the multifactorial disorders will have far wider medical implications. Some examples include

- Diabetes (insulin dependent)
- Psychiatric illness
- Dementia
- Cancer
- Coronary artery disease
- Hypertension
- Mental retardation
- Congenital malformations, e.g., cleft lip or palate

Considerable effort will go into trying to understand the molecular basis for the multifactorial disorders in the coming decade. This is an important goal because (1) The ability to detect those who are genetically predisposed will allow more targeted preventative programs to be developed. (2) Genetic information will identify new therapeutic targets or strategies. (3) As the genetic component to the multifactorial disorder is understood, the environmental contributions become easier to identify and manipulate in terms of prevention.

As discussed earlier, a major barrier to progress at the molecular level is the use of strategies such as linkage analysis that are less effective for gene discovery when diseases have subtle phenotypes or a multifactorial basis

Table 4.2 A comparison of type I and type II diabetes mellitus

Type I diabetes mellitus (also called juvenile diabetes, insulin dependent diabetes mellitus—IDDM)	Type II diabetes mellitus (also called non insulin dependent diabetes—NIDDM)
Pancreas makes little or no insulin.	Body cannot use the insulin that is made.
Treatment involves the taking of insulin.	Treatment involves the taking of drugs to allow the insulin to be used, or more insulin is produced. Weight loss and exercise help.
Disease arises from autoimmune destruction of the pancreatic β islet cells. Usually presents before the age of 30 years, most often in childhood or teens.	Disease of late onset with a significant genetic component. MODY (maturity onset diabetes of the young) is a distinct subtype of NIDDM with autosomal dominant inheritance and onset <25 years of age.
About 10 million affected worldwide.	About 150 million affected worldwide.
Associated with strong genetic predisposition + environment effect.	Risk factors include (1) aged >45 years, (2) overweight, (3) family history diabetes, (4) gestational diabetes during pregnancy.

for inheritance. More sophisticated computer analysis programs have been suggested as the way forward in these complex traits. A brute-strength approach directed against the vast bank of information available as a result of the Human Genome Project may, in the longer term, provide the answers. Alternatively, new strategies will be required. It is expected that information about normal physiological processes (e.g., memory) may emerge as by-products of research into the polygenic and multifactorial traits.

DIABETES

The challenges ahead in the multifactorial disorders are well illustrated by type 1 diabetes (insulin dependent diabetes mellitus or IDDM). IDDM is the third most prevalent chronic disease of childhood, with 0.3% of the population affected by age 20. The lifetime risk for IDDM is about 1% (i.e., 10–20 million people worldwide). The basic defect in IDDM involves an autoimmune destruction of the pancreatic β islet cells so that insulin is not available to regulate glucose metabolism. Concordance rates in IDDM twins are 5–10% for DZ twins and 30–40% for MZ twins, which suggests a significant genetic component to this form of diabetes. A comparison between IDDM and the more common type II diabetes (non insulin dependent diabetes mellitus or NIDDM) is given in Table 4.2.

A lot is now known about the genetic effects involved in IDDM, but less about the environmental risks apart from the association with viral infections such as coxsackie B4. The strongest genetic effect in IDDM involves the HLA locus (IDDM1). Both class I and class II genes are implicated (Figure 4.2). Not much is known about the actual HLA effect except that it is a major determinant contributing around 42% to development of IDDM. Other loci implicated in IDDM include the VNTR polymorphism found in association with the insulin gene (IDDM2). This polymorphism has three different alleles,

Table 4.3 Genetic defects associated with maturity onset diabetes of the young (MODY)

Class of MODY	Underlying gene
MODY1	Hepatic nuclear factor 4 alpha
MODY2	Glucokinase
MODY3	Hepatic transcription factor 1
MODY4	Insulin promoter factor 1

with allele I showing an increased risk for IDDM and allele III a reduced risk. It is proposed that the type III allele, which correlates positively with higher levels of insulin mRNA, has its effect through immune tolerance. However, the IDDM2 effect is relatively small (10%) compared to the IDDM1 effect. Many more genetic loci have been implicated (up to IDDM15) although little is known about actual genes within these loci.

Less is known about the genetics of type II diabetes mellitus except for a subtype of this disorder known as MODY (maturity onset diabetes of the young). MODY shows an autosomal dominant mode of inheritance with high penetrance in multigenerational families. Mutations in several genes are now known to cause MODY (Table 4.3). Not surprisingly, there are many mutations within each gene, and this limits the use of DNA screening to detect those at risk. One practical reason for DNA testing to confirm MODY in a young patient with diabetes is that these individuals, unlike IDDM patients, do not usually require treatment with insulin.

DEMENTIA

The dementias comprise an increasingly important group of disorders, particularly in countries with an ageing population. Aggregates of dysfunctional proteins or peptides in the brain are the pathologic hallmark of the demen-

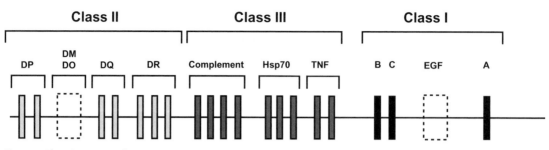

Fig. 4.2 The HLA genes and IDDM.
The HLA genes reside within the MHC (major histocompatibility) locus on chromosome 6p21. The MHC spans a region of 3.6 Mb (1 megabase = 1 million base pairs) and contains about 200 genes arranged in three groups: Class I (HLA-A, B, C, E, G and F), Class II (HLA-DP, DQ, DR, DO and DM) and Class III. To simplify the HLA locus, only key genes are shown as bars. The dotted box represents a cluster of genes. The highest risk for IDDM is associated with the DR3.DQ2 and DR4.DQ8 haplotypes. However, about 10% of Caucasians with IDDM do not carry these markers, and other HLA haplotypes carry susceptibility in these individuals. Although the genetic risk factors are being characterised, it is still far from clear how they precipitate or facilitate the autoimmune destruction associated with type I diabetes.

tias. These aggregates are considered to have arisen from abnormal folding of proteins. One source of knowledge about the dementias has come from transgenic mouse models (Box 5.1). Similarly, understanding the rare genetic types of dementias has contributed further information about the more common sporadic forms. For example, mutations in the amyloid precursor protein (*APP*) gene leading to Alzheimer's disease are exceedingly rare. Yet they have identified an abnormality involving β amyloid that also occurs in the sporadic condition. No therapies directed to the specific underlying pathological process are presently available in the dementias although therapies targeting the abnormal β amyloid are being developed. In the longer term, an understanding of the molecular basis for dementia will enable at-risk individuals to be accurately identified and expand the scope for novel therapies.

Alzheimer's Disease

Alzheimer's disease makes up about two thirds of dementia cases. It has a prevalence of around 1% in those aged 65–69 years and increases to 40–50% in those 95 years or older. Early onset dementia (arbitrarily defined as less than 60–65 years of age) makes up about 6–7% of Alzheimer's disease. In less than 5% of cases, Alzheimer's disease is inherited as an autosomal dominant disease. Mutations in three genes are found in genetic Alzheimer's disease. They are β amyloid precursor protein (*APP*) on chromosome 21, presenilin 1 and presenilin 2 genes.

In Alzheimer's disease abnormal accumulations contain two different proteins: (1) plaques—made up of extracellular fibrillar aggregates of β amyloid peptide—particularly the peptide that is 42 amino acids in length ($A\beta_{42}$) and (2) neurofibrillary tangles comprising the protein tau. Mutations in the *APP* gene are considered to

produce Alzheimer's disease by increasing the number of $A\beta_{42}$ peptides generated, and these peptides are highly amyloidogenic. The development of Alzheimer's disease in Down's syndrome is thought to reflect the additional copy of *APP* because of the trisomy 21. Mutations in the presenilin 1 and 2 genes are thought to have their effect through proteolysis of the β amyloid precursor protein, thereby leading to increased production of the $A\beta_{42}$ peptide.

The basis for the more common, late onset, sporadic form of Alzheimer's disease is unknown. However, a genetic association involves a variant of the *APOE* (apolipoprotein E) gene and risk for Alzheimer's disease. The *APOE* gene has three variants—$\varepsilon2$, $\varepsilon3$, $\varepsilon4$. Carrying one *APOE* $\varepsilon4$ allele nearly doubles the lifetime risk for Alzheimer's disease. Not carrying an *APOE* $\varepsilon4$ allele reduces the risk by 40%. How this works is not known, and it should also be noted that many in the population who are $\varepsilon4$ positive do not develop Alzheimer's disease, and many (40–70%) with late onset Alzheimer's disease are not $\varepsilon4$ positive. Routine DNA testing for *APOE* subtypes is not indicated because of this uncertainty, and DNA testing for the three Alzheimer's disease genes described earlier should be restricted to the appropriate circumstances, e.g., early onset cases or cases with a positive family history. The same precautions described for Huntington's disease predictive testing apply to Alzheimer's disease predictive testing (see Chapters 2, 3).

Prion Disease

Another form of dementia is prion disease. Prion disease may be sporadic, inherited, iatrogenic or transmissible from animal to human via infected tissue. This form of dementia, also called transmissible spongiform encephalopathies, includes Creutzfeldt-Jakob disease

(CJD) in humans, bovine spongiform encephalopathy (BSE), and scrapie in sheep and goats. In 1996, the emergence of variant CJD (vCJD) is thought to have arisen from transmission across the species of the BSE agent. The common abnormality in all prion diseases involves a post-translational change from a normal protein—host prion protein PrP^c—to an abnormal "infectious" form—PrP^{Sc} (P^c—cellular; P^{Sc}—scrapie).

The function of the normal prion protein is unknown. It is coded for by a gene called *PRNP*. Mutations in this gene are associated with familial cases, which comprise about 10% of the prion diseases in Europe. However, in the vast majority of sporadic cases, there are no detectable DNA mutations, and the change from PrP^c to the abnormal PrP^{Sc} is thought to occur because of somatic mutations or other, as yet unknown, genetic or environmental factors. Risk factors for developing vCJD include young age, residence in the UK especially between 1985 and 1990, and intriguingly, homozygosity for a methionine 129 polymorphism in the *PRNP* gene. At this position there is either a methionine or a valine. In normal individuals, the different combinations of met/met, met/val and val/val are all present. However, in patients with vCJD, 100% are met/met, suggesting that this may lead to genetic predisposition. If this is correct, it is hypothesised that a second wave of vCJD will occur at some later date from the met/val heterozygotes because they have only one copy of the met polymorphism.

Prion disease remains a challenge for the future, particularly to explain how the infectious forms occur without any apparent conventional infectious agents being involved. It is postulated that the abnormal PrP^{Sc} protein itself functions as the "infectious" agent (see also Chapter 8 and Figure 8.8 for more discussion on CJD).

SCHIZOPHRENIA

Schizophrenia is a common and debilitating psychiatric disorder affecting 1% of the population. Schizophrenia usually begins during adolescence or early adult life, and can be difficult to diagnose because the phenotype is not always easy to define. Diagnostic features of schizophrenia include (1) Positive ones such as auditory hallucinations, delusions (characteristically paranoid), disorganised speech, grossly disorganised or catatonic behaviour; (2) Negative ones such as inability to pay attention; loss of will, drive and the sense of pleasure; flattened affect and social withdrawal. In addition, most patients have some history of behavioural dysfunction such as social and learning difficulties.

Genetics

Despite earlier controversy, there is now strong evidence for a major genetic component in schizophrenia. Risks

---

> **Box 4.1 The usefulness of twins in estimating genetic versus environmental effects in multifactorial traits.**
>
> Monozygotic (MZ) or identical twins develop following division of a single fertilised ovum. Although each twin will start off with the same DNA content, they may not be exactly 100% identical genetically because of post-zygotic changes in the DNA, and epigenetic effects that might influence overall gene expression. In contrast, dizygotic (DZ) twins result from the fertilisation of two ova by different sperms. Thus, on average, DZ twins share half of their nuclear genes, which is comparable to non-twin siblings. Generally, the environment shared by DZ and MZ twins is similar. Thus, twins are a popular model to assess the relative contributions of genes and environment in disease. This type of approach in schizophrenia research has been productive. Concordance (both of the twins are affected or unaffected) has shown that about 45% of MZ twins will both develop schizophrenia. In contrast, the same risk for DZ twins is about 15%. These results suggest that there is a genetic component in pathogenesis of schizophrenia. To counter the criticism that MZ and DZ twins may not necessarily be exposed to a comparable environment (e.g., MZ twins share a single placenta *in utero* and are more likely to be "closer" than their DZ counterparts), a study in schizophrenia has looked at a small number of MZ twins who were raised apart. Concordance for the development of schizophrenia was not reduced, thereby strengthening the genetic association.

---

in first (5–15%), second (2–6%) and third (2%) degree relatives of individuals affected with schizophrenia are higher than the general population. The genetic component in schizophrenia is considered to be about 70–80% based on family, twin and adoption studies, with a concordance rate in monozygotic twins around 45% compared to about 15% in dizygotic twin pairs (Box 4.1).

Molecular genetic analysis in schizophrenia has been constrained for a number of reasons: (1) The difficulty in defining a reliable phenotype with overlap into other psychiatric disorders. (2) An inability to determine a mode of inheritance. Autosomal dominant with reduced penetrance, autosomal recessive, multifactorial, unstable DNA triplet repeats and a number of other combinations have all been proposed. (3) The confounding effects of drug or alcohol abuse and the potential for some neurological disorders to produce schizophrenic-like features so complicating inclusion and exclusion criteria for linkage studies.

In this discussion schizophrenia is considered a multifactorial disorder as the inheritance pattern is complex, and it is likely to involve a number of genes, each contributing a relatively small effect. Gene/environment interactions then determine the complete phenotype. Suspected environmental effects include perinatal or childhood brain injury and psychological stress. Since clinical and biochemical studies had not provided clear explanations for the genetics or pathogenesis of schizophrenia, the next option was positional cloning.

Positional Cloning

An early clue where to start looking in the genome for schizophrenia genes was provided when it was observed in one family that two schizophrenic males were partially trisomic for chromosome 5q11.2-q13.3. Therefore, DNA polymorphic markers for this region were used in linkage analysis. In 1988, LOD scores between 3–6 were obtained in two British and five Icelandic families with schizophrenia. The LOD score is a statistical measure of the strength of a linkage and is described further in Figure A.4. Thus, DNA studies were consistent with the cytogenetic observation and pointed to a schizophrenia gene in association with chromosome 5q11-q13. However, subsequent pedigrees have been investigated, and they have failed to confirm the chromosome 5 findings. One initial explanation was that schizophrenia is caused by a number of mutations that involve other loci. Reassessment of the chromosome 5 linkage studies with more families and additional DNA probes has indicated that the positive LOD score may have been a chance finding, and the association between schizophrenia and chromosome 5 was probably not significant.

Subsequently, many other loci have been implicated in schizophrenia. The most promising have been on chromosomes 1, 6, 8, 12, 13 and 22. Genes at these loci have now been shown to be strong candidates for the development of schizophrenia (Table 4.4). Some success in a complex multifactorial disorder such as schizophrenia has resulted from a strategy that combined the traditional positional cloning by linkage analysis with association studies that focus on candidate genes in identified loci.

Table 4.4   Genes implicated in schizophrenia (from Gerber *et al* 2003; Harrison and Owen 2003)

Gene, metabolic pathway	Locus	Earlier linkage study focused attention to the gene region	Confirmed by others	Mouse model
RGS4 (regulator of G-protein signalling-4). Shown to be down-regulated on microarray studies and is an inhibitor of the glutamate receptors (these are the predominant excitatory neurotransmitter receptors in the mammalian brain).	1q21–22	Yes	Yes	No
DTNBP1 (dysbindin). Acts through the NMDA (N-methyl-D-aspartate) receptors (NMDA represents one particular class of glutamate receptors).	6p22	Yes	Yes	No
NRG1 (neuregulin 1). Present in glutamatergic synaptic vesicles and involved with NMDA receptors. One haplotype with this gene increased two-fold the risk for schizophrenia.	8p12–p21	Yes	Yes	Yes
PPP3CC. Gene that encodes the calcineurin γ catalytic subunit. Involved in both the dopaminergic pathway and the NMDA receptors, hence, a link between the dopamine and glutamine theories for schizophrenia.	8p21.3	Yes	Recently described	Yes
DAAO (D-aminoacid oxidase). See *G72* for mode of action. Although both are associated with increased risk for schizophrenia, a combination of risks for *G72* and *DAAO* was shown to be synergistic.	12q24	No	No	No
G72. This gene and *DAAO* have key effects on the NMDA receptors.	13q34	Yes	Yes	No
COMT (catechol-O-methyltransferase). Although this gene directly affects monaminergic neurotransmission, it is likely to have an indirect effect also on other synaptic populations, including glutamatergic ones.	22q11	Yes	Yes/No[a]	Yes
PRODH (proline dehydrogenase). This gene potentially affects glutamatergic synapses via several mechanisms.	22q11	Yes	No[a]	Yes

[a] Both *COMT* and *PRODH* were suitable candidate genes for study because they resided within the region that is deleted in the genetic disorder velocardiofacial syndrome. In this syndrome there is a high rate of psychosis.

Pathogenesis

An early focus on the dopaminergic system as the basis for schizophrenia (dopamine hypothesis) was based on the observations that (1) Amphetamine abusers could develop schizophrenia-like illnesses due to excess dopamine release and (2) D_2 (dopamine) antagonists were effective in treating schizophrenia. The emphasis on the dopaminergic system may, on retrospection, have had a confounding effect. A refocusing was required when recently discovered genes were shown to be involved with the glutamatergic system. The glutamate hypothesis for schizophrenia now implies that there is reduced activity within the glutamate neurotransmitter system, particularly due to malfunction of the NMDA (N-methyl-D-aspartate) receptors. Glutamate transmission plays a key role in neuronal development, neuroplasticity and neuroexcitotoxicity, and each of these functions is impaired in schizophrenia. Inhibition of the NMDA receptors in normal volunteers induces a schizophrenia-like illness. Although the dopamine hypothesis has not been excluded (and indeed it may still play a role in some forms of schizophrenia), molecular analysis in schizophrenia has identified another, and potentially more important, pathway.

Despite the difficulties in discovering genes for complex genetic disorders, the example of schizophrenia illustrates that it is still appropriate to utilise molecular strategies to study these conditions since success will eventually follow. Although there is still a lot to learn about schizophrenia, the first major breakthroughs have emerged in providing a better understanding of the molecular basis for this important psychiatric disorder. No doubt therapies that are directed to the glutamatergic system will soon be forthcoming to test further the importance of this pathway in schizophrenia.

NOVEL MECHANISMS FOR DNA AND DISEASE

INTRODUCTION

Clinical genetics in the pre-molecular medicine era was more straightforward, with a focus predominantly on descriptive syndromes. Observations about genetic disorders could usually be grouped within the three traditional forms of inheritance: (1) autosomal dominant, (2) autosomal recessive and (3) X-linked. The molecular medicine era has progressed clinical genetics into a more scientifically exact as well as a dynamic speciality, with new modes of genetic inheritance being described. This has been accompanied by an exponential increase in knowledge of genetics and the DNA mechanisms involved in pathogenesis. The traditional book and journal resource is giving way to the computer and the Internet to allow the new discoveries to be understood, as well as accessed rapidly (Figure 4.3).

In addition to the QTLs in the inheritance of the complex genetic disorders, other mechanisms need to be considered. They involve novel DNA changes that can influence the way a gene functions. Some examples include mitochondrial DNA (mtDNA), genomic imprinting and mosaicism. Epigenetics, another mechanism, will be discussed in greater detail under Oncogenesis.

MITOCHONDRIAL INHERITANCE

The nucleus is not the only organelle in eukaryote cells that contains DNA. Mitochondria have their own genetic material in the form of a 16.6 kb double-stranded circular DNA molecule. mtDNA is characterised by a high mutation rate (5–10 times that of nuclear DNA), few non-coding (intron) sequences, a slightly different genetic code and maternal inheritance. The last occurs because spermatozoa make a negligible contribution to the conceptus in terms of mitochondrial DNA (Table 4.5).

Mutations in mtDNA produce their effect through a deficiency in the respiratory chain that comprises five enzyme complexes within the inner mitochondrial membrane. Defects in the respiratory chain affect a number of cellular processes, including (1) production of ATP, (2)

Table 4.5 Differences between nuclear DNA and mitochondrial DNA

Mitochondrial DNA (compared to nuclear DNA)	Comments
High mutation rate	Due to (1) lack of protective histones, (2) few DNA repair systems, and (3) the generation of oxygen-free radicals from the respiratory chain
Genetic structure	mtDNA does not have introns or display asexual segregation (i.e., recombination does not occur between mtDNA)
Unique genetic code	UGA = tryptophan (stop codon in nuclear DNA) AGA and AGG = stop codons (arginine in nuclear DNA) AUA = methionine (isoleucine in nuclear DNA)
Input to mtDNA function occurs from both nuclear and mtDNA genes	This makes inheritance patterns for mtDNA disorders difficult to predict

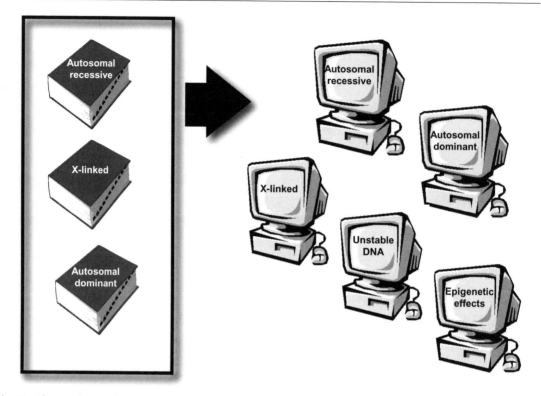

Fig. 4.3 The complex ways in which DNA can influence genetic inheritance and the phenotype have necessitated a move from the traditional paper source of information to the computer.

apoptosis, (3) production of reactive oxygen species and (4) cellular oxidation and reduction. Each of the five respiratory chain complexes is made up of several protein subunits encoded by both nuclear DNA and mtDNA. Therefore, the inheritance patterns for mtDNA disorders can be complex because they can be autosomal recessive, autosomal dominant or X-linked (if the key respiratory chain gene is coded by nuclear DNA) or maternal in inheritance (if the respiratory chain gene is entirely mtDNA in content). As well as respiratory chain specific mutations, genes involved in mtDNA replication and transcription as well as protein import and assembly are encoded by nuclear DNA, but mutations in them can produce a mtDNA phenotype.

Mitochondrial DNA and Genetic Disease

It is only since 1988 that some genetic disorders—particularly those affecting organs with high energy requirements such as the brain and skeletal and heart muscles—have been proven to result from mutations in mitochondrial DNA. Although it was suspected that mitochondria were involved on the basis of maternal inheritance, biochemical abnormalities and abnormal

morphology on microscopy, definitive proof required DNA characterisation.

Features that suggest a mitochondrial DNA origin for an underlying disease are (1) Maternal inheritance; i.e., both males and females can be affected, but the disorder is transmitted only by females. (2) The pathophysiology involves defects in respiratory chain function (i.e., energy production); therefore, likely diseases for which there might be a mtDNA origin include encephalopathies, myopathies and cardiomyopathies. (3) There can be variable expression in affected individuals. This is explained on the basis that each mitochondrion contains 2–10 DNA copies, and in each cell there can be 1000–10 000 mtDNA copies. The mtDNA molecules in each cell are usually identical (called homoplasmy). However, if there are mutated mtDNA species, different proportions of the wild-type to mutant mtDNA can be found in each cell and tissue. This is called **heteroplasmy**—the finding of a mixture of mutant and wild-type mitochondrial DNA species in the same cell. (4) Tissues will be affected differentially on the basis of their energy requirements. Thus, the central nervous system, skeletal and cardiac muscle fibres are at highest risk. In addition, tissues with a high mutant-to-normal mtDNA ratio are more likely to be affected.

Table 4.6 Some examples of mtDNA genetic disorders (Graff *et al* 2002)

mtDNA defect	Disease	Clinical phenotype	DNA mutation(s)
Complex V respiratory chain mutations	Leigh's syndrome	Severe progressive encephalopathy in children (milder in adults)	Similar phenotype whether caused by mtDNA or nuclear DNA mutation—common feature involves energy metabolism. Severity related to the proportion of mutant/normal mtDNA species, i.e., heteroplasmy.
Complex I respiratory chain mutations	Leber's hereditary optic neuropathy	Causes blindness, predominantly in young males (although reduced penetrance; i.e., most carriers never become blind)	Three common missense mutations involved. Generally associated with homoplasmy (i.e., these are mild mutations).
mt-tRNA gene (tRNA^{lys})	Myoclonus epilepsy and ragged red fibres syndrome (MERRF)	Myoclonus epilepsy, mental retardation, ataxia, tremor, muscle atrophy	Inhibits mitochrondrial protein synthesis and causes a respiratory chain deficiency.
mt-tRNA gene (tRNA$^{Leu(UUR)}$)	Myopathy, encephalopathy, lactic acidosis, stroke-like episodes	Seizures, episodic vomiting and repeated cerebral episodes causing hemiparesis, hemianopia or cortical blindness	Heteroplasmic A3243G missense change common defect.
mtDNA rearrangements	Kearns-Sayre syndrome	Ophthalmoplegia, ptosis, retinal degeneration, ataxia, heart block	Deletions/duplications in mtDNA. Usually heteroplasmic and include at least one tRNA gene.

The types of mutations in mtDNA range from deletions to duplications to single base changes. It is interesting that the more severe mutations demonstrate heteroplasmy since they would otherwise be lethal. Because of their effect on reproductive fitness, these mutations are very heterogeneous, suggesting independent origins. On the other hand, the milder point mutations can be found in all cells, i.e., homoplasmy. Examples of some genetic disorders that arise from mtDNA defects (as well as nuclear DNA defects that mimic the mitochondrial phenotype) are given in Table 4.6.

The list of mitochondrial DNA-associated defects has grown considerably as PCR allows the mitochondrial genome to be studied with greater ease. Previously, the heterogeneity of the clinical phenotypes and the difficulty in using conventional biochemical approaches for study of mitochondria meant that there was little understanding of aetiology or pathogenesis. Even DNA technology before the availability of PCR was demanding since large quantities of a tissue rich in mitochondria (e.g., the placenta) were required to enable sufficient DNA to be isolated for use in a technique such as Southern blotting (see the Appendix). However, all this has changed with PCR, and strategies can be developed that allow the 16.6 kb genome to be amplified in segments using blood as a source of mitochondrial DNA. These segments are then screened for mutations by methods such as dHPLC (see the Appendix). Fragments showing differences in nucleotide bases are confirmed to be abnormal by DNA sequencing. Alternatively, amplified fragments can be sequenced directly.

Mitochondrial DNA and Ageing

The mutation theory of ageing is based on DNA (particularly mtDNA with its higher mutation rate than nuclear DNA) undergoing continuous damage from environmental agents such as UV light, radiation and chemicals, as well as endogenous agents such as free radicals and reactive oxygen species that are by-products of the normal metabolic processes in cells. This damage in nuclear DNA is repaired by a number of repair systems (see Table 3.6), but there are fewer of them in mitochondria. Therefore, ageing is considered to result from a net accumulation of mutations with time, and these mutations eventually cause permanent damage. mtDNA is particularly susceptible to this form of damage, and so contributes to the ageing process as it gradually accumulates mutations. High energy–dependent organs such as the brain are particularly at risk. Unlike the genetic effects described earlier, these changes are somatic in origin, and not passed on to the next generation; i.e., they would appear as sporadic disorders.

Although mtDNA is implicated in the ageing process, it has only recently become apparent that mtDNA genes can also prolong life. Using the *C. elegans* worm model, a genomics approach comprising microarrays and RNA interference (RNAi) was utilised to identify longevity genes (see RNA Interference in Chapter 2; Microarrays in Chapter 5) considered to reside within the worm's chromosomes I and II (a total of more than 5000 genes). RNAi was introduced into the *C. elegans* when the worms ate bacteria containing the RNAi. The effects of gene inhibition brought about by the RNAi was monitored through microarrays. A number of worms had their life span extended with the RNAi treatment. From these, it could be shown that mtDNA genes were implicated in the induced longevity. For example, a mutation turning off one of the tRNA synthetase genes had a marked effect on the worm's life span.

Interest has now shifted to the insulin-like signalling pathway and *daf-2*, a key gene in promoting longevity in *C. elegans* (Box 4.2). To investigate further the role of *daf-2*, the same RNAi and microarray approach was used to inactivate about 93% of the *C. elegans* genes. Downregulated genes were detected through microarrays. For

the worms with an extended life span, it was possible to classify the genes involved into three categories: (1) Stress genes; i.e., in prolonging life, this pathway works through enhanced repair mechanisms. (2) Antimicrobial response genes. This makes sense since worms feed on bacteria, and eventually they are killed by the bacteria. (3) A whole range of genes involved in metabolism. Probably the most important observation was that the three distinct pathways contributing to the worm's life span could all be linked back to the *daf-2* gene, which is now shown to be a key regulator of longevity but working through many different downstream effects.

GENOMIC IMPRINTING

Contrary to Mendel's original theory that genes from either parent have equal effect, it is now clear that expression of a few genes is dependent on their parent of origin. The mechanism involved is called genomic (or genetic) imprinting, which refers to the differential effects of maternally and paternally derived chromosomes or segments of chromosomes or genes. Genomic imprinting implies that during a critical time in development, some genetic information can be marked temporarily so that its two alleles undergo differential expression. The critical period is considered to be the time of germline formation. Imprinting can be erased or re-established in the germ cells of the next generation (Figure 4.4). Approximately 50 imprinted genes have now been identified in the human and mouse.

Going hand in hand with imprinting is the concept of epigenetics (introduced in Chapter 2). Epigenetics refers to the regulation of gene expression without altering the nucleotide sequence. Epigenetic regulation is important in a number of human diseases, including multifactorial diseases, complex syndromes and cancer. The allele-specific expression of imprinted genes is accomplished by allele-specific epigenetic modifications (for example, cytosine methylation, methylation or acetylation of histones). An imprinted locus will be inherited along mendelian lines, but this will not be apparent until it is seen that the expression pattern for that locus is dependent on the parent of origin (Figure 4.5). Imprinting plays a fundamental role in normal development during embryonic and postnatal life (see Chapter 7 for further discussion). In addition, imprinting is involved in brain function and behaviour. In cancer, the imprinting pattern in tumours can be disturbed. Since imprinting means that one of two alleles is normally inactive (imprinted), it follows that a mutation in the remaining allele can lead to genetic disease because neither gene is now expressing. However, if the mutation affects the imprinted allele, there will be no clinical consequence because the imprinted allele does not express. In the latter scenario, a mutated imprinted allele causes no immediate problem, but it may do so in subsequent generations if

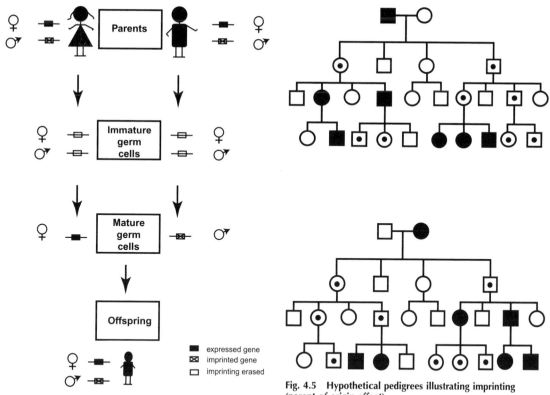

Fig. 4.4 Imprinting of genes in males and females.
Not all genes are equally expressed in males and females. The example given shows a gene that is active only when it is inherited from the female (the gene derived from males is inactive or imprinted). The imprint is erased in the immature germ cells and then re-established in a parent-specific manner as the germ cells mature. The imprint established in the germline is then maintained with each cell division. In somatic cells this imprint is reflected by differential levels of expression between the two alleles. Progeny inherit the appropriate imprinting pattern. Read more about imprinting during germ cell development in Chapter 7.

Fig. 4.5 Hypothetical pedigrees illustrating imprinting (parent of origin effect).
An imprinted locus is inherited in a normal manner, but the expression of the two alleles will depend upon the parent of origin. **Top:** The paternal allele is inactive (imprinted). There will be no expression of the mutant allele when transmitted by the father. For the mutant gene to express its phenotype, it must pass through the maternal line. **Bottom:** In this case, the maternal allele is inactive (imprinted), and the disease phenotype becomes apparent only after paternal transmission of the mutant allele. In both cases, carriers (indicated with a dot in the circle or square) have normal phenotypes but can transmit the trait depending on their sex. There are equal numbers of affected and unaffected males and females in each generation.

the imprint is reset because it has now been transmitted by a parent of the other sex.

Disorders Involving Imprinting

Two rare syndromes with overlapping but different phenotypes have been localised by cytogenetic analysis to the same region on chromosome 15q11-q13. The two are Prader-Willi syndrome and Angelman's syndrome. Features of these disorders are summarised in Table 4.7. The aetiology of both remained unknown until cytogenetic and then molecular analysis identified atypical modes of genetic inheritance. It is now considered that distinct but

adjacent segments of chromosome 15q11-q13 are critical for normal development. Loss of the **paternal** segment of this chromosome region affecting a number of paternally expressed genes (including *SNRPN*, *NDN*, *MAGEL2* as well as a cluster of paternally expressed small nucleolar RNAs) leads to the Prader-Willi syndrome. In contrast, loss of the **maternal** segment (containing the *UBE3A* gene, which is expressed from the maternal allele in certain parts of the brain) produces Angelman's syndrome; i.e., the two syndromes exhibit oppositely imprinted chromosomal segments.

Table 4.7 Clinical, cytogenetic and DNA features of Prader-Willi syndrome and Angelman's syndrome

Feature	Prader-Willi syndrome	Angelman's syndrome
Clinical	Obesity, short stature and mild mental retardation, behavioural problems. Characteristic facies, small stature, hands and feet. Hypo-gonadism. Floppy and feeding problems in the newborn period.	Ataxia, hyperactivity, epilepsy, lack of speech, severe mental retardation but happy disposition. Characteristic facies. Can be floppy at birth.
Incidence	1 in 10 000 to 1 in 15 000.	1 in 10 000 to 1 in 40 000.
DNA	Paternal deletion 73%, maternal uniparental disomy 25%, imprinting mutation 2%, other rare.	Maternal deletion 70%, paternal uniparental disomy 2%, imprinting region mutation 5%, *UBE3A* mutation 10%, and unknown 13%.

Imprinted genes are not randomly distributed across the genome. Indeed, many such genes are clustered, suggesting that the regulation of imprinting may be based on a domain-like structure. This is now confirmed with an imprinting control element located in the chromosome 15q11-q13 region, and controlling in *cis* (*cis*—a region of a gene or nearby locus that affects its function) the nearby Angelman's syndrome and Prader-Willi syndrome genes via methylation, chromatin folding and gene expression. More recently, a number of non-translated RNA species have been identified within the imprinted regions, suggesting that they may also be playing a role in the regulation of imprinting. One such non-translated RNA is snoRNA (sno—small nucleolar).

Molecular (DNA) Diagnosis and Counselling

Imprinting is more accurately detected, and its implications have become better understood with the utilisation of molecular technology. This has enabled accurate assessment of the parental origin for chromosomal abnormalities such as deletions, aneuploidies or uniparental disomies (see the next section for further description of uniparental disomy). For DNA diagnostics in disorders such as Prader-Willi and Angelman's syndromes, the initial DNA test determines the methylation status of the imprinting region. This highly reliable test does not define the underlying defect. Further DNA analysis is required for this purpose. Counselling issues in both Prader-Willi syndrome and Angelman's syndrome are complex but important because the parents will invariably want to know the risk of recurrence of these conditions. Generally, in Prader-Willi syndrome the risk is low if the primary defect is a *de novo* deletion or uniparental disomy. In Angelman's syndrome, counselling is more complex because a wider range of genetic defects can be involved, including cases in which the underlying defect is not known. Generally, risks are low if there is a *de novo* deletion or uniparental disomy. Because both syndromes are rare and involve complex genetic abnormalities, counselling and DNA diagnosis should be undertaken by specialised health professionals.

Uniparental Disomy

Uniparental disomy occurs when two copies of a chromosome are inherited from one parent, and nothing is inherited from the other parent. There are two types of uniparental disomies: (1) isodisomy—both chromosomes from the one parent are identical copies and (2) heterodisomy—the two chromosomes represent different copies of the same chromosome. Uniparental disomy has been reported in both Prader-Willi syndrome and Angelman's syndrome. Cytogenetic analysis will not detect uniparental disomy because the chromosomal numbers are preserved. It requires molecular analysis to show that the two chromosomes originated from the same parent. Since the chromosomal (and gene) content is not changed in uniparental disomy, disease will occur only if the chromosome involved in the disomy contains imprinted genes. Thus, Prader-Willi syndrome will develop if uniparental disomy leads to the presence of two maternal copies for chromosome 15. Another mechanism leading to genetic abnormalities with uniparental disomy involves isodisomy if the two identical chromosomes each carry the same recessive mutation (see cystic fibrosis following).

Uniparental disomy is not unique to chromosome 15. Cystic fibrosis occurring in the case of a carrier mother but normal father has been explained by uniparental disomy. In this situation, non-paternity was excluded, and it was shown that affected children had inherited two copies of the mutant chromosome from their carrier mothers; i.e., isodisomy had occurred. It is noteworthy that the cystic fibrosis phenotype was also associated with developmental abnormalities, e.g., moderate to severe intrauterine and postnatal growth retardation. Thus, it is possible that paternally derived gene(s) located on chromosome 7 are required for normal development.

There are three explanations for uniparental disomy. They involve fertilisation with disomic (diploid content) or nullisomic (no chromosomal content) gametes (Figure 4.6). The pathway leading to a trisomic conceptus is considered to be the more likely since chromosome 15 is

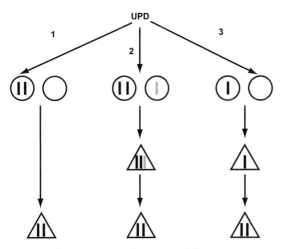

Fig. 4.6   Three ways in which uniparental disomy can occur.
Gametes are depicted as circles; zygotes, as triangles. A chromosome is shown as a bar; it is present as one copy (monosomy—the normal situation), two copies (disomy) and no copies (nullisomy). **(1)** One gamete has two copies of a chromosome, and the other has no copies. This situation can arise following non-disjunction. Fertilisation between these two gametes would produce the normal diploid number, but both chromosomes have come from the one parent, i.e., either iso- or heterodisomy. **(2)** Fertilisation in this case is between a disomic gamete and a normal monosomic one. The zygote is trisomic and is unlikely to survive unless one of the three chromosomes is lost. By chance (33% of the time), the one lost will have come from the normal gamete; i.e., the zygote is again diploid for that chromosome, but both originate from the same parent. **(3)** A third scenario involves fertilisation between a normal gamete and a nullisomic one. One way for the zygote to survive involves duplication of the single chromosome. In this case, uniparental isodisomy results. The mechanism depicted in **(2)** is considered the most likely since trisomy has been reported in chorionic villus samples, but the newborn has Prader-Willi syndrome secondary to uniparental disomy. The initial trisomic situation is corrected, which allows the fetus to survive, but at a cost because disomy results.

one of the more frequent trisomies associated with spontaneous miscarriages. If trisomy were to follow from non-disjunction (uneven division of chromosomes during meiosis), a viable disomic conceptus is possible only if one of the three chromosomes is lost. If this were to occur, one third of the concepti will have the two remaining chromosomes originating from the same parent. Thus, one mechanism for the Prader-Willi syndrome is maternal non-disjunction giving a trisomic conceptus, which is rescued when the paternal chromosome is lost.

MOSAICISM

Mosaicism refers to the presence in an individual (or a tissue) of two or more cell lines that differ in genotype or chromosomal constitution but have been derived from a single zygote. Mosaicism is the result of a mutation that occurs during embryonic, fetal or extrauterine development. Mosaic cellular populations can arise from mutations in nuclear or mtDNA in post-zygotic cells, epigenetic alterations in DNA and numerical or structural abnormalities in chromosomes. All these alterations can proceed from normal to abnormal or vice versa. The time at which the defect arises will determine the number and types of cells (somatic and/or germ cells) that are affected. It is likely that mosaicism will be found in all large multicellular organisms to some degree. Mosaicism is now able to be studied in a greater number of circumstances and in more depth because DNA techniques allow an accurate genotypic assessment of multiple tissues. In this way the identity of individual cells can be established.

In contrast to mosaicism is chimaerism. This refers to the presence in an organism of two or more cell lines that are derived from different zygotes. During embryogenesis, cells from two distinct embryos can mix; e.g., dizygotic twins could do this because of intra-uterine transfusion of cells from one twin to the other. It would be expected that this type of chimaerism would lead to immunological tolerance following a graft from one twin to the other. A more common example of chimaerism is an allogeneic organ transplant.

Chromosomal Mosaicism

Females are examples of chromosomal mosaicism since there will be random inactivation of one of the two X chromosomes in all tissues. As mentioned in Chapter 3, this can complicate carrier testing for X-linked genetic disorders. It should also be noted that X-inactivation is not complete since regions on the X chromosome are spared, e.g., the pseudoautosomal region on the end of the short arm. Thus, normal females do not have the phenotype of Turner's syndrome in which one of the two X chromosomes is missing. Both Turner's syndrome (45,X) and Down's syndrome (trisomy for chromosome 21) have had chromosomal mosaicism demonstrated by cytogenetic analysis of cultured lymphocytes. In fact, a conceptus with Turner's syndrome probably survives to term only when a coexistent normal cell line is also present. Thus, chromosomal mosaicism explains why an aneuploid fetus can survive to term if a normal cell line is present in the placenta. The common autosomal trisomies involving chromosomes 13, 18 and 21 are sometimes found as somatic mosaics. In nearly all cases, the zygote is initially completely trisomic, but the loss of one of the trisomic chromosomes produces a normal cell line that persists in the embryo.

A number of explanations for chromosomal mosaicism are observed at prenatal diagnosis: (1) maternal contamination of sampled tissue, (2) laboratory artefact,

(3) confined placental mosaicism and (4) true fetal mosaicism. Chromosomal mosaicism usually results from non-disjunction occurring in an early embryonic mitotic division, leading to the persistence of more than one cell line. With early fetal sampling made possible by chorionic villus biopsy, it has become apparent that chromosomal mosaicism affecting the placenta occurs more frequently than previously considered (~1–2% of samples). Chromosomal mosaicism confined to the placenta can produce false diagnostic results particularly in karyotypes obtained from chorionic villus sampling. Retarded intrauterine growth in a fetus with a normal karyotype may result from aneuploidy (the addition or subtraction of single chromosomes) confined to the placenta.

Somatic Cell Mosaicism

Mitotic errors at the DNA copying stage can give rise to mutations in single human genes. The clinical effect of somatic mosaicism depends on when the mutation arose and in what cell types. Somatic mutations that occur as early events in development will give rise to a more generalised disease phenotype. On the other hand, a late onset will manifest by localised or segmental disease because only some cell lines are affected. Clues to the presence of mosaicism may come from the finding in sporadic genetic disorders of marked tissue dysplasia, which is patchy in distribution. Alternatively, mild phenotypic manifestations in a person with an apparent spontaneous single gene mutation or a mild phenotype in an individual whose offspring or parent is severely affected may represent examples of somatic cell mosaicism (Table 4.8).

Germinal Mosaicism

From animal studies it has been estimated that the proportion of mosaicism in germ cells can vary from a few percent to 50%. Animal studies have also shown that germline cells separate from somatic cells at an early stage in development. Therefore, post-zygotic mutations that give rise to mosaicism can affect either somatic cells, germ cells or occasionally both.

Germinal mosaicism is one explanation why parents, who are apparently normal on genetic testing, can have more than one affected offspring with an X-linked or dominant genetic disorder (e.g., X-linked: Duchenne muscular dystrophy, haemophilia A or B; and autosomal dominant: osteogenesis imperfecta, tuberous sclerosis, achondroplasia, neurofibromatosis type 1). Therefore, the suspicion of germ cell mosaicism means that recurrence of a genetic disorder needs to be considered when individuals are being counselled. Germinal mosaicism affecting sperm has been sought using PCR. Normal DNA

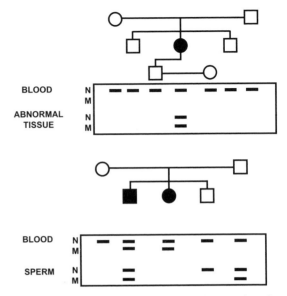

Fig. 4.7 Hypothetical pedigrees demonstrating somatic and germinal mosaicism identified by DNA testing.
N = normal; M = mutant. Open symbol = normal phenotype; filled symbol = affected phenotype. **Top:** DNA testing shows in the peripheral blood lymphocyte that all individuals have only the normal DNA marker. However, biopsy of abnormal tissue (e.g., skin) in the affected person shows that the DNA pattern here is different, and a mutant band is also present, i.e., somatic mosaicism. **Bottom:** This pedigree illustrates two affected individuals with an autosomal dominant disorder but phenotypically normal parents. DNA markers in the peripheral blood confirm that the two offspring have the genetic disorder. Examination of sperm DNA from males shows that the father of the two affected individuals has germinal mosaicism, because some of the sperm have the DNA mutation. The proportion of affected sperm could be estimated by comparing the intensities for the normal and mutant DNA bands.

Table 4.8 Some Mendelian disorders associated with somatic mosaicism (Youssoufian and Pyeritz 2002)

Class of disorder	Examples
Metabolic	Lesch-Nyhan syndrome, tyrosinaemia type I
Immune dysfunction	Wiskott-Aldrich syndrome, adenosine deaminase deficiency
Clotting	Haemophilia A and B
Skeletal	Marfan's syndrome
Muscular	Duchenne muscular dystrophy, congenital myotonic dystrophy
Tumour suppressor	Neurofibromatosis types 1 and 2, tuberous sclerosis
Nervous system	Friedreich's ataxia

patterns obtained from somatic cells, such as peripheral blood, are compared with sperm DNA patterns. The latter would show both normal and mutant DNA forms if there is germline mosaicism. From the frequency of the mutant form, a theoretical recurrence risk can be estimated (Figure 4.7).

ONCOGENESIS

INTRODUCTION

"Cancer is essentially a genetic disease at the cellular level" (W Bodmer, 1994). Evidence for a genetic component in cancer includes

- Increased risk for tumour development is seen in those with a genetic defect in DNA repair.
- Chemicals or physical agents that mutate DNA also elicit tumours in animals.
- Some structural chromosomal rearrangements in the germline predispose to tumour development.
- Somatically acquired mutations in the DNA of tumour cells can resemble those seen in familial cancers.
- The malignant phenotype can be conferred on normal cells by gene transfer studies with oncogenes.
- The malignant phenotype can be induced *in vivo* by gene manipulation in transgenic mice.

For many years, tumourigenesis was hypothesised to represent a **multistep** process. However, it was only with the application of recombinant DNA techniques that evidence for this became available. It is now possible to identify molecular (DNA) changes responsible for the initiation, promotion and progression of cancers. The ability to define mutations at the DNA level has enhanced the accuracy of diagnosis. Therapeutic options based on our knowledge of the DNA changes in cancer are now being attempted.

An individual's response to cancer is complex. It involves a number of factors such as the state of the immune system, nutrition and well-being, the extent of disease, response to treatment, and the development of drug resistance. In addition, genetic changes involve oncogenes, tumour suppressor genes, apoptotic genes, repair genes and epigenetic modifications to DNA. Just as was described for the genetic disorders, cancers demonstrate two distinct inheritance patterns: (1) rare but high penetrance cancers, for example, the familial adenomatous polyposis described in Chapter 3, Mendelian Genetic Traits, and (2) common but low penetrance cancers likely to represent multifactorial disorders. It is the latter category that will be discussed in this chapter.

Cancer is very heterogeneous in its presentation, clinical types, biologic progression and treatment options. These features would make anyone pessimistic that it will ever be completely controlled. However, there is increasing optimism that at the molecular level, there are starting to be seen shared pathways in some cancers, thereby allowing a better understanding of pathogenesis and new opportunities for drug design. For example, there are a number of common breakdowns in normal cell function in cancers. They include (1) genomic instability, (2) uncontrolled and sustained proliferation, (3) lack of differentiation, (4) defective apoptosis and (5) potential to invade (metastasise). As well as the common defects, a series of genes play a normal role in the cells' functions. It is only when these genes are mutated through genetic inheritance or somatic changes that they become carcinogenic. Although there seems to be a large number of genes, they fall into a limited number of classes, and the same genes play a role in many cancers.

Some of the nomenclature in molecular oncology needs to be defined. **Oncogenes** (*onkos*—the Greek word for mass or tumour) are genes associated with neoplastic proliferation following a mutation or perturbation in their expression. Oncogenes have normal counterparts in the genome called **proto-oncogenes**. They play an essential physiological role in normal cellular growth control. v-onc and c-onc are abbreviations for oncogenes found in retroviruses (v) and their cellular equivalents (c). **Tumour suppressor genes** are genes that also play a role in normal cellular proliferation and differentiation. Loss or inactivation of tumour suppressor genes can lead to neoplastic changes. **Transformation** refers to the acquisition by normal cells of the neoplastic phenotype. **Signal transduction** describes the transfer of signals from extracellular factors and their surface receptors by cytoplasmic messengers to modulate events in the nucleus.

ONCOGENES

Retroviruses

The initial observation implicating viruses and cancer came in 1910 when P Rous demonstrated that a filterable agent (virus) was capable of inducing cancers in chickens. Fifty-six years later, he was awarded a Nobel Prize for this work. Retroviruses have since been identified as the cause of cancers in many species, including mammals, although much less frequently in humans. The relevance of the retroviruses to human cancers remained unclear until appropriate investigative tools became available. These tools included the availability of gene transfer studies and more recently the potential to

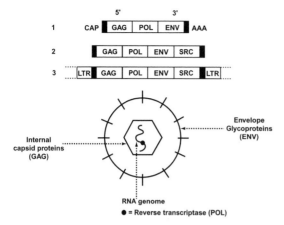

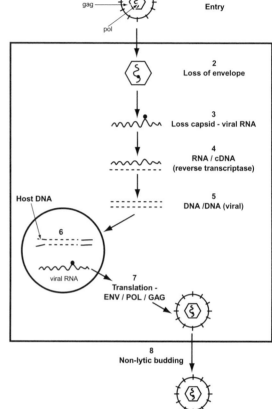

Fig. 4.8   The structure of a retrovirus.
(1) RNA tumour viruses (retroviruses) have an RNA genome. This RNA has two features of eukaryotic mRNA, i.e., a capped 5′ end and a poly-A tail at the 3′ end. Retroviral RNA codes for three viral proteins: (i) a structural capsid protein (*gag*), which associates with the RNA in the core; (ii) the enzyme reverse transcriptase (*pol*); and (iii) an envelope glycoprotein (*env*), which is associated with the lipoprotein envelope of the virus. **(2)** Transforming retroviruses have an oncogene. In the example here, the oncogene is that of the Rous sarcoma virus (*src*). **(3)** Retroviruses are so named because they have an RNA genome and are able to replicate through formation of an intermediate (provirus), which involves integration of the retroviral genome into that of host DNA. The provirus has LTRs (long terminal repeats) on either side of the RNA genes. The LTRs are several hundred base pairs in size and insert adjacent to smaller repeats derived from host DNA.

Fig. 4.9   Life cycle of a retrovirus.
(1) The envelope protein enables the retrovirus to bind to the surface of host cells on infection. **(2–5)** Double-stranded DNA derived from viral RNA and the action of reverse transcriptase is required before the retroviral genome can be integrated into that of the host. **(6–8)** The provirus formed replicates to produce mature viral particles, which are extruded from the cell by non-lytic budding.

identify mutations affecting DNA or chromosomes, as well as the development of transgenic mouse models.

The RNA tumour viruses (retroviruses) provided the first proof that genetic factors can play a role in carcinogenesis. Retroviruses have three core genes (*env*, *gag*— coding for structural proteins—and the third, *pol*, which codes for reverse transcriptase; Figure 4.8). As discussed earlier (Chapter 2, Figure 2.2) reverse transcriptase is an enzyme that allows RNA to be converted into cDNA. In this way, the retrovirus can make a DNA copy of its RNA, which can then become incorporated into the host's genome (Figure 4.9). A fourth gene gives retroviruses the ability to induce tumour growth *in vivo* or to transform cells *in vitro*. In the latter situation, cells lose their normal growth characteristics and acquire a neoplastic phenotype. An example of this would be loss of contact inhibition so that instead of growth in a single cell layer *in vitro*, there is unregulated proliferation into clumps. Retroviral DNA sequences responsible for transforming properties are called viral oncogenes (v-onc). Names are derived from the tumours in which they were first described—for example, v-*sis*, S̲imian s̲arcoma virus; v-*abl*, murine A̲belson l̲eukaemia virus; v-*mos*, M̲o̲loney

sarcoma virus; v-*ras*, r̲at s̲arcoma virus; but v-*src* is from virus s̲arcoma-producing.

Viral oncogenes have cellular homologues called cellular oncogenes (c-onc). The term "proto-oncogene" was coined to describe cellular oncogenes that do not have transforming potential; i.e., they do not form neoplasms in their native state. There are many proto-oncogenes. Their roles in terms of cellular growth control are complex and may involve interactions between a number of the proto-oncogenes. The sites or modes of action of proto-oncogenes can be divided into (1) growth factors, (2) receptor tyrosine kinases, (3) signalling—through membrane tyrosine kinases or transduction and (4) transcription factors (Table 4.9).

Table 4.9   Functional classes of oncogenes and tumour suppressor genes

Class	General function	Examples[a]	Specific action
Growth factors	Act via cell surface receptors to induce cellular division.	SIS	Codes for the β chain of the platelet derived growth factor (PDGF).
Receptor tyrosine kinases	Binding to their membrane receptors is the first step in delivery of mitogenic signals to the cell's interior to initiate cell division.	erbB family	ERBB1—epidermal growth factor receptor 1; ERBB2 (or HER-2/neu)—epidermal growth factor receptor 2.
		FMS	Receptor for colony stimulating factor 1.
Signalling	Membrane tyrosine kinases: Structural changes increase the kinase activity and from this the membrane receptors or signal transduction.	SRC	Activates mitogenic signalling pathways.
	Transduction: Method by which the extracellular growth factor at the cell surface receptor transfers (transduces) its signal to the nucleus by a number of ways, including via G proteins.	RAS	Membrane associated G protein and activates signalling pathways.
Transcription factors	Proto-oncogenes can encode nuclear binding factors and in this way control gene expression.	MYC	Plays major role in control of cell proliferation and apoptosis.
Cell surface proteins	Adenomatous polyposis gene (familial colon cancer).	APC	Interacts with β catenin.
Cell cycle factors	Inhibitors of cell cycle progression.	TP53	TP53 gene—cell cycle regulator as well as apoptosis.
		RB	Retinoblastoma gene—cell cycle regulator.
Apoptosis	Programmed cell death.	BCL2	Opposite effect to TP53, i.e., blocks apoptosis and so prolongs a cell's life.
DNA repair	DNA repair.	ATM	Ataxia telangiectasia gene.
	Mismatch repair.	MLH1	Hereditary non polyposis colorectal cancer gene.

[a] The last four classes involve tumour suppressor genes.

Oncogene Activation

From molecular studies, a number of mechanisms have been shown to disrupt proto-oncogenes, leading to uncontrolled cell division. These mechanisms include

(1) Chromosomal translocation: Changing the positions of oncogenes within the human genome can alter their function. A way to do this is through a translocation, i.e., the movement of a segment of a chromosome to another chromosome. Chromosomal breakpoints that produce the Philadelphia chromosome found in chronic granulocytic leukaemia involve a translocation of the c-ABL oncogene on chromosome 9 to a gene on chromosome 22, which is called the breakpoint cluster region or BCR (Figure 4.10). The hybrid BCR-ABL transcript produces a novel protein with tyrosine kinase activity. Transgenic mice that express the BCR-ABL fusion gene develop lymphoblastic leukaemia or lymphoma.

(2) Gene amplification: Amplified c-myc sequences have been described in a number of human cancers. In neuroblastoma, a childhood tumour, the N-MYC gene may become amplified up to 300 times.

(3) Point mutations: Single base changes in the DNA sequence of proto-oncogenes have been consistently observed in the different ras genes. A wide range of malignancies demonstrate ras mutations. It is noteworthy that the sites for mutations in ras are limited to codons 12, 13 and 61, which are located within the regions that code for binding of GTP/GDP. Therefore, failure by ras to convert the active complex (GTP-ras) to inactive (GDP-ras) would produce an excess of stimulatory activity leading to unregulated cellular proliferation.

(4) Viral insertion: Another mode by which proto-oncogene function can be perturbed is through insertion of viral elements. An example of this is thought to occur with the hepatitis B virus and hepatocellular carcinoma.

TUMOUR SUPPRESSOR GENES

The identification of proto-oncogenes and oncogenes in the pathogenesis of cancer was an exciting development in molecular medicine. However, results from in vitro and in vivo studies were not always consistent, and only

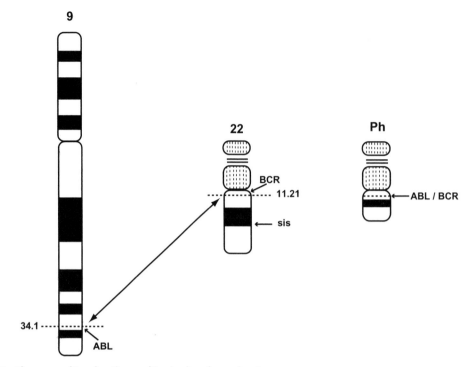

Fig. 4.10 Chromosomal translocation resulting in altered gene function.
A reciprocal translocation between chromosomes 9 and 22 produces the Philadelphia (Ph) chromosome in chronic granulocytic leukaemia.----- = breakpoints. The Ph chromosome comprises the portion of chromosome 22 above ----- and the small segment of chromosome 9 below the -----. This results in juxtaposition of *ABL* from chromosome 9 with *BCR* from chromosome 22. The *sis* proto-oncogene is not considered to have a functional effect from this translocation because it is located at some distance from the actual chromosome 22 breakpoint (22q11.21). See also Figure 2.11, which shows a FISH image of this translocation, and Figure 4.18, which illustrates the fusion gene.

about 20% of human tumours demonstrated changes in these genes. Oncogenes were not abnormal in the inherited cancer syndromes. Thus, other molecular explanations were sought.

Perturbations in genes can produce gain-of-function (stimulatory) or loss-of-function (inhibitory) signals to cell growth. Examples of gain-of-function effects are the proto-oncogenes described earlier. In these situations, mutations produce positive signals leading to uncontrolled proliferation. On the other hand, tumour formation can result through loss-of-function. This is associated with another group of genes known as the **tumour suppressor genes**. Evidence for these in the pathogenesis of cancer has come from three observations: (1) somatic cell hybrids and transgenic mice, (2) inherited cancer syndromes and (3) loss of heterozygosity in tumours.

(1) Somatic cell hybrids and transgenics: In the late 1960s, murine cell hybrids formed by fusions between normal and tumour cells were found to revert to the normal phenotype. Subsequently, as the hybrid clones

were propagated in culture, the tumour phenotype became re-established. This effect was seen in a wide range of tumour lines and was considered to indicate the influence of tumour suppressor genes derived from the normal cells. Subsequent loss of chromosomes, which occurred on serial passage of cell lines, enabled reversion to the neoplastic phenotype when the tumour suppressor genes were lost. Today, micro-cell transfers allow one or a few chromosomes within a reconstituted membrane from normal cells to be delivered to recipient tumour cells. In this way, it has been possible to identify tumour suppressor genes in a variety of chromosomes. Sophisticated molecular techniques such as gene knockout enable specific genes to be inactivated in transgenic mice. This shows definitively that a particular gene functions as a tumour suppressor (see *TP53* on page 101 and Chapter 5 for further discussion on gene knockouts).

(2) Inherited cancer syndromes: Familial recurrence is seen in most tumours although the proportion is often low. One exception is the tumour retinoblastoma, for which the inherited form comprises up to 40% of cases.

The remainder are sporadic. Although rare, retinoblastoma is the most common intraocular malignancy in children, with an incidence of about 1 in 14 000 live births. In familial cases, about half of the children can be affected, i.e., an autosomal dominant mode of transmission (Box 4.3).

In the early 1970s, epidemiological studies of both retinoblastoma and Wilms' tumour led A Knudson to propose his **two-hit model of tumourigenesis**. This model required, in either the sporadic or genetic forms of retinoblastoma, the tumour cells to acquire two separate genetic changes in DNA before a tumour developed. The **first** or predisposing event could be inherited either through the germline (familial retinoblastoma), or it could arise *de novo* in somatic cells (sporadic retinoblastoma). The **second** event occurred in somatic cells. Thus, in sporadic retinoblastoma both events arose in the retinal (somatic) cells. In familial retinoblastoma the individual had already inherited one mutant gene and required only a second hit affecting the remaining normal gene in the somatic cells. The frequency of somatic mutations was sufficiently high that those who had inherited the germline mutation were likely to develop one or more tumours (Figure 4.11).

The mode of inheritance for retinoblastoma is autosomal dominant with incomplete penetrance. However, at the cellular level, it is recessive since loss or inactivation of both alleles is required to change a cell's phenotype. A cell containing only one of its two normal alleles will usually produce enough tumour suppressor gene product to remain normal. Loss of the cell's remaining wild-type allele exposes it to uncontrolled proliferation. Incomplete penetrance reflects the requirement for the second (somatic) mutational event and explains the skipping of some generations. On the other hand, sporadic forms of the tumour involve two separate somatic events. The second hit must occur in the same cell lineage that has experienced the first or predisposing hit. The probability of this is relatively rare, and so sporadic forms of the tumour occur later in life and have the additional features of being unifocal and unilateral. The two hits proposed by Knudson were confirmed in the form of the two alleles for the retinoblastoma gene when it was finally cloned and characterised.

(3) Loss of heterozygosity (usually abbreviated to LOH): In the inherited tumours, the somatic event that affects the second (normal) allele and so exposes the recessive mutation can involve chromosomal loss or molecular abnormalities such as a deletion. Chromosomal rearrangements, losses or aneuploidy (abnormal chromosome numbers) can be detected cytogenetically by studying constitutional cells (e.g., lymphocytes or fibroblasts) and comparing these cells to tumour cells. However, small DNA deletions (microdeletions) will be difficult to find. Cytogenetic analysis does not usually allow the parental origin for chromosomes to be determined.

Box 4.3   Two rare tumours demonstrating important genotypic changes in cancer.

Retinoblastoma: Retinoblastoma is an important intraocular malignancy in children. It is usually detected within the first three years of life. Since treatment with surgery or radiotherapy is potentially curative, it is essential to make an early diagnosis. With increasing survival of treated patients, it has become apparent that those with the heritable form are at increased risk for other tumours, e.g., osteosarcoma. An important breakthrough in understanding retinoblastoma came with the finding of a deletion involving chromosome 13q14 in the blood of patients with the heritable type, and in the tumour cells of both the sporadic and heritable forms; i.e., loss of heterozygosity had occurred. Subsequently, DNA polymorphic markers showed that the normal allele on the chromosome 13 inherited from the unaffected parent was the one lost in the tumour cells, indicating that loss of heterozygosity produced its effect by uncovering a mutation in the germline. This was consistent with the two-hit model for cancer. Using positional cloning, the retinoblastoma (*RB1*) gene was identified. The protein from this gene was expressed in most tissues, not only in the retina. Nevertheless, confirmation of *RB1* as a likely candidate gene came with the finding that the normal 4.7 kb mRNA transcript was not present in retinoblastoma cell lines. Mutations in the gene (i.e., deletions, point mutations) were also detected in DNA from tumours. An apparent increased frequency of osteosarcoma in patients with retinoblastoma was strengthened when similar changes in *RB1* were found in these bone tumours. Surprisingly, sporadic cases of osteosarcoma occurring independent of retinoblastoma also had mutations involving *RB1*. From DNA studies, it is estimated that in ~10% of cases, the disease can arise following inheritance of a germline mutation from a carrier parent or through the occurrence of a *de novo* germline mutation (approximately 30% of cases). This leads to bilateral heritable disease. Progeny of those with bilateral heritable disease have a 50% risk of developing retinoblastoma. **Li Fraumeni syndrome:** This rare autosomal dominant disorder predisposes to a range of tumours, including sarcoma, leukaemia and breast and brain tumours. Affected family members have one mutant *TP53* gene in their germline. Malignant cells, on the other hand, have abnormalities affecting both alleles. These data reinforce the two-hit hypothesis for tumourigenesis and implicate the *TP53* tumour suppressor gene as one component in the pathogenesis of the Li Fraumeni syndrome. What other genetic defects are involved in this disorder remain to be determined. Recently, it has been shown that germline *TP53* mutations can occur without the classical Li Fraumeni syndrome. This is found in individuals who develop multiple malignancies.

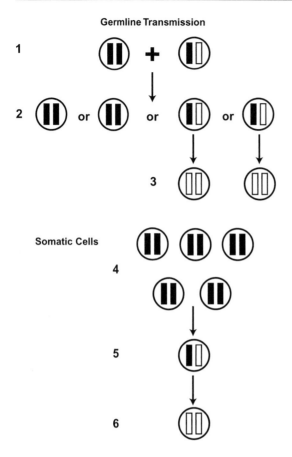

Germline Transmission

Fig. 4.11 Two-hit hypothesis for cancer.
The first or predisposing event can be inherited through the
germ cells (**1–3**) or arise in somatic cells (**4–6**). In both cases,
a second somatic event is required to inactivate the remaining
normal allele. **Germline:** (**1**) First event occurring in the
germline. The wild-type tumour suppressor gene for a
particular locus is shown as ■. The first hit affects one of the
germ cells to inactivate one tumour suppressor gene □.
(**2**) Offspring have a 50% chance of inheriting the mutant
tumour suppressor gene (dominant inheritance). (**3**) A second
hit is required to inactivate the remaining tumour suppressor
gene. Because the germ cells were initially involved, all
cells will be predisposed. Therefore, the second hit is more
likely to happen in the appropriate cell, e.g., retinal cells in
retinoblastoma. This produces multifocal tumours that present
at an earlier age. Other members of the family can become
affected. **Somatic cell:** (**4**) Sporadic tumour formation requires
two separate somatic events because now the germ cells have
wild-type (normal) tumour suppressor genes. (**5**) If the first hit
affects a somatic cell, it inactivates the tumour suppressor
gene in that cell and its progeny. To get both tumour
suppressor genes inactivated, the second hit (**6**) must involve
the predisposed line. This is less likely to occur than the
situation in (**3**). Therefore, tumour formation is delayed. If it
occurs, it is sporadic and unifocal.

The availability of DNA polymorphic markers has
meant that individual parental contributions in normal
and tumour cells can now be distinguished (Figure 4.12).
DNA markers flanking a locus containing a tumour
suppressor gene will thus demonstrate loss of heterozygosity if that segment is missing. The same approach is
useful in studying sporadic tumours with the base line or
constitutional genotype determined from the affected
individual's normal tissues, and the two parental contributions from analysis of the pedigree. They can then be
compared to DNA patterns in the tumour.

DNA studies to detect loss of heterozygosity in
tumours have been used to identify putative tumour suppressor loci, and from these, attempts to clone and characterise candidate genes have been made. Two tumour
suppressor genes—*TP53* on chromosome 17p and *DCC*
(<u>d</u>eleted in <u>c</u>olon <u>c</u>arcinoma) on chromosome 18q—
were detected in this way. Loss of heterozygosity has
been described in many forms of cancer, and the same
tumour suppressor regions may be involved in more than
one malignancy.

Tumour Suppressor Genes and Cancer

The various roles played by tumour suppressor genes in
cancer are still being defined. Ways in which these genes
in their wild-type or normal forms can prevent the development of cancer include (1) inhibiting cell proliferation,
(2) inducing differentiation or cell death and (3) stimulating DNA repair. The first two mechanisms will be
described in more detail in the section Cell cycle and
apoptosis. DNA repair was summarised in Chapter 3.
However, it should be noted that the genes involved in
regulating the cell cycle and apoptosis, for example,
TP53, can also indirectly contribute to DNA repair since
they slow down the cell cycle (or stimulate apoptosis)
and so assist the DNA repair enzymes to correct any
defects.

Caretakers and Gatekeepers

The two-hit model works well with a tumour such as
hereditary retinoblastoma. However, not all tumour suppressor genes play the same role in tumourigenesis. This
led to K Kinzler and B Vogelstein in 1997 to propose two
classes of tumour suppressor genes—caretakers and gatekeepers (Table 4.10). Basically, gatekeeper tumour suppressor genes play a central role in regulation of cellular
proliferation, and mutations in these genes can directly
lead to tumours developing. In contrast, caretaker tumour
suppressor genes play a more global role, maintaining
genome integrity. Mutations in this class of genes set up
the genetic environment for tumours to develop through
mutations in the gatekeepers or proto-oncogenes. From
observing retinoblastoma, a two-hit model involving a
gatekeeper tumour suppressor gene emerged. With care-

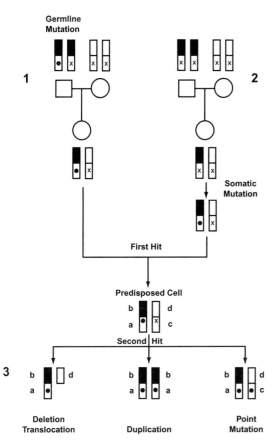

Fig. 4.12 Identifying parental alleles by molecular testing and detecting molecular (DNA) defects that produce a second (somatic) loss at a tumour suppressor gene locus. The mutation in the tumour suppressor gene is depicted by •; the normal allele, by x. **(1)** In the first family, there is inheritance of a germline mutation. **(2)** In the second family, the germline is normal but a somatic mutation occurs. In both situations the result is a predisposed cell. DNA polymorphism markers for the two alleles in the predisposed cell are indicated by b/a (mutant) and d/c (normal). **(3)** Loss of the remaining (normal) tumour suppressor gene via a second hit can occur by a number of different mechanisms, e.g., from left to right: deletion or an unbalanced translocation; a duplication or a more discrete abnormality such as a point mutation that occurs on the second (normal) allele. **DNA detection**: Loss of heterozygosity can be detected if it is assumed that four informative DNA markers distinguish the two parental contributions, and they are situated at the tumour suppressor gene location. For simplicity, the polymorphic markers are depicted as "a" (mutant) or "b, c, d" (normal). These give two haplotypes of b,a/d,c. Loss of heterozygosity is seen as b,a/d– (deletion, translocation) and b,a/b,a (duplication). Loss of heterozygosity will be masked in the point mutation situation since the normal locus (identified by the "c" marker) is still present although it is now non-functional because it has acquired a discrete mutation.

taker tumour suppressor genes, the number of hits is considerably increased because genes to be mutated include a caretaker followed by at least one other. It is likely in sporadic tumours demonstrating a positive correlation with increasing age that six or more genetic mutations will be necessary. This led to a multistep hypothesis for tumour development, which was introduced in Chapter 3 using the example of familial adenomatous polyposis (see Figure 3.7). In colon cancer, mutations in *K-RAS* and loss of heterozygosity on chromosome 5q (i.e., the tumour suppressor gene—adenomatous polyposis coli or *APC*) are the initial events in tumourigenesis. They are then followed by loss of heterozygosity on chromosome 18q (*SMAD4* gene) and chromosome 17p (*TP53* gene).

Other ways in which tumour suppressor genes can perform their role is by limiting a cell's proliferative capacity through inducing it to undergo differentiation. In this way, the relatively greater mitotic activity seen in the undifferentiated cell gives way to an end-cell that divides less frequently. As well as preventing the formation of tumours, an additional role for tumour suppressor genes lies in normal development. Finally, tumour suppressor cells can reduce the risk of tumour formation by ensuring that a cell that has undergone a mutational event is induced to die by the process of apoptosis.

CELL CYCLE AND APOPTOSIS

Cell Cycle

Cellular and tissue integrity requires an exquisite balance between cell proliferation and cell death. The balance between a series of positive and negative signals determines whether cells will continue to live or die. The cell cycle consists of a series of highly ordered events that lead to duplication and division of a cell. The process requires production of new DNA, segregation of chromosomes, mitosis and then division. Extracellular signals control entry into, exit from and progress of the cell cycle. At key points in the cell cycle, signalling pathways monitor the progress of upstream events prior to a cell progressing further. These monitoring stages in the cell cycle are often called checkpoints. The cell cycle is divided into five components (Figure 4.13): G_0—resting phase with cells having their 2n (diploid) DNA content; G_1—cell growth phase (2n); S—DNA synthesis phase (4n); G_2—cell growth phase (4n); M—mitotic phase (4n → 2n). A critical step in control of the cell cycle comes at the G_1 to S transition. After this point, the cell is irreversibly committed to the next cell division.

At the molecular level, observations based on studies from yeast and the frog *Xenopus* have enabled the complex steps involved in the cell cycle to be better understood. Key components that have a stimulatory effect on the cell cycle are the cyclins that work in

Table 4.10   Gatekeeper and caretaker tumour suppressor genes (TSG)

Gatekeeper TSG	Caretaker TSG
This class of TSG plays a central checkpoint role in the cell cycle and so cellular proliferation. Examples are retinoblastoma, neurofibromatosis 1, familial adenomatous polyposis	TSG involved in the maintenance of genome integrity. Failure produces genome hypermutability. Examples are the DNA repair genes.
A specific tissue distribution for the tumour is found (see above examples)	Unlike the gatekeeper TSG, inactivation of this class does not promote tumour development directly. Tumours arise because of the genome instability leading to mutations in other genes (including gatekeepers and proto-oncogenes). Tissue specificity not a feature of this class.
Inactivation of this class is rate limiting for tumour formation. This class follows Knudson's two-hit hypothesis for tumour development in both hereditary and sporadic forms of tumours.	Mutations in caretaker TSGs would be less likely to be associated with sporadic cancers because four hits would be involved—two initiating hits on the caretaker TSG followed by the two hits with the gatekeeper genes required for eventual tumour development.

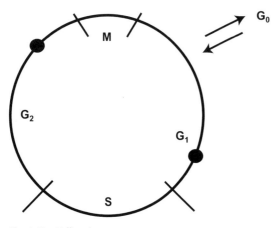

Fig. 4.13   Cell cycle.
The cell cycle has five distinct stages. G = gap; S = DNA synthesis; M = mitosis. The • indicate the position of two key checkpoints, although there are others. At these checkpoints the cell evaluates progress and can arrest the cycle if repair is needed. Checkpoints can also lead to activation of apoptosis if cell damage cannot be repaired. Different cyclins are produced at various stages of the cell cycle. Growth factors and mitogenic signals induce cells to leave the quiescent (G_o) phase and progress through G_1. Oncogenes promote growth and are particularly involved with the G_1 phase; tumour suppressor genes inhibit the cell cycle, promote apoptosis and focus on the S phase. Repair genes do their work during S and G_2. In a rapidly proliferating somatic cell, the entire cell cycle can take up to 24 hours to complete. G_1 is the longest phase (about 12 hours); S phase, about 7 hours; $G_{2,}$ 4 hours; and mitosis, 1 hour.

concert with their catalytic partners (cyclin dependent kinases—CDK) to hyperphosphorylate the products of the retinoblastoma (Rb) tumour suppressor family. As a result, transcription factors are released, and they lead to upregulated expression of genes that are crucial for cell cycle progression. Thus, the retinoblastoma pathway is fundamental to tumour formation, and not surprisingly,

retinoblastoma proteins play a role in a range of tumours apart from the classic example of genetic retinoblastoma (see Box 4.3).

In contrast, negative influences on the cell cycle come from a series of checkpoints that respond to various stimuli. The tumour suppressor genes can reduce the potential for tumour formation by interfering with the progress of the cell cycle until damaged DNA is repaired. One of the key players here is *TP53*, which responds to DNA strand breaks by stimulating the expression of p21, *GADD45* and cyclin G, leading to cell cycle arrest so there is time for DNA repair to take place (more about *TP53* below). During the S and G2 phases, DNA repair systems are essential to make sure that DNA has been copied correctly before undergoing mitosis.

The *APC* gene (involved in familial adenomatous polyposis described in Chapter 3) provides a useful example of the positive and negative effects on the cell cycle. In Figure 3.6, the structure/function relationships of the *APC* gene are given. They show how *APC* is important for maintaining normal cell division through its effect in binding β catenin. However, a mutation in *APC* leads to accumulation of β catenin, which then works through the G_1-S checkpoint region and myc + cyclin D1 + retinoblastoma to stimulate the cell cycle, and eventually tumours develop.

Apoptosis

Development as well as ongoing maintenance of many adult tissues relies on the balance between proliferation, differentiation and cell death. Cell death can occur in a number of ways. (1) **Necrosis.** Features of this include an accidental initiation, e.g., severe injury, associated with degenerative and inflammatory responses. Nearby tissues or cells can be damaged, and the final cellular debris are removed by phagocytes. Necrosis occurs rapidly after the initial insult and does not have the cellular morphologic changes found with apoptosis. (2) **Apoptosis.** This occurs in normal tissues and involves a type of cell suicide requiring the interaction of genetic events and producing a non-inflammatory demise of the cell. At the DNA level, the inter-nucleosomal degradation occurring in DNA can be seen in the form of small double-stranded fragments that migrate in a ladder pattern (composed of multiples of about 185 bp). This route for cell death requires time to take place after the initial insult.

Apoptosis is a highly regulated multistep process comparable to what is seen with the cell cycle. Indeed, a discussion of the cell cycle and apoptosis as separate events is artificial since both share many of the key regulators. Virtually all cells have an inbuilt apoptotic program triggered by a variety of stimuli (growth factor withdrawal, genotoxic insults, UV irradiation), thereby ensuring maintenance of cellular integrity. There are two major apoptotic pathways: (1) extrinsic and (2) intrinsic. Extrinsic stimuli include tumour necrosis factor ligands that induce trimerisation of cell-surface death receptors. The intrinsic pathway involves the mitochondria and signals from pro-apoptotic genes such as *BAX* (BCL2-associated X protein). *BAX* produces a protein that alters the mitochondrial membrane permeability. This releases cytochrome c into the cytoplasm, which stimulates the final step in apoptosis through the production of proteolytic proteins called caspases. At the molecular level, apoptosis involves a number of genes. In humans, two that are well characterised are *BCL2* and *TP53*.

BCL2 belongs to a multigene family, and it inhibits some forms of apoptosis; i.e., it protects cells from death. *BCL2* and *BAX* genes share considerable homology despite having opposite effects on apoptosis. The proteins encoded by these genes form heterodimers, and the ratio of Bcl-2 to Bax protein determines whether life or death results; i.e., when the *BCL2* effect is predominant, apoptosis is inhibited ("life"), whereas a predominance of *BAX* leads to an acceleration of programmed cell death. *BCL2*'s effect on apoptosis results from developmental stimuli, cytokine withdrawal or cytotoxic stress condi-

tions. *BCL2* is not involved with responses to death receptor–induced apoptosis occurring via the tumour necrosis factor receptor pathway. Lymphoid cells exposed to an activated *BCL2* gene following a chromosome 14:18 translocation eventually develop into a malignant lymphoma because spontaneous mutations that occur in these cells are unable to be contained by the cell dying, and so they accumulate.

In contrast to *BCL2* is the effect of the *TP53* gene, which can induce a damaged cell to undergo apoptosis and so remove a potential focus for tumour formation. Cells damaged by chemotherapeutic agents will stimulate the production of *TP53* and so undergo apoptosis. This additional anti-tumour effect may explain why cancers that have wild-type *TP53* genes respond better to treatment. On the other hand, a mutant *TP53* gene cannot function in this way, and damage to the cancer cell produced by the chemotherapy will accumulate in cells that have not been directly killed by the treatment. These cells will then form a new clone of more malignant, treatment-resistant tumour cells.

TP53

TP53 (also written p53 or P53) has been described as the most important cancer-related gene. This tumour suppressor gene is located on chromosome 17p13.1, and is implicated in both inherited and sporadic cancers. Indeed, it is the most frequently altered tumour suppressor gene in human non-haematopoietic malignancies, with more than 50% of solid tumours showing a loss of wild-type *TP53*. The gene has 11 exons and codes for an mRNA of about 2.5 kb, producing a protein of 393 amino acids (Figure 4.14). The gene's importance is suggested by its evolutionary conservation; e.g., mouse and human proteins are about 80% homologous. The gene is expressed in all cells.

TP53 functions as a tumour suppressor gene since it inhibits the transformation of cells in culture by onco-

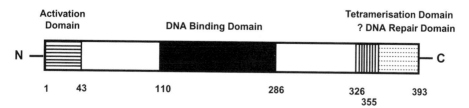

Fig. 4.14   Structure of the *TP53* gene.
TP53 has four functional domains. From N to C terminals, these are Activation Domain, DNA Binding Domain, Tetramerisation Domain and a Basic Domain that appears to be involved in DNA repair regulation. The Activation Domain can bind DNA and is important for *TP53*'s role in regulating transcription of target genes. Most of the mutations in *TP53* occur in the DNA binding domain, identifying this as an important functional region. *TP53* in its tetrameric form has an enhanced activity for interacting with DNA and proteins, and this conformation relies on the third domain. The last domain has nine basic amino acids and appears to be involved in *TP53*'s role in DNA repair. The numbers indicate the approximate amino acid in relation to the functional domains.

Table 4.11   Key proteins involved in the cell cycle and apoptosis

Protein[a]	Cell cycle	Apoptosis
p53	Stops the cell cycle following DNA damage. Does this through various pathways including p16 (retinoblastoma) and p21 (*WAF1*, *CIP1*).	Key regulator of apoptosis—does this through transcription and non-transcription–dependent processes. Result is stimulation of apoptosis.
pRb	Rb strong inhibitor of the cell cycle by interacting with transcription factors such as E2F. When pRb is phosphorylated, transcription factors are released, and they stimulate proliferation, particularly G_1 to S.	Cleaved by caspase leading to stoppage of cell cycle during apoptosis.
Bcl-2	Increases the length of G_1 and can promote exit into G_0. Retards entry back into the cell cycle.	Key role is in inhibiting apoptosis (prolonged uncontrolled exposure to bcl-2 leads to leukaemia).
Cyclins and CDK	Growth factors stimulate cyclin D leading to it associating with Cdk4 and Cdk6 in early G_1. This allows Rb to become phosphorylated, and then the cell can move past the G_1 checkpoint into S phase.	Apoptotic effect not well defined.
CDK inhibitors	Either block CDK/cyclin assembly or CDK activities leading to arrest of cells at different points in the cell cycle (particularly G_1 arrest) until conditions are suitable for the cell to progress through the cell cycle.	Apoptotic effect not well defined.
myc	Nuclear phosphoprotein that functions as a transcription factor stimulating the cell cycle, particularly its early control (transition from G_0 to G_1 to S).	Cell proliferative effect balanced by its role in stimulating apoptosis. Involved pathways not well defined but thought to be different to cell cycle effect.
ras	Stimulates the cell cycle through phosphorylation of genes involved in cell cycle progression, and at the same time inhibits apoptosis.	Working through another pathway, ras stimulates apoptosis.

[a] Abbreviations: Rb—retinoblastoma protein; CDK—cyclin dependent kinase.

genes and the formation of tumours in animals. Transgenic mice that have both *TP53* genes inactivated by gene knockout studies are normal at birth, but by 6–9 months of age, 100% develop a range of cancers. In humans, inheritance of a mutated *TP53* gene can produce the serious multi-organ cancer syndrome called Li Fraumeni (see Box 4.3). A key feature of tumour suppressor genes (i.e., loss through chromosomal or DNA rearrangement) is seen with the *TP53* locus on chromosome 17p.

Cancers in which there have been mutations affecting the *TP53* gene include colon, lung, brain, breast, melanoma, ovary and chronic myeloid leukaemia in blast crisis. Defects observed lead to loss of both alleles in 75–80% of cases, with one defect often a deletion and the second a missense point mutation. The latter leads to production of an abnormal protein. Another way to interfere with *TP53* is through the binding of exogenous viral antigens or cellular oncogenes to the normal p53 protein.

TP53 plays a key role in inhibiting tumour development through the mechanisms mentioned earlier: (1) checkpoint control of the cell cycle, (2) induction of apoptosis and (3) stimulation of the DNA repair mechanism. There is also evidence that *TP53* may have a negative effect on angiogenesis, an essential property for

solid tumours to progress. When DNA is damaged, *TP53*-mediated pathways attempt to repair the injury through arrest of the cell cycle and stimulation of the DNA repair mechanisms. When repair is not successful, *TP53* stimulates the apoptotic pathway to remove the damaged cell. The *TP53* and pRb/E2F (see Table 4.11) pathways are also interconnected since free E2F with its stimulatory effect on the cell cycle (it becomes free when pRb is phosphorylated) induces *TP53* transcription, which has an inhibitory effect on the cell cycle.

In normal cells, the level of *TP53* is low because of its 20-minute half-life. However, following exposure of the cell to DNA-damaging agents (e.g., irradiation or certain chemicals), hypoxia, the level of the p53 protein, dramatically increases. The 53 kDa protein encoded by *TP53* is a transcription factor that can regulate a number of genes at the DNA level. p53 blocks progression of the cell cycle in the G_1 phase (see Figure 4.13). This allows DNA repair to occur prior to entry into the S phase. The cell cycle effect of p53 ensures that damaged DNA is not allowed to replicate; hence, it has been called the "guardian of the genome."

Mutant *TP53* forms demonstrate altered growth regulatory properties and can also inactivate normal (wild-type) p53 protein. The latter is called a **dominant**

negative effect since inactivation of one of the two tumour suppressor loci can produce what appears to be a dominant effect if the mutant protein inhibits or interferes with the product from the remaining normal allele.

EPIGENETICS

The concept of epigenetics, or the influencing of gene expression without altering the DNA sequence, was introduced in Chapter 3. Here, the discussion of epigenetics is broadened to include the development of cancer. One way this might occur is through the hypermethylation of tumour suppressor genes, producing in effect an "epimutation." An interesting report by Suter *et al* in 2004 describes two patients with multiple cancers and inactive *MLH1* DNA repair genes secondary to hypermethylation rather than the traditional DNA mutation. This suggests that epigenetic changes may be more important in the pathogenesis of cancer than previously considered.

Gene Silencing

Most of what has been described in molecular oncogenesis relates to mutations in various genes. However, another powerful way in which to inhibit genes (particularly tumour suppressor genes) is by epigenetic means. DNA methylation (as well as histone methylation and acetylation) are recognised mechanisms by which gene function can be altered without changing the DNA sequence. Whether methylation patterns represent primary or secondary effects has yet to be determined, but at least there is a clear association with methylation and down-regulated gene expression, and demethylation and active gene expression. In tumours, DNA methylation provides confusing data. Loss of methylation has been observed in CpG dinucleotides (normally, most of them are methylated) and increased methylation in CpG islands associated with gene promoters (normally, demethylation would be found here).

Knudson's two-hit hypothesis requires in genetic tumours that a germline mutation (first hit) is followed by a second hit (somatic mutation). Another way this can occur is for the second hit to inactive the gene through methylation of the promoter. It is also possible that in sporadic cancers, hypermethylation of the promoter might down-regulate both alleles. Imprinting might also be a mechanism by which the "second hit" occurs. If a locus is imprinted, only one of the two alleles is functional. In this circumstance, it would require a single "hit" to inactivate the one functional allele.

Loss of DNA Methylation

Loss of methylation can potentially impact on tumour development since normally repressed genes might be induced to express or over-express in the case of a chromosomal translocation that brings into proximity: (1) a proto-oncogene or (2) distant promoters or other regulatory sequences able to induce expression of the demethylated gene. Loss of methylation might also affect the functional stability of chromosomes as the pericentromeric region of the chromosome requires DNA methylation for stability and replication of DNA.

Recently, loss of imprinting has been observed in some cancers; e.g., both maternal and paternal *IGF2* alleles are expressed in Wilms' tumours. In normal tissue, it is the paternal *IGF2* allele alone that is functional. *IGF2* (insulin growth factor 2) is a gene that has, as its name implies, growth stimulatory effects. Hence, the normal output from a single gene is increased when the maternal allele also expresses. The effect of "relaxation" of imprinting is not clear, but it may predispose to tumour formation since a gene that is not normally expressed is now functional. The story of imprinting and carcinogenesis is still in its early days. As well as explaining how tumours develop, the loss of imprinting opens the potential for a future line of treatment since re-establishing the imprint, if this were possible, would allow the additional gene that is expressing to be turned off.

Imprinting

Reference was made earlier in this chapter to parent of origin effects (imprinting) and the possible role these effects might play in genetic disorders. It is now evident that similar effects are seen in some cancers. For example, the source of two unusual human tumours is parental specific. Complete hydatidiform mole has the normal number of 46 chromosomes, but all are **paternal** in origin (for example, duplication of a haploid sperm). Ovarian teratoma is the opposite in that all 46 chromosomes are **maternally derived** (failure of extrusion of the second polar body). Some cancers that form part of the inherited syndromes (e.g., osteosarcoma, Wilms' tumour, Beckwith-Wiedemann and hereditary paraganglioma) demonstrate features consistent with imprinting. At the clinical level, it is important to consider the possibility of imprinting so that counselling given to families with inherited or familial cancers is accurate. The potential for imprinting to confuse the inheritance pattern may explain why the genetic component of some tumours has remained obscure.

Carotid body tumours (also called hereditary paraganglioma) are rare tumours arising in the chemoreceptor structures in the head and neck. Linkage analysis has, for some time, identified a locus for these tumours on chromosome 11q23-qter. A number of family studies have shown a parent of origin effect; i.e., the disorder is transmitted only through males. Therefore, a mutation in the paternally derived (functional) gene produces a deficit leading to tumour formation. On the other hand, trans-

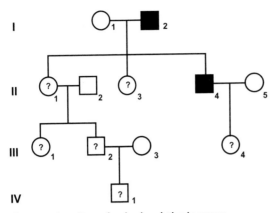

Fig. 4.15 A pedigree showing imprinting in cancer.
Four generations of a family are illustrated. Two males (I-2, II-4) presented with bilateral carotid body tumours when aged in their 30s. The consultand (II-4) has asked advice concerning risk to his daughter and two sisters. Although the tumour is transmitted as an autosomal dominant trait, risk assessment is more complex since the tumour demonstrates imprinting with full expression only occurring when there is transmission through the **male** germline. Therefore, the risk for the consultand's daughter is 50%, and she will develop the tumour if she inherits the mutant allele from her father. The same situation holds for the consultand's two sisters (II-1 and II-3). The *a priori* risk for III-1 and III-2 is 25%. However, neither individual will develop a tumour if he/she inherits a mutant gene from the mother, but he/she will still be a carrier. The *a priori* risk for IV-1 is 12.5%, and tumour development would be expected since there has been transmission from a male.

mission of the mutant gene through the maternal line does not lead to tumour formation because of maternal imprinting; i.e., the gene is not active when it is transmitted by the mother, and so a mutation in the inactive gene will have no effect. However, offspring will still be carriers and at 50% risk of having affected children if they are males (Figure 4.15).

Recently, mutations in a gene *SDHD* that codes for the small subunit (cybS) of cytochrome b in mitochondrial complex II have been found in hereditary paraganglioma. Interestingly, the gene shows biallelic expression in brain and lymphocytes, but it has not been studied in the carotid body or other paraganglioma cells. However, the imprinted gene *UBE3A* also showed biallelic expression in Angelman's syndrome, but in the brain only the maternal allele is expressed normally.

BREAST CANCER

The reader might wonder why breast cancer has been included in Chapter 4, and familial adenomatous polyposis, another genetic cancer, in Chapter 3. This is to reflect the relative complexities of the two disorders both

clinically and at the molecular level. The rare genetic disorder familial adenomatous polyposis has a clear autosomal dominant inheritance pattern in all families investigated. This is consistent with the 100% penetrance observed with *APC* gene mutations despite the fact that colon cancer in familial adenomatous polyposis has a multistep pathogenesis. There still remain unanswered questions about the *APC* gene, but from the clinical perspective, the counselling of patients and families with *APC* mutations is relatively straightforward, as are the therapeutic options; i.e., colectomy is required, with the only question being what is the best time to do this. Breast cancer provides a contrast. Even 10 years after *BRCA1* was found, its role in pathogenesis is still being determined. Issues related to genetic counselling remain complex.

Clinical Features

After non-melanoma skin cancer, breast cancer is the most commonly diagnosed cancer in women, and the commonest cause of cancer death in this group. It is second only to lung cancer as a cause of death from cancer. By 75 years of age, nearly 1 in 10 women in the USA will develop this disease. Pathogenesis of breast cancer is complex, involving physiological, environmental, lifestyle and genetic factors.

In contrast to the **inherited** cancer syndromes such as retinoblastoma or familial adenomatous polyposis described previously, **familial** cancers refer to neoplasms that cluster in families. However, because of inadequate markers to detect the predisposed phenotype, it is difficult to ascertain who is at risk. Many types of familial cancers have been reported, but the sites most commonly involved are breast, ovary, melanoma, colon, blood and brain. Clinical features that suggest a familial cancer include (1) two or more close relatives affected, (2) multiple or bilateral cancers in the same person and (3) early age of onset and clustering (e.g., occurrence of both cancer of the breast and ovary). Some facts about breast cancer, particularly of relevance to molecular medicine, are given in Table 4.12.

BRCA1 and BRCA2 Genes

Historical developments in the breast cancer DNA story include the following:

- During the late 1980s, loss of heterozygosity in breast cancer tissue was reported for a number of chromosomes.
- In 1990, breast cancer was localised to chromosome 17q21 by linkage analysis.
- During 1990–1992, the chromosome 17q location was confirmed by demonstrating loss of heterozygosity in both breast and ovarian cancer samples.

Table 4.12 **Molecular medicine facts about breast cancer**[a]

Feature	Explanations
Genetic inheritance	Of women, 5–10% have a mother or sister with breast cancer; 10–20% have a first- or second-degree relative with breast cancer. The mode of inheritance is usually autosomal dominant, and the two usual genes involved are *BRCA1* and *BRCA2*. However, mutations in these genes account for only about 2–3% of all breast cancers. Although these two genes are considered as high penetrant, women with mutations in *BRCA1* or *BRCA2* have a lifetime risk of developing breast cancer that varies from 36–85%; i.e., penetrance is far from complete in many.
Genes	*BRCA1*, *BRCA2*, Li Fraumeni syndrome (*TP53* gene), Cowden's syndrome (*PTEN* gene). Breast cancer may be found more frequently in ataxia telangiectasia (*ATM* gene), but this is controversial.
BRCA3[b] locus	There is at least one other high penetrance breast cancer gene, but to date the actual gene(s) remains elusive; hence, BRCA3 presently refers to non-*BRCA1* and non-*BRCA2*.
Low penetrance genes	In contrast to *BRCA1* and *BRCA2* that are high penetrance genes (that is how they were found; i.e., these genes made a significant, albeit confusing, impact on the development of breast cancer), there is evidence that there are additional low penetrance QTL-like genes: (1) They are more common in the population; (2) They are difficult to find because there are multiple; (3) Each has a small additive effect on the phenotype. A number of reports based on case-control (association) studies have suggested possible candidate genes, but these reports are speculative.
Features to suggest a genetic basis for breast cancer	(1) Autosomal dominant inheritance—this may not be very apparent especially in males with *BRCA1* mutations who might not develop disease. (2) Bilateral breast cancer, or both breast and ovarian cancer. (3) Early onset breast cancer. In addition, a family history of one or more males with breast cancer or Ashkenazi Jewish background suggests underlying *BRCA1* or *BRCA2* mutations.
Other risk factors	Age, previous breast disease, reproductive history, menstrual history, oestrogen therapy, radiation exposure, lifestyles, diet and alcohol intake.

[a] Information derived from and Wooster and Weber 2003. [b] The convention is for the symbols for human genes to be written in italicised capitals. Genes for animals are italicised small case. Proteins encoded by genes can have p in front of them, and/or are written in small/uppercase but not italicised. Hence, BRCA3 is not an actual gene but a location.

- In 1994, the *BRCA1* gene was cloned, and the BRCA2 locus on chromosome 13q12-q13 was identified. The potential for a third locus (BRCA3) was raised although in 2004 this remained to be confirmed by the finding of a gene(s).
- By 1995, it was shown that some sporadic ovarian cancers had mutations in the *BRCA1* gene, but no sporadic breast cancers had abnormalities affecting this gene.
- In 1995, the *BRCA2* gene was isolated.
- In 2002, microarrays demonstrated how breast cancer patients could be stratified into high and low risk. There was also some evidence that *BRCA1* and *BRCA2* mutation carriers might be distinguishable from the microarray fingerprint.

The successful finding of the first breast cancer–specific gene *BRCA1* was announced in the October 1994 issue of *Nature Genetics* as "The glittering prize." It took almost four years of intensive work, including input from the biotechnology sector, before the locus identified as BRCA1 revealed the underlying gene. By contrast, the *BRCA2* gene was found in a considerably shorter time because the discovery process was able to make use of genome sequence information; i.e., *in silico* positional cloning became possible (see Chapter 3, Appendix and Figure A.4 for more discussion about positional cloning).

A key step with *in silico* positional cloning is the annotation of the raw DNA sequence data, i.e., using sophisticated computer programs to study the raw data and then attempt to predict the likely location of genes and other important DNA landmarks such as regulatory control regions (Figure 4.16).

Subsequently, intense efforts have been directed to both *BRCA1* and *BRCA2* genes to understand the molecular issues of relevance to breast cancer. These issues include (1) the complex DNA changes that are likely to be involved in cancer, (2) the significance of DNA mutations in the pathogenesis of breast cancer, (3) the resource-intensive efforts (clinical, counselling and education) necessary to utilise effectively the information derived from these genes and (4) the additional ethical, social and educational issues that have appeared, in a relatively short time frame, in the management of a woman with breast cancer, particularly if there is a positive family history. The first two points will be dealt with in this chapter, and the remaining two issues are covered in Chapter 10.

At the molecular level, the breast cancer story is still unfolding. Both are large genes (*BRCA1* 24 exons, 1863 amino acid protein; *BRCA2* 27 exons, 3418 amino acid protein). They function as tumour suppressor genes; i.e., breast/ovarian cancer is an autosomal dominant condition with inherited mutations in either gene. The second

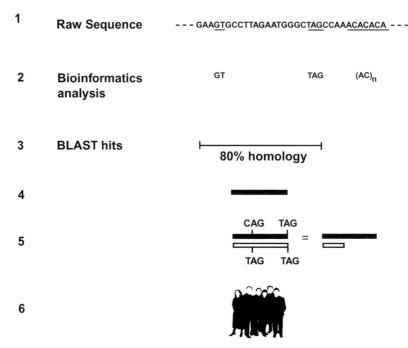

Fig. 4.16   From DNA sequence to gene discovery—DNA annotation.
(1) Raw sequence data—in this form it makes little sense. **(2)** Bioinformatics analysis of the sequence identifies a possible splice site (GT), a possible termination codon (TAG) and a run of AC repeats $(AC)_n$. The GT and TAG findings are clues that this segment of raw DNA sequence might contain the end portion (i.e., 3′ segment) of a gene. **(3)** Further bioinformatics analysis using a program such as BLAST (see Chapter 5) indicates that the region in question is about 80% homologous to a known gene—an important finding because this provides further evidence that this segment of DNA might contain a gene, and later on the sequence homology allows function to be implied through knowledge of the known gene's function. **(4)** The entire gene is now isolated (from knowing a little about the 3′ segment of the gene). **(5)** The gene is sequenced in normal and affected individuals (■ normal, □ affected). The sequencing confirms the gene to be the cause of the disease because, in the centre of the gene in the affected individuals, the CAG codon (normally, coding for glutamine; see Table 2.1) has a point mutation that converts it to TAG, i.e., a premature stop codon. This will produce a truncated protein. **(6)** The final step in positional cloning is to investigate large groups of patients to determine the frequency of this mutation and its role in pathogenesis through phenotype/genotype studies.

(normal) gene is either inactivated by a dominant-negative effect, or, as is found in breast and ovarian cancer tissue, the second (normal) allele is often deleted, leading to complete loss of function. There is some evidence that the tumour suppressor effect occurs through DNA repair. However, the story remains incomplete because mutations in the two genes do not necessarily lead to cancer. Microarray fingerprints in breast cancer patients (see Genomics in Chapter 5) suggest that those with *BRCA1* and *BRCA2* mutations might be distinguishable on the basis of their gene expression profile. These early preliminary data open up another approach to studying the role of these genes in breast cancer.

A confusing aspect of the breast/ovarian cancer story centres on the relative risks as well as penetrance for mutations in the two genes. The state of knowledge at this time implicates *BRCA1* in breast cancer as well as breast and ovarian cancer families. *BRCA2* is involved in breast cancer, but there are also associations with male breast cancer, ovarian cancer, prostate cancer and pancreatic cancer. The penetrance for *BRCA1* and *BRCA2* mutations is estimated to be between 36–85% for breast cancer and 16–60% for ovarian cancer. For the individual, these figures clearly show the mutation-positive woman to have a risk above the background population, but whether the risk is closer to 36% or 85% would be critical when considering radical treatment options such as bilateral mastectomy. However, this information is not available in most circumstances, and so important decisions become personal choices. Attempts are being made to use computer-based algorithms to estimate risk, but ultimately the value of this approach depends on accurate epidemiologic data that need to be based on a thorough understanding of the molecular pathology (see

Bioinformatics–Medical Informatics in Chapter 5 for further information about risk estimation).

Not surprisingly, considerable effort has gone into DNA mutation detection in *BRCA1* and *BRCA2*. In mid-2004, nearly 1000 mutations were listed in the Human Gene Mutation Database in Cardiff (http://www.hgmd.org/). The expected molecular heterogeneity is found, with DNA changes involving point mutations (missense, nonsense, frameshift), splicing defects, small deletions or insertions and rearrangements. The great majority of mutations produce a truncated protein. Unless there are founder effects (discussed below), only rarely does a mutation recur in unrelated individuals. Thus, the potential for widespread screening in the general population was fast disappearing, as the information on the underlying mutations accumulated. However, for individual families DNA testing for breast cancer predisposition became a realistic option if the mutation was known.

A founder effect was first observed in those of Ashkenazi Jewish origin for three mutations (*BRCA1* 185delAG, 5382insC, and *BRCA2* 6174delT) with carrier frequencies of around 0.9%, 0.3% and 1.3%, respectively; i.e., these mutations account for about 25% of the early-onset breast cancer in Ashkenazi Jewish women. Founder mutations have also been described in the Netherlands, Iceland and Sweden (www.cancer.gov/cancerinfo/pdq/genetics/breast-and-ovarian).

An interesting observation to emerge from DNA mutation testing and genotype/phenotype correlations involved *BRCA2*. It now seems likely that there is a specific region in this gene (ovarian cancer cluster region in exon 11 between ~nt 3000–nt 6600) where mutations increase the risk for developing ovarian cancer and reduce the breast cancer risk.

Mutation DNA testing was undertaken in many laboratories until 2002–2003 when problems arose in relation to the US company Myriad and its enforcement of the patent it held on *BRCA1* (the company also claims a patent for *BRCA2*). This problem resulted in some confusion whether Myriad would allow DNA testing outside its own laboratory or would licence other laboratories (see Chapter 10 for more discussion about this issue). On a more positive note, Myriad showed that with the right technology and the high-throughput DNA sequencing capability of a large laboratory (contrasting to small DNA laboratories with limited facilities), it was possible to sequence the *BRCA1* and *BRCA2* genes directly rather than use the more tedious SSCP or dHPLC scanning or even the PTT test (see the Appendix for descriptions of these tests). This had the potential to produce more effective mutation detection with a faster turnaround time, but it was expensive, and so many could not afford this test.

Sporadic and Other Forms of Breast Cancer

Early expectations following the discovery of the *BRCA1* and *BRCA2* genes were soon moderated when it became apparent that (1) Penetrance was variable. (2) Multiple mutations were making population screening impossible in most communities. (3) Testing of the individual's DNA was time consuming and expensive. (4) Finally, the disappointing finding was made that sporadic breast cancer (the more important and common form of breast cancer) was not associated with germline (i.e., blood) mutations. Somatic mutations were not detected in breast cancer tissue, and only few in ovarian tissue. Therefore, like familial adenomatous polyposis, germline mutations in rare genetic cancers did not explain the more common tumour types. What remains to be excluded is that epigenetic changes might function in the sporadic forms and through methylation of promoters, or some other mechanism, interfere with the expression of these genes. There is some indirect evidence that this may be so; i.e., mRNA for *BRCA1* is present in lower levels in breast cancer tissue.

Other genes increasing the risk for breast cancer in association with familial cancer syndromes include (1) Li Fraumeni syndrome: *TP53* mutations (see Box 4.3), (2) Cowden's syndrome: involves the gene *PTEN*, a protein tyrosine phosphatase, and is associated with tumours in the breast, gut and thyroid and (3) Ataxia telangiectasia: mutations in the *ATM* gene are associated with haematologic malignancies and perhaps a higher risk for breast cancer. In addition, there are DNA polymorphisms in genes involved in metabolism of oestrogen or environmental carcinogens, for example, the cytochrome P450 enzymes (see Chapter 5). In the latter group, the risks of breast cancer remain to be proven. Overall, when *BRCA1*, *BRCA2* and the above genes or DNA polymorphisms are considered together, they still account for only about 20% of cases associated with familial risk.

An observation with potential clinical significance relates to *TP53* and *ATM* and the fact that individuals with mutations in these genes are considered to have an increased sensitivity to ionising radiation. This is relevant since a reason for undertaking DNA mutation analysis is to enable closer surveillance of at-risk women. Presently, the usual form of surveillance, apart from self-examination of the breasts, involves mammography (i.e., repeated ionising radiation), but should this be recommended in those with changes in *TP53* and *ATM*? It would seem that each interesting development in molecular medicine leads to new knowledge but at the same time raises more questions. For the researcher, this is a challenge. For the health professionals and involved families, each new piece of the puzzle will eventually lead to more effective prevention and therapeutic options.

SOMATIC CELL GENETICS

Somatic cell genetics now joins the single gene, polygenic, multifactorial and chromosomal disorders as a distinct group of genetic disorders although the DNA changes are not inheritable. The role played by mutations in the genetic material of somatic cells is well exemplified by cancer, e.g., the familial adenomatous polyposis example described in Chapter 3 in which a series of mutations affecting somatic cells enable progression from the localised tumour to a malignant, metastasising cancer.

The potential for mutation at the somatic cell level has recently been highlighted by a novel molecular defect producing **anticipation** (increasing severity or earlier age at onset of a genetic disease in successive generations; see Figure 3.4). In the case of Huntington's disease, the fragile X syndrome and myotonic dystrophy (discussed in Chapter 3), it has been shown that anticipation is associated with a heritable and unstable DNA nucleotide triplet able to increase in size from one generation to another. The triplet repeat demonstrates both somatic instability as well as instability following inheritance through one of the parental germ cells. This novel mechanism for instability remains to be fully characterised. Somatic cell changes associated with ageing, autoimmune disease and congenital malformations are other areas of research now starting.

HAEMATOPOIETIC MALIGNANCIES

Solid tumours are initiated by two or more mutations in DNA followed by a multistep progression. In contrast, leukaemias do not generally demonstrate the random genome instability seen in the solid tumours, and they are often associated with a single non-random reciprocal chromosomal translocation event. Usually, translocations lead to tumour formation through inactivation of a tumour suppressor gene or activation of a proto-oncogene. Haematopoietic malignancies present in the first instance as an aggressive disorder or become more malignant during the course of their natural history. Access to abnormal cells in the peripheral blood or bone marrow is possible. Therefore, tumours of the haematopoietic cells have provided convenient models with which to study DNA changes during various stages of a malignancy.

Immunoglobulin Genes and T Cell Receptor Genes

Lymphocytes are unique cells since they are able to undergo somatic rearrangements of their immunoglobulin or T cell receptor genes. This is essential to generate molecules of sufficient diversity to enable recognition of the vast array of antigens to which an organism will be exposed. Thus, gene families encoding the immunoglobulin and T cell receptor genes are arranged in two configurations: (1) functionally inactive or germline state and (2) functionally active or rearranged state, each of which is unique and contributes to the polyclonal response (Figure 4.17). Immunoglobulin diversity in the B lymphocytes reflects rearrangements of the heavy chain region on chromosome 14, followed by rearrangements in the κ light chain genes and, if necessary, the λ light chain genes. The repertoire is further diversified by the addition of somatic mutations, including random insertions of nucleotides at the V-D and D-J junctions. Similar scrambling of genes occurs to form the T cell receptor repertoire.

The process of gene rearrangements is error prone. Therefore, it is possible that the immunoglobulin or T cell receptor genes can be accidentally spliced next to or into other genes, including the proto-oncogenes. One way for this to occur is from a chromosomal translocation. Following this, the cells containing the rearranged immunoglobulin or T cell receptor genes can now be driven by the associated proto-oncogene, and eventually a malignant clone arises. Should a lymphoid cell form one such clone, all sister cells will carry the hallmark of its unique gene rearrangement. This is utilised when investigating patients with haematopoietic malignancies (discussed further below).

Chronic Myeloid Leukaemia

Chronic myeloid leukaemia, which affects young adults, is a malignant clonal disorder involving a pluripotential haematopoietic stem cell. It usually presents in chronic phase, and within three to four years develops into an accelerated and then acute phase called blastic transformation. More than 95% of cases have the Philadelphia (Ph) chromosome, the result of a reciprocal translocation involving chromosomes 9 and 22 (Table 4.13, Figures 4.10 and 4.18). The fusion gene product formed from the translocation (*BCR-ABL*) contains *ABL*, a proto-oncogene with tyrosine kinase activity. However, the fusion protein demonstrates unregulated tyrosine kinase activity, leading to a number of downstream effects including increased cell proliferation, reduced apoptosis, adhesion abnormalities and genomic instability. The translocation is sufficient to produce the leukaemia. During development of blastic transformation, additional DNA changes develop. These changes can involve *TP53* or *RB1* (retinoblastoma).

Until recently, the only curative treatment for chronic myeloid leukaemia was an allogeneic stem cell bone marrow transplant. However, this procedure is not avail-

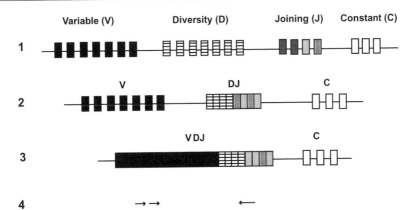

Fig. 4.17   Immunoglobulin genes in the germline and how they rearrange.
During development of a stem cell into a B or T lymphocyte, there are rearrangements of the germline immunoglobulin genes (which number in the hundreds). This rearrangement generates the diversity in immune proteins necessary for effective antigen recognition. **(1)** The different immunoglobulin heavy chain genes are V—variable; D—diversity; J—joining; C—constant. **(2)** The first recombination in the heavy chain locus involves a D to J step. **(3)** This step is then followed by V to D-J recombination. **(4)** To detect these rearrangements, DNA primers for PCR are based on regions that are known to be conserved (→ ←). Similar rearrangements occur with the immunoglobulin light chain genes (λ, κ) and the T cell receptor genes.

Table 4.13   Translocations and gene changes in the haematologic malignancies (from Gribben 2002)

Disorder	Translocation detectable by cytogenetics	Gene change detectable by PCR
Chronic myeloid leukaemia (CML)	t(9,22)(q34;q11)[a] Occurs in about 95% of CML, 25% adult ALL,[b] occasional childhood ALL and rarely in AML	Juxtaposes the *BCR* gene (chromosome 22) and the *ABL* proto-oncogene from chromosome 9.
Acute promyelocytic leukaemia (APML)	t(15;17)(q21;q21)	Juxtaposes *RARα* gene (chromosome 17) and the *PML* gene (chromosome 15).
Follicular lymphoma (85%) and diffuse lymphoma (30%)	t(14;18)(q32;q21)	*BCL2* proto-oncogene on chromosome 18 becomes juxtaposed to IgH locus on chromosome 14.
B cell CLL, myeloma, mantle cell lymphoma	t(11;14)(q13;q32)	*BCL1* proto-oncogene on chromosome 11 is juxtaposed to IgH.
Burkitt's lymphoma, B cell ALL	t(8;14)(q24;q32)	Juxtaposes exons 2 and 3 of proto-oncogene *MYC* and IgH.

[a] This terminology to describe a cytogenetic rearrangement means that there is a translocation (t) between chromosomes 9 and 22. The position on 9 is q34 (long arm band 34), and on 22, it is q11 (long arm band 11). [b] Abbreviations: CML—chronic myeloid leukaemia, ALL—acute lymphoblastic leukaemia, AML—acute myeloid leukaemia, APML—acute promyelocytic leukaemia, IgH—immunoglobulin heavy chain, CLL—chronic lymphocytic leukaemia.

able for a large number of patients, and there is significant mortality and morbidity associated with transplantation. During the late 1990s, a new drug was designed specifically to interfere with tyrosine kinase activity. The drug (imatinib) is one of the first designer drugs based on molecular knowledge of aetiology and has proven to be very successful. Imatinib could become the treatment of choice before marrow transplantation. Studies are presently under way to assess the long-term efficacy of this drug, and whether it alone can cure this disease, or it needs to be used in conjunction with other therapeutic options for chronic myeloid leukaemia, a condition that was once considered an incurable disease.

Acute Promyelocytic Leukaemia

Acute promyelocytic leukaemia is a rare variant of acute myeloid leukaemia involving the promyelocyte cells. In addition to the usual leukaemia-related problems, patients with acute promyelocytic leukaemia are at risk of severe bleeding due to clotting factors being deficient, which exacerbates the effect of thrombocytopenia

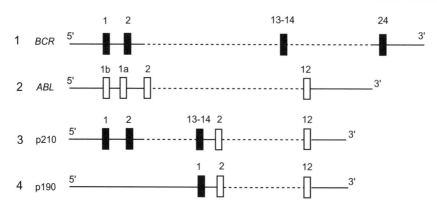

Fig. 4.18 Molecular (DNA) changes involving proto-oncogenes from the translocation producing the Philadelphia chromosome.
(1) Depicted is the *BCR* (breakpoint cluster region) gene (■). It has 24 exons. In chronic myeloid leukaemia, the break in this gene is around exon 13 or 14. **(2)** The *ABL* proto-oncogene (□) has the tyrosine kinase activity, and following the translocation, it is interrupted 5′ to exon 2. **(3)** The fusion gene for chronic myeloid leukaemia encodes for a 210 kd protein (p210) and comprises exons 1 to 13–14 of *BCR* (■) and exons 2–12 of *ABL* (□). **(4)** A smaller fusion protein (p190) is found with acute lymphoblastic leukaemia occurring in adults with the Ph chromosome. This fusion gene involves exon 1 of *BCR* (■) and exons 2–12 of *ABL* (□).

normally found in acute leukaemia. In terms of molecular medicine, acute promyelocytic leukaemia is a useful model to illustrate how molecular understanding of pathogenesis can be translated into the development of more appropriate therapy regimens and new designer drugs. Like chronic myeloid leukaemia, acute promyelocytic leukaemia is associated with a particular translocation (see Table 4.13). In the great majority of cases, this translocation disrupts two genes and leads to the formation of a fusion protein PML-RARα (*PML*—a putative transcription factor implicated in a number of cellular processes, including apoptosis, growth regulation, tumour suppression, RNA processing; *RARα* is the retinoic acid receptor alpha gene). *RARα* can fuse with a number of other genes apart from *PML* (<5% of cases), but for the purpose of this discussion, we will focus only on *PML*. A key activity of the *RARα* gene involves the neutrophil differentiation pathway, so inactivation of this gene's function through the translocation-produced fusion protein not surprisingly leads to maturation arrest at the promyelocyte stage. The fusion protein is considered to have a number of actions, including a dominant negative effect on the normal gene product. Unlike the Ph chromosome that appears to be solely involved in the development of chronic myeloid leukaemia, there is some evidence from transgenic animal studies showing incomplete penetrance and long latency periods that other genetic factors may be important. Nevertheless, the translocation is a key component in pathogenesis.

Knowledge of the molecular pathology has now been applied in developing novel treatments for this leukaemia. These treatments comprise inhibitors of the PML-RARα fusion protein. One particularly effective

drug is ATRA (all *trans*-retinoic acid). ATRA works through binding to PML-RARα, thereby inhibiting its downstream effects, as well as inducing degradation of this fusion protein.

Like the chronic myeloid leukaemia example, a new therapeutic has been designed specifically for the mutant gene products in this leukaemia. The remission rate for acute promyelocytic leukaemia has now dramatically improved, particularly when ATRA is used in combination with chemotherapy (complete remission of about 90%). However, ATRA is not without complications, and so molecular or cytogenetic confirmation is essential to be sure that the appropriate translocation has occurred in this leukaemia. Another future challenge is the monitoring and post-treatment management to ensure a long-term remission or cure. For this, molecular testing is needed to detect minimal residual disease, with evidence accumulating that the longer term outlook may be improved if treatment is started in early relapse (detected molecularly) instead of waiting for a full haematologic relapse.

Minimal Residual Disease

Minimal residual disease refers to submicroscopic disease, i.e., disease that remains occult within the patient but eventually leads to relapse. A patient's response to treatment for acute leukaemia is influenced by many factors (Box 4.4). At the time of diagnosis of leukaemia, the tumour burden is high (up to 10^{12} leukaemic cells), and even at complete remission, the traditional microscopic approaches have limited capability to detect residual disease, and even a complete

(1) Host: Age, physical state of well-being and genetic background (pharmacogenetic response to drugs). Apart from age, these factors are difficult to measure but can make important contributions to the treatment outcome. **(2) Leukaemia:** The type of leukaemia, the tumour burden (there can be 10^{12} leukaemic cells at the time of diagnosis) and the development of drug resistance in these cells. **(3) Treatment:** The drug dosage (influenced by the genetic ability to metabolise drugs—see Pharmacogenetics in Chapter 5), as well as any drug-drug interactions that may occur. **(4) Minimal residual disease:** There is growing evidence in a number of haematologic malignancies (including myeloma) of benefit that comes from post-treatment monitoring for minimal residual disease. In children with acute lymphoblastic leukaemia receiving identical treatment, poor prognostic features include the following: male, white blood count $>50 \times 10^9$/litre, aged less than 1 year or older than 10 years, the Ph chromosome and slow initial response to treatment. Minimal residual disease monitoring can identify good and bad prognostic indicators that are independent of known clinical and biological prognostic markers. For example, detectable minimal residual disease at any stage following initial treatment is a poor prognostic indicator in acute leukaemia. Similarly, patients undergoing allogeneic bone marrow transplants for leukaemia have a poorer long-term outlook if minimal residual disease is detected post-transplant. In adults with chronic myeloid leukaemia, a progressive ongoing reduction in tumour mass (determined by PCR detection of the *BCR-ABL* fusion gene) can still be observed even a year after achieving complete remission with imatinib treatment. Patients with such a decline in tumour burden do better, in terms of disease-free survival, than those who do not show this trend.

remission based on microscopy can be associated with a residual 10^8 to 10^{10} leukaemic cells. In this circumstance, it is understandable that relapse can occur. Therefore, to improve response to treatment (and even to select the treatment of choice), minimal residual disease monitoring has become an important component of modern therapy for leukaemia. Minimal residual disease monitoring is not readily available with the solid malignancies because with the leukaemias, the blood (or bone marrow) provides a source of accessible tissue for monitoring.

Minimal residual disease detection was first attempted with polyclonal or monoclonal antibodies. However, some of the antigens detected by these antibodies were also present on normal or precursor cells, and so better methods were needed. Today, two useful approaches for

detecting minimal residual disease are (1) PCR and (2) flow cytometry. Each method has its own strengths and weaknesses in particular situations. PCR is valuable because it is very sensitive (for example, it can detect 1 leukaemic cell in 10^3–10^8 normal cells). PCR primers can be designed to detect fusion transcripts or immunoglobulin/T cell gene rearrangements. More recently, quantitation based on real-time Q-PCR (see the Appendix) has proven to be a valuable new addition to the monitoring of minimal residual disease.

A variation of the minimal residual disease theme involves the monitoring of progress following bone marrow transplantation from another donor (allogeneic marrow). This type of treatment is used in a number of leukaemias and bone marrow aplasias. Early identification of engraftment (presence of donor cells) or relapse (presence of host cells) is important to optimise post-transplantation treatment. However, the small number of cells usually present makes conventional diagnostic approaches difficult. PCR is ideally suited to this situation provided primers can be designed to detect DNA sequences that will distinguish host and donor cells. The extreme sensitivity of PCR means that only a few cells are necessary for assay, and so the dilemma of engraftment versus relapse can be resolved early, and appropriate treatment started.

As discussed previously, the exciting novel therapies for chronic myeloid leukaemia and acute promyelocytic leukaemia have made a dramatic impact on the initial remission rate. However, what to do next remains an open question, and to answer this, minimal disease monitoring will be important in ongoing clinical trials. Only about 50% of acute leukaemias have clearly identifiable chromosomal breakpoints, and so new targets for minimal residual monitoring need to be identified, particularly for the acute myeloid leukaemias.

DIAGNOSTIC APPLICATIONS

Microscopy

The standard approach in anatomical pathology is microscopic examination of a tissue section with one or more stains. A diagnosis is made on the basis of cell morphology and staining characteristics. A greater level of resolution is possible with electron microscopy. Another development is immunophenotyping, which allows the identification of specific antigens by staining with monoclonal or polyclonal antibodies. More recently, to these investigative approaches can be added *in situ* hybridisation using DNA probes to detect specific sequences. An illustration of the value of *in situ* hybridisation is provided by the human papillomaviruses. A description of these viruses, their detection and association with genital tract lesions is found in Chapter 8. Diagnosis and typing of such viruses have become simplified and more accurate with tests such as DNA amplification by the polymerase chain reaction (PCR). In the research laboratory, the availability of *in situ* hybridisation has allowed the tissue and cellular localisation of mRNA transcripts from these oncogenic viruses to be determined.

Fine Needle Aspiration

A useful source of tissue for histopathological examination is obtainable through fine needle aspiration. This procedure is more rapid and less traumatic than conventional biopsy and carries less risk of tumour dissemination. The problem of whether sufficient material can be obtained for histological assessment has now been overcome to some extent by PCR. Another advantage of PCR is that archival materials such as formalin-fixed, paraffin wax–embedded tissue blocks are suitable for DNA testing. Laser-capture microscopy is a new development that allows individual cells in a sample to be studied, thereby avoiding the contaminating effect of nearby normal cells.

Solid Malignancies

As indicated in Chapter 2, there are different classes of DNA tests (see Table 2.8). In terms of the solid malignancies, the ones considered most often in the clinical scenario are familial adenomatous polyposis, hereditary nonpolyposis colon cancer and breast cancer. In addition, there are many examples of rare tumours for which DNA testing options are available. Having described the complexities involved in (1) finding the relevant germline mutations in the above cancers and (2) understanding the clinical implications of the various mutations, it is not surprising that it is universally recommended that these tests should be undertaken with considerable care and thought.

Molecular testing in solid malignancies generally involves **predictive** DNA testing because germline mutations are involved. Therefore, this class of tests has clinical, social, ethical and legal implications for the patient (and family). In many centres, the ordering of these tests is restricted to specialised clinics that can provide the necessary experience as well as appropriate genetic counselling. Guidelines have also been developed to assist the health professionals in deciding on what are the risk scenarios or risk family structures, and what should be done, or when to refer these problem cases. In other centres, these tests are available more on a user-pay basis, and this approach can lead to problems if the necessary counselling and support are not provided. It should also be noted that knowledge of these genes and the significance of any mutation may change as experience builds. This means that the indications for testing are not always straightforward, and long-term follow-up may be required.

The implications for DNA testing for germline *TP53* mutations are more complex than with other genes that predispose to cancer (e.g., *APC*, HNPCC in colon cancer) because *TP53* does not produce tumours of a certain type. The broad range of neoplasms found with *TP53* makes clinical screening and follow-up difficult. Hence, the balance between resources, costs and clinical effectiveness needs to be considered with even greater care. *TP53* mutation analysis continues to be an essential research tool and increasingly has a role to play in epidemiological studies. In terms of individual disorders, the merits of each case should be considered.

Leukaemias and Lymphomas

A similar strategy to that described in anatomical pathology is usually followed in the leukaemias and lymphomas, i.e., morphologic appearance, staining characteristics and immunophenotyping. Knowledge of cytogenetic changes, particularly translocations, provides additional information about diagnosis and prognosis. DNA changes can also be sought (see Table 4.13). In contrast to the DNA tests for solid malignancies described earlier, the haematological tests fall within their own class since they have implications only for the patient who already has established disease, and there are no direct (genetic) consequences for family members (see Table 2.8).

Compared to conventional cytogenetic tests, the sensitivity of the molecular approaches (particularly PCR) is high; e.g., 1 cell in 100 000 with the Philadelphia chromosome can be identified. This was discussed earlier under Minimal Residual Disease. PCR is particularly suitable for the identification of translocation breakpoints since many of them have specific, invariant sequences associated with them, and so the appropriate amplification primers are easier to design. The sensitivity of PCR in this situation is high; e.g., a leukaemic cell mixed with 10^5 to 10^6 normal cells is distinguishable.

In the lymphoproliferative disorders (lymphocytic leukaemias or lymphomas), an additional molecular approach is available through examination of immunoglobulin and T cell receptor genes. The finding of gene rearrangements usually means a clonal population is associated with a haematologic or immunologic disorder. However, the finding of a clonal population does not necessarily mean it is malignant. The latter conclusion will be made on the basis of gene, haematologic and clinical criteria. A disadvantage of the PCR technique in this situation is the false negative result. This result does not reflect a deficiency in the technology, but the failure of the PCR primers to detect the underlying gene rearrangements.

The designing of DNA primers for PCR involving immunoglobulin or T cell receptor genes is particularly difficult for a number of reasons: (1) the complexity of the germline structure, (2) the occurrence of DNA rearrangements, (3) additional somatic mutations once rearrangements have occurred, (4) the possibility of incomplete or aberrant gene rearrangements and (5) clonal evolution in some leukaemias. To avoid some of these problems, primers to conserved sequences within

V_H and J_H are selected. Amplification will not be obtained from germline sequences since they are too far apart (see Figure 4.17). Most monoclonal populations found in the lymphoproliferative disorders are eventually detectable, but the yield is not as high as PCR-based detection of known translocations.

FUTURE

MULTIFACTORIAL DISEASES

As indicated earlier, the important public health issues that are now being faced by many communities involve the multifactorial disorders. Resources and effort in epidemiology, molecular medicine and bioinformatics will be needed to ensure that molecular medicine–based approaches enable strategies to be developed that will lead to more effective prevention and treatment. Obesity will be used as an example of a contemporary challenge involving this complex group of conditions.

Obesity

The World Health Organisation has proclaimed obesity to be a global epidemic. In the USA, 61% of adults are considered to be overweight (body mass index 25–29.9), and 26% are obese (body mass index ≥30). Obesity is no longer a disease of Western society but afflicts many communities. Increased morbidity and mortality reflect the associated disorders that go with obesity, including diabetes, hypertension, coronary artery disease and arthritis. Apart from surgical approaches to obesity, no definitive forms of treatment will lead to consistent and significant weight loss. Indeed, the common approaches of diet and increased exercise have, in general, relatively small effects, and in most cases they are only temporary measures.

The aetiological basis for obesity is similar to what has been described for many of the multifactorial disorders. This involves a core genetic component to which are added various environmental and lifestyle influences. For obesity, twin studies would suggest that the genetic component is substantial (50–90%) since concordance rates for body mass index between DZ and MZ twins are 32% and 74%, respectively. Human genome maps for obesity have implicated many genes and numerous loci. The 2003 update (Snyder *et al* 2004) lists 43 mendelian syndromes relevant to human obesity, with 24 of these syndromes having relevant genes or suitable candidates identified. However, the mendelian syndromes associated with obesity are rare, and so in terms of complex genetic inheritance, the 208 human QTLs in this map are particularly relevant. These QTLs are important because they are likely to be involved in the common, sporadic

forms of obesity. Overall, the 2003 human obesity gene map lists more than 430 genes, markers and chromosomal regions thought to be associated with the obesity phenotype.

Why has the human evolved with many genes that are proving to be detrimental to health in our society? One explanation is that over a long evolutionary period, the environment has generally been inhospitable in terms of food availability. Therefore, natural selection would provide a greater survival advantage for humans who were able to eat ravenously during times of food being available and/or able to store ingested calories more effectively to protect against times when food was scarce. However, in modern society the environment has changed, allowing much easier and prolonged access to high caloric food, and humans are less active physically. This environment is less compatible with the gene pool that has evolved, and the consequence is obesity.

A hypothesis for obesity, known as the lipostatic model, involves the central nervous system as the key regulator of food intake (appetite) and energy expenditure (thermogenesis). Afferent signals in this pathway come from insulin and leptin, with the plasma levels in these hormones reflecting the body fat content. Acting via the central nervous system, insulin and leptin exert a catabolic effect; i.e., they lead to a reduction in food intake by suppressing appetite, and they promote thermogenesis by increasing energy expenditure. Overall, this has a negative effect on weight gain (Figure 4.19). The insulin and leptin pathway acts via hypothalamic melanocortin signalling, and the catabolic output results from release by the hypothalamus of thyrotropin releasing hormone (TRH), corticotropin releasing hormone (CRH) and oxytocin. Evidence that the melanocortin pathway is important comes from the rare Mendelian disorders associated with mutations in single genes that lead to gross obesity phenotypes (Table 4.14).

In practical terms, how will molecular medicine contribute to the control of the obesity epidemic? The demonstration that complex pathways and mechanisms are likely to be involved will make it necessary to revisit the contemporary view that obesity is a product of overeating and lack of exercise. While these factors are crucial contributors, they alone may not provide the answer to the control of obesity. From the therapeutic

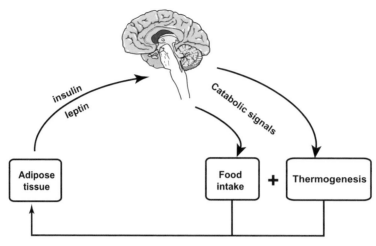

Fig. 4.19   Schematic representation of central (brain) control of body weight (the lipostatic model of body weight regulation; Cummings and Schwartz 2003).
Depicted is a simplified version of catabolic pathways involved in body weight regulation. Two key afferent signals are insulin and leptin. In the blood, the level of these hormones reflects the body fat content, and both act by decreasing appetite and increasing thermogenesis. Although leptin appears to be the more important of the two, it was disappointing to find in obesity that leptin was increased (as expected), but this did not produce an anti-obesity effect; i.e., obesity is a state of leptin resistance. In the hypothalamus, leptin works through the melanocortin signalling pathway. Key genes in this pathway include *POMC* (proopiomelanocortin). The protein from this gene is cleaved by *PC1* (prohormone convertase 1) into α melanocortin (*αMSH*), which works via melanocortin-4 receptors (*MC4r*). The final (catabolic) output from the hypothalamus comes from the release of thyrotropin-releasing hormone (TRH), corticotropin-releasing hormone (CRH) and oxytocin. The catabolic pathway described is also matched by an anabolic one; i.e., this pathway increases food intake and reduces thermogenesis. Leptin also acts on this pathway and inhibits it.

Table 4.14   Human obesity syndromes caused by single gene mutations involving the leptin-melanocortin signalling pathway

Gene	Mutations	Features
LEP	6 mutations described (rare)	Leptin gene. *LEP* and the next three all give rise to extreme hyperphagia and severe childhood onset obesity. Autosomal recessive mode of transmission.
LEPR	3 mutations described	Leptin gene receptor.
POMC	2 mutations described	Proopiomelanocortin (POMC) is associated with catabolic activity and is located in the hypothalamic arcuate nucleus.
PC1	1 mutation described	POMC is cleaved by prohormone convertase 1 (PC1) into αMSH, which exerts its effect via melanocortin-4-receptors (MC4R).
MC4r	>70 mutations described	See PC1. *Mc4r* is located within the hypothalamic paraventricular nucleus (also called the satiety centre) because lesions here produce obesity and hyperphagia. Autosomal dominant mode of inheritance. Phenotype is typical of more common forms of obesity.
SIM1	1 mutation described	Produces a phenotype like *Mc4r* deficiency. *SIM1* is important for formation of the paraventricular nucleus.

viewpoint, little of importance has emerged in the form of drugs that have a significant impact on weight gain or loss. Surgical intervention is used, but there is considerable evidence that this simply leads to deposition of fat at other sites, because the impaired central control leading to overeating and reduced thermogenesis has not been corrected. The molecular analysis of obesity will provide a new opportunity to understand the molecular pathogenesis and, from this, design drugs that specifically target genes such as those in the leptin-melanocortin signalling pathway that have been described.

NOVEL MECHANISMS FOR DISEASE

Epigenetics

Our understanding of how normal genes function and how genetic disorders arise has been very much influenced by the single gene Mendelian disorders; i.e., gross quantitative differences in terms of genes working or not working are used to explain phenotypic changes. In contrast, human development involves a complex interplay of genes that express at specific times. Apart from actual changes in DNA sequence, other mechanisms for controlling gene expression involve epigenetic DNA effects such as methylation and the differential expression of maternal and paternal alleles (imprinting). A number of fundamental questions about imprinting remain to be answered. For example, how and when does imprinting occur? The mouse model, including transgenic mice, has provided some information through the identification of imprinted genes and chromosomal loci that can now be studied. Methylation has also been implicated as an important factor modifying the phenotype in the imprinting process. Answers on imprinting are still being sought, and it is expected that new knowledge will provide greater insight into the complex regulation of human genes. Epigenetics and its effect on human development are discussed further in Chapter 7.

Epigenetic control of cancer cells (as well as loss of this control) adds an additional level of complexity in the multistep hypothesis used to explain carcinogenesis. For knowledge in this area to proceed further will require more basic knowledge of epigenetic control of gene expression, as well as better bioinformatics tools to allow sophisticated analysis that takes into account complex combinatorial models that include epigenetic regulation. Gene silencing, through epigenetic changes, provides an interesting contrast to gene silencing secondary to a DNA mutation, since the former mechanism is potentially reversible. Hence, further understanding of epigenetic changes in cancer may provide alternative targets for therapy.

ONCOGENESIS

Genes to Watch

The next few years will be accompanied by a better understanding of the various molecular events involved in the cancer pathway. An interesting recent observation has suggested that *TP53* may play a role in preventing teratogenic effects during development. This focuses on the importance of understanding the normal physiological functions of tumour-associated genes, as well as the changes that occur following mutations in their DNA. Many more cancer genes need to be isolated, and even more work is needed to understand their significance. For

example, a number of genes have been implicated in prostate cancer, and an even greater number of susceptibility loci have been identified in this common disease of males. However, the significance of these changes for diagnostic, prognostic and therapeutic purposes is still far from being understood.

A recent observation linking *BRCA2* to sporadic breast cancer involves a novel protein product of the gene *EMSY*. *EMSY* maps to chromosome 11q13.5, a locus that is amplified in 13% of sporadic breast and 17% of sporadic ovarian cancers. The connection with *BRCA2* occurs via exon 3, which binds the EMSY protein. Overexpression of *EMSY* leads to inactivation of *BRCA2*, which would be analogous to a mutation in the latter. This interesting finding needs further study.

The ends of chromosomes (telomeres) comprise TTAGGG repeats bound to proteins. This protects the chromosome from degradation. As cells divide, the telomeres shorten because of loss of TTAGGG repeats. When telomere shortening reaches a critical length, the cell dies. To avoid this, the cell has a telomerase enzyme that adds TTAGGG onto pre-existent telomeres. In cancers, telomeres are usually short, and so these abnormal cells would die in the normal course of events, except that cancers can up-regulate telomerase activity. There are other intriguing ways in which telomerase might promote tumourigenesis, including giving a survival advantage to these cells independent of telomere length. As knowledge is gained about telomeres, it is likely that this will have implications for cancer, including novel targets for therapy (for example, inhibiting telomerase), as well as ageing, since telomere shortening is also observed in the normal ageing process.

Metastasis

Most cancer patients die as a result of metastatic effects, rather than the primary tumour. Hence, it is disappointing that relatively little is known about the molecular basis for metastasis, particularly as there is some evidence that metastases occur earlier in tumour development than previously was thought, but for some reason the metastatic cells have a poorer growth potential than normal cells. Indeed, the presence of tumour cells at distant sites to the primary does not necessarily mean that they will develop into metastatic foci. The relative inefficiency for metastasis was initially considered to reflect the early phase in metastasis when tumour cells invaded blood vessels and travelled to distant sites. Now, it is proposed that the rate-limiting step occurs in the growth of the tumour once it reaches the secondary site.

Metastasis suppressor genes are potential genetic mechanisms that explain how the growth of metastatic cells is regulated at distant sites. Unlike the convention tumour suppressor genes, the metastasis suppressor genes are more likely to be turned off by epigenetic modifications than mutations, and although these genes can suppress the *in vivo* growth of metastases, they do not affect primary tumour growth. There are only about seven metastatic suppressor genes described, and there is a long way to go before the biology of metastasis is clearly understood. From this knowledge will come better ways in which to understand the natural history of each cancer and new drugs to target specifically the metastatic cells.

The Forgotten Tumour

Although the number one killer from cancer, lung cancer does not have the same high public profile compared to breast cancer (as a cancer of women) and prostate cancer (as a cancer of men). Nevertheless, its association with cigarette smoking should make molecular analysis easier because the environmental contributor to carcinogenesis is well defined, unlike most other solid cancers. Despite this, it is disappointing that relatively little is known about causation in this tumour. The usual loss of heterozygosity at different loci (suggesting somatic loss or inactivation of tumour suppressor genes) and a long list of proto-oncogenes shown to be mutated in lung cancer provide no major clues to what specifically is happening at the molecular level.

Research in lung cancer is more difficult because there are no examples of genetically acquired disease like the rare genetic cancers found with bowel and breast, and so there are fewer clues where to look. Nevertheless, the challenge to defeat lung cancer is there, and it will be interesting to see where the breakthroughs will emerge. It might be said—why bother, when there is an environmental carcinogen involved? Admittedly, the public health preventative measures must remain a priority in this tumour. But, there are also questions that need to be addressed, such as why can cancer develop years after one stops smoking? What is the contribution from and how does passive smoking inhalation lead to lung cancer? Why can some people smoke heavily and never develop lung cancer? As shown in the chronic myeloid leukaemia and acute promyelocytic leukaemia examples, knowledge of molecular pathogenesis can also lead to novel drug therapies.

Chromosomal Translocations

The predictable association between a number of haematopoietic malignancies and chromosomal translocations has been an important development in understanding cancer. The availability of PCR has enabled minimal residual disease to be sought by looking for these translocations. However, PCR has also shown that normal individuals with no haematological findings of leukaemia or lymphoma can have the same chromosomal translocations. It remains possible that these indi-

viduals will develop a lymphoproliferative disorder some time in the future, but for the present, this finding challenges the premise that translocations are sufficient for the development of a tumour. Another explanation is that PCR is not detecting these changes in functional cells but in fragments of extracellular DNA that are not biologically active.

A similar finding has recently been observed in children with SCID (severe combined immunodeficiency) treated by gene therapy and subsequently complicated by the development of T cell leukaemia (see Chapter 6). In two children, an insertional event leading to activation of the proto-oncogene *LMO2* is thought to have caused the leukaemia, although it is likely that there are other contributing factors in aetiology (see Table 6.11). In the follow-up of other treated children, one has been shown to have what is likely to be *LMO2* inactivation, but to date this child has not developed leukaemia. More information is needed about these chromosomal and gene rearrangements, and why most produce a leukaemia or lymphoma, but in rare cases this does not appear to have happened.

FURTHER READING

Introduction

Botstein D, Risch N. Discovering genotypes underlying human phenotypes: past successes for Mendelian disease, future approaches for complex disease. Nature Genetics 2003; 33(suppl):228–237 (*provides a detailed look at the ways in which Mendelian disease–causing genes were discovered and the challenges ahead in attempting to discover genes for the complex genetic disorders*).

Ewens WJ, Spielman RS. Locating genes by linkage and association. Theoretical Population Biology 2001; 60:135–139 (*provides a more mathematically based discussion of linkage and association studies*).

Lewis CM. Genetic association studies: design, analysis and interpretation. Briefings in Bioinformatics 2002; 3:146–153 (*gives a nice, simple overview of an association study*).

Multifactorial Diseases

Cloninger CR. The discovery of susceptibility genes for mental disorders. Proceedings of the National Academy of Sciences USA. 2002; 99:13365–13367 (*brief but useful summary of major advances in the molecular genetics of schizophrenia*).

Gerber DJ et al. Evidence for association of schizophrenia with genetic variation in the 8p21.3 gene, *PPP3CC*, encoding the calcineurin gamma subunit. Proceedings of the National Academy of Sciences USA. 2003; 100:8993–8998 (*original report for the most recent of the schizophrenia genes discovered*).

Harrison PJ, Owen MJ. Genes for schizophrenia? Recent findings and their pathophysiological implications. Lancet 2003; 361:417–419 (*excellent review that critically analyses the likely significance of the various schizophrenia candidate genes*).

Kelly MA, Rayner ML, Mijovic CH, Barnett AH. Molecular aspects of type 1 diabetes. Molecular Pathology 2003;5 6:1–10 (*gives overview of molecular associations in IDDM*).

Nussbaum RL, Ellis CE. Alzheimer's disease and Parkinson's disease. New England Journal of Medicine 2003; 348:1356–1364 (*useful molecular overview of these two important diseases*).

Pedersen NS, Smith E. Prion diseases: epidemiology in man. APMIS 2002; 110:14–22 (*brief but useful review of the different prion diseases*).

Wong AHC, van Tol HHM. Schizophrenia: from phenomenology to neurobiology. Neuroscience and Biobehavioral Reviews 2003; 27:269–306 (*an extensive overview of schizophrenia including clinical, molecular and treatment options*).

Novel Mechanisms for Disease

Arantes-Oliveira N, Berman JR, Kenyon C. Healthy animals with extreme longevity. Science 2003; 302:611 (brief overview of daf2 *C. elegans* mutants and longevity).

Finch CE, Ruvkun G. The genetics of aging. Annual Reviews of Genomics and Human Genetics 2001; 2:435–462 (*comprehensive review of ageing including the genetics and metabolic pathways involved*).

Graff C, Bui T-H, Larsson N-G. Mitochondrial diseases. Best Practice & Research Clinical Obstetrics and Gynaecology 2002; 16:715–728 (*succinct review of mtDNA and its mutations*).

Lee SS, Lee RYN, Fraser AG, Kamath RS, Ahringer J, Ruvkun G. A systematic RNAi screen identifies a critical role for mitochondria in *C. elegans* longevity. Nature Genetics 2003; 33:40–48 (*original article describing how RNAi was used to detect mtDNA mutants, which prolonged life*).

Murphy CT *et al.* Genes that act downstream of DAF-16 to influence the lifespan of *Caenorhabditis elegans*. Nature 2003; 424:277–284 (*easy-to-follow original report of how pathways and genes influencing life span can be discovered by microarrays*).

Walter J, Paulsen M. Imprinting and disease. Seminars in Cell & Developmental Biology 2003; 14:101–110 (*excellent overview of imprinting*).

Youssoufian H, Pyeritz RE. Mechanisms and consequences of somatic mosaicism in humans. Nature Reviews Genetics 2002; 3:748–758 (*succinct and readable review of what can be a difficult topic to understand*).

www.mitomap.org/ A Human Mitochondrial Genome Database (*web address for MITOMAP, an Internet site that provides information as well as catalogues the various mutations in mtDNA*).

www.mgu.har.mrc.ac.uk/imprinting/imprinting.html The web site for the Mammalian Genetics Unit in Harwell UK (*provides information on imprinting, including resources and references*).

Oncogenesis

Baak JPA, Path FRC, Hermsen MAJA, Meijer G, Schmidt J, Janssen EAM. Genomics and proteomics in cancer. European Journal of Cancer 2003; 39:1199–1215 (*comprehensive review of genes and proteins in cancer*).

Bodmer WF. Cancer genetics. British Medical Bulletin 1994; 50:517–526 (*early description of cancer as a genetic disease*).

Coultas L, Strasser A. The role of the Bcl-2 protein family in cancer. Seminars in Cancer Biology 2003; 13:115–123 (*overview of bcl2 and apoptosis*).

Herman JG, Baylin SB. Gene silencing in cancer in association with promoter hypermethylation. New England Journal of Medicine 2003; 349:2042–2054 (*nice review on DNA methylation and its potential effects in cancer*).

Ponder BAJ. Cancer genetics. Nature 2001; 411:336–341 (*provides overview of genetic and non-genetic contributions to tumour development*).

Schultz DR, Harrington WJ. Apoptosis: programmed cell death at a molecular level. Seminars in Arthritis and Rheumatism 2003; 32:345–369 (*useful review of apoptosis*).

Suter CM, Martin DIK, Ward RL. Germline epimutation of *MLH1* in individuals with multiple cancers. Nature Genetics 2004; 36:497–501 (*interesting study showing how epimutations can produce cancer*).

Vermeulen K, Berneman ZN, Van Bockstaele DR. Cell cycle and apoptosis. Cell Proliferation 2003; 36:165–175 (*excellent overview of overlapping proteins and pathways that control the cell cycle and apoptosis*).

Wooster R, Weber BL. Breast and ovarian cancer. New England Journal of Medicine 2003; 348:2339–2347 (*excellent up-to-date review of this subject*).

www.cancer.gov/cancerinfo/pdq/genetics/breast-and-ovarian (*National Cancer Institute educational web site–useful source of information*).

http://www.ncbi.nlm.nih.gov/entrez/query.fcgi?db=OMIM (*The OMIM site. Type 168000 in search OMIM for . . . and this will provide information on carotid body tumours and inheritance*).

Somatic Cell Genetics

Campana D. Determination of minimal residual disease in leukaemia patients. British Journal of Haematology 2003; 121:823–838 (*overview of this topic particularly in the acute leukaemias*).

Goldman JM, Melo JV. Chronic myeloid leukemia—advances in biology and new approaches to treatment. New England Journal of Medicine 2003; 349:1451–1464 (*comprehensive summary of chronic myeloid leukemia*).

Gribben JG. Monitoring disease in lymphoma and CLL patients using molecular techniques. Best Practice & Research Clinical Haematology 2002; 15:179–195 (*easy-to-read overview of the B lymphoproliferative diseases*).

Mistry AR, Pedersen EW, Solomon E, Grimwade D. The molecular pathogenesis of acute promyelocytic leukaemia: implications for the clinical management of disease. Blood Reviews 2003; 17:71–97 (*detailed summary of APML*).

Future

Blasco MA. Telomeres and cancer: a tale of many endings. Current Opinion in Genetics and Development 2003; 13:70–76 (*a succinct and informative summary of telomeres and their possible role in cancer*).

Cummings DE, Schwartz MW. Genetics and pathophysiology of human obesity. Annual Reviews of Medicine 2003; 54:453–471 (*provides a nice overview of obesity with the various molecular abnormalities well described*).

Hugh-Davies L *et al.* EMSY links the *BRCA2* pathway to sporadic breast and ovarian cancer. Cell 2003; 115:523–535 (*original research report providing preliminary evidence linking BRCA2 to sporadic forms of breast and ovarian cancer*).

Janz S, Potter M, Rabkin CS. Lymphoma and leukaemia associated chromosomal translocations in healthy individuals. Genes, Chromosomes and Cancer 2003; 36:211–223 (*provides interesting insights into what can be found in apparently normal subjects with PCR*).

Kauffman EC, Robinson VL, Stadler WM, Sokoloff MH, Rinker-Schaeffer CW. Metastasis suppression: the evolving role of metastasis suppressor genes for regulating cancer cell growth at the secondary site. The Journal of Urology 2003; 169:1122–1133 (*provides a useful insight into how metastases spread and grow*).

Sekido Y, Fong KM, Minna JD. Molecular genetics of lung cancer. Annual Reviews of Medicine 2003; 54:73–87 (*lists all the molecular changes that have been described in lung cancer*).

Snyder EE *et al*. The human obesity gene map: the 2003 update. Obesity Research 2004; 12:369–439 (*interesting article summarising all the reports that describe a gene or locus involved in obesity*).

5

GENOMICS, PROTEOMICS AND BIOINFORMATICS

INTRODUCTION

Prior to the Human Genome Project, the focus of molecular medicine was genetics. With the completion of the Human Genome Project, genetics has evolved into genomics (see Figure 1.3). The new term implies a more global approach to capturing the wealth of knowledge about human and other DNA sequences now stored in databases. The goal is to analyse the complete genome (genomics) in terms of gene function, or define the protein repertoire associated with the genome (proteomics). For this, more sophisticated computer programs need to be developed (bioinformatics) that will allow DNA or protein-based information to be converted into knowledge of gene or protein function. The various genomic and proteomic endeavours are now well under way, with new technologic advances ensuring faster and cheaper analytic potential. A limitation to progress remains bioinformatics, an area of science that will provide exciting developments for the future. An extension of genomics is pharmacogenetics and pharmacogenomics. This area of molecular medicine will ensure that the right drug can be found for the right patient, and new drugs will be developed on the basis of DNA knowledge. Decision-making will also be enhanced at the consulting room or bedside as nanotechnology allows genomic and proteomic analytic tools to be miniaturised.

GENOMICS

The genomics era has been made possible through two major developments: (1) an exponential increase in knowledge and information about human and other genomes as a consequence of the Human Genome Project and (2) technology developments allowing high-throughput analysis of DNA. An example of the latter is the microarray, a new development that moves genetics from focusing on one or few genes or gene sequences, to genomics with its emphasis on the analysis of hundreds to hundreds of thousands of genes simultaneously.

MICROARRAYS

Microarrays are orderly, high-density arrangements of nucleic acid spots ($\sim 10^2$ to 10^6). Each spot represents a DNA probe that is attached to an immobile surface. Probes include cDNA or oligonucleotides. The latter are smaller, and so allow a larger number of spots per microarray; for example, computer-aided deposition of spots can produce a cDNA density of up to 10000 cDNA per cm^2 and 1000000 oligonucleotides per cm^2. This is offset by the greater potential for non-specific hybridisation with the oligonucleotides because of their smaller size. Materials onto which the spots (DNA probes) can be attached include glass, silicon, nylon and nitrocellulose membranes. DNA that is hybridised to the microarray can be labelled with radioactivity or fluorescein, with the latter now the choice because multiple colours can be detected with a laser (see the Appendix for technical details about microarrays).

Microarrays allow a snapshot to be taken of gene activity in the cell. This information can be compared between control (normal cells) and cells that are abnormal. High-throughput screening of gene expression possible with microarrays can reveal molecular signatures of what is occurring at the cellular level. This knowledge can be exploited in the clinic for diagnostic purposes, or in the research environment to understand disease initiation and progression. The development of microarrays has required improvements in bioinformatics. This development has been necessary for designing probes for the hybridisation conditions required and to analyse the complex data sets generated with microarrays. Analysis includes the comparison of the various hybridisation signals to set standards, thereby providing the ability to identify different colours—i.e., what genes are being expressed—as well as assessing the intensity of the different colours—i.e., the level of up or down regulation for each gene (Table 5.1).

There are different types of gene microarrays (Table 5.2). The most useful to date has been the expression DNA microarray that enables the transcriptome (all the RNA species in a given cell) to be studied and compared with the transcriptome in another cell. In this type of analysis, it is possible to measure any number of mRNA species from 100s–100000s. For example, what is the difference at the genomic level between a cell line that is growing normally and the same cell line that has become cancerous? A way in which to make this comparison is to hybridise the mRNAs from the two different cell lines, i.e., normal (control) versus cancer cell lines against a glass slide or silicone chip onto which has been spotted genes of relevance to carcinogenesis (Figure 5.1). These expression arrays allow the identification of genes that help to explain the biology of tumours, or detect tumour-specific targets for better diagnostics or new drug development. Commercially produced microarrays are now available covering a wide range of interesting genes (*TP53*, CYP450) or genetic pathways (apoptosis) or organisms (*E. coli* gene array).

DNA can be spotted onto a microarray in different ways: (1) High-density microarrays—the oligonucleotide microarray described above, and allowing >10^6/cm^2 density of gene dots, but offset by the time (>12 hr) required for processing. (2) Spotted microarrays—here the microarray density is less (>10^4/cm^2), but they are cheaper to produce. In development are: (3) Bead arrays—beads containing DNA comprise the hybridisation surface. (4) Electrically addressable arrays—hybridisation of DNA to the array leads to a tiny electric current being generated immediately. The advantage of these is instant recognition, but this is offset by a very low density of spots (perhaps in the hundreds of genes; Figure 5.2). See also Chapter 9 for a description of these different microarrays in the scenario of bioterrorism.

An example of how microarrays will impact on future clinical decision-making is illustrated by work from van

Table 5.1   Colour coding in 2-colour hybridisation microarrays

Colour of hybridisation dot	Significance
Green	This colour is produced when DNA is labelled with a dye called Cy3, which is usually used with the normal (control) tissues. If the microarray dot is coloured green, it means that the expressed gene in that microdot represents normal (control) tissue exclusively. The intensity of the green gives some indication of the level of expression of that particular gene.
Red	This colour is produced when DNA is labelled with a dye called Cy5. DNA labelled with Cy5 usually comes from the abnormal (experimental or unknown) tissues. If the microarray dot is coloured red, it means that the expressed gene in that microdot has been contributed from abnormal (unknown or experimental) tissue exclusively. The intensity of the red gives some indication of the level of expression of that particular gene.
Yellow	This colour means the relevant gene in both normal (control) and abnormal (unknown or experimental) tissues is expressing (red + green = yellow); i.e., the gene may be constitutively expressed in that tissue.
Black	This colour means neither gene is expressed.

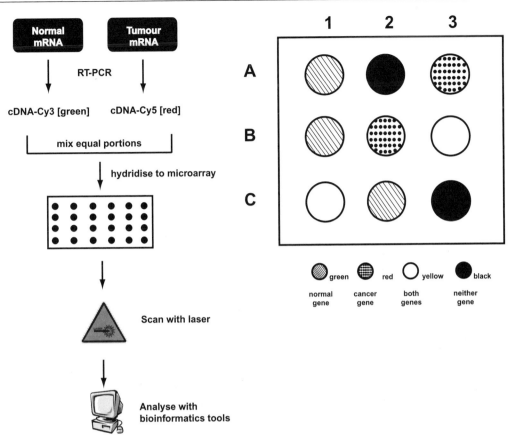

Fig. 5.1 Comparing gene expression in normal versus cancer tissue with a DNA microarray.
The diagram illustrates how a microarray can indicate which genes are important in a particular cancer tissue. (**Left**) Total mRNA from both normal and cancer tissue is made into cDNA. The normal tissue cDNA is labelled with a green dye (Cy3) and the cancer tissue cDNA with a red colour (Cy5). The cDNAs are mixed in equal proportions and hybridised to the microarray that has spotted onto it DNA probes (either in the form of cDNAs or oligonucleotides) for genes with relevance to cancer. Following hybridisation, the excess cDNAs are washed off, and the microarray plate is scanned with a laser to detect various colour changes (red, green, yellow and black). Using appropriate software and the results from control DNA samples, it is possible to identify the intensity of each red and green colour to estimate the level of the gene being expressed. (**Right**) The output from a microarray is based on four colour changes—green (normal tissue expressing), red (cancer tissue expressing), yellow (both normal and cancer tissues expressing) and black (neither tissue expressing). In this particular example, it is evident that, in the tumour, the genes that are predominantly expressing in that tissue are 2B and 3A, while genes 1A, 1B and 2C are not expressing in tumour because the green colour indicates that it is the normal tissue which is contributing to the hybridisation. Both normal and tumour tissue genes are contributing to the dots for 1C and 3B; i.e., these genes are constitutively expressed in both tissues. Finally, gene microdots represented by 2A and 3C are not expressed at all in either tissue. This microarray shows that cancer-specific genes can be identified by the DNA probes in 2B and 3A. See an actual microarray profile in Figure 5.2.

de Vijver and colleagues (2002) involving a study of 295 women with breast cancer. The work was initiated because breast cancer patients with the same disease staging have different treatment outcomes and survival. Current prognostic indicators rely on lymph node status as well as histological grade of the tumour. Treatment options vary including adjuvant chemotherapy and anti-oestrogen agents such as tamoxifen. In many circumstances it is difficult for patients to choose from the various treatment options offered, particularly when it is known that a large number of women will respond well without additional treatment that has side effects of its own.

In developing a microarray for breast cancer, the researchers first took mRNA from 98 primary breast tumours coming from women under 55 years who were lymph node negative. Of these, 34 patients subsequently developed metastases within five years; 44 remained

Table 5.2 Types of gene microarrays

Function of microarray	Applications
Expression microarray, i.e., compares the expression profile between two different cells	The most successful microarrays to date have been those based on comparing expression profiles in various tumours.
Open expression arrays, e.g., SAGE—Serial Analysis of Gene Expression	In contrast to the expression array described above, SAGE is an open system because it allows expression of all mRNA species, not just those that are spotted onto the chip (see the Appendix for details about SAGE).
Detecting single base changes, e.g., mutations or SNPs	These microarrays are not routinely used because of their cost. However, the technology is rapidly improving, e.g., the 10K (10 000) SNP chip released by the Affymetrix company has now become a 100K SNP chip, with expectation that it will be a 250–500K chip in the next few years. Access to so many SNPs will drive down costs. The potential to measure this large number of SNPs (or DNA mutations) in the one chip is enormous.
Comparative genomic hybridisation (CGH) or microarrays used to look for genomic gains or losses	DNA probes on the microarray are actually genomic DNA derived from known chromosomal loci. Hybridisation against normal and tumour tissue will allow the identification of genomic regions that are amplified (genomic gain due to gene amplification) or lost (deleted) in tumours. From this, it becomes possible to identify regions involved in the development of cancer through a gain or loss of gene function.

disease free after five years; 18 were shown to have *BRCA1* germline mutations; 2 had *BRCA2* germline mutations. The mRNAs were hybridised against 25 000 human genes. It was shown that about 5000 of these genes demonstrated consistent changes between the tumour tissues, and on this basis, it was possible to stratify women into high- and low-risk groups. Subsequently, it was shown that only 70 of the 5000 genes were necessary for determining risk status. These 70 genes were spotted onto another microarray, and a larger cohort of women studied. In that study, 295 consecutive patients with primary breast cancer were analysed with the 70 gene array. These patients were followed for a median time of 6.7 years. Based on the RNA expression profile, the group was able to be divided into high risk (180 patients) and low risk (115 patients) for metastasis. The RNA profile was shown to be a more powerful predictor of outcome than the standard histologic grade, surgical stage and node status.

The potential for a more objective predictor to guide the selection of treatment would be invaluable for managing breast cancer. This study was a landmark because it illustrated the applicability of microarrays in clinical medicine. More studies are now needed to confirm and expand the preliminary findings. The cost of microarrays would also need to come down to allow access for more women. Another change that will be needed is a move from the traditional formalin-preserved tissue to fresh tissues (or banks of frozen tissues) since non-preserved tissues are required to prepare the mRNAs.

Despite the challenges described, microarrays are progressing through the research phase of molecular medicine. In clinical medicine, they will be particularly valuable in areas such as

(1) Diagnostic testing and disease classification,
(2) Treatment selection through sophisticated pharmacogenetic analysis of the individual, as well as the underlying pathologic tissue,
(3) More accurate prognostic indictors based on algorithms derived from the individual's genetic makeup, as well as the tissue's genetic profile.

An example to illustrate 1–3 above is the use of a gene microarray to study children with acute lymphoblastic leukaemia (ALL). Holleman *et al* (2004) were able to identify gene profiles in the leukaemic cells that were associated with resistance to the four antileukaemic agents used in treatment. The children with these profiles had a poor prognosis. The investigators also concluded that a relatively small number of genes were associated with drug resistance. Whether the genetic profile identified by microarray was superior to the conventional prognostic indicators remains to be seen. However, the identification of genes involved in drug resistance provides targets for developing new antileukaemic agents that would be particularly valuable in the poor response subgroup of patients identified.

DNA Banks

There are a number of reasons given why DNA might be stored in a bank. They include (1) The rapid advances in gene discovery can quickly change what is a genetic defect of unknown aetiology today, to a disease with a known DNA marker in the future. (2) The necessity in linkage analysis to have key family members available for testing, particularly if these individuals might die. (3) Altruistic participation in research. Some commercial

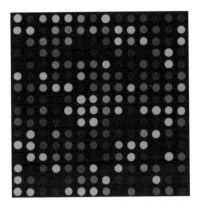

Fig. 5.2 A gene microarray. (*reproduced with the permission of Dr C Roberts and Dr S Friend, Rosetta Inpharmatics LLC, Seattle, USA*).
Top: This is a 1-inch × 3-inch glass slide onto which has been imprinted 25 000 genes in the form of oligonucleotides. The oligonucleotides are 60mers (i.e., 60 bases in size), and they are imprinted *in situ* onto the glass slide using a conventional inkjet printer head. The longer oligonucleotides used for hybridisation (as discussed previously, a 20mer oligonucleotide is generally sufficient to detect a specific region of the genome) provides this particular microarray with a technical advantage because, when constructing the array, any mistakes or errors in the inkjet printing will not be as critical as would be the case with a 20mer array. On the periphery of the slide and diagonally are grid lines. They are used for alignment when a scanner reads the array. The scanner would be faced with the picture shown, i.e., 25 000 spots with various colours and intensities, including green, red, yellow (both red and green have hybridised), black (neither red nor green has hybridised). **Bottom:** This is a small section of the microarray magnified to show the individual spots and the colours generated. At this stage, the scanner needs to identify the various colours as well as their intensities, and these data need to be stored. Bioinformatics-based tools are needed for this aspect of the microarray, as well as the next step, which is not as technically demanding as the actual microarray but equally challenging because the vast amount of data needs to be interpreted in terms of their potential biologic significance. (see colour insert)

banks (including DNA banks being established or proposed by pharmaceutical companies) have proposed that individuals (depositors) can add their DNA to the company's bank. The samples are used for research, and indirectly there is the potential benefit to the depositor who might then have access to new drugs. This would be particularly useful in circumstances where new drugs are unaffordable. In a clinical trial setting, the depositors' DNA markers will be known; this information will be valuable to the company (this issue is discussed further under Pharmacogenomics). (4) Dubious reasons for a DNA bank include the potential for human cloning at some future date.

Individuals or families in which there is a genetic component to disease (particularly rare or esoteric diseases) should be informed that future technological developments might allow further definitive study of the disorder, even though nothing can be offered at present. In these circumstances, it is important that health professionals are aware of the potential of recombinant DNA technology so that counselling given to individuals or families is accurate and permits them access to future developments. The result is DNA (in the form of a tissue or DNA itself or immortalised cell lines) stored in a DNA bank so that it is accessible, if required, at some future date.

A number of professional societies have proposed guidelines for DNA banks. They cover actual physical facilities; the relationship between depositors, their families and health professionals; confidentiality of information; safety precautions and quality assurance measures. The word "depositor" rather than "donor" is used because the individual giving the sample maintains ownership and is not acting as a donor in the broadest sense. Depositors need to have clear statements on the length of banking, the potential problems and their rights in respect of the banked DNA.

DNA can be stored in the form of whole blood or as DNA for years at −20°C or −70°C. Lymphocytes transformed with Epstein-Barr virus (EBV, see Chapter 8) produce an immortalised cell line that can be cryopreserved over many years in liquid nitrogen. Aliquots can be thawed and propagated as required. The advantage of immortalising lymphocytes is the availability of an unlimited amount of DNA which is suitable for a technique such as RT-PCR that can be used to identify mRNA transcripts in peripheral blood (see Chapter 2 and the Appendix). On the other hand, a disadvantage of immortalised cell lines is the technical demands required to prepare and maintain them. Material that is suitable for DNA banking includes blood, hair follicles, liver/spleen and other tissues from a deceased individual, abortus specimens, buccal cells from mouth washes and dried blood spots (Guthrie spots) that are collected as part of neonatal screening programs (see Chapter 7).

A DNA bank is not simply a diagnostic laboratory that keeps a number of DNA specimens in the refrigerator for future purposes. It is a planned activity with very strict operating guidelines. There are legal requirements, defined above, that the depositors and curators of the bank will need to understand. What can be done with the DNA, particularly in terms of research, requires careful thought and the appropriate consent processes. Other banks—for example, bone marrow, brain, various other tissues—can overlap the DNA bank because tissue from these banks is also useful as a source of DNA. Security to ensure that all samples are adequately protected means sophisticated measures are needed. In some commercial banks, various computer-based approaches are developed to code samples, which means the donors can never be identified (i.e., the samples are de-identified). Alternatively, samples are potentially identifiable, but security is enhanced by having a third independent party keep a check on privacy and confidentiality, or holding the key that links a sample with a particular individual. In Chapter 10 there is discussion on national DNA databases (banks) that have been set up in the UK and Iceland.

ANIMAL MODELS

Unlike humans, animals can be manipulated experimentally and bred under specific conditions. Therefore, animal models for human diseases have been extensively developed. These models can arise spontaneously, but a more effective approach is to produce experimentally the animal model required, thereby mimicking the human condition better and allowing the natural history of a disorder to be followed progressively over a number of generations. As discussed previously in Chapter 4, some of the more complex multifactorial diseases in humans may best be resolved by research using a genetic breeding approach coupled with DNA analysis.

Traditional Models

Over the years, inbred strains, particularly the laboratory mouse, have been the mainstays for studies involving a wide range of human disorders. Inbred mice are produced by repeated sister-brother matings over about 20 generations. The result is a syngeneic mouse that will be identical (e.g., homozygous) at every genetic locus, and to other mice of the same strain. Another type of inbred mouse is the congenic one. Although derived from one strain, selective breeding allows this animal to have genetic material from a second strain at a single locus. Naturally derived animal models provide considerable information, but they have limitations; e.g., the mutation may not be representative of that found in the human disorder. More importantly, for many diseases, a suitable animal model does not exist.

Recombinant DNA (rDNA) approaches provide a means to create new animal models or manipulate existing ones to test the function of genes. The rDNA approaches can broadly be divided into two strategies: (1) reverse or genotype-driven animal models and (2) forward or phenotype-driven models. The reverse strategy is essentially the transgenic animal; i.e., manipulating a specific gene in a mouse will provide more information about the disease itself. In contrast, the forward strategy focuses on the disease (phenotype), and from this, knowledge of the underlying genomic changes can be gained. An example of the forward approach is the ENU mouse, described in more detail below.

Transgenic Mice

Transgenic mice have become an invaluable resource to further our understanding of human disease. They are produced by microinjection of DNA into the pronucleus of a fertilised oocyte, which is then inserted into a pseudopregnant foster mother. Transgenics are useful for studying a wide range of disorders. Although the injected transgene is randomly inserted into the genome, it can still function, and its expression will produce a new phenotype. The next level of sophistication for the transgenic mouse is the animal created by gene knockout. This involves homologous recombination between an introduced gene and the corresponding gene in the animal; i.e., integration into the genome is no longer random. In this way, the normal gene is replaced by one with a known mutation or vice versa. Gene function can be inhibited or the effect of a specific mutation observed (see Figure 6.10 and Chapter 6 for further discussion of homologous recombination).

Embryonic stem cells have been critical for developing knockout transgenics. Since these types of stem cells are totipotential, they can be genetically manipulated and then reintroduced into the blastocyte of a developing mouse to produce a chimaera. Foreign DNA that has become integrated into the germline of the chimaera will enable the gene to be transmitted to progeny. Appropriate matings will produce homozygotes containing the transgene (Figure 5.3). Embryonic stem cells allow a gene to be targeted to its appropriate locus and replace its normal wild-type counterpart by homologous recombination, thereby producing a null allele. Using this approach, a better understanding of genetic inheritance or disease pathogenesis becomes possible (Box 5.1). The utility of knockout studies to define the function of unknown genes is illustrated by the mouse *hox-1.5* gene, which was inactivated by homologous recombination. Homozygous mutants for this defect developed a phenotype similar to the human DiGeorge syndrome, i.e., absent parathyroid and thyroid glands with defects of the heart, major blood vessels and cervical cartilage. Until the mouse data became available, all that could be said about the complex phenotype associated with this syndrome was that it was produced by a deletion in chromosome 22q11. From the animal work, it became possible to identify candidate genes for further study in the human.

Despite their value, the two types of transgenics described represent an all or nothing effect, and with widespread expression of the transgene in many tissues, it is difficult to investigate subtle gene effects or distinguish primary from secondary effects. The uncontrolled

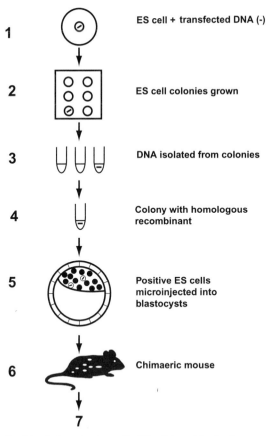

1 ES cell + transfected DNA (-)

2 ES cell colonies grown

3 DNA isolated from colonies

4 Colony with homologous recombinant

5 Positive ES cells microinjected into blastocysts

6 Chimaeric mouse

7

Fig. 5.3 Embryonic stem cells (ES cells) used for *in vivo* expression of recombinant DNA.
This method produces transgenic mice, which can be used to test the function of genes *in vivo*. (1) ES cells are transfected with foreign DNA. Many ES cells will take up the DNA, but this will involve different sites in the mouse genome because of random integration. In a very rare case, the integration will involve the correct part of the genome by a process of homologous recombination. (2) Colonies of ES cells are grown. (3) DNA is isolated from pools of colonies. (4) The colony that has DNA integrated into the correct position in the genome by homologous recombination can be identified by PCR. (5) ES cells that have the homologous recombined DNA are injected into mouse blastocysts. (6) Using different coloured mice as sources of ES cells (e.g., white mouse) and blastocysts (e.g., black mouse) will enable chimaeric mice to be distinguished. (7) If the transgene has also integrated into the germline, it will be possible to obtain a homozygote animal by appropriate matings (see Chapter 6 for further discussion of ES cells).

expression of the transgene during embryonic development could also be lethal if it is not normally expressing at this time. Earlier attempts at controlling gene expression in transgenics were fairly primitive and involved the use of inducers, for example, a heavy metal. This was possible because the inserted gene had a promoter

Box 5.1 Transgenic animal models of genetic disease.

Abnormal folding and aggregation of proteins in the central nervous system are considered to be key factors in the development of neurodegenerative disorders such a Alzheimer's disease (Chapter 4), Huntington's disease (Chapter 3) and Parkinson's disease (degeneration of dopaminergic neurons in the substantia nigra by this process leads to Parkinson's disease). In a rare familial form of Parkinson's disease, mutations have been detected in α synuclein (a synaptic protein), suggesting that this could be the protein that is misfolded and accumulates. To study this, the human α synuclein gene was inserted into the fruit fly (*D. melanogaster*) and the mouse. In the fruit fly, this produced a selective death of dopaminergic neurons, indicating that α synuclein exerted its toxic effect on specific sets of neurons. Transgenic mice created by inserting a human α synuclein gene demonstrated a number of the characteristics of Parkinson's disease, including motor deficits and inclusion formation. The mice also showed that amyloid β peptide 1–42 (a substance implicated in Alzheimer's disease) and oxidative stress exacerbated the toxic effects of the α synuclein protein. A second strain of transgenic mice was created to express human β synuclein (a homologue of α synuclein) to test the hypothesis that the β synuclein could reduce the protein accumulation and misfolding. When the two strains were crossed (to form double transgenics), it was shown that the severity of the Parkinson's disease was reduced. The mechanism by which β synuclein works remains to be determined, but the transgenic mouse model will be essential for answering this question. Thus, the transgenic model confirmed the importance of α synuclein in pathogenesis of Parkinson's disease and identified a novel way in which accumulation of this protein might be treated. There is also the potential that common pathways will be discovered for both Alzheimer's and Parkinson's diseases, thereby identifying targets for new drugs or cellular therapies (Hashimoto *et al* 2003).

element that was inducible when exposed to heavy metals. When the latter were placed in the drinking water, the transgene would be activated. However, this would not allow subtle or physiologic regulation of the transgene, and heavy metals are potentially toxic. A better approach was needed. The next development in the transgenic mouse allowed targeting to tissues as well as more effective regulation of the transgene's expression. This is called a conditional knockout, which means the inserted gene can be switched on or off "conditional" to a specific stimulus. One approach to make a conditional transgenic utilises what is called the Cre-lox system (Figure 5.4).

Cre-lox system to generate a conditional transgenic mouse

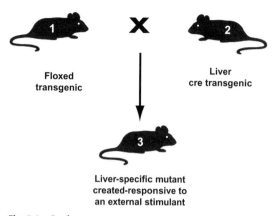

Fig. 5.4 Cre-lox system to generate a conditional transgenic mouse.
Cre (causes recombination) recombinase enables recombinations to be made where there are loxP (locus of recombination) sites. **(1)** The floxed transgenic (flanked by lox) is produced by the usual embryonic stem cell homologous recombination approach, but in this case the gene of interest is constructed so that it is flanked by loxP sites. Mice with this transgene are bred to homozygosity, but have no phenotypic changes because the Cre recombinase is needed. **(2)** To introduce the Cre recombinase requires breeding to a Cre transgenic. This transgenic has Cre under the control of a promoter that can be tissue or time specific. **(3)** Offspring of the Cre/Floxed mating on exposure to a stimulant that turns on the promoter, will develop tissue specific (or time specific) recombination to inhibit gene function.

ENU Mice

The availability of the mouse genomic sequence has made the ENU mouse a particularly valuable resource. ENU (N-ethyl-N-nitrosourea) is a potent germline mutagen generating single nucleotide mutations in DNA. Using this chemical, it is possible to create random mutations in the mouse's DNA and then observe the various phenotypes that result. Phenotypes that resemble human diseases are then investigated at the mouse genome level to identify the underlying gene. From this knowledge, the human homologue can be isolated. Compared to the transgenics described earlier, this model is phenotype driven; i.e., the phenotype is first identified, and then this allows the genotype (gene) to be determined. ENU is very efficient at producing mutations. The difficult step with this mouse model is detecting the various phenotypic changes, particularly subtle ones. Variations of the ENU mouse are now being developed, and some of them have a lesser requirement for the more difficult phenotypic end point and rely more on genome screens looking for mice with mutations in genes of particular interest. The ENU mouse has high-

lighted the potential for *forward genetics*; i.e., mutagenesis of the genome produces different phenotypes. From these, it becomes possible to detect the underlying genes, i.e., no prior knowledge of genes is needed.

Zebrafish

Although zebrafish have been used since the early 1970s as a tool to understand development and genetics, a relatively new addition to the genomics laboratory is the fish tank. This tank is used to house the zebrafish (*Danio rerio*), a model organism now attracting much attention because of its small size, short life cycle and ease of culture. Since its genome is only half that of the human or mouse, it is easier to work with in terms of gene identification, and a complete sequence of the zebrafish genome will soon be available. It is a particularly good model when studying development because embryos are transparent and they develop outside the mother's body (Table 5.3). Sequencing of the zebrafish genome started in 2001 and is expected to be completed by 2005 using the two different approaches described earlier for the Human Genome Project (see Chapter 1): i.e., (1) a whole genome shotgun approach and (2) the more traditional cloning to generate overlapping clones that are then sequenced. Comments from the Sanger Institute in Cambridge imply that the former approach (used by Celera in sequencing the human, bacteria and mouse genomes) is faster, but the quality of the result is inferior to the traditional but slower strategy (see Zebra Fish Information Network on www.zfin.org).

In the zebrafish, antisense approaches to gene manipulation (see Chapter 6) have been used successfully to knock out genes and then observe the effects on the phenotype. Recently, the utility of the zebrafish for drug evaluation has been highlighted since it is possible to expose directly the embryos (by placing the drug in the fish tank) and so observe potential toxic effects of drugs. In the review of zebrafish by Langheinrich (2003), an interesting section details the effects of drugs of addiction (cocaine, nicotine and ethanol) on zebrafish. Exposure to cocaine in the zebrafish leads to a decrease in visual sensitivity and a conditioned behavioural change consistent with the model that cocaine addiction works through midbrain dopamine. In a second experiment, zebrafish were exposed to ENU. Mutants produced by ENU included three with insensitivity to cocaine, altered dark adaptation and memory capacity. Putting this together, genes involved in dopaminergic signalling in the retina and brain were able to be implicated in the cocaine addictive effect.

PHARMACOGENETICS

There are no consistent definitions for the terms "pharmacogenetics" and "pharmacogenomics." One defini-

Table 5.3 Some comparisons of model organisms (see Zebra Fish Information Network on www.zfin.com)

Trait	Zebrafish	Other model organisms
Vertebrate	Yes, so useful to study developmental processes, as well as behavioural traits by forward genetics, i.e., genome mutagenesis from which phenotypes are used to identify genes.	*Drosophila* (fruit fly), yeast and *C. elegans* (worm) are less useful because they are invertebrates but still valuable at the cellular or biochemical level. Behavioural traits in the mouse are difficult because of breeding limitations.
Mammal	No.	Mouse is a mammal and therefore closer to the human.
Breeding	Ideally placed here. Embryos are transparent, develop outside the mother, so can be followed easily. Large numbers of eggs are laid by females, and within 24 hours can be visualised as tiny fish.	Mice take 21 days to develop, and access *in utero* is limited.
Physical manipulation of embryo	Very easy because external to mother. Can expose to chemicals simply by releasing them into the tank water (absorbed by skin and the gills of embryos).	More difficult.
Genetic manipulation	Although possible to manipulate genes by approaches such as antisense, cannot produce a knockout by homologous recombination.	Knockout mice well established. Experimental and basic knowledge of *Drosophila* and mice superior because these models have been around a lot longer.
Whole genome sequenced	Started in 2001; expected to be completed by 2005.	Starting in 1996, many model organisms have had genome sequenced, including many bacteria, viruses, yeast, *C. elegans*, *Arabidopsis thaliana*, rat and mouse.

tion for pharmacogenetics describes the effect that the genotype has on drug response, i.e., inherited differences in drug effects. In contrast, pharmacogenomics involves the application of genomic information to drug design, discovery and clinical development. Another attempt at distinguishing the two is based on technology; i.e., in pharmacogenomics, genome-wide strategies (e.g., microarrays) are used to identify the inherited basis for differences between individuals in their response to drugs. A third, and likely correct view, is that pharmacogenetics and pharmacogenomics are the same. In this chapter, an attempt will be made to distinguish pharmacogenetics and pharmacogenomics because they both continue to appear in the literature. In keeping with the distinction between genetics and genomics first introduced at the beginning of this chapter, pharmacogenetics will be used to explain the *in vivo* differential effects of drugs in patients due to inherited gene variants. Pharmacogenomics will be used to explain the differential effects of compounds *in vivo* or *in vitro* on the expression of all genes (see Table 5.4, which attempts to compare a number of definitions). In the longer term, pharmacogenetics will ultimately be replaced by the broader concepts possible in pharmacogenomics.

A 1998 study by Lazarou and colleagues reported that the overall incidence of serious adverse drug reactions in hospital practice was 6.7%, while the incidence of fatal adverse drug reactions was 0.32%. The extrapolation from this was that more than 100 000 deaths occur annually in the USA because of adverse drug reactions, and they are potentially avoidable deaths. While acknowl-

edging that the study design was open to criticism, the authors concluded that adverse drug reactions in hospitals were a critical clinical issue that needed resolution.

Variables determining differences in drug metabolism include age, sex, weight, body fat, function of various organs (liver, kidney, heart and lung), smoking, alcohol intake, nutrition, environmental factors and concomitant disease. Drug–drug interactions are another important cause of adverse drug events. The final variable is genetic, and this can work at various levels of the drug's life, i.e., metabolism, transport, target and disease pathways. A genetic contribution to drug metabolism has been known for some time when it became evident that drug levels in the blood or urine were variable, and this variability was inherited. However, it was not until the molecular era that the genetic contribution could be clearly attributed to various drug metabolising genes and the basis for these gene effects identified.

Cytochrome P450 Enzymes

Inherited variations in our ability to metabolise drugs are common (Table 5.5). An important group involved in drug metabolism is the cytochrome P450 enzymes (CYPs). They are the major phase I drug metabolising enzymes (phase I involves oxidation, reduction and hydrolysis). CYPs that share at least 40% DNA sequence homology are grouped within families denoted by an Arabic number. A letter after this denotes a subfamily, and members within subfamilies are numbered sequentially, for example, *CYP3A4*. Although humans have

Table 5.4 Some comparisons between pharmacogenetics and pharmacogenomics

Pharmacogenetics	Pharmacogenomics	Reference
Used in the clinical setting to identify the drugs and drug doses most likely to be optimal for the patient, i.e., more targeted, more effective medicines for patients	Used in the pharmaceutical research setting to find the best drug candidate from a given series of compounds under investigation, i.e., finding new medicinals quicker and more efficiently	Lindpainter 2003
Study of differences among a number of individuals with regard to clinical response to a particular drug, i.e., one drug, many genomes	Study of differences among a number of compounds with regard to the gene expression response in a single genome/transcriptosome; i.e., many drugs, one genome	
Involves the study of patient populations	Entails the use of an increasing number of databases to identify genes for screenable targets that are not yet known to be genetically related to disease	Paul and Roses 2003
The differential responses of patients to drug compounds based on genetic polymorphisms, e.g., SNPs	Seeks to identify disease-relevant drug targets at the molecular level and to target drugs to clinical populations with specified haplotypes	
One drug for many patients	Many drugs for many patients	
From phenotype-to-genotype	From genotype-to-phenotype	Weinshilboum 2003
The role of inheritance in the individual variation in drug response, i.e., the right drug and dose for each patient	The influence of DNA sequence variation on the effect of a drug	

more than 55 P450 genes, only about half a dozen are involved in the metabolism of the great majority of prescribed and over-the-counter drugs. The P450 2 family comprises at least five subfamilies, designated A through E. Enhanced activity of the *CYP2D6* gene has been associated with a number of cancers (bladder, liver, pharynx, stomach and cigarette-induced lung cancer). The explanation is that increased metabolism of various environmental toxins by *CYP2D6* leads to the accumulation of toxic intermediates that are carcinogenic.

CYP2D6 also illustrates the broad effects that a gene can have on drug metabolism, ranging from an inability to metabolise it to ultra-fast metabolism of the drug. A large number of drugs (close to 20% of those commonly prescribed) are metabolised by *CYP2D6*. About 5–10% of Caucasians have a deficiency in their metabolising potential, and so drug effects are exaggerated. This deficiency is inherited as an autosomal recessive trait, so those with a homozygous mutation (one mutant allele from each parent) are at most risk. Once the gene was cloned, it was possible to show that mutations impairing the gene's function were heterogeneous, including single base changes and deletions. About 75 mutations have now been described (see http://www.imm.ki.se/cypalleles—a DNA mutation database for P450). This heterogeneity in the number of mutations would make it difficult to undertake routine DNA screening for *CYP2D6* (compare with 6-mercaptopurine, which follows) with conventional technology, although DNA chips would be ideal to screen for variants in this gene. Another interesting mutation with *CYP2D6* is the pres

ence of multiple gene copies with up to 12 being described; i.e., patients with this abnormality would be super-metabolisers, and so drug doses would need to be increased to achieve a comparable therapeutic effect. This variant seems to be particularly common in East Africans.

CYP2C9 is the important gene involved in warfarin metabolism. Two genetic variants are associated with impaired metabolism of warfarin, and so increased plasma concentrations of this drug would be expected. Thus, patients with these genetic variants would require less warfarin to remain anticoagulated and would be at greater risk of bleeding complications if normal doses were used. The poor metabolising genetic variants are found in about 10% of Caucasians but, in contrast, only in 2% of black Africans.

CYP2C19 is involved in metabolism of the proton-pump inhibitor omeprazole, which is used in combination with ampicillin for treatment of *H. pylori*. It could be predicted that poor metabolisers (associated with genetic variants in two of the gene's exons) are more likely to achieve eradication of *H. pylori* infection because for the same drug dose higher plasma levels of omeprazole could be achieved. This was found in one study looking at the effect of these genetic variants in Japanese. Again, racial differences are observed with the *CYP2C19* slow metabolising variants in Europeans showing that ~4% have the slow metaboliser genotype, whereas in Japanese the variants have a much higher frequency (~20%). Therefore, Japanese would be expected to respond better to treatment. As is evident from the

Table 5.5 Pharmacogenetics of some drugs (from Weinshilboum 2003)

Drug metabolising enzyme	Frequency in population of poor metabolism phenotype	Examples of drugs (or chemicals) metabolised	Drug effects in association with poor metabolism polymorphism
Cytochrome P450 2D6 (*CYP2D6*)	9% United Kingdom 1% Arabs 30% Hong Kong Chinese	Nortriptyline, Codeine	Enhanced Decreased
Cytochrome P450 2C9 (*CYP2C9*)	10% Europeans	Warfarin Phenytoin	Enhanced Enhanced
Cytochrome P450 2C19 (*CYP2C19*)	2.7% White Americans 3.3% Sweden 15% China 18% Japan	Omeprazole	Enhanced
Dihydropyrimidine dehydrogenase	1% population	Fluorouracil	Enhanced
Butyrylcholinesterase (pseudocholinesterase)	~1 in 3500 Europeans	Succinycholine	Enhanced
N acetyltransferase 2	52% White Americans 17% Japanese	Isoniazid Hydralazine Procainamide	Enhanced
Thiopurine S-methyltransferase	~1 in 300 Caucasians ~1 in 2500 Asians	Mercaptopurine Azathioprine	Enhanced
Catechol O-methyltransferase	~1 in 4 Caucasians	Levodopa	Enhanced
Paraoxonase 1 (Gln192 variant)	75% Northern Europeans 3% Some Asians	Paraoxon[a] Diazoxon[a] Sarin[b] Simvastatin	No effect[c] Enhanced[c] Enhanced[c] No effect

[a] Insecticide. [b] Nerve toxin. [c] The Gln192 PON-1 variant is associated with a higher catalytic efficiency for hydrolysis. This leads to more rapid clearance of an agent like sarin and the insecticide diazoxon, but has little effect on the insecticide paraoxon, suggesting that the latter compound may be metabolised *in vivo* by another system distinct from PON-1.

descriptions of risk populations given above, as well as the data in Table 5.5, the common molecular mechanism that would explain clinical variability with drugs is a genetic polymorphism influencing drug metabolism or drug effect, with the inter-individual and inter-racial effects simply reflecting the distribution of the polymorphism, and not a specific trait found in any particular population.

Statins

A study by Chasman and colleagues (2004) focuses once again on the genetic determinants that might influence responses to important pharmacologic agents. In this clinical trial, 1536 individuals taking pravastatin for cholesterol lowering were investigated to identify if genetic markers could explain inter-individual variation in drug response. Statins work by inhibiting 3 hydroxy 3 methylglutaryl coenzyme A (HMG-CoA) reductase. This leads to a lowering of total cholesterol, particularly the LDL cholesterol, and so statins are potentially very effective

in reducing cardiovascular risk. The study design involved the use of 148 SNPs from 10 candidate genes known to be associated with cholesterol lowering. Patients' responses to a standard dose of pravastatin over a 24-week period were compared to their genetic (SNP) profiles. It was shown that two SNPs in the HMG-CoA gene (the gene targeted by statins) were strongly associated with a 22% lower reduction in cholesterol and a 19% smaller effect in LDL lowering; i.e., patients with these two SNPs demonstrated a significantly impaired response to pravastatin compared to others taking the same drug. As the authors rightly noted, the study now needs to be repeated to confirm its conclusions. How the SNPs exert their effects is unknown since they are both within the gene's introns. Nevertheless, it will not be long before pharmaceutical companies make use of genetic information to provide further guidance on drug doses, as well as define those patients who are less likely to respond (the pravastatin example) or those at more risk of side effects (the 6-mercaptopurine example to follow).

6-mercaptopurine

Although the importance of genes in drug metabolism is now well established, the application of pharmacogenetics into clinical practice has been slow. This is well exemplified by 6-mercaptopurine, a purine antagonist drug that was introduced more than 30 years ago to treat malignancies such as leukaemia, although it is now also used as an immunosuppressive agent for serious inflammatory diseases. For over two decades, it has been known that in some patients severe toxic effects, involving the bone marrow predominantly, can occur despite standard doses being used. Metabolism of 6-mercaptopurine occurs through thiopurine S-methyltransferase, and the gene for this is called *TPMT*. Caucasian populations can be divided into three groups based on their level of activity of *TPMT* in red blood cells. The molecular basis for the three groupings is now understood through gene analysis. Mutations within the *TPMT* gene can produce changes in two amino acids—Ala154Thr and Tyr240Cys. These mutations can be inherited together or separately. Those who are heterozygous for the mutations are at increased risk of side effects from this drug, whereas homozygous-affected for both the above (approximately 1 in 300 Caucasians are in this category) are at significant risk of toxicity if normal doses are used, since there is complete loss of catalytic activity. Although other variants of *TPMT* influence metabolism, the two missense changes described above are found in about 75% of affected patients, making it easier to set up a DNA testing protocol.

In mid-2003, the USA's Federal Drug Administration (FDA) formed a committee to consider whether it should specifically recommend that DNA testing for *TPMT* mutations become mandatory before the drug is used. Perhaps surprisingly, the committee recommended against this, although it wanted more information to be included on the drug's container label about the genetic risks with *TPMT* variants. The lack of enthusiasm for what would be the first compulsory pharmacogenetic DNA test was based on several issues, including (1) the costs of the DNA test, (2) the difficulty that some physicians might have in interpreting the test results, (3) a possible delay in starting treatment, (4) the potential that this might reduce drug doses and so suboptimal treatment, which could have serious consequences in a potentially fatal disorder such as leukaemia and (5) a final justification given by those not wanting a mandatory DNA test was the ease with which the drug toxic effects could be monitored by serial blood counts.

Of the various reasons, the potential to cause more harm if the drug dose is reduced would seem to be the most persuasive argument, but the others are less so. The costs of the DNA test would seem insignificant compared to the costs of the drug itself or hospitalisation if complications such as neutropenia or thrombocytopenia

prolonged hospital stay. It would also seem reasonable to do the test after treatment was started so that at least those predisposed could be monitored more regularly. Health professionals must understand what a DNA test is and its implications, and *TPMT* would be a straightforward example to gain this knowledge.

For the present, some health professionals continue to be reluctant to add pharmacogenetics to their patient workup. However, this is unlikely to be justifiable in the longer term, and no doubt once the FDA or equivalent regulatory bodies mandate pharmacogenetic testing for drugs with potentially severe side effects, others will follow. The first step in this direction was taken in November 2003 when the FDA released a draft document titled: *Guidance for Industry: Pharmacogenomic Data Submissions* (www.fda.gov/cder/guidance/). When the consultation process if completed, this will provide guidance on how pharmacogenomics data: (1) can be used by sponsors to inform regulatory decisions; (2) will be utilised by the FDA for decision making. Pharmaceutical companies are steadily building up their databases that compare DNA markers and drug responses in the many clinical trials undertaken. In future, the companies will be indicating which tests require pharmacogenetic assessment (1) to streamline their drug development process (discussed further under Pharmacogenomics) and (2) as a form of medico-legal protection.

PHARMACOGENOMICS

Information about the potential for adverse drug reactions is obtained at two stages in a drug's development. The first is the pre-marketing randomised clinical trials mandated by regulatory authorities. The second is the post-marketing experiences, i.e., the results of using the drug on patients for therapeutic purposes. Pre-marketing is stringently controlled and expensive because it involves random trials. However, it cannot answer all questions about toxicity or provide all permutations and combinations of genes, drug-drug interactions, environment and ethnicity that will influence the overall drug effect. Post-marketing monitoring does not involve formal randomised studies but generally relies on observations. These observations will not easily detect rare or unusual effects. Ultimately, costly and long-term cohort studies are needed to fully evaluate the efficacy and side effects associated with drugs. Not surprisingly, it can be expected that new drugs will continue to produce unexpected side effects unless other strategies, in particular pharmacogenomics, can be added to the regulatory and marketing steps.

One opportunity for pharmaceutical companies to streamline their pre-marketing randomised clinical trials is to stratify patient populations based on genetic markers that are likely to influence metabolism of the drug. In this way, patients most at risk of adverse events, or equally

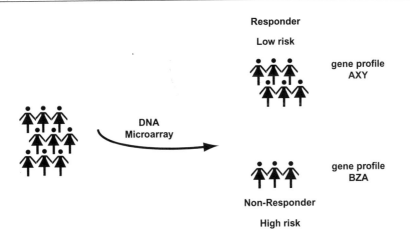

Responder

Low risk

gene profile
AXY

DNA
Microarray

gene profile
BZA

Non-Responder

High risk

Fig. 5.5   Pharmacogenomic analysis of an experimental drug for the treatment of hypertension in pre-eclampsia.
A DNA microarray can be developed based on the pharmacology and toxicology experimental data for a new antihypertensive drug. The microarray would contain many genes of relevance to pathways required for the drug to act and be metabolised. A phase I/II clinical trial (which essentially means the trial is being undertaken to assess toxicity and dosage rather than efficacy) is then conducted, and the patients' genetic responses are documented with the microarray. At the end of the trial, patients can be divided into two main groups: (1) good responders, or low risk of side effects and (2) poor responders, or high risk of side effects. The data from microarray profiles would then be examined to distinguish the two groups; for example, Group 1 has the AXY profile based on various SNPs, and Group 2 has the BZA profile. From the genetic information, the large and expensive phase III study would specifically select patients with the AXY gene profile. The pharmaceutical company might also go back and look more carefully at why patients with the BZA genetic profile did not respond or had more side effects. The genetic information might then allow the company to redesign the drug (to increase the target population) or use the genetic information to warn some individuals against using the drug or advising a modified treatment regimen because of an underlying genetic susceptibility.

important, patient populations less likely to respond to the drug, will be excluded. This stratification will increase the chance of the drug making it to the marketplace because it will reduce the number of side effects and enhance the potential for efficacy to be demonstrated. There is also the potential for genomic information obtained during the earlier phase I/II studies to be used when designing the more expensive and larger phase III trials, particularly in relation to inclusion and exclusion criteria (Figure 5.5). Trials that are stratified in this way should be less expensive and faster to conduct. Savings in the regulatory process might be expected to be passed on in the form of cheaper drugs. A potential downside to this approach is patient "outliers" who will not have access to new drugs because they will be specifically excluded by the pharmacogenomics strategy used to market the drugs. This poses an ethical dilemma that will need to be addressed.

Preclinically, many lead compounds are produced with therapeutic potential. Genomic analysis would provide a further filter to identify those compounds most likely to work and least likely to produce side effects. This knowledge would be based on *in vitro* and *in vivo* (animal) data as the genomic profile would be used to identify likely pathways involved in drug effects, as well as the drug's metabolism. *In silico* evaluation (i.e., drug modelling with bioinformatics) would be another tool for

the preclinical workup. Another strategy might be to select a compound whose action and metabolism rely on genes with little genetic variability, and so the compound is more likely to demonstrate greater consistency between patients (and different populations), thereby giving that drug an expanded target population. The complexities associated with drugs, their effects and metabolism and the way genomic information can be used in this context are illustrated by reference to the human enzyme paraoxonase-1 (PON-1).

Paraoxonase-1

Paraoxonase-1 or PON-1 (the gene is called *HUMPONA*) is a high-density lipoprotein-associated serum enzyme. It demonstrates three distinct activities: (1) Protecting low-density lipoproteins from oxidative modifications. Thus, individuals with low PON-1 enzyme activity have higher cardiovascular risks related to progression of atherosclerosis. (2) Metabolism of some drugs including the diuretic spironolactone and the cholesterol-lowering agents such as simvastatin. (3) Metabolism of organophosphorus compounds including insecticides that act by inhibiting acetylcholinesterase activity and toxic nerve agents such as sarin that are also organophosphorus compounds with anti-acetylcholinesterase activity.

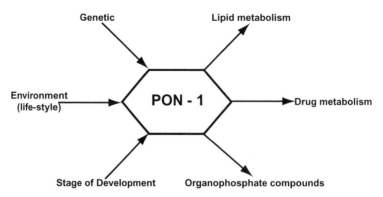

Fig. 5.6 Variables affecting paraoxonase-1 (PON-1) activity.
Both genetic and environmental factors influence metabolism of this drug. Genetic effects can result from a change at the catalytic site because of a Gln192Arg variant. There are also 5′ promoter changes that influence the expression of the gene. Environmental and lifestyle effects include **Reduced PON-1**: pregnancy, renal disease, diabetes, cirrhosis, smoking, alcohol but daily moderate intake of alcohol increases PON-1. **Elevated PON-1**: some dietary fats, simvastatin, phenobarbital. During development, the level of PON-1 increases from birth to about 15–25 months of age, when it reaches its maximum level. This may explain why young children are more susceptible to environmental toxins such as insecticides compared to adults.

Early biochemical analysis of serum PON-1 levels showed stratification into three groups in humans—high, intermediate and low. At the genomic level, a number of different variants for this enzyme have been defined. One codon variation (Gln192Arg) influences catalytic efficiency in relation to hydrolysis of different substrates; for example, the Gln192 allele allows the insecticide diazoxon to be hydrolysed more rapidly, but this differential effect is not seen with the insecticide paraoxon. The Gln192 allele also allows a neurotoxic agent such as sarin to be metabolised more effectively. In addition to the catalytic genetic variation, 5′ promoter polymor-

phisms influence the gene's expression. Finally, there are environmental, life style and developmental influences on the overall activity of this gene (Figure 5.6). Hence, PON-1 genetic markers are less valuable in predicting response than biochemical measurement of the enzyme level. Nevertheless, knowledge of the molecular controls has allowed a better understanding of the importance of the environment and other factors with PON-1. An understanding of the PON-1 pathway has also proven useful in poisoning with sarin, a nerve toxin used in bioterror (see Box 9.3).

PROTEOMICS

Proteomics is the analysis of the total proteins (proteome) expressed by a cell, tissue or organism. Although this does not fall within the definition of molecular medicine (see Chapter 1), the Human Genome Project has transformed the rigid discipline mentality in clinical medicine, research and industry into one that increasingly must consider the "big picture." We are now in the functional genomics era; i.e., there is a move away from structural analysis such as the location of genes based on DNA sequence data to a study of gene function. For this, bioinformatics will need to be more broadly based and not just focus on DNA sequence data. Therefore, genome analysis has to be complemented by understanding the proteome and the transcriptome. The latter refers to the analysis of mRNA species produced in the cell, tissue or organism. The integrated approach to studying biological

systems and the effect of stresses or stimuli is called **systems biology**.

Prior to the Human Genome Project demonstrating that the human genome has far fewer genes than originally anticipated (from 100 000 or more to about 30 000), it was often said that the most direct way to understand our complex proteome (millions of proteins versus tens of thousands of genes) was to characterise genes and, from this, understand the proteins. Methods to identify new genes in the genome and then sequence them have improved sufficiently to make this an achievable target. However, the gene to protein approach now needs rethinking because the smaller number of genes implies there must be something else occurring at the level of the genome to account for the complex proteome (see Table 1.6). Whether this can be entirely explained by alternative splicing that was described in Chapter 2 would seem unlikely. RNA species discussed in Chapter 2 or some other as yet to be discovered mechanism means that the genomic approach is more complex than originally anticipated. Epigenetic changes as well as post-translational modifications such as glycosylation further complicate a direct link between DNA sequence and protein function. Hence, effort is increasingly being directed back to the study of proteins, so molecular medicine must now include proteomics. Recent new technologies for protein analysis, particularly mass spectrometry, have revitalised the potential for protein analysis.

MASS SPECTROMETRY

Although the term "proteomics" was coined in 1994, a limitation to its development has been the difficulty in sequencing a protein. This limitation became even more apparent as DNA sequencing methods improved dramatically as the Human Genome Project progressed. Although the sequencing of ESTs (expressed sequence tags) gave rapid information about transcripts present in various cells and tissues, it was incomplete, particularly for low-abundance mRNA species. In the late 1980s, methods were developed to allow the ionisation of proteins and peptides, thereby opening up the potential for techniques such as mass spectrometry to be used for protein sequencing and characterisation. For mass spectrometry, the mass-to-charge ratio (m/z) of a molecule is measured. To do this, the substance must be ionised, and it is then transferred to a high vacuum system where it is exposed to a laser beam. The laser blasts off the ionised protein, and it flies down a vacuum tube towards an oppositely charged electrode. From its transit time, the mass can be calculated very accurately. MALDI-TOF (MALDI—matrix assisted laser desorption ionisation; TOF—time of flight) mass spectrometers now enable the mass of peptides or proteins to be determined rapidly and accurately. The result is a spectrum based on the various m/z ratios generated, with the height of each peak in that spectrum reflecting the abundance of that particle. Bioinformatics is then needed to take these mass data and, through algorithms, convert them into likely protein sequences and identity. Once high-throughput methods became available to characterise proteins accurately, it was necessary to develop databases (just like the ones used to store DNA data). Despite these developments, the proteomic databases remain inferior to the genome databases, because they are limited by substrate access since proteins need to be isolated from relevant tissues (compared to the universality of genomic DNA, which is identical in all tissues). In this respect, laser-capture microscopy has been an important development since it allows the obtaining of a pure cell population from a tissue section containing many different cells.

DRUG DEVELOPMENT

The prohibitive costs in developing new drugs and the relatively low success rate in taking the drug to the marketplace have required the pharmaceutical industry to rethink its strategies. Proteomics has now become an integral step in new drug discovery and development. Ultimately, it is the protein in health or during disease to which drugs need to the targeted. High-throughput proteomics allows a wide range of proteins to be identified during disease, and they become potential targets for new drugs. As well, the proteomic analysis can identify biomarkers in accessible biological fluids that can be used during all steps in drug development to monitor efficacy and toxicity. Although a gene (and so its protein) may be structurally normal, it can still be dysfunctional through modifications such as glycosylation or phosphorylation in the protein. This, as well as protein-protein interactions, can also be sought during the proteomic workup. The next step in drug development requires some form of validation *in vivo*, and this means a genomics approach, perhaps through a transgenic animal in which the candidate protein and its perturbations can be modelled. Once the protein is confirmed as a good candidate, it is necessary to return to sophisticated proteomic and bioinformatics tools to identify the protein's three-dimensional structure and, from this, predict likely targets for the drug.

Systems Biology Approach

Although impressive new drugs have been developed to target specific problems—for example, the statins to treat high cholesterol mentioned earlier—the challenge now is how to develop drugs for more complex diseases in which there is multiorgan pathology, for example, diabetes, heart disease, even ageing. The traditional approach would be to use either genomics or proteomics or transcriptomics-based strategies in the complex diseases. A more powerful option now available is to utilise bioinformatics to integrate all the information that these different modalities can provide. Once a drug is developed, it also becomes possible to use the same systems biology approach to model the product within an integrated genomics, proteomics and transcriptomics framework to predict efficacy, likely toxicities or other unexpected effects.

CLINICAL PROTEOMICS

An example of how proteomics might be applied in clinical medicine is given by ovarian cancer. Because this cancer develops insidiously with few signs or symptoms, most patients are diagnosed at an advanced stage. One approach to improve outcomes in ovarian cancer is the development of protein-based biomarkers that allow early identification of this tumour. The biomarkers must be accessible (e.g., blood, urine or saliva), and they must demonstrate an appropriate level of specificity and sensitivity. In the case of the ovarian cancer example given, a mass spectrometer profile involving five separate m/z values provided 100% sensitivity and specificity in 52 normal and 92 stage I–III ovarian cancer patients, including 15 out of 15 who were stage I (Petricoin and Liotta 2003).

Like the landmark DNA microarray study of breast cancer described earlier, a similar proteomic-based analysis of lung cancer has demonstrated the potential for this approach. In this study MALDI-TOF was used to produce a protein spectrum from a 1 mm diameter section taken from a frozen tissue biopsy. Not surprisingly, a lot of "noise" was obtained; for example, more than 1600 protein peaks were observed. However, with bioinformatics it became possible to look for differential expression between tumour samples and reduce the number of peaks to workable data. From this, a profile based on 15 distinct peaks was used to distinguish patients with poor prognosis non-small cell lung cancer from those with good prognosis (Yanagisawa *et al* 2003). As the values of the various protein profiles are confirmed, they could be used as adjuncts to diagnosis, determining prognosis and assisting with treatment regimens. Ultimately, determining the identity of the various protein peaks will provide invaluable insights into understanding pathogenesis of the tumour.

A word of caution would be appropriate at this point. The ovarian cancer proteomics study described above was first published in 2002. By 2004, doubts were being raised about the results. In a news feature in *Nature* titled "Running before we can walk," some questions have been raised about the reproducibility of the proteomic data. In particular, there is a suggestion that the difference in the m/z values between the normal and ovarian cancer groups was generally less than 500. With mass spectrometry, values below 2000 for m/z are regarded with suspicion because they could be artefacts. At this stage, it remains to be resolved whether the data are genuine or artefacts of the technology. However, it is important to note that cutting-edge science, which is often the case with molecular medicine, must go through the usual peer review and rigorous clinical trial assessment steps before it can be safely taken from the laboratory to the bedside. Sometimes, there is pressure for this process to be shortened through lobbying by consumers, politicians or health professionals. This is particularly relevant to a high-profile area such as molecular medicine and must be avoided at all costs.

BIOINFORMATICS

HISTORICAL

Bioinformatics is the application of computational tools and analysis used to capture, store and interpret biological data. In modern biological research, bioinformatics is essential for managing and analysing data. The computer is also having on impact on medical practice through the availability of sophisticated Internet databases accessible to patients, the community and health professionals. There is the potential for computers to assist in decision-making. This form of bioinformatics is sometimes called **medical informatics**, but for convenience, the generic term "bioinformatics" will be used to cover both clinical and research applications. The importance of bioinformatics has closely paralleled the growth in molecular medicine. Two key catalysts for the developments in bioinformatics were the Internet's arrival (Box 5.2) and the Human Genome Project (Chapter 1).

The profile for bioinformatics changed dramatically in the 1980s when DNA sequencing data began to accumulate. These data had to be stored, and the traditional paper methods were inadequate for the volume of information generated. So the sequences were deposited electronically in various databases such as GenBank, EMBL. Information about proteins was placed in databases such as PIR (Protein Information Resource). As well as storage, programs were required to analyse the data. Initially, the major interest was in DNA sequences usually obtained as part of a gene discovery search. Computational analysis was required to determine if the sequence had any homology with other genes. In other words, did this sequence represent a gene, and if so, what was its likely function? For this, DNA sequence was interrogated with a program such as FASTA (abbreviation for <u>Fast</u>—<u>all</u>) to see if it had similarity (the molecular term for this is "homology") to known DNA sequences in the databases. Finding 100% homology was bad news because it was likely that the gene had already been discovered since its sequence was in the database. Finding 0% homology was equally depressing because the gene had not been discovered, but it would be problematic for the researcher to take the finding further without having some clue to a likely function for this gene. The best result was to find some homology. This meant the gene was still novel. Since it had some sequence similarity to another gene that had already been characterised, knowledge of the latter's function could be put to use in understanding how the newly discovered gene might work.

In the past decade, the Human Genome Project has increased exponentially the output of DNA-based data

Box 5.2   Origin of the Internet (Al-Shahi *et al* 2002).

The Internet is a revolution in communication, compatible with the impact made by radio, telephone and television. The Internet comprises a worldwide network of cables and computers that facilitate the transmission of information. It was first developed by the US military in the 1960s to enable rapid information transfer. In 1969 ARPANET, which comprised four host computers connected together, was launched, and in 1972 there was the first public demonstration of the technology, including the first electronic message. The original ARPANET grew into the Internet, which now hosts many services of relevance to molecular medicine, including the WWW (World Wide Web), email and FTP (File Transfer Protocols), which allow documents and software to be rapidly transferred over the Internet. Although the terms "Internet" and "WWW" are used interchangeably, it is relevant to stress that the WWW is only one of the many services available through the Internet.

Apart from the military gains from having a secure means of communication, the motivation for the original ARPANET was the sharing of expensive computers. A secondary goal was electronic mail although this subsequently became a key issue both in science and society. The WWW was launched in 1989 for the sharing of information in physics research. It became public in 1991, and in 1993 a browser was developed that made the WWW a lot more user friendly; hence, its use rapidly expanded (see Chapter 1 for more information on the WWW). To identify the individual Internet servers, unique IP (Internet Protocol) addresses were developed based on the format 111.222.333.444, which gives about

4.2 billion locations (these are now starting to run out!). Browsers allowed the Internet to be used without requiring sophisticated computer programming knowledge and also introduced the options for graphics. The development of search engines such as Google (www.google.com) allowed rapid access to a vast amount of information about any topic. Internet capacity is described as bandwidth, and many service providers now allow the use of broadband to speed up the transmission of information.

As indicated in the text, the Internet is intimately tied up with future developments in molecular medicine. In terms of medical practice, three main concerns about the Internet are (1) security, (2) corruption by viruses and (3) accuracy of information. Security is less of an issue to the individual than it might be to corporations. However, in terms of patient data, the Internet must be adequately protected, and this becomes a major challenge for health facilities, e.g., electronic records, results of patients' investigations. Corruption by viruses is increasingly a problem because of the potential effects on individual computers as well as the downtime experienced by networks. The third concern reflects the vast amount of information on the Internet, the quality of which is variable. The easy access to this information by health professionals, patients and the community means that harm is possible if the wrong information is obtained. For a fascinating tour of the Internet, details on how it started and evolved, go to www.isoc.org/internet/history/brief.shtml, which is the URL for the Internet Society.

---

not only in relation to the human genome, but also many model organisms (see Chapter 1, Table 8.2). Today, nearly 200 whole genomes have been sequenced. Together, they provide data sets that are over 10 times more complex than the human genome. The pressure is now on bioinformatics to produce better software to interrogate this information for two main purposes: (1) Very large segments of DNA sequences must be annotated; i.e., the primary DNA sequence is analysed closely using sophisticated computer programs so that mistakes can be edited, and regions in the sequence likely to contain genes, or other potentially interesting changes, identified (see Figure 4.16). (2) Comparisons or homology searches with the various data sets for model organisms need to be undertaken. This approach is particularly valuable in the design of new vaccines based on unique DNA sequences and so proteins that are specific for various pathogens.

More sophisticated software had to be developed to cope with the increasing complexity in data analysis. The old workhorse FASTA has given way to a program called BLAST (Basic Local Alignment Search Tool), which pro-

vides more rapid and better information about DNA sequences and gene characterisation. As discussed in Chapter 1, Celera's strategy for sequencing whole genomes from human to model organisms was based on breaking up the genome into random fragments and then sequencing each of them individually. The puzzle was reassembled by relying on computer analysis to identify the many overlapping fragments, and in this way the sequence for the whole genome could eventually be reconstructed. The computer power required for this process was formidable.

Today, the level of complexity continues to rise with the requirement for bioinformatics in DNA microarrays. This technology allows gene expression data to be generated from many thousands of genes. The information must be stored and analysed. It must also be cross-compared so that data from a number of experiments can be standardised, and data generated across laboratories should also be comparable. The initial requirement for bioinformatics to provide understanding of simple one-dimensional objects such as a DNA sequence has significantly changed with microarrays since the information now required is related to networks and the relationship between various genes. An interesting experiment is presently under way to meet the hardware (computer power) challenge for bioinformatics. This involves the development of a grid of supercomputers that connect major research institutions. One such initiative funded by the European Union is called the DataGrid Project. Its aim is to build the next generation of computing infrastructure by linking computer and database resources across widely distributed scientific communities. With this type of computer power, homology searches that today take days to a few weeks to complete will be undertaken within seconds to hours.

For proteomics, bioinformatics programs need to search and analyse various protein databases, and ultimately this information will make some prediction about a protein's structure. Complex proteomics data generated from technologies such as X-ray crystallography, nuclear magnetic resonance and electron microscopy are often required to create three-dimensional models of molecules. Fifty years ago, the collective intellects of Francis Crick and James Watson took many months to come up with the double helical model for DNA. Now, modern bioinformatics analysis would have been able to produce a three-dimensional model of DNA based on the X-ray crystallographic pictures of Rosalind Franklin and Maurice Wilkins within a fraction of this time.

IN SILICO *POSITIONAL CLONING*

Positional cloning for gene discovery has been described extensively (Chapters 1, 3, 4, Appendix). The modern version of positional cloning is called *in silico* positional cloning to highlight that gene discovery increasingly relies on computer-based strategies and information found in databases rather than the traditional "wet laboratory" approach. Indeed, the latter may gradually disappear, and all new genes will be found and characterised *in silico*. Discovering genes for multifactorial genetic disorders is more difficult than the single gene disorders because the gene effects in the former are more subtle, they may be additive and the modes of transmission are difficult to identify because of the relatively small effects. Confounding environmental interactions are also found. Therefore, different strategies are needed, and a direct analysis of DNA sequences in databases called *in silico* positional cloning is now being pursued.

In silico positional cloning describes the analysis of candidate genes obtained directly from the databases. Another shortcut involves the identification of functional SNPs in candidate genes. Most SNPs are in non-gene regions of the genome, and so they provide no direct information about function. However, some SNPs are found within a gene's coding sequence (they are sometimes called cSNPs or coding SNPs) or the gene's regu-

latory regions. This class of SNPs could have functional implications because of their locations, and so they are potentially more valuable compared to SNPs that are found in regions that do not contain gene sequences. cSNPs provide a shortcut to identifying a gene that might be important because it is relatively easy to conduct an association study by testing one particular cSNP with some prior knowledge that different alleles of that cSNP are likely to show functional variability in terms of gene expression (Figure 5.7).

MEDICAL INFORMATICS

In a rapidly moving field such as genomics, information needs to be regularly updated. The Internet is the best route for accessing databases and journals that provide the health practitioner (and often patients and families) with relevant information. In terms of genetic disorders, one of the most useful databases is OMIM—*Online Mendelian Inheritance in Man*. This regularly reviewed database is an extensive resource, with more than 15 000 entries. For each clinical condition described, it provides links with relevant publications as well as the related DNA or protein data and comes in a historically formatted summary. Other useful databases are listed in Tables 5.6 and 5.7.

Computer software is increasingly being used by health professionals to enhance their practice. Patient information recorded electronically provides the start for electronic medical records or computer-generated prescriptions that are useful to highlight potential adverse drug combinations. Commercial programs such as Cyrillic (see the software company FamilyGenetix on http://www.familygenetix.com/) allow pedigrees to be drawn. More interesting for future practice in genomics will be the availability of computer-generated algorithms for decision-making. A simple example of this is given in the study of Emery *et al* (2000), in which 18 simulated cases involving familial breast and ovarian cancers were reviewed by general practitioners in the United Kingdom using three alternative approaches: (1) a computer-based decision support package designed for general practice (i.e., primary care physicians), (2) a commercial pedigree drawing program and (3) the traditional pen and pencil to draw a pedigree and determine risks. Not surprisingly, (1) was shown to be superior based on criteria such as correct referrals, correct drawing of pedigrees and the time taken to complete each case. The company mentioned above is presently validating the program GRAIDS (Genetic Risk Assessment by Internet with Decision Support) in the general practice environment to enhance risk assessment, decision-making and communication for genetic cancers. The National Cancer Institute has also developed a useful Internet-based program that allows the doctor or counsellor to input clinical information of relevance to breast cancer risk such as family history and

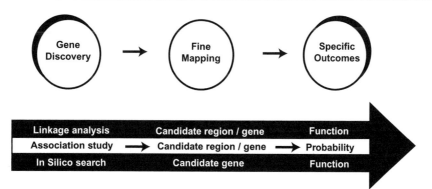

Fig. 5.7 Gene discovery pipeline.
Strategies: The gene discovery pathway can follow traditional linkage analysis in the single gene mendelian disorders, or in the case of more complex diseases, an association study is needed. Alternatively, a direct *in silico* search of the genomic databases can be made by computer for both single gene and complex genetic disorders. **Narrowing the region of interest**: Once a genomic region of interest is defined, it needs to be narrowed in size by fine mapping to reduce the number of genes in that region. The narrowing step can form part of the positional cloning or can utilise the identification of likely candidate genes. **Outcomes**: In the linkage study, predicting function in the recently discovered gene might require a transgenic animal to be made, or could be determined from the type of underlying mutations found in disease. The outcome from an association study generally involves a probability that a gene is involved, and additional work is required to translate this into function. The *in silico* gene discovery pathway is potentially more flexible since candidate genes form the basis of the study, and the identification of cSNPs in these genes allows function to be considered at an early stage in the pathway.

Table 5.6   Some useful bioinformatics web sites

Name	URL (Uniform Resource Locator)[a]	Comments
National Center for Biotechnology Information (NCBI)	www.ncbi.nlm.nih.gov	Contains a repository for many bioinformatics tools and databases
BioInform	www.bioinform.com	Provides bioinformatics news but requires subscription for access
International Society for Computational Biology	www.iscb.org/	Provides useful links in bioinformatics and is a site more likely to be of interest to the reader with a strong commitment to bioinformatics
GeneCards	http://bioinformatics.weizmann.ac.il/cards/	Provides a vast amount of data (38338 entries) on genes
Cytochrome P450 nomenclature committee home page	www.imm.ki.se/cypalleles	Interesting to click on this and observe the considerable heterogeneity (although not unexpected) with DNA changes in the various P450 genes
Human Gene Mutation Database	http://archive.uwcm.ac.uk/uwcm/mg/hgmd0.html	Contains an international database of genetic mutations causing a wide range of diseases
Zebrafish information network	www.zfin.org	Provides a lot of information directed predominantly to high school students and their use of the zebrafish for school-related activities
Clinical proteomics program data bank	http://clinicalproteomics.steem.com	Discusses clinically relevant issues in proteomics; gives examples and further links
Proteomics and Cancer	http://www.cancer.gov/newscenter/pressreleases/proteomicsQandA/	Discusses clinically relevant issues in proteomics; gives examples and further links

[a] URLs without the http:// will have this added automatically when the address is written.

Table 5.7 Clinically relevant resources available online

Name	URL (Uniform Resource Locator)	Comments
Online Mendelian Inheritance in Man (OMIM)	www.ncbi.nlm.nih.gov/entrez/query.fcgi?db=OMIM	A must for any clinician dealing with genetic diseases. Reputable and regularly updated. Links to DNA and protein information and databases.
The National Cancer Institute (USA) information service	www.cancer.gov/cancer_information/	Summarises evidence-based medicine with information on genetic basis for various cancers, including breast, ovarian, colorectal. Comprehensive site with lots of useful facts.
Breast cancer risk assessment tool	http://bcra.nci.nih.gov/brc/	Provides an Internet-based interactive tool for measuring a woman's risk of invasive breast cancer.
OMNI (UK Internet resource in health and medicine)	www.omni.ac.uk/	Identifies more than 8000 resource items.
Canadian Diabetes Association web site	www.diabetes.ca	Provides wide-ranging information for patients and health professionals dealing with diabetes.
GeneClinics	www.geneclinics.org	Discusses information relevant to diagnosis, management and counselling for specific genetic disorders.
National Centre for Biotechnology Information (NCBI)	www.ncbi.nlm.nih.gov/About/primer	Provides a science primer with useful summaries of many topics in genomics.
Your disease risk	www.yourdiseaserisk.harvard.edu/	Interactive site allowing an individual's risks to be assessed for cancer, diabetes, heart disease, osteoporosis and stroke.
Genomics and Population health	www.cdc.gov/genomics/activities/ogdp/2003	Impressive site from the USA's CDC dealing with a range of clinically relevant public health issues in genomics.
Gene Therapy Clinical Trial Site	www.wiley.co.uk/genmed/clinical/	Lists about 636 gene therapy studies undertaken worldwide.
GeneTests	www.genetests.org	Provides educational material and a directory to many DNA testing laboratories, particularly in the USA.

previous breast pathology. This information is then returned to the health professional in the form of a relative risk. The program also has succinct information about various options available for the at-risk patient (see Table 5.7).

The Internet will be essential for educating patients, families and the community at large. A potentially receptive target to direct concepts of molecular medicine and the implications for future health are schoolchildren, many of whom have been brought up with computers and the Internet from their earliest days in school. The genomics era will continue to be a strong driving force for further development and sophistication of bioinformatics. Apart from the obvious research value of bioinformatics, an opportunity now exists to develop Internet resources for education and clinical decision-making.

e-Counselling and e-Consultations

Professional counselling services provided in conjunction with genetics clinics are increasingly faced with greater demands and more complex scenarios. This trend will continue as new genes and genetic risks are defined. In this environment, the traditional one-to-one, face-to-face counselling needs reassessment, with the potential that computer-based education might provide better ways in which to deliver counselling, which can be repetitive and not necessarily requiring the same detail for each patient. Following up and ensuring that the patient has understood what are difficult concepts of risk and even a basic understanding of disease and inheritance are further challenges in this environment. While it is acknowledged that each individual patient (or couple) has particular problems or questions that need to be addressed, and ultimately this may be possible only with a personal approach, there is no reason why computer-based educational resources cannot be developed to assist in counselling, i.e., the concept of e-counselling. These resources could be made available to those requiring counselling, and this is then followed by the more formal counselling session. This approach provides the patient (or couple) with the opportunity to consider the various issues before the formal consultation, and so questions and concerns can be more specific. The counsellor also has the opportunity to gauge the level of knowledge or comprehension based on the previous e-counselling and so structures the counselling sessions and any follow-up visits accordingly.

An interesting concept to consider as the Internet reaches more homes is the e-consultation, i.e., the online doctor-patient consultation. Apart from the privacy and confidentiality issues related to Internet traffic, this approach has potential advantages for the patient—and even the doctor when it comes to simple problems. However, there will always remain the medico-legal issues since it becomes even more difficult to establish the patient's ability to comprehend the information provided, and the doctor may not have the whole picture from a brief email message. However, as access to the Internet increases, it is likely that this medium will be used as part of the consultative process. "Guidelines for the Clinical Use of Electronic Mail with Patients" is the interesting title for a document in the *Journal of the American Medical Informatics Association*. This article provides guidelines for email exchanges between health professionals and patients.

FUTURE

. . . OMICS

The evolution into the . . . omics (genomics, proteomics, transcriptomics) refers to the nontargeted identification of all genes, proteins or mRNA transcripts in a particular biological specimen or organism. To the . . . omics could also be added epigenomics (from epigenetics), phenomics (from the phenotype) and metabolomics. The big picture view possible with the . . . omics sets the stage for future research in molecular medicine and distinguishes the above from previous activities. For example, metabolic studies are not new, but now with modern technologies such as nuclear magnetic resonance (NMR) and mass spectrometry (MS), it becomes possible to obtain a global metabolic profile in any cell, tissue or organism (metabolomics). Previously, the study of metabolites was focused to a limited number of products. There are three limitations to the promised great leap forward in . . . omics. Two, involving high costs and technical issues, are resolvable, as is occurring in genomics. The third is bioinformatics, an important area for directing future resources as well as building more human capacity.

In parallel with increasing automation will come more effective and efficient genomics strategies. In particular, the present focus on a mutation in a DNA sample (genetics era) will be replaced by DNA chips that allow many hundreds or thousands of DNA mutations to be assessed simultaneously (genomics era). Technologic developments will facilitate DNA analysis, but at the same time, the amount of information available will provide a challenge in terms of protecting privacy confidentiality and developing better education strategies. The issues of high costs and complexity found with the microarrays will not take long to resolve. An important driver for this is the long lag time and the high costs involved in developing new drugs. For example, it is estimated that it can take up to 10 years and about a billion dollars to get a new drug into the marketplace, and along the way there are many failures. Pharmaceutical companies are also concerned that the drug discovery pipeline is slowing, with fewer drugs now emerging. Therefore, the industry is placing a lot of interest in microarrays as a means by which costs can be reduced as described earlier in pharmacogenetics (see also Figure 5.5), and more targets for drug development are identified. This will ensure that developments in microarrays are fast tracked, leading to important practical spin-offs at the clinical interface.

An interesting paper by Zambrowicz and Sands (2003) looks at the 100 best-selling drugs and identifies the role played by mouse knockouts in the drugs' development and regulatory processes. The authors comment on the pharmaceutical industry and the pressure to identify the next 100 winners as quickly, cheaply and effectively as is practical. In this, the potential to alter the phenotype (for example, the ENU mouse) or the genotype (the knockout) will play a major role.

Health care systems are struggling with the rising costs of drugs. In some countries, second- and third-generation drugs are unaffordable. As well as paying for drugs, the health care system needs to devise ways in which the safety of drugs can be monitored and adverse effects minimised. Therefore, there is considerable interest in the potential for pharmacogenomics, with the promise that this technology will reduce adverse events, identify new drug targets and reduce the costs required in the regulatory processes by allowing smaller but genetically more selective cohorts to be used in clinical trials.

NANOTECHNOLOGY

As well as the promising developments that will come with bioinformatics, an equally exciting field that will impact on molecular medicine is nanoscience. Nanoscience is defined as the study of the fundamental principles of molecules and structures with at least one dimension approximately between 1 and 100 nanometres (1 nanometre [nm] is 1×10^{-9} of a metre). To put this size into perspective, a human hair measures about 50 000 nm, and the smallest particle visible to the human eye is 10 000 nm.

The nano dimension is not simply a reflection of smallness, but an entirely new concept that sits somewhere between human-made devices and molecules found in living things. As such, the usual restrictions and rules such as conductivity, hardness, melting point and so on, need to be redefined. Like bioinformatics, which crosses a number of disciplines (biology, engineering, computer science and information technology), nanoscience involves many disciplines—chemists, physicists, engineers (materials, electrical and chemical).

One application of nanotechnology in molecular medicine will be the availability of DNA biosensors. In Chapters 2 and 3, the potential to construct a small oligonucleotide and hybridise this to its unique corresponding DNA sequence in the genome was shown to be the basis for amplification by PCR. Once a DNA sequence is amplified, its presence can be recognised in various ways, for example, by electrophoresis and so size determination, or by DNA hybridisation. For the latter, primers are labelled with fluorescein, and the presence of hybridisation is detected with a laser beam. An alternative approach that will be possible through biosensors will involve the use of gold or silver nanodots that can be attached to the DNA. The hybridisation of oligonucleotide-labelled nanodots will be instantly recognisable with calorimetric detectors without the requirement for DNA amplification by PCR. This miniaturisation of the detection process will allow DNA testing to be moved out of the traditional "wet" laboratory to the consulting office or the bedside.

Drug delivery and bioavailability remain key determinants of a drug's effectiveness. In terms of gene therapy, the inability to deliver sufficient amounts of DNA into the nucleus is a key issue that remains unresolved. Nanotechnology will open up new strategies to target drugs to places where they are needed and get them there efficiently, for example, by encapsulating them within nanoscale cavities inside polymers. Lipid encapsulation has been used to deliver DNA (i.e., genes) into cells, but like the viral vectors, the amount transfected in this way is very small. An intriguing possibility with nanotechnology is the use of magnetic nanoparticles that can be bound to the drug (or gene) to be delivered, and then magnetic fields outside the body used to guide the drug (or gene) to the location required.

FURTHER READING

Introduction

Campbell AM, Heyer LJ. Discovering genomics, proteomics and bioinformatics. 2003. Cold Spring Harbor Laboratory Press, San Francisco (*useful introductory textbook*).

Guttmacher AE, Collins FS. Genomic medicine—a primer. New England Journal of Medicine 2002; 347:1512–1520 (*overview of genomics in medicine, including many examples and a glossary*).

Genomics

Aitman TJ. DNA microarrays in medical practice. British Medical Journal 2001; 323:611–615 (*overview of microarrays in relation to medicine with nice illustrations*).

Chasman DI, Posada D, Subrahmanyan L, Cook NR, Stanton VP, Ridker PM. Pharmacogenetic study of statin therapy and cholesterol reduction. Journal of the American Medical Association 2004; 291:2821–2827 (*well-written and informative study how pharmacogenetics can identify patient subgroups responding suboptimally to treatment with statins for cholesterol lowering*).

Costa LG, Cole TB, Jarvik GP, Furlong CE. Functional genomics of the paraoxonase (*PON1*) polymorphisms: effects on pesticide sensitivity, cardiovascular disease, and drug metabolism. Annual Reviews of Medicine 2003: 54:371–392 (*various functions of paraoxonase and the associated molecular variants*).

Cox RD, Brown SDM. Rodent models of genetic disease. Current Opinion in Genetics & Development 2003; 13:278–283 (*summary of the various rat and mouse models used in modern genomics research*).

Evans WE, McLeod HL. Pharmacogenomics—drug disposition, drug targets and side effects. New England Journal of Medicine 2003; 348:538–549 (*overview of the various factors involved in drug metabolism*).

Hashimoto M, Rockenstein E, Masliah E. Transgenic models of α synuclein pathology: past, present and future. Annals New York Academy of Sciences. 2003; 991:171–188 (*description of transgenic mice models for Parkinson's disease*).

Holleman A et al. Gene expression patterns in drug resistant acute lymphoblastic leukemia cells and response to treatment. New England Journal of Medicine 2004; 351:535–542 (*shows potential value of microarray analysis of tumour cells*).

Langheinrich U. Zebrafish: a new model on the pharmaceutical catwalk. BioEssays 2003; 25:904–912 (*shows the contribution of the zebrafish to knowledge of human development and behaviour*).

Lazarou J, Pomeranz BH, Corey PN. Incidence of adverse drug reactions in hospitalised patients: a meta-analysis of prospective studies. Journal of the American Medical Association 1998; 279:1200–1205 (*provides estimates of drug side effects in hospitalized patients*).

Lindpainter K. Pharmacogenetics and the future of medical practice. Journal of Molecular Medicine 2003; 81:141–153 (*overview of pharmacogenetics particularly in the pharmaceutical industry*).

Marshall E. Preventing toxicity with a gene test. Science 2003; 302:588–590 (*summarises the events around the FDA's decision not to make TPMT DNA testing compulsory*).

Maughan NJ, Lewis FA, Smith V. An introduction to arrays. Journal of Pathology 2001; 195:3–6 (*brief but useful introduction to microarrays. This issue of the* Journal of Pathology *is directed to microarrays—well worth reading*).

Meisel C *et al*. Implications of pharmacogenetics for individualizing drug treatment and for study design. Journal of Molecular Medicine 2003; 81:154–167 (*considers basic pharmacologic principles behind drug metabolism as well as the genetic effects*).

Paul NW, Roses AD. Pharmacogenetics and pharmacogenomics: recent developments, their clinical relevance and some ethical, social and legal implications. Journal of Molecular Medicine 2003; 81:135–140 (*distinguishes pharmacogenetics and pharmacogenomics*).

Van de Vijver MJ *et al*. A gene expression signature as a predictor of survival in breast cancer. New England Journal of Medicine 2002; 347:1999–2009 (*a key paper describing the application of an expression microarray in breast cancer management*).

Weinshilboum R. Inheritance and drug response. New England Journal of Medicine 2003; 348:529–537 (*useful overview of CYP genes*).

Wood AJJ. Racial differences in the response to drugs—pointers to genetic differences. New England Journal of Medicine 2001; 344:1292–1396 (*editorial illustrating the molecular mechanism for racial susceptibility to drugs*).

Proteomics

Check E. Running before we can walk? Nature 2004; 429:496–497 (*a response to the Petricoin and Liotta paper on proteomics and ovarian cancer*).

Davidov EJ, Holland JM, Marple EW, Naylor S. Advancing drug discovery through systems biology. Drug Discovery Today 2003; 8:175–183 (*describes the integrated . . . omics approach now being used in drug development*).

Patterson SD, Aebersold RH. Proteomics: the first decade and beyond. Nature Genetics 2003; 33(suppl): 311–323 (*detailed but comprehensive overview of how proteomics has developed and where it is going*).

Petricoin EF, Liotta LA. Clinical applications of proteomics. Journal of Nutrition. 2003; 133:2476S–2484S (*the focus here is how proteomics, including new developments, will be used in oncology*).

Walgren JL, Thompson DC. Application of proteomic technologies in the drug development process. Toxicology Letters 2004; 149:377–385 (*gives an overview of proteomics strategies in drug development*).

Yanagisawa K *et al*. Proteomic patterns of tumour subsets in non small cell lung cancer. Lancet 2003; 362:433–439 (*example of how proteomic profiles are likely to become valuable adjuncts in cancer management*).

Bioinformatics

Al-Shahi R, Sadler M, Rees G, Bateman D. The internet. Journal of Neurology, Neurosurgery and Psychiatry 2002; 73:619–628 (*succinct overview of the Internet's development*).

Bayat A. Bioinformatics. British Medical Journal 2002; 324:1018–1022 (*readable overview directed to medical applications*).

Botstein D, Risch N. Discovering genotypes underlying human phenotypes: past successes for Mendelian disease, future approaches for complex disease. Nature Genetics 2003; 33 (suppl):228–237 (*very sophisticated review explaining where gene discovery has been and where it is going with the use of SNPs, DNA databases and bioinformatics*).

Emery J *et al*. Computer support for interpreting family histories of breast and ovarian cancer in primary care: comparative study with simulated cases. British Medical Journal 2000; 321:28–32 (*relatively simple but illustrative example of how computer-based information helps with clinical decision-making*).

Kane B, Sands DZ. Guidelines for the clinical use of electronic mail with patients. Journal of the American Medical Informatics Association 1998; 5:104–111 (*provides guidelines for email exchanges between health professionals and patients*).

Kanehisa M, Bork P. Bioinformatics in the post-sequence era. Nature Genetics 2003; 33(suppl):305–310 (*provides historical overview of bioinformatics development*).

Internet based resources are summarised in Tables 5.6, 5.7.

http://eu-datagrid.web.cern.ch/eu-datagrid/ (*web address for the EU's initiative to building computing infrastructure for the future*).

Future

Emilien G, Ponchon M, Caldas C, Isacson O, Maloteaux J-M. Impact of genomics on drug discovery and clinical medicine. Quarterly Journal of Medicine 2000; 93:391–423 (*nice review of molecular medicine in the genomics era*).

Ratner M, Ratner D. Nanotechnology. A gentle introduction to the next big idea. 2003. Prentice Hall, New Jersey ISBN 0-13-101400-5 (*provides an exciting glimpse into the future*).

Robertson JA, Brody B, Buchanan A, Kahn J, McPherson E. Pharmacogenetic challenges for the health care system. Health Affairs 2002; 21:155–167 (*attempts to forecast the effect of pharmacogenetics on industry, regulation, medical practice and the patients*).

Zambrowicz BP, Sands AT. Knockouts model the 100 best-selling drugs—will they model the next 100? Nature Reviews Drug Discovery 2003; 2:38–51 (*an interesting paper summarising the best sellers (in drugs) and the role played by transgenics in their development*).

http://www.royalsoc.ac.uk/document.asp?id=2023 (*address for the UK's Royal Society and the Royal Academy of Engineering report–"Nanoscience and nanotechnologies: opportunities and uncertainties"—published in 2004. Very comprehensive overview of the topic*).

GENETIC AND CELLULAR THERAPIES

RECOMBINANT DNA–DERIVED DRUGS

FACTOR VIII

A review of the treatment options in haemophilia A demonstrates the potential for molecular technology in drug design and production of recombinant therapeutics. The haemophilia model also demonstrates the use of PCR technology to ensure greater safety in the supply of blood products (see Chapters 3 and 8 for clinical details about haemophilia and molecular infection control measures).

The factor VIII gene provides many challenges in producing a drug by recombinant DNA (rDNA) means. It is large (26 exons) with a genomic structure extending over 186 kb. The protein comprises 2332 amino acids and is synthesised as a single chain that is processed as it moves from the rough endoplasmic reticulum to the Golgi apparatus. As this occurs, the mid-portion (B subunit) of the molecule is excised since it is not required for haemostatic function (see Figure 3.16). The heterodimers formed are held together by calcium. Post-translational processing is needed.

Plasma-derived Products

A historical summary of developments in the treatment of haemophilia is given in Table 6.1. Landmarks include the isolation in 1964 of a specific factor VIII–enriched product known as cryoprecipitate. In the 1970s, antihaemophiliac factors with higher concentrations and improved stability led to more effective treatment programs utilising home therapy. In this way patients were able to treat themselves the instant a bleeding episode was noted. However, complications occurring with the plasma-derived antihaemophiliac factors remained significant (Table 6.2). The initial problems associated with hepatitis B infection from blood products were thought to have been resolved once blood banks introduced donor screening programs, and viral-inactivating steps were incorporated into commercial production. Nevertheless, the subsequent recognition of other viruses (e.g., HIV, hepatitis C and parvovirus) again highlighted the problems of human-based biological products. The increasing number of haemophiliacs infected with HIV illustrated further limitations of plasma products, even those that had undergone viral-inactivation steps. These processes included various combinations of heating and/or organic solvent exposure. These processes increased the production costs but gave no guarantee that all viruses (both known and unknown) would be neutralised. For example, parvovirus can withstand temperatures to 120°C, and viruses without lipid envelopes are not sterilised by organic solvents or detergents.

Table 6.1 Milestones in the treatment of haemophilia A

Year	Discovery
1840	Bleeding episode treated with normal fresh blood.
1920s	Plasma rather than whole blood shown to be effective.
1930s–1950s	Fractionation of plasma → various components with antihaemophiliac activity. 1937: Factor VIII implicated as the cause of haemophilia A.
1964	Cryoprecipitate isolated. Produced by allowing frozen plasma to thaw. When it is thawed, a cold insoluble precipitate (cryoprecipitate) remains. This precipitate contains a high concentration of factor VIII.
1970s	High-potency freeze-dried factor VIII concentrates available, allowing prophylactic home therapy that is both feasible and effective.

1980s	The low point in haemophilia treatment occurs in the early 1980s with the finding that HIV and HBV in the blood supply infect many haemophiliacs. This leads to the following:
	  Biotechnology companies Genentech (San Francisco) and Genetics Institute (Boston) clone and express the factor VIII gene (1984) with the aim to produce a recombinant DNA product.
	  More effective viral-inactivation steps incorporated in the manufacture of factor VIII products.
	  Monoclonal antibody-purified factor VIII becomes available.
	Alternative haemostatic pathways used to bypass the effect of inhibitors.

1988	Clinical trials of rh factor VIII start; encouraging results emerge.
1990s	Early 1990s: First generation rhFVIII and rhFIX products licensed for clinical use. rhFVII used to bypass factor VIII inhibitors. Novel means of expressing rhFIX tried, e.g., transgenic expression in sheep milk. 1999—First B domain–depleted rhFVIII released. 1999—Gene therapy clinical trials in haemophilia A and B start in USA.
2001	Transient but significant increases in FVIII or FIX levels reported after gene therapy.
2003	Second generation rhFVIII well established. Unlike first generation recombinant products, these do not use bovine albumin or other human proteins as stabilisers although there is exposure to albumin during some steps in manufacture. However, no reports of infections occurring with recombinant products. Third generation rhFVIII (these have no exposure to exogenous animal or human albumin or plasma proteins during any stage in manufacture) developed. Viral inactivation steps also added to manufacturing process. Trials under way. The first (ADVATE) was approved by the FDA in 2003.
2004	Third generation rhFVIII shown to be as effective as earlier recombinant products with less risk of infections both known and unknown being transmitted.

Liver disease as a cause of morbidity and mortality in haemophilia is well documented. An association between hepatitis B virus, hepatitis C virus and liver disease is established. Immunosuppression involving both B and T cell lymphocyte function and independent of infection with HIV was also found in haemophiliacs. A mechanism proposed for immunosuppression implicated the many contaminating plasma proteins present in the factor VIII or factor IX products.

Finally, a critical consideration with any human-derived product is its availability, which will always depend on a regular supply of donors. This supply can never be guaranteed. Most developed countries have moved away from plasma-derived products, although for financial reasons, they may still be used by the older haemophiliacs already infected with HIV or hepatitis. There is little option but to use plasma-derived factors

if recombinant products are not available or are too expensive.

Monoclonal antibodies to factor VIII were included in production protocols during the 1980s. The resulting product had a higher potency and greater purity. The reduction in contaminating proteins was considered to be an added bonus if this effect was significant in the immunosuppression observed in haemophiliacs.

rDNA-derived Products

In 1987, the first patient was treated with a recombinant DNA–derived human factor VIII. The steps involved in preparing this product are summarised in Figure 6.1. The use of mammalian cell lines such as CHO (Chinese hamster ovary) enabled complex post-translational steps (e.g., glycosylation) to be undertaken. Recombinant

Table 6.2 Problems associated with use of plasma-derived haemophilia treatment products

Problem	Details
Infection	Lipid enveloped viruses: Hepatitis B, hepatitis C, HIV, West Nile virus. 60–70% of haemophiliacs with severe disease in the 1980s infected with HIV. Higher infection rate for HCV. Non-lipid enveloped viruses: Hepatitis A, Parvovirus B19 Others: Slow virus infections; a range of organisms including non-viruses that are suspected but not proven to be pathogenic; unknown organisms or ones yet to emerge.
Liver disease	Progressive and potentially fatal liver disease in 10–20% of those with chronic HBV or HCV developing cirrhosis. Risk for hepatocellular carcinoma 30 times higher than general population.
Immunosuppression	Contaminating proteins in factor concentrates (including pure ones) implicated as the cause for immunosuppression. Both T and B cell function impaired.
Inhibitors	Exposure to neoantigens produces a risk of antibodies developing against coagulation factors (10–20% of haemophilia A, 3–5% of haemophilia B) with some correlation to the underlying molecular defect. This complication associated with all types of factor VIII concentrates.
Cost	Antihaemophilic factor expensive, particularly when costs of purification and viral inactivation are added.
Availability	Plasma-derived products always limited. Less of an issue now that there are the recombinant products, but the availability of the latter is limited by cost.

Table 6.3 Relative purities of various factor VIII preparations

Source	Activity (units/mg protein)
Plasma	0.01
Cryoprecipitate	0.3
Concentrates 1970s	0.5
Concentrates 1980s	1.5
Immunopurified	2000
Recombinant	2000

factor VIII produced this way was shown to have fewer protein contaminants although some were invariably present. The activity of the recombinant product was equivalent to monoclonal antibody-purified factor VIII, and the potential to develop inhibitors is comparable to other products (Table 6.3).

Expression studies of the recombinant DNA–derived factor VIII product have shown that optimal stability requires association with the von Willebrand factor. Therefore, one commercial product was obtained by co-transfecting cDNAs for both factor VIII and the von Willebrand factor. It was also expected that removal of the B-domain (not required for haemostasis) would facilitate expression of recombinant factor VIII for commercial production. The first multicentred human trials of recombinant human factor VIII were started in 1988, and results soon showed promising laboratory and clinical responses. These studies also confirmed that recombinant and plasma-derived factor VIIIs were biologically identical.

The value and efficacy of recombinant DNA–derived factor VIII are now well established, and since introduced, there have been no reports of infections resulting from its use. The first generation recombinant products were soon replaced by second generation factors, and now there are third generation products that contain no human or non-human protein products during manufacture. The availability of a regular and controllable supply possible with recombinant products will allow better planning and more effective treatment of bleeding problems. Balancing this are the high costs for recombinant therapeutics, and so they are not affordable to many. However, the costs will eventually fall, and then plasma-derived substances for the treatment of coagulation disorders will no longer be needed.

Inhibitors in Haemophilia

Apart from treatment issues in haemophilia, another application of molecular medicine is to provide a better understanding of the development of inhibitors (antibodies) to factors VIII or IX and expand the treatment options for this complication. The finding of inhibitors in haemophilia is a serious development occurring in approximately 10–20% of patients with haemophilia A, or 3–5% of patients with haemophilia B. Inhibitors occur as a consequence of exposure to new antigens (neoantigens) to which immunological tolerance has not developed. These individuals are placed at a much greater risk of dying from an uncontrollable bleeding episode since conventional factor replacement becomes ineffective. Genotype/phenotype correlations show that inhibitor formation is more likely to occur in those with factor VIII mutations that produce a truncated protein, or

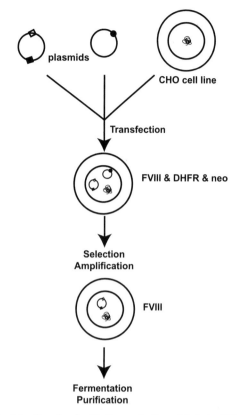

Fig. 6.1 Steps involved in the production of rh factor VIII.
The recombinant plasmid has genes (in the form of cDNAs) for factor VIII (■) and dihydrofolic acid reductase (DHFR = □) as well as their respective promoters (not drawn to scale). The factor VIII gene requires a mammalian cell line such as CHO (<u>C</u>hinese <u>h</u>amster <u>o</u>vary) to enable post-translational changes to occur. Another plasmid containing a gene for resistance to the antibiotic neomycin (neo^r) is also co-transfected (●) into CHO cells. The result is CHO cells containing factor VIII, DHFR and neo^r genes that remain as extrachromosomal DNA or, for more stable expression, are integrated into the CHO genome. Specific selection for CHO cells containing the neo^r gene is possible by culturing in a medium containing G418, a neomycin analogue that is lethal to cells unless they carry a gene for neomycin resistance. Amplification for the factor VIII containing CHO cells is then obtained by culturing the CHO cells in medium with increasing concentrations of methotrexate. This will select CHO cells that have the DHFR gene (and so also the factor VIII) since DHFR-deficient CHO cells will not grow in the presence of methotrexate. The factor VIII-expressing CHO cells are finally fermented in large, commercial volumes. Protein isolated from these cells is purified by immunoaffinity chromatography with monoclonal antibodies to factor VIII. The end product is checked for contaminants (mouse, CHO proteins, DNA) and its functional activity assessed. Recombinant human factor VIII is then available for clinical use.

significantly disrupt production of factor VIII protein as would occur with the flip tip inversion (see Table 3.15 for a summary of mutations in haemophilia A). In contrast, missense mutations are less disruptive of the protein's structure and so less likely to be associated with inhibitor formation.

Activated plasma-derived coagulation factors (mixtures of activated factors X and VII and tissue factor) were produced to overcome the block on factor VIII activation resulting from the development of antibodies. However, they were expensive, had all the potential complications associated with factor VIII and, in addition, were reported to increase the risk for thrombosis. Activated factor VII on its own appeared to be a useful way to bypass the factor VIII block, but isolation of this substance from plasma was commercially impractical. This problem was resolved with the development of a recombinant DNA–derived product. Recombinant human factor VII has now been successfully used in a number of patients with factor VIII antibodies, including one report involving a haemophiliac with antibodies to both human and porcine factor VIII who was also a Jehovah's Witness. This individual developed a life-threatening cerebral haemorrhage but refused treatment with human blood products. However, recombinant DNA–derived substances were acceptable, and recombinant factor VII eventually stopped his bleeding.

GROWTH HORMONE AND GONADOTROPINS

Human growth hormone, a protein of 191 amino acids, is essential for growth. Because this hormone is species-specific, its only biological source is human. Following the successful treatment of a pituitary dwarf in 1958 with human growth hormone, programs were established to isolate this substance from pituitaries obtained from human cadavers. However, the programs were ceased in the mid-1980s when a number of recipients died from Creutzfeldt-Jakob disease, a fatal slow virus infection of the central nervous system (see Chapter 8).

Growth hormone has now been replaced with a recombinant product following the cloning and expression of the gene in 1979. Because the mature protein does not require sophisticated post-translational modifications, it can be prepared using a relatively simple bacterial expression system. However, two potential disadvantages are associated with this expression system: (1) the necessity for extensive purification to remove impurities of bacterial origin, particularly endotoxins, and (2) the presence of an additional methionine amino acid at the start of the protein. The latter occurs because the eukaryotic start codon (ATG) is translated in the prokaryotic system into a methionine. Clinical trials during the mid-1980s demonstrated the efficacy of the recombinant growth hormone, and it has remained in contin-

Table 6.4 **Problems inherent in urinary-produced gonadotropins**

Logistics of collecting large urine volumes.

Pooling required produces significant batch-to-batch variability in terms of potency.

Increasing worldwide demand difficult to meet.

Contamination invariable including urine proteins, infectious agents and excreted drugs.

Difficult to avoid luteinising hormone (LH) in the preparations, so hormonal purity difficult to achieve.

uous use since. No significant side effects have become apparent. The additional methionine does not, as originally feared, lead to a major increase in antigenicity of the product.

Similar developments occurred with the gonadotropins, a glycoprotein hormone family used in infertility treatment and comprising FSH (follicle stimulating hormone), LH (luteinising hormone) and HCG (human chorionic gonadotropin). These gonadotropins were first prepared from animal products in the 1930s. Subsequently, human pituitary glands were used, but this source was discontinued for the reason given earlier with growth hormone. Urinary-derived gonadotropins then remained the only available products, and despite a good record for safety, there were concerns about the human source (Table 6.4). The problem was solved when the genes for the above hormones were isolated and expressed in CHO cells because, like haemophilia, post-translational glycosylation was required for activity. rhGonadotropins are now available. They demonstrate 99% purity, leading to higher specific activity and lower immunogenicity. Risks of infections or being exposed to other foreign proteins in urine are now eliminated. The relative costs of urinary derived and recombinant product remain controversial depending on how the cost analysis is undertaken. However, there is little doubt that the recombinant hormones will become the first choice for infertility treatment.

HAEMATOPOIETIC GROWTH FACTORS

Bone marrow haematopoietic (blood forming) cells are derived from the proliferation and differentiation of progenitor cells that form specific lineages following their interaction with cytokines. The pluripotential stem cell is the ultimate source of the lymphoid and myeloid precursor cells. The latter differentiates into the platelet, erythroid, neutrophil and macrophage lineages. As the progenitors mature into committed stem cells, they become responsive to a more limited range of cytokines although a number are often required for optimal effect. The major cytokine groups are the colony stimu-

lating factors (CSFs) and the interleukins (ILs). These substances are involved in the regulation, growth and differentiation of a variety of cellular components. Haematopoietic growth factors include the CSFs, the ILs and erythropoietin.

The CSFs stimulate the formation and activity of white blood cells including granulocytes and macrophages. The prefix indicates the target cell, i.e., G-CSF (granulocyte); M-CSF (macrophage or monocyte). GM-CSF is less lineage-restricted since it can affect both granulocytes and macrophages. GM-CSF also stimulates the precursor cells in the erythroid (red blood cell) and megakaryocyte (platelet) lineages. Many cell types including fibroblasts and endothelial cells produce one or more of the CSFs. Base-line production is low but rises rapidly following exogenous stimuli such as bacterial endotoxins. In turn, the CSFs provide the proliferative signals for cellular differentiation.

Knowledge that haematopoietic growth factors existed has been around since the 1960s. However, the minute amounts isolated and the complex interactions and target cells associated with these factors limited major progress (Figure 6.2). In the 1970s, traditional biochemical methods enabled small quantities of M-CSF and GM-CSF to be produced, followed by G-CSF in the early 1980s. However, nothing further could be done because of the amounts available. The *in vivo* significance of these products remained unclear until the relevant genes were cloned. The first to be cloned by AMGEN, a US biotechnology company, was G-CSF in 1986. The others soon followed. Expression of the cloned genes enabled the products to be tested using *in vivo* animal studies, then primate and finally human trials. The FDA approved specific clinical uses of G-CSF and GM-CSF in 1991.

Examples given earlier in haemophilia, growth hormone and gonadotropins have clearly illustrated the value of recombinant DNA–derived therapeutics. The haematopoietic growth factors continue this theme, although they also demonstrate how the various permutations and combinations possible through rDNA technology can complicate the evaluation and rational use of these products. Recombinant haematopoietic growth factors have been commercially produced using a variety of vector systems, e.g., *E. coli*, yeast and mammalian cell lines such as CHO. The expressed proteins in each case have structural limitations; e.g., the bacterial expression system produces a recombinant protein that is not glycosylated; and the yeast system, a partially glycosylated growth factor. The fully glycosylated substance can be made only with the more sophisticated mammalian expression systems. An altered antigenic structure for a recombinant DNA–derived product can have two consequences: (1) Antibodies might be induced, particularly if there is long-term use of the product and (2) Immunogenicity can be impaired if the recombinant DNA product is to be used as a vaccine.

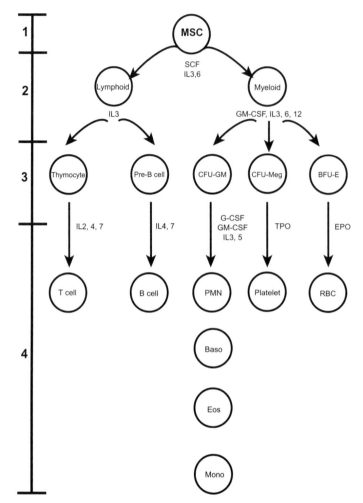

Fig. 6.2 Pathways for the differentiation of haematopoietic cells and the associated cytokines and growth factors.
Four compartments are defined: **(1)** multipotential stem cell (capable of self-renewal), **(2)** multipotential progenitor cells, **(3)** lineage-committed progenitor cells and **(4)** mature cells. The ultimate source of all haematopoietic cells is the multipotential stem cell (MSC). The two multipotential progenitor cells produce lymphoid cells or the other blood cells. Further along the pathway are more differentiated progenitor cells including the lymphoid precursors and CFU-GM (CFU—colony forming unit; GM—the four types of cells from CFU-GM are granulocyte [PMN], basophils, eosinophils and monocytes [macrophages]). The CFU-Meg (colony forming cells megakaryocyte) will produce platelets under the influence of thrombopoietin (TPO). The BFU-E (burst forming unit erythroid) will produce red blood cells under the influence of erythropoietin (EPO). Differentiation into the various pathways occurs in response to cytokines. SCF—stem cell factor; IL—interleukin; CSF—colony stimulating factor.

rhG-CSF and rhGM-CSF

G-CSF is a non-glycosylated protein and so can be produced in the more basic bacterial expression system. In contrast, GM-CSF is glycosylated and requires either a yeast expression system or CHO cells if this post-translational modification is to be included in the product. GM-CSF has also been produced as a non-glycosylated form through the bacterial system. There are two major indications for using G-CSF or GM-CSF. First

is the treatment of febrile episodes related to neutropenia (secondary to disease, drug therapy or bone marrow transplantation). The second is in the mobilisation of stem cells from peripheral blood, for example, to stimulate recovery after bone marrow transplantation. Variations on the latter include the leukopheresis step involved before marrow transplantation when a greater yield of stem cells becomes possible by treating the donor with the CSFs. *In vitro* uses of CSF will be discussed under gene therapy.

Although G-CSF is the preferred drug, there remain many unanswered questions about the relative values of G-CSF and GM-CSF because different products are commercially available. For example, is the glycosylated or non-glycosylated form of GM-CSF superior? One reason for preferring G-CSF is the apparent fewer side effects with this product. But this also needs to be confirmed. Both products are very expensive, and so their use has been limited to specific indications described above. Many clinical trials have been reported, but over a decade after they were first approved by the FDA, the relative values of G-CSF and GM-CSF (indications, efficacy, toxicity and costs) remain to be fully determined.

ERYTHROPOIETIN

Anaemia is an important complication of renal failure and accompanies a number of chronic conditions including malignancy, AIDS, aplastic anaemia and rheumatoid arthritis. The anaemia of chronic disease has a significant effect on morbidity. Components of this anaemia include (1) decreased production of the hormone erythropoietin (usually abbreviated to EPO), (2) iron deficiency (in the presence of normal iron stores) and (3) chronic infection or illness, which can impair bone marrow function. Erythropoietin is the primary regulator of red blood cell production. The majority of the hormone is produced in the kidney. Its level is regulated by the tissue oxygen tension present in the kidney. Erythropoietin feeds back to the bone marrow where it acts on the committed erythroid progenitors and precursors. In the presence of hypoxia, the level of erythropoietin increases and the red blood cell mass expands. Once hypoxia is corrected, erythropoietin production in the kidney is suppressed.

Erythropoietin was discovered at the beginning of the last century. It had been suspected for some time that reduced production of this hormone played a major role in the anaemia of chronic renal failure. However, it was not possible to take this any further in the investigative or therapeutic sense since the amount of erythropoietin that could be isolated from kidneys or the urine was insignificant. Thus, the potential for erythropoietin was limited until the gene was cloned in 1985. Recombinant human erythropoietin is now produced by expression of its cDNA in a mammalian cell line system similar to that described for haemophilia. The necessity for a sophisticated expression system reflects the requirement for glycosylation.

Clinical Applications

Recombinant human (rh) erythropoietin given to anaemic patients being dialysed for end-stage renal failure is capable of raising the haemoglobin level, and so reduces or removes the necessity for blood transfusions. Thus, the potential complications associated with blood transfusions (infection, immunosensitisation to HLA antigens, iron and circulatory overload) can be avoided. The improvement in the quality of life for this chronically ill population has been used to justify the costs of rherythropoietin. Other indications for the recombinant product include the specific normocytic normochromic anaemia found in cancer, severe infection or chronic inflammatory disorders such as rheumatoid arthritis. Although this form of anaemia had been identified for many years, it was difficult to determine what role erythropoietin played since an accurate plasma assay was not available. However, once the gene was cloned, it became possible to measure plasma erythropoietin very accurately.

In cancer, there is evidence that the level of erythropoietin is reduced and so contributes to the co-existent anaemia. A number of trials have now shown that the anaemia can be corrected in ~60% of cancer patients who are treated with recombinant erythropoietin. The availability of rhEPO also means that blood transfusions can be avoided. Although a lot of data have been generated about the value of rhEPO treatment for anaemia secondary to a broad range of clinical conditions, there remains some controversy about its value, particularly in relation to the quality of life and justification of the costs involved for this product.

Side effects reported with recombinant human erythropoietin include local reactions to the injection as well as more serious complications such as hypertension, seizures and venous thromboses. Some of these effects may reflect an inappropriate increase in the red blood cell mass secondary to the EPO effect. rhEPO is given three times per week, and it is known that a greater sialic acid content on the carbohydrate component of EPO leads to a longer serum half-life. Therefore, a new synthetic product (Darbepoetin alfa) was produced. It is like EPO but has two extra N-linked carbohydrate addition sites. This makes the drug 10 times as powerful, and it can be given once weekly. Although this product was approved by the FDA for treating anaemia of chronic renal failure, it is also used to treat the anaemia associated with cancer chemotherapy.

VACCINES

Successes with rDNA-derived therapeutic drugs summarised above have not been matched in vaccines. With the exception of the hepatitis B virus vaccine, the results with rDNA vaccines have been disappointing. This is despite the considerable financial commitment driven in part by the urgency to find vaccines for infectious diseases such as AIDS, TB and malaria. Commonly used conventional vaccines such as poliomyelitis, measles and rubella are, with few exceptions, composed of **live** (infectious) and **attenuated** (non-pathogenic but immunogenic) organisms. As such, they have proven to be highly

Table 6.5 Components of the hepatitis B virus (from Keating and Noble 2003)

Component	Antigens	Serum markers and their significance
Surface (envelope) components	Small or major S protein (hepatitis B surface antigen)— the most important component Middle protein (comprises S and pre S2 components) Larger protein (comprises S, pre S2 and pre S1 components)	The presence of antibodies to HBsAg (anti HBs) and disappearance of HBsAg indicates immunity. Chronic carriers have persisting HBsAg and no anti HBs.
Inner core	Hepatitis B core antigen (HBcAg)	Anti HBc are the first antibodies to appear and indicate recovery or chronic infection.
Hepatitis B e antigen	Hepatocytes infected with HBV secrete HBeAg derived from the core gene	HBeAg indicates high level of infectivity.

effective, relatively cheap and so affordable by most communities. Other conventional vaccines are made up of **inactivated (killed) microorganisms** (e.g., Salk poliomyelitis vaccine). Another variable with vaccines lies in their antigenic composition; i.e., one or more antigenic components (i.e., **subunit vaccines**) may be available as with influenza and recombinant hepatitis B vaccines.

Despite the efficacy of the conventional vaccines, it should be noted that some would not have reached clinical use with the stringent licensing regulations now in force. For example, the oral poliomyelitis (Sabin) vaccine can revert on rare occasions to the wild-type (neurotoxic) strain and so produce poliomyelitis. Subacute sclerosing panencephalitis is a very rare neurological complication following infection with the measles virus including some vaccine-derived strains. It can be fatal or lead to permanent neurological sequelae including mental retardation. It is unlikely that the two vaccines mentioned would have been marketed with these potential risks in today's litigation-conscious society. Modern production techniques require more stringent quality control steps during manufacture as well as better assessment of toxicity. In terms of standardisation and quality control, recombinant DNA–based vaccines have a lot to offer.

Hepatitis B Vaccines

Hepatitis B virus (HBV) is a DNA virus. It contains surface and core components that are distinct (Table 6.5). Approximately two billion people are infected with HBV, and of these about 350 million become chronic carriers. Approximately one million chronic carriers die annually from complications such as cirrhosis and hepatocellular carcinoma (statistics from Keating and Noble 2003). Endemicity rates vary greatly (Table 6.6).

In 1982, a hepatitis B vaccine became available using plasma from known chronic hepatitis B carriers. In this circumstance, stringent purification and inactivation procedures became mandatory. Thus, the vaccine was expensive, and the amount that could be produced was

Table 6.6 Endemicity of HBV (from Keating and Noble 2003).

Endemicity[a]	Geographic regions	Predominant mode of transmission
High (8–20% chronic carriers)	East and South East Asia, sub-Saharan Africa	Vertical (i.e., childbirth) or horizontal (child to child)
Low (<2% chronic carriers)	Northern Europe, North America	Parental and sexual modes of transmission

[a] Areas of high to intermediate endemicity include South and Central Asia, Middle East, North Africa, Southern and Eastern Europe.

limited by the availability of infected plasmas. The vaccine was not well received by the public in view of the theoretical risk that other viruses (e.g., HIV) might be transmitted despite the inactivation processes undertaken. Because of these problems and the importance of hepatitis B virus as a cause of liver disease (see Chapter 8), a recombinant DNA–derived vaccine was released in 1987.

The HBV story illustrates the usefulness of recombinant DNA technology in vaccination programs leading to a declining incidence over the past decade. In some countries, it has been possible to show a reduction in hepatocellular carcinoma rates. Nevertheless, HBV remains a global problem even in countries with low endemicity. Routine vaccination programs for newborns, infants, children and high-risk groups (e.g., health workers, prisoners) have been implemented to reduce the spread of this virus. More still needs to be done, including better vaccination rates for other high-risk groups such as intravenous drug users and young homosexual males.

A number of HBV vaccines are now available. They include plasma-derived as well as various subunit recombinant products. The latter vary in their surface antigen component, with several having only the S component, whereas others also have pre S2 proteins. Because glycosylation is required, the recombinant vaccines have been produced in genetically engineered yeast expression systems. In 1991, the FDA approved a combined HBV and hepatitis A virus vaccine.

The recombinant vaccines have now been used extensively and confirmed to be as effective as the plasma-derived product, although there appears to be variability in immunogenicity depending, to some extent, on the subunits incorporated into the vaccines. For many vaccines it has been possible to show more than 10 years of immunity although the aim on a global basis would be for lifelong protection. It has also been possible to identify categories of individuals who demonstrate poorer vaccination responses based on their anti HBs responses. They include the elderly, males, those who are obese, smokers and those with chronic diseases. For these people, revaccinations or the use of higher antigen-based vaccines is indicated.

The recombinant hepatitis B vaccine has many advantages, which include (1) An unlimited supply is available.

(2) Production can be standardised more effectively. (3) Greater flexibility is possible with the type of structure that is produced. (4) The vaccine will become cheaper and safer with time. The cost is not a small consideration since the communities that have the highest carrier rate for hepatitis B are often those that can least afford expensive vaccines. In these communities plasma-derived vaccines may be the only option. With increasing competition the hepatitis B vaccine's cost has come down. As more becomes known about HBV infection, it will be possible to alter the recombinant vaccine to enhance its immunogenicity.

Although the only successful human recombinant DNA–derived vaccine is that for the hepatitis B virus, more successes with recombinant DNA vaccines have been obtained in veterinary practice as well as the meat and livestock industry.

CELLULAR THERAPIES

This section includes a broad range of therapies with the common theme being biological-based treatments involving the manipulation or preparation of cells. This is in contrast to the conventional pharmaceuticals. The relevant examples in terms of molecular medicine are the gene therapies. Stem cells and xenotransplants are also discussed because in some circumstances the DNA for these cells will be manipulated, making them examples of both cellular and gene therapies.

SOMATIC CELL GENE THERAPY

Gene therapy can be defined as the transfer of genetic material (DNA or RNA) into the cells of an organism to treat disease or for marking studies. Because of the latter option, a better description would be gene **transfer** rather than gene **therapy** since a therapeutic intent is not necessary. However, for convenience the term "gene therapy" will be used to cover all applications. When first proposed as a therapeutic option, gene therapy was considered only in the context of genetic disorders. Today, gene therapy has broader treatment applications, particularly for cancer and infectious diseases. Disorders for which gene therapy has been tried or considered include.

Genetic Diseases

- Immunodeficiencies, e.g., adenosine deaminase deficiency
- Cystic fibrosis
- Familial hypercholesterolaemia
- Storage disorders, e.g., Gaucher's disease
- Coagulopathies, e.g., haemophilias A, B

- Haemoglobinopathies, e.g., β thalassaemia, sickle cell disease

Acquired Diseases

- Cancer, e.g., melanoma, brain and renal tumours
- AIDS
- Vascular disease
- Neurological disorders, e.g., Parkinson's disease, Alzheimer's disease

A number of criteria have been proposed to identify the types of genetic disorders for which gene therapy would be indicated. They include (1) a life-threatening condition for which there is no effective treatment, (2) the cause of the defect is a single gene and the involved gene has been cloned, (3) regulation of the gene need not be precise and (4) technical problems associated with delivery and expression of the gene have been resolved (see below). Similar considerations would hold for acquired disorders such as cancer, although in these circumstances a cure might not be the prime goal, and so the same stringent criteria would not necessarily apply.

Because of the potential risks with manipulating the human genome, the regulation and monitoring of gene therapy protocols by various government and institutional biosafety committees have been intense. It was not until September 1989 that the USA National Institutes of Health (NIH) approved the first **marker** study involving transfer of DNA into patients with melanoma, a malignant skin cancer. In September 1990, the first **therapeutic** transfer of a genetically engineered cell was undertaken in a four-year-old child with the potentially

Table 6.7 Range of gene therapy proposals reviewed by the NIH's Recombinant DNA Advisory Committee (RAC)[a]

Type of study	Number of protocols	Disease (number)[b]
Marking	41	Autologous stem cell transplantation studies involving a range of malignancies including breast cancer, leukaemia, lymphoma
Therapeutic	607	Infectious diseases (42)[c] Monogenic gene disorders (59) Cancer (430) Miscellaneous diseases (76)

[a] A more comprehensive overview of gene therapy trials worldwide can be found on http://www.wiley.co.uk/genmed/clinical. [b] The numbers do not equal 607 since some protocols were withdrawn or replaced by others. [c] Of these studies, 39 of 42 protocols involved HIV.

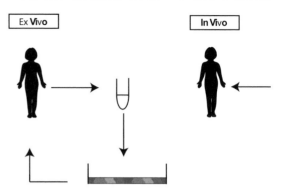

Fig. 6.3 Transferring DNA into cells.
Ex vivo: This approach involves the removal of cells from the patient. DNA (or RNA) is next introduced into the cells, which are then cultured to obtain adequate numbers. The introduction of DNA by infection using a viral vector is called transduction. The genetically altered cells (which may also be physically or antigenically altered following the *ex vivo* manoeuvres) are finally returned to the patient. In some circumstances, *ex vivo* transfer is the only feasible option, e.g., haematopoietic cells. In terms of safety, there is more confidence with *ex vivo* transfer since only the appropriate cells will take up the DNA/RNA. *In vivo*: A more physiological approach and challenge for the future is *in vivo* transfer, which involves direct entry of DNA (or RNA) into the patient. Targeting is necessary in this form of transfer.

fatal genetic disorder, adenosine deaminase deficiency. By late 2004, 669 gene therapy protocols had been reviewed by the NIH's Recombinant DNA Advisory Committee (RAC). They covered both marking and therapeutic trials (Table 6.7).

Gene therapy in the human refers to **somatic cell gene therapy**, which means the target is not a germ cell, so that transmission to future generations cannot occur. Germline gene therapy (an example of which would be the transgenic animals described in Chapter 5) has been prohibited for various reasons, which are discussed in Chapter 10.

Strategies for Gene Delivery

There are two ways to transfer DNA (RNA) into cells—*ex vivo* or *in vivo* (Figure 6.3). A prerequisite for *ex vivo* transfer is the necessity to culture cells *in vitro*. Therefore, not all cells will be useful targets for this type of gene therapy. Another requirement is the ability to return the genetically altered cells to the patient; i.e., the cells need to be transplantable. The above considerations have meant that a lot of the work involving *ex vivo* transfer has focused on haematopoietic cells. Apart from the fact that *ex vivo* transfer may be the only suitable approach available in many cases, it has another advantage in terms of safety; i.e., there is more control over which cells will take up the foreign DNA. However, *in vivo* transfer is considered to be more physiological and may be the only option in some circumstances, e.g., disseminated cancer. *In vivo* transfer remains a priority awaiting further developments to ensure that the right cells express the transferred DNA, and they do so in adequate numbers. The concept of **targeting** becomes a real issue when *in vivo* transfer is considered (discussed further below).

The ultimate aim in gene transfer is to get DNA into specific tissues. Again there are two major approaches—**physical** and **viral** (biological) means. The cell and

nuclear membranes can be made more permeable to DNA following co-precipitation of DNA with calcium phosphate, or an electric shock—called electroporation (Table 6.8). Using micropipettes, it is possible to inject DNA into the cell's nucleus. More novel approaches to facilitate movement of DNA into a cell include (1) injection of DNA directly into muscle cells; (2) insertion of DNA via cationic liposomes in the process known as lipofection, i.e., synthetic spherical vesicles that have lipid bilayers and so are able to cross the cell membrane and (3) coating of DNA with proteins and the "gene gun"—DNA-coated microprojectiles. Physical methods can be relatively inefficient when it comes to cells taking up DNA. More importantly, DNA inserted into the host genome in this way is usually present as multiple copies; i.e., there is no control over the sites of insertion, and so the function of normal genes could be affected. Finally, the expression of the introduced gene is only transient.

The current preferred method for gene transfer involves the use of viruses, particularly the retroviruses. Wild-type retroviruses can convert their RNA into double-stranded DNA, which can then integrate into the host's genome (see Figure 8.2). Viral proteins encoded by the *gag*, *pol* and *env* genes make up approximately 80% of the retroviral genome. These RNA segments can be deleted and replaced by a foreign gene, e.g., human adenosine deaminase (ADA). Now the recombinant retrovirus is no longer

Table 6.8   Delivery systems for gene transfer
A number of approaches can be used to get DNA into cells to allow genes to be expressed. The large number of options available would suggest that no single method is ideal. Even within the viral vectors, there are advantages and disadvantages found with each.

Type of approach	Delivery method
Physical	**Greater membrane permeability to DNA:** calcium phosphate coprecipitation, electroporation (electric shock) **Microinjection:** into the cell nucleus, into muscles **Other methods:** insertion via liposomes, coating DNA with proteins, gene gun, microencapsulation
Viral	**Integrated into host's genome:** retrovirus, adeno-associated virus, lentivirus **Not integrated into host's genome:** adenovirus, herpes simplex 1 virus,[a] smallpox virus

[a] The herpes simplex 1 based vector does not integrate (and so a transient effect would be expected), but the wild-type virus displays latency. Although the molecular basis for latency remains to be defined, it means that this virus could potentially be made to express its inserted gene over a longer time frame without the necessity for integration to occur.

infectious because it cannot make its own structural proteins. This is a prerequisite for gene therapy. Persistent infection by the genetically engineered retrovirus would not be permissible since it might lead to neoplastic change, the wrong cells expressing the gene, or the germ cells becoming infected and so passing on the gene or any genetic defects created to future generations. To become a useful vector for DNA transfer, the retrovirus must infect in a controlled way. This can be done with **packaging cells**. These cells contain a **helper** retrovirus that has also been genetically manipulated to produce empty virions; i.e., structural proteins are present, but a complete infectious virion cannot be made. However, the retroviral vector with its inserted ADA gene can utilise the structural proteins produced by the helper virus in the packaging cells to form a complete (infectious) virion, which can undergo one round of infection. This would be enough to get the genetically engineered retroviral RNA into the target cells' DNA. Advantages of the retroviral vector for DNA transfer include (1) A single virus infects one cell. (2) The virus is usually non-immunogenic. (3) Integration into the host genome means there is the potential for long-term expression of the inserted gene. Disadvantages are (1) The target cell must be dividing before the retrovirus can integrate into the cell's genome. (2) Transduction efficiency is usually inadequate. (3) DNA insert size is limited, which can be a problem if a large gene is involved. (4) Since retroviral vectors are produced from living cells, there is the worry that contaminants derived from these cells will be present.

Risks are involved with using retroviruses as gene transfer vectors: (1) Integration is random, and so there is

always the worry that a normal gene is inactivated or an oncogene is activated. (2) Retroviruses have the potential to revert to replication-competent organisms and so induce cancer (Figure 4.8). Because of these issues, a number of other viruses have been developed for gene therapy. Two of them (adeno-associated virus and lentivirus) will integrate into host DNA and so can lead to long-term expression of the transduced gene; i.e., a cure is possible. The latter goal must be balanced by long-term side effects if the integration has interfered with the function of a normal gene. The other viral vectors (as well as the physical means) do not lead to integration, and so the associated genes are expressed for only a short term; i.e., treatment will need to be repeated (Table 6.9).

Target Cells

Another consideration in gene therapy is the target cell. If a retroviral vector is used for transduction, an important prerequisite for the target cell is for it to be dividing so that the retrovirus can integrate into the host genome. The target cell should also be appropriate to the type of expression required. For example, a neurologic disorder may derive no benefit from the transfer of genes into haematopoietic cells. Finally, the target cell needs to be long-lived to prolong the effects of gene therapy. The ideal target cell would be pluripotential stem cells since integration of a gene into this cell should produce a cure, or at the very least a long-term effect. Because of the potential availability of stem cells and the considerable experience gained with bone marrow transplantation, a lot of the work to date has focused on the haematopoietic stem cells as targets for gene transfer.

The bone marrow pluripotential stem cell is a rare cell that to date has been satisfactorily isolated and characterised only in the mouse. Gene transfer into human bone marrow–derived stem cells has been possible because of the infectious capability of the retroviruses. Nevertheless, expression observed in these instances has been low and of short duration. Thus, gene therapy in this circumstance would not be effective in disorders such as the β thalassaemias for which significant gene expression is required to produce an adequate supply of protein. This may be overcome in the near future with new developments in molecular technology. These developments include (1) the potential to stimulate division of the pluripotential stem cells with the recombinant human growth factors, thereby making these cells move out of the G_o phase of the cell cycle and so become more accessible to infection with a retrovirus; (2) the use of monoclonal antibodies to identify surface antigens found on primitive cells, e.g., CD34[+] cells; and (3) the availability of DNA sequences that can significantly up-regulate (i.e., increase) gene expression.

Table 6.9 A comparison of different methods for introducing DNA in gene therapy

Method[a]	Target cell	Chromosomal integration (gene insert size in kb)	Other advantages	Other disadvantages
Retrovirus	Limited to dividing cells	Yes (medium-sized insert ~8 kb)	Easy to manipulate. Considerable experience because used in many trials. Potential for long-term gene expression.	Known to cause disease in humans, i.e., risk if recombination occurs as non-random integration means endogenous genes can be disrupted.
Adenovirus	Both dividing and non-dividing cells	No, episomal (medium-sized insert ~8 kb)	Easy to manipulate. Can achieve high titres in production. Used in many trials. Does not integrate, i.e., no risk of insertional mutagenesis	Potentially immunogenic and can provoke toxic response in host (current batches of this virus less immunogenic). Short-term expression.
Adeno-associated virus	Both dividing and non-dividing cells	Yes (small insert size ~5 kb)	Not known to cause disease in humans. Non-immunogenic.	Relatively new. Difficult to purify. Lack of expressed viral protein means limited immune responses provoked. Potential for insertional mutagenesis. Detected in semen but unclear if in sperm or tissues or fluids.
Lentivirus	Both dividing and non-dividing cells	Yes (medium-sized insert ~8 kb)	Non-immunogenic. Easily manipulated. Long-term gene expression.	Derived from HIV and so concern that wild-type recombinant is formed. Potential for insertional mutagenesis. Difficult to manufacture.
Herpes simplex virus	Both dividing and non-dividing cells	Extra-chromosomal (large insert size >30 kb)	Does not disrupt endogenous genes, i.e., remains extrachromosomal. Potential for latency, i.e., long-term expression.	Limited use if patient already has developed immunity to this virus.
Naked plasmid DNA	Most cells	No (relatively large inserts >10 kb)	Can be inserted into many cells by various means. Easy to make. Minimal biosafety risks.	Larger than oligonucleotides and so requires some packaging. May provoke immune response. Generally DNA vaccines are safe but ineffective.

[a] The transmission of genetic material from one cell to another by viral infection is called **transduction**. Acquisition of new genetic markers by incorporation of added DNA into eukaryotic cells by physical or viral means is called **transfection**.

A recent report would suggest more optimism for gene therapy in thalassaemia. The success in this case was based on the use of a lentiviral vector considered to be more efficient in gene transfer. The study by Puthenveetil *et al* (2004) describes a complete correction of the *β* thalassaemia major defect following gene therapy in *in vitro* and xenotransplant models. This now needs to be reproduced in human clinical trials but provides some hope for treatment of a serious disease in many communities.

Cells with potential value for gene therapy are summarised in Table 6.10. They display a number of features that make them useful targets for the transfer of genetic information. However, none promises a cure even if effects are long-lived because they are not stem cells; i.e., life span is finite as the cells do not have an unlimited potential for self-renewal and do not differentiate into various types of secondary cells.

Gene Marking

Gene therapy protocols are of two types: those which are (1) expected to produce a therapeutic result and (2) used to mark a cell with a gene so that it can be followed or identified as part of a research protocol. One marking-based research interest is to follow cells in the transplantation scenario, for example, to determine in cancer if relapses following treatment occur in host (patient) or donor cells. This use of gene transfer will provide answers to a number of interesting biological questions (Box 6.1).

Tumour-infiltrating lymphocytes are lymphoid cells that invade solid tumours. They can be grown in culture and have tumour-killing potential. For example, removal of these cells from a melanoma and growth in the presence of an interleukin (IL2) enable the tumour-infiltrating lymphocytes to be cultured and then reinfused back into patients. The lymphocytes target to the melanoma where

Table 6.10 Target cells for gene transfer
Target cells need to be long-lived and of a suitable type for the expression required.

Cell	Utility
Haematopoietic stem cell	Sources are bone marrow and umbilical cord blood. The possibility for gene transfer into the pluripotential stem cell means a cure is feasible. Haematological and immunological defects are the types of disorders that could be corrected. Particularly suited to *ex vivo* transfer.
Lymphocyte	The major advantages of the lymphocyte are its role in immunity, its relatively long life and its ease of access in the blood. This cell has been the target for gene transfer in melanoma and the immunodeficiency disorders. Also suited to *ex vivo* transfer.
Respiratory epithelium	Not suited for *ex vivo* transfer. Slowly dividing respiratory epithelial for *in vivo* transfer with retroviruses. At present good transfer is possible with adenoviruses and lipofection.
Hepatocyte	*Ex vivo:* Cultures can be obtained from liver tissue. Cells transduced with retrovirus can be reimplanted in the liver via the portal circulation. *In vivo:* Novel methods to get DNA into cells involve coating DNA with a protein receptor recognised by hepatocytes.
Fibroblast, skin cells	Easy to access and grow in culture. Can produce biologically active compounds, e.g., coagulation factor IX. The main problem is their short-term effect, which may be due to graft rejection. Suitable for *ex vivo* or *in vivo* transfer.
Skeletal muscle	A problem with skeletal muscle is that it is post-mitotic (and so retroviral vectors less effective) and multinucleate. *In vivo:* Injection of DNA in plasmid form into muscle cells enables expression of the DNA without it incorporating into host genome. *In vitro:* Adenoviral vectors have been used to insert a mini dystrophin gene into cells. Approaches require further assessment.

Box 6.1   Gene transfer for marking: Determining the cause of relapse in leukaemia following intensive chemotherapy and autologous bone marrow transplantation.

Autologous bone marrow transplantation remains an option for treating some leukaemias with a poor prognosis. In this situation, marrow is harvested from the patient prior to intensive chemotherapy that will lead to marrow aplasia. The patient is rescued from the otherwise fatal aplasia by reinfusing his/her own marrow that has either been purged of leukaemic cells by various *in vitro* manoeuvres or is a remission marrow; i.e., leukaemic cells are not present to any extent by conventional assessment. The combination of intensive chemotherapy and autologous marrow transplantation will cure leukaemia in some individuals, but others will eventually relapse. An important but unanswered question relates to the source of this relapse. Relapse could have occurred in **transplanted cells**; i.e., purging of leukaemic cells from the marrow had been inadequate or the remission marrow contained a significant number of leukaemic cells. On the other hand, the source of

relapse could have been **leukaemic cells** that had persisted in the patient's bone marrow even after intensive chemotherapy. The two possibilities could not be investigated by conventional means since (1) Cells used for transplantation are identical to those in the patient (i.e., autologous) and so could not be distinguished and (2) The population of cells to be studied is extremely small. Both problems have now been overcome by molecular technology. First, PCR enables very few cells to be studied. Second, autologous cells can be marked with a gene (e.g., the neomycin gene) prior to re-infusion into the patient. PCR is then used to identify whether relapsed cells had the neomycin gene. Results from such marking studies have confirmed that relapse can involve the transplanted cells, and so purging protocols have been inadequate, and/or remission marrows contain sufficient leukaemic cells to produce a relapse.

---

they can induce regression of the tumour. The first human gene transfer involved a marking study with the addition of a neomycin gene via a retroviral vector into tumour-infiltrating lymphocytes obtained from patients with advanced melanoma. The transduced lymphocytes were

then reinfused into the patients and their duration of survival and any toxic effects noted. The results were satisfactory in that the neomycin-containing tumour-infiltrating lymphocytes persisted in the circulation; they were able to target to the melanoma cells, and there were

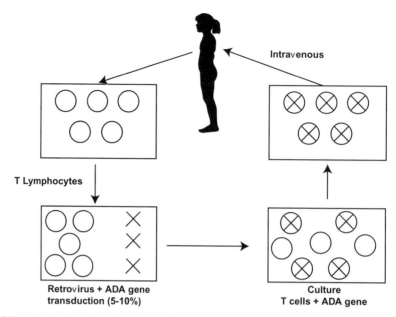

Fig. 6.4   Protocol for adenosine deaminase (ADA) gene therapy.
T lymphocytes (○) are removed from the patient's circulation and infected (transduced) *ex vivo* with a retrovirus containing the wild-type ADA gene (x). Transduction efficiency at about 5–10% is not sufficient, and so cells will need to be cultured. The patient is re-infused with her own lymphocytes after sufficient numbers of the transduced cells are grown in culture. Since pluripotential stem cells have not been used, a cure is unlikely, and so the procedure must be repeated, depending on the life span of the transduced lymphocytes.

no apparent side effects directly attributable to the retroviral vector.

Genetic Diseases—Introducing New Genes

(I) Severe combined immunodeficiency (SCID) due to adenosine deaminase (ADA) deficiency: In 1990, a four-year-old child with the potentially fatal autosomal recessive disorder, adenosine deaminase deficiency, received an infusion of her own lymphocytes that had been genetically altered by a retrovirus containing a normal adenosine deaminase gene (Figure 6.4). Adenosine deaminase deficiency was chosen for a number of clinical reasons: (1) It constituted an important cause of the severe combined immunodeficiency syndromes in children. (2) Death within the first 1–2 years of life was common. (3) Medical treatment was suboptimal. This included a recently released drug called PEG-ADA that comprised the natural product (ADA) coupled to polyethylene glycol (PEG) to increase its half-life. PEG-ADA was very expensive, and follow-up time had not been sufficient to confirm its long-term efficacy. Another therapeutic option for adenosine deaminase deficiency is bone marrow transplantation. This can produce a cure if an HLA-identical sibling donor transplant is successful. However, less than 20% of patients had a matched sibling. Alter-

native bone marrow sources such as a haplo-identical parent or a matched unrelated donor result in much poorer outcomes after transplantation.

The four-year-old girl described above had been treated with PEG-ADA but had responded inadequately to this form of therapy. In this circumstance, approval was given to attempt gene therapy. Features at the DNA level that made adenosine deaminase deficiency a good candidate for gene therapy included (1) The target or affected cells were lymphocytes and so accessible through the blood. (2) T lymphocytes have a relatively long life span. (3) The gene had been cloned and was relatively small at 3.2 kb in size. (4) It was expected that a moderate level of gene expression would be sufficient to reduce mortality in this condition.

At first, stem cells isolated from the bone marrow of the four-year-old girl were proposed as the targets for gene transfer. Subsequently, it was found that mature T lymphocytes isolated from peripheral blood were more practical alternatives to the elusive stem cells (estimated to comprise between 1 in every 10 000 to 100 000 cells). Multiple infusions of genetically engineered autologous T lymphocytes were given at 1–2 monthly intervals. Since the target cell in this case was no longer the pluripotential stem cell, a cure became unlikely. By 1995, 10 patients with adenosine deaminase deficiency had undergone various forms of gene therapy. In some cases it has been possible to identify transduced cells in the patients' peripheral blood or bone marrow for six months or more following the last treatment course. Cellular and humoral immune responses in the patients improved, and this has been accompanied by a better quality of life for these children. For ethical reasons, none had PEG-ADA treatment stopped, although in some cases the dose has been reduced without ill effects. Unfortunately, the therapeutic potential for gene therapy in these children was never able to be fully determined because of the concomitant use of PEG-ADA.

In 2002, a report was published describing how one of six patients with ADA and treated with gene therapy was taken off PEG-ADA treatment. The rationale for this was based on laboratory evidence that the PEG-ADA might be inhibiting the T lymphopoiesis. As predicted, the patient started to improve with transduced T lymphocytes gradually replacing non-transduced lymphocytes (because the former would have a selective advantage). This promising result was reported about the same time as the SCID-X1 gene therapy successes described below.

(II) Severe combined immunodeficiency–X-linked type 1 (SCID-X1): As well as deficiency of the adenosine deaminase enzyme, a more common cause of SCID results from mutations in the common γ chain of the interleukin 2 receptor (IL2Rγc). The gene is also found in interleukins 4, 7, 9, 15 and 21 and is essential for the development of T lymphocytes and natural killer (NK) cells from early lymphoid precursors. Thus, infants with SCID-X1 lack T and NK cells, and their functionally deficient B cells are probably secondary to the T cell deficiency (Box 6.2). This type of SCID is fatal within the first year of life unless immune function can be restored.

SCID-X1 was considered a better candidate for gene therapy than ADA deficiency because it was a disease that primarily involved the T and NK (natural killer) lymphocytes. Other considerations in favour of using gene therapy for SCID-X1 included

- An HLA-identical sibling donor for bone marrow transplantation was not usually available.
- The γc gene was expressed all the time, and so regulation was not an issue.
- Since T cells are long lived, it could be expected that whatever effect was obtained would last for a reasonable duration.
- A final and important consideration was that the reinsertion of a functioning γc gene into a cell could be expected to give that cell a proliferative advantage; i.e., in the patient, lymphocytes with the transduced normal γc gene would have a greater chance of surviving, and so low transduction efficiency seen in gene therapy (i.e., not enough of the gene could be inserted into a cell) would be less of a problem.

> **Box 6.2   Case study: SCID-X1 (severe combined immunodeficiency–X-linked type 1).**
>
> X-linked SCID is a rare genetic disease (about 1 male in 75 000 live births) produced by a deficiency in T cells and natural killer (NK) cells associated with abnormal B cell function. Affected boys will die within a year if the T and NK deficiencies are not corrected. The molecular basis for SCID-X1 is a mutation within the gene known as γc (γ common gene). A mutation in this gene causes a severe disorder because the gene codes for a subunit found on six different cytokine receptors (interleukins 2, 4, 7, 9, 15 and 21 receptors). Definitive treatment for SCID-X1 (90% chance of cure) involves bone marrow transplantation from an HLA-matched sibling. Unfortunately, less than 20% of affected infants have such a donor, and so considerably more risky marrow transplants are possible (mortality now can approach 30% because the matching is less ideal). Even with transplantation, B cell function may not be restored, and these children might require lifelong supplementation with immune globulin. Therefore, an alternative approach is gene therapy. Other considerations in favour of gene therapy are (1) A clearly defined mutation in a relatively small gene and (2) the involvement of long-lived lymphocytes. Therefore, in 1999 a normal γc gene was transduced into the lymphocytes of children with SCID-X1.

In 1999, the first children with SCID-X1 were treated by gene therapy. The protocol was similar to that described earlier for ADA deficiency, but now the normal γc gene was inserted into a subset of lymphocytes known as CD34$^+$ cells, i.e., more primitive lymphocytes that could be expected to differentiate into the missing T and NK lymphocytes. CD34$^+$ cells were harvested from the children (usually less than 12 months of age), and these cells were transduced *ex vivo* with the γc gene inserted in a retroviral vector.

In 2002, the first formal report of the study was published in the *New England Journal of Medicine*. Of five boys treated in this way and followed up to 2.5 years, four were well with normal growth (i.e., apparently cured) and one continued to improve. This was an exciting development and appeared to justify a frustrating decade of hard work in gene therapy with no obvious success stories. By 2002, a follow-up report indicated that 9 of 10 treated SCID-X1 patients had responded very positively. The first two treated were now in their fourth year, and doing well. However, in September 2002, almost three years after the gene therapy, one of the children developed an acute T cell leukaemia, and a few months later a second treated child was shown to have the same complication. Tragically, in the midst of the first successful application of gene therapy, 2 of 14 SCID-X1

Table 6.11 Contributing factors to the development of acute T cell leukaemia in two SCID-X1 children treated with gene therapy

Event	Explanation
Insertional mutagenesis—the retroviral vector containing the γc gene inserted in or near the proto-oncogene *LMO2*	The retroviral γc gene insert also contained a promoter element to drive the γc gene. When the γc gene construct has inserted in or near the *LMO2* gene, it is thought to have activated this proto-oncogene and so produces leukaemia. The *LMO2* gene is associated with the development of T acute lymphoblastic leukaemia in transgenic mice, and human leukaemias associated with translocations.
CD34+ stem cells and status of stem cells in young (<12 months old) children. Relative dose of transduced cells.	Some evidence that stem cells (including the CD34+ precursor cells) in young children may be more susceptible to insertional mutagenesis events because they are still immature and have a greater proliferative capacity (the 2 patients with acute leukaemia were the youngest of the first 14 treated). Although the numbers of transduced cells were comparable to what had been used in adults, on a per kg basis there remained the possibility that these children had received excessively large numbers of transduced CD34+ cells.
Selective advantage provided by the IL2Rγc gene.	The IL2Rγc gene is essential for normal T cell formation but less critical with other cells. Hence, the major selective advantage for transduced cells would be those that are T cell developing. The IL2Rγc gene is also a component of six interleukins, all of which act as T cell growth factors.
Replication advantage of the CD34+ cells in a disease such as SCID	A number of other gene therapy protocols had used CD34+ cells, and leukaemia had not developed. One explanation is that SCID itself, because of the important selective advantage of the transduced CD34+ cells (compared to all other endogenous lymphocytes), was a contributing factor to the leukaemia. As part of this, the relatively "empty" marrow devoid of endogenous T cells because of the SCID defect would be a fertile environment for the proliferation of the inserted cells.
Insertional mutagenesis may not be entirely random	Early evidence is available that viruses that integrate have a predilection for certain regions of the genome; e.g., they are more likely to insert in or near genes than in the intergenic regions. Some even seem to prefer regions associated with transcriptionally active oncogenes. Retroviruses are not the only risk vectors; lentiviruses and adeno-associated viruses also integrate.
Unknown effects	A third treated SCID child has had an insertion into his *LMO2* gene, but to date acute leukaemia has not developed.

children treated had developed an unusual form of acute leukaemia involving T cells.

The next 12 months saw a number of changes in gene therapy protocols from both regulatory (Box 6.3) and scientific viewpoints. What caused the development of acute leukaemia is still debated. There is consensus that a number of contributing factors were involved, and these are summarised in Table 6.11. It is now considered that insertional events from transduced viral vectors that integrate into chromosomes are more common than initially thought, but most do not involve important genes such as proto-oncogenes. In addition, cells with these insertional events do not have a proliferative advantage, and so have a finite life span. In terms of SCID, a proto-oncogene known as *LMO2* is now well established as being causative because both children have been shown to have the γc gene inserted in or very near that proto-oncogene. However, this alone is unlikely to be sufficient to explain the development of leukaemia. More knowledge is needed and no doubt will come from further follow-up and study of the SCID-treated children. The clinical holds placed on some SCID gene therapy trials by various regulatory bodies have now been lifted with greater precautions being required. They include the treating of older children (perhaps their bone marrow

environment is less susceptible) and the use of smaller doses.

From a regulatory perspective, the bar for justifying a clinical study of gene therapy has been raised, with greater emphasis being placed on a risk/benefit assessment, including what alternative treatments are available. Nevertheless, it is important to emphasise that the first successful gene therapy (rather than gene transfer) results have now been reported.

(III) Haemophilia: The second edition of *Molecular Medicine—An Introductory Text* included a section on gene therapy of cystic fibrosis. However, nothing significant has developed in this particular genetic disease, and so the section has been replaced with a discussion of haemophilia, which has shown recent promising results. Haemophilia is considered a good candidate for gene therapy because (1) No significant regulation of the inserted factor VIII or factor IX genes would be required, and even an over-expressed gene would not be a problem since normal plasma levels have considerable variability with the upper limit being 150% of normal. (2) A small increase in the factor plasma levels would be sufficient to convert a severe disease into a relatively milder form. (3) It would be relatively easy to access blood cells for *ex vivo* transduction, and these would be

an adequate source of factors VIII or IX. (4) Good animal
models are available for pre-clinical studies.

A number of clinical trials completed or under way
involving both factor VIII and factor IX gene therapy have
utilised a variety of strategies including (1) *in vivo* (intravenous, intramuscular or intrahepatic) administration of
a viral vector containing the normal gene and (2) *ex vivo*
transduction studies (using fibroblast cells implanted into
omentum after *ex vivo* transduction). Interesting results
have emerged, including the finding of retroviral
sequences in semen following intravenous injection. This
latter study was stopped in view of the potential for accidental germline spread although it is likely that the vector
sequences were in the tissues or fluids biopsied rather
than the sperm itself. A few patients demonstrated a persistently elevated factor VIII level (one lasted 10 months),
and intrahepatic injection of factor IX allowed one
patient a transient rise in his factor IX to 13%. However,
all treated patients eventually failed to show persistent

gene expression, a problem common to most other gene
therapy studies.

In mid-2004, the Avigen company discontinued its
haemophilia gene therapy trials following the development of minor abnormalities in liver function studies associated with a lowering of the factor IX levels. Presently, this
is thought to represent an immune response to the adenovirus associated vector. Where this leaves haemophilia
gene therapy remains to be determined, but it is a temporary disappointment for a promising model.

Cancer Gene Therapy

Gene therapy for cancer can be considered in a number
of categories. Protocols are designed to

- Stimulate the natural immunity to cancer cells
- Kill or interfere with the growth of cancer cells with
  drugs
- Insert a wild-type tumour suppressor gene, e.g., *TP53*
- Increase tolerance to high doses of chemotherapy or
  delay drug resistance

1. Gene therapy to stimulate the immune system: Important cells in the host's immune response to tumour
   antigens include lymphocytes, macrophages and neu-
   trophils. Cytokines are proteins involved in key host
   defence mechanisms. Examples include interleukins
   1, 2, 6, 8; interferon γ, and TNF-α (tumour necrosis
   factor α). Other substances that can be used are
   immune co-stimulatory molecules such as B7.1. The
   presence of these substances is required to activate
   cytotoxic T lymphocytes leading to tumour rejection
   and a wider activated immune response to deal with
   metastatic tumour. Therefore, it is not surprising that
   a number of gene therapy protocols in cancer utilise
   various permutations of the above cytokines or B7.1.
   These genes are inserted into tumour cells or autologous fibroblasts, which are then injected back into the
   patient. In this way it is hoped to stimulate an immune
   response specifically to the tumour cells. Another
   approach that has involved a more direct local
   destruction of cancer cells was initially attempted
   by inserting the gene for tumour necrosis factor, a
   potent cytolytic agent, into the tumour-infiltrating
   lymphocytes.

2. Local delivery of cytotoxic drugs: To overcome the
   problem of non-selectivity with cytotoxic treatment, a
   gene therapy approach that utilises the conversion of
   an inactive compound (prodrug) to an active metabolite has been devised. For example, thymidine kinase
   (TK), an enzyme from the herpes simplex 1 virus
   (HSV-1), is harmless to mammalian cells. However, TK
   converts the antiviral agent ganciclovir to an active
   substance that damages DNA and leads to apoptosis
   of that cell. Similarly, 5 fluorocytosine (the prodrug)
   can be changed to 5 fluorouracil (a cytotoxic agent)

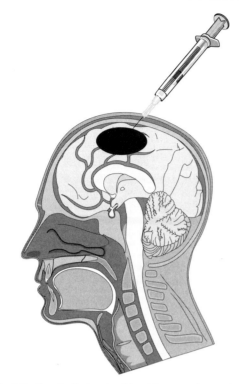

Fig. 6.5 Hypothetical protocol involving gene therapy for glioblastoma multiforme.

Glioblastoma multiforme is a malignant brain tumour that responds poorly to conventional treatment. Using imaging techniques, it would be possible to inject the tumour with a herpes simplex virus vector containing its promoter sequence and the TK (thymidine kinase) gene. The patient is next given ganciclovir, and only those cells that have TK will be able to convert the ganciclovir to its toxic metabolite. As well as the tumour cells that express HSV-TK, it has become apparent that neighbouring cells are also susceptible to damage. This bystander effect contributes to the tumour killing, and is thought to occur because of leaking of the active metabolite through cellular gap junctions to nearby cells without the transfected gene. Another benefit may come from use of HSV-1 rather than a retrovirus since the herpes virus in its natural infection can establish latency within the central nervous system (as shown by recurrence of cold sores). If the genetically engineered HSV-TK can be induced to behave in this manner, a longer course of controlled treatment might become possible.

by reaction with cytosine deaminase. Following gene transfer, cells which express the 5 fluorocytosine will be destroyed when exposed to cytosine deaminase, while the remaining cells will be safe. Hence, this form of gene therapy has been described as involving a "suicide gene." An example of "suicide gene" therapy is given in Figure 6.5 using as a model glioblastoma multiforme, a difficult-to-treat brain tumour.

3. Replacement of abnormal tumour suppressor cell activity: *TP53* is a key tumour suppressor gene. It is important for normal cell functioning and protects the cell from DNA damage by inducing cell cycle arrest and stimulating apoptosis. Mutations in *TP53* are found in more than 50% of cancers. Therefore, *in vivo* transfer protocols have been attempted to replace *TP53* in difficult-to-treat tumours such as lung cancer.

4. Modulating drug sensitivity and resistance: Two key problems associated with chemotherapy in cancer are bone marrow toxicity and the development of drug resistance. To protect the bone marrow, chemoprotection gene therapy protocols have been designed to target stem cells and introduce into them genes such as *MDR1* (multidrug resistance 1). This gene codes for P-glycoprotein and provides cells with resistance to a wide range of cytotoxic drugs.

Although there are many gene therapy approaches to the treatment of cancer, none to date has produced unequivocal cures or major improvements. There are many explanations, in particular the relatively low efficiency in delivering the desired gene that is possible with current viral vectors. More work is being carried out to address this issue.

RELATED GENE THERAPIES

Earlier, a definition for somatic cell gene therapy was proposed. However, as novel approaches are developed to manipulate DNA or RNA, a broad enough definition becomes elusive. Hence, gene and "related" therapies are sometimes used to capture new developments. As indicated earlier in Chapter 2, RNA has lived in the shadow of DNA. Apart from its traditional role in transcription and translation, a whole range of new activities for RNA has recently been identified, including the formation of RNA-RNA, RNA-DNA or RNA-protein interactions. These interactions, as well as the observation that RNA can have a catalytic effect, open up the potential for RNA in therapeutics.

Antisense Oligonucleotides

The rationale behind antisense oligonucleotides as a form of gene therapy is straightforward. An oligonucleotide about 12–20 bases in size will specifically bind (hybridise) to a region within DNA or RNA. This is a critical primer binding step in the polymerase chain reaction (PCR—see Chapter 2). *In vitro*, when the DNA polymerase and the four nucleotide bases are added, the segment of DNA defined by the two primers will be amplified to complete the PCR. However, *in vivo*, the attachment of an oligonucleotide to its corresponding partner (in DNA or RNA) based on A-T (U in RNA) and C-G pairing will produce a duplex that is susceptible

Table 6.12 The various stages at which antisense oligonucleotides can inhibit protein synthesis

Step in protein production	How the antisense oligonucleotide can inhibit
Transcription of DNA	The antisense oligonucleotide can form a triplex with the double helix. This three-layered helical formation cannot be transcribed.
Processing of the mRNA transcript	The initial mRNA transcript contains all the genomic information, but it must be processed into a mature mRNA by splicing out the introns. The splicing step can be interfered with if there is antisense oligonucleotide also available. Antisense RNA binding to mRNA forms an RNA duplex that is degraded by RNAse H.
Transport of the mRNA from nucleus to cytoplasm	Antisense oligonucleotides can interfere with this and so reduce the mRNA concentration and half-life.
Translation of the mRNA into protein	Antisense oligonucleotide can interfere with the protein initiation signal for translation.

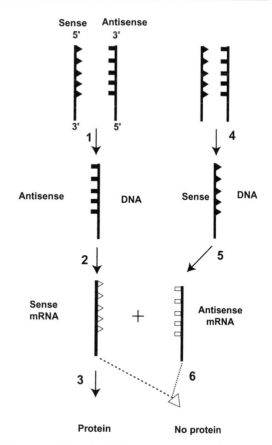

Fig. 6.6 Antisense technology as a form of gene therapy. Antisense RNA can inhibit the expression of a gene and would be effective in diseases in which unwanted gene expression occurs, e.g., HIV infection. The underlying mechanisms that produce the antisense effect are not fully known. One explanation involves a binding of the antisense RNA (or an antisense oligonucleotide) to sense mRNA and so inhibition of the latter during translation (protein synthesis). The normal transcription → translation pathway is illustrated on the left of the diagram (**1, 2, 3**) with the antisense DNA strand providing the template for transcription by RNA polymerase to make (sense) mRNA. On the right (**4, 5, 6**), the DNA sense strand has been copied. This could be a normal response by an organism to foreign DNA or a genetically engineered gene which has been "flipped" around so that the sense sequence becomes the template for mRNA synthesis. Alternatively, oligonucleotides with the antisense sequence are introduced into the cell. The result is the same, with the antisense mRNA or oligonucleotide binding to the sense mRNA and inhibiting its activity.

to degradation by various enzymes (Figure 6.6). The inhibitory effect of oligonucleotides can occur at a number of steps required for protein production (see Table 6.12).

Because of the way they work, antisense oligonucleotides would not be useful for replacing a missing gene or gene product. On the other hand, they have been used in therapeutic trials as a form of gene therapy to inhibit genes producing unwanted proteins in situations such as (1) cancer, when oncogenes have been activated; (2) infections, especially HIV because there will be viral-specific mRNA sequences that can be targeted for inhibition; and (3) post-operatively after coronary artery stenting when excessive proliferation can lead to blockage.

In vivo, antisense oligonucleotides exist, but they would not have any significant effect because it would be difficult to deliver sufficient to the nucleus (or even cytoplasm), their half-life is very short and they would quickly be degraded. However, various chemical changes have been made to the manufactured antisense oligonucleotides that give them a longer action, and they are more resistant to degradation *in vivo*. The potential for antisense technology has already been demonstrated in agriculture; e.g., genetically engineered tomatoes have been produced to make them mush-resistant because the gene producing polygalacturonase, which breaks down the cell wall, has been inhibited by antisense RNA. A number of plants have now been genetically engineered in similar fashion to give them resistance to particular viruses.

Antisense strategies have partly been driven by developments in technology, allowing automated synthesis of oligonucleotides. These have enabled the industrial production of relatively cheap high-quality oligonucleotides. Successful chemical modification of the oligonucleotides (the third generation antisense oligonucleotides are now being produced) reduces their potential for endogenous breakdown by cellular and nuclear nucleases. Considerable technical problems still remain to be resolved. For example, antisense molecules are not catalytic. Thus, binding to a target that could be reversible would reduce effectiveness. Delivery of antisense compounds also needs to become more efficient. Membrane-based vehicles (e.g., liposomes, similar to those described in gene therapy) are able to internalise antisense oligonucleotides into cells. Once in the cell, the oligonucleotide can act within the cytoplasm or gain access to the nucleus if DNA is the target. The pharmacokinetic properties and potential toxicity of oligonucleotides are still to be fully determined. Studies to date have utilised non-physiological conditions since saturating quantities of oligonucleotides have been required to achieve adequate intracellular and nuclear concentrations. Nevertheless, the novelty of this technology and the successful *in vitro* studies observed to date are very promising.

From a regulatory perspective, oligonucleotides are considered to be traditional drugs and not gene therapy products because they have a very short half-life, and they do not alter the sequence of the DNA. An example is the antisense oligonucleotide Vitravene®. It has been designed to inhibit proteins required for replication of the human cytomegalovirus (CMV) in a situation such as underlying HIV and CMV retinitis. It is applied locally and is effective. More recently, the 18 base antisense oligonucleotide Genasense G3139 has been manufactured to inhibit the mRNA for the human gene *BCL2*. This gene normally inhibits apoptosis (see Chapter 4). However, the inhibitory effect becomes a problem when the gene is over-expressed in a number of tumours including leukaemia, lymphoma, lung, breast, myeloma and melanoma. Perturbation of *BCL2* also induces drug resistance (anticancer drugs and radiotherapy rely on apoptosis for their effects). Thus, blocking of *BCL2* expression would seem a useful strategy for treating a number of malignancies. G3139, used in combination with cytotoxic agents, is now undergoing various clinical trials to determine the value of this antisense oligonucleotide in cancer treatment.

In a few cases, oligonucleotides have been designed in a way that will allow them to change the DNA sequence, and in this circumstance, they are classified as biological agents (i.e., gene therapy). At some future time, long-acting oligonucleotides might need to be monitored to ensure that there are no non-specific effects on the actual DNA structure.

Ribozymes

Another form of RNA-based gene therapy involves the use of catalytic RNA molecules known as ribozymes. They are naturally occurring RNA species that cleave

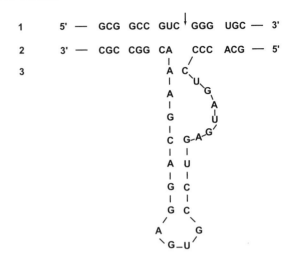

Fig. 6.7   The structure of a ribozyme.
(**1**) An mRNA molecule. (**2**) A ribozyme showing nucleotide sequences that are complementary to those in the mRNA. (**3**) The ribozyme's catalytic domain. The mRNA and ribozyme bind on the basis of Watson and Crick base pairing, i.e., A with T (U in RNA) and G with C. This is followed by sequence-specific cleavage, which occurs to the right of a GUC (↓) on the mRNA. The ribozyme is then free to cleave other mRNA species.

RNA at specific sequences. Ribozymes would have similar applications to those described for antisense oligonucleotides, i.e., RNA species from tumours or infectious agents that can be specifically inhibited. A potential advantage of ribozymes over antisense molecules lies in the former's catalytic activity so that following binding there is cleavage of target RNA. Specificity of the ribozyme rests with the hybridising (antisense) arms located on either side of the molecule's catalytic activity domain (Figure 6.7). Clinical trials undertaken using ribozymes are still at the phase I level. The ribozymes in these studies are being used to inhibit RNA produced by infectious agents such HIV, hepatitis C virus and aberrant gene expression in various cancers.

Technological constraints for ribozymes include their design, which makes production more difficult. Ribozymes are also susceptible to degradation by RNAses. Just as has been described for most other gene therapy approaches, more efficient methods for delivery of ribozymes into cells will also need to be developed, and the *in vivo* effect has to be of greater duration. A combination of ribozyme and antisense technology is a promising future development. In this strategy, ribozymes are incorporated into antisense oligonucleotides, thereby providing the latter with catalytic activity. These catalytic antisense molecules have twice the efficiency of the conventional substances. Another possibility is the use of combination therapy with ribozymes and chemotherapy, thereby reducing the target load and demand on the ribozyme.

DNA Vaccines

Nucleic acid (DNA) vaccines involve the direct injection of genes (in the form of naked, plasmid DNA or DNA delivered by another vector system) expressing viral proteins into a patient to produce a sustained antigenic stimulus and so generate an ongoing immune response. DNA vaccines, which have now been in use about 10 years, provoked initial interest because of the relatively simple way in which they could be prepared to deliver a range of protein antigens for immunisation. In animal studies these vaccines stimulate both humoral and cell-mediated immune mechanisms (similar to what occurs with live attenuated vaccines) without integrating into host DNA. Thus, they are an alternative but safer approach to live viral vaccines and are better than inactivated (dead) vaccines in the breadth of the immune response they elicit. Routes of administration are flexible, e.g., parenteral, mucosal or the gene gun, which delivers tiny amounts of DNA-coated gold or tungsten beads. DNA vaccines are presently being assessed in AIDS, malaria and a variety of cancers.

Infectious agents that are difficult or dangerous to produce by conventional culture techniques (e.g., rabies virus) could also be better developed through recombinant DNA means. Genetic manipulation would also be useful to reduce the likelihood of reversion to wild-type strains (e.g., poliomyelitis) or to increase the antigenicity of a particular component derived from the infecting organism.

It is evident from results now emerging that DNA vaccines are safe, but this is offset by their low potency, which reduces their effectiveness. To correct this, it will be necessary to increase the efficiency with which the DNA is being delivered and/or enhance the antigenicity of the protein produced. DNA vaccines also pose a regulatory challenge since they lie somewhere between a conventional vaccine (which has as one of its purifying steps the removal of any nucleic acids) and the traditional gene transfer vector. Hence, for the purpose of this discussion, they have been grouped under "related gene therapies." Issues that remain unresolved include the unequivocal demonstration that integration into DNA does not occur, and even if it does, what are the consequences? Anti-DNA antibodies to injected DNA are known to develop in animal studies, but their development and potential for autoimmune disease in humans remains unknown.

STEM CELLS

Stem cells are of two classes: (1) embryonic—found within the embryo's inner cell mass (Figure 6.8) and (2) adult—found within differentiated tissues such as the bone marrow. Stem cells have become topical as potential novel ways in which to repair or replace damaged tissues, or even organs. For this, stems cells have two important properties: (1) self-renewal, with the capacity to make more stem cells and (2) ability to give rise to different progeny when exposed to the appropriate signals. In doing this, a progenitor cell is first formed. This is the precursor to the specialised cell (called a differentiated cell). Stem cells have varying potential to differentiate into other cells: (1) Unipotent stem cells develop into one cell type. (2) Multipotent stem cells can form multiple cells types. (3) Pluripotent stem cells can differentiate into most, if not all, of the adult cell types in the body. (4) Totipotent stem cells will form all cell types, both adult and others, such as the placenta.

Embryonic Stem Cells (ES Cells)

In the embryo's blastocyst stage before implantation (about day 5–7 embryo), the inner cell mass contains all the cells that will make up the fetus. Some of these cells are pluripotential because they will give rise to all types of somatic cells as well as the germ cells. When these pluripotential stem cells are grown in vitro, they are called embryonic stem cells or ES cells (see Figure 6.8). When maintained under appropriate culture conditions, embryonic stem cells can be cultured indefinitely in an undifferentiated state. When differentiated, embryonic stem cells give rise to the three major cell lineages (endodermal, mesodermal and ectodermal). Mouse embryonic stem cells were isolated in 1981, and the human equivalents were found in 1998.

There are a number of potential applications for embryonic stem cells:

1. Research—understanding disease pathogenesis: Transgenic mice are useful animal models to study human disorders (see also Chapter 5). They are produced by microinjection of DNA into the pronucleus of a fertilised oocyte. Although the gene of interest is not inserted into its correct position in the genome, it still remains possible to add new genes that can function in vivo. Thus, expression of the mutant transgene will produce the clinical phenotype. An extension of this is the transgenic mouse, which has been created by gene knockout. This process involves homologous recombination between an introduced mutant gene and the corresponding wild-type gene. Now gene function can be inhibited or the effect of a specific mutation observed.

   Embryonic stem cells have been critical for developing knockout transgenics. Since embryonic stem cells are totipotential, they can be genetically manipulated and then reintroduced into the blastocyte of a developing mouse to produce a chimaera. Foreign DNA that has become integrated into the germline of

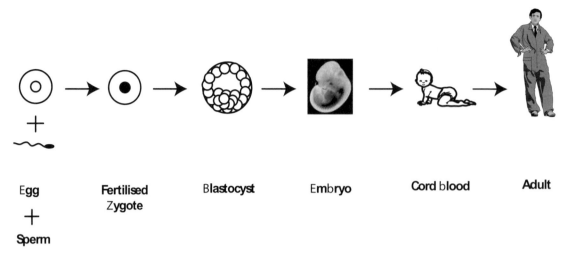

Egg	Fertilised	Blastocyst	Embryo	Cord blood	Adult
+	Zygote				
Sperm					

Fig. 6.8 Sources of stem cells.
Embryonic stem cells are derived from the 5–7 day embryo known as the blastocyst. The outer layer of cells depicted as circles is the trophoblast, which will go on to form the placenta. The cells at the bottom forming the inner cell mass are the embryonic stem cells. Cord blood stem cells obtained from the placenta at birth and various tissues in the adult provide the source of adult stem cells.

the chimaera will enable the gene to be transmitted to progeny. Appropriate matings will produce homozygotes containing the transgene. Embryonic stem cells allow a gene to be targeted to its appropriate locus and replace the normal wild-type counterpart by homologous recombination. Using this approach, a better understanding of genetic inheritance or disease pathogenesis becomes possible.

2. Human transplantation of tissues or organs: The pluripotential and immortal qualities of embryonic stem cells make them ideal candidates for use in transplantation to repair damaged tissues or replace tissues that have undergone degenerative changes. There is now evidence in mouse work that embryonic stem cells might prove useful in conditions such as Parkinson's disease, myocardial infarction and spinal cord injuries. Significant technical challenges are needed before the promises of the embryonic stem cells are achieved in humans. These challenges include growing large numbers of pure stem cells for the cell type required. It remains to be determined what would be the best type of cell to use in particular situations, i.e., what degree of differentiated embryonic stem cell is needed for various scenarios? Another important consideration with embryonic stem cells is the question of rejection since the source of the tissue is not normally the recipients. This could be addressed by anti-rejection treatment similar to what is already used in allografts. An alternative way in which the rejection of embryonic stem cells might be controlled is through therapeutic cloning (somatic

cell nuclear transfer or SCNT), which is discussed further below. A comparison of embryonic stem cells and adult stem cells is given in Table 6.13.

A controversial aspect of embryonic stem cells concerns their source, which presently comes from excess embryos obtained during *in vitro* fertilisation procedures (discussed further in Chapter 10). While these excess embryos would eventually be destroyed, their use for human research has provoked controversy. In response to this, governments have in some cases limited access to embryonic stem cell lines. For example, in the USA, only 15 lines created before 9 PM EDT on August 9, 2001, can be used for research that is funded through the NIH (see http://stemcells.nih.gov/registry/index.asp). Cell lines produced after the above date are available, but NIH research money cannot be used for experimentation with these lines. The regulatory framework for human embryonic stem cell research is very much in flux with the US position described above flanked by countries such as the UK and Singapore that allow therapeutic cloning (see cloning on page 166), and others such as Germany and Italy that do not allow any embryo research, including the production of embryonic stem cells.

Adult Stem Cells

Although embryonic stem cells have the greatest potential to differentiate into various cell types, the disadvantages mentioned earlier as well as the ethical and moral questions raised about their source of derivation make adult stem cells attractive alternatives for transplantation

Table 6.13   Comparing properties of embryonic and adult stem cells

Property	Embryonic stem cell	Adult stem cell
Source	Apart from technical considerations, complicated by ethical, legal and social issues.	Limited only by technical issues.
Immunogenicity	Rejection a concern if source is another human, i.e., an allograft (so immunosuppression likely to be required). Antigenicity unresolved. A way around this is therapeutic cloning to obtain the embryonic stem cell from the patient.	Not relevant if autologous, i.e., sourced from patient. Becomes an issue if insufficient cells produced by the patient or the problem is acute and an allogeneic source is needed.
Ability to form multiple tissues, i.e., plasticity	There is no dispute that embryonic stem cells have the capacity to differentiate into any cell type.	There is still debate about the plasticity of adult stem cells. Not proven beyond doubt that adult stem cells are as versatile in this particular role.
Production	Although these stem cells are pluripotential and immortal, there remains a problem with the selection and expansion of pure populations of required cells types.	Easier to grow almost indefinitely in culture, but in terms of a mass market, adult stem cells inferior to embryonic stem cells.
Telomerase	Telomerase, the enzyme restoring the telomeric ends of chromosomes following cell division, is found in high levels in immortal cells such as the embryonic stem cells.	Telomerase levels in haematopoietic stem cells are insufficient to maintain the telomere length. This would be a concern for long-term growth.
Gene expression and tumour formation	Control of gene expression may not be tightly regulated. In mice, teratomas can develop alongside various differentiated cell types. Risk of tumour formation in human cell lines not known.	Tumours not reported.
Success in clinical trials to date	Proof of principle has been shown in animal studies, but data from human work are still very preliminary.	Already successful—e.g., bone marrow and cord blood transplants—although not necessarily showing a pluripotential stem cell effect as distinct from a haematopoietic stem cell.

provided their potential to differentiate into different cells and tissue types (described as plasticity) can be proven. Therefore, the debate centres around the degree of plasticity possible, with claims that adult stem cells can differentiate into a wide range of tissues. Others are more sceptical, wanting to know if the adult stem cell is actually changing its function, or if this apparent plasticity is due to a coexistent or itinerant stem cell that has been carried along with, say, a haematopoietic stem cell that now appears to be producing a brain cell. Another explanation of the adult stem cell's apparent plasticity involves cell fusion between something like a haematopoietic stem cell and the host's target cell, thereby making it appear as though the haematopoietic stem cell has differentiated into a distinct cell.

The traditional hierarchical model for haematopoiesis has, at the top, a self-renewing pluripotential stem cell. Below this is the committed progenitor cell. The next level is the lineage restricted precursor cell. If it is proven that bone marrow (adult) stem cells can differentiate into tissues such as blood, bone, muscle, neural cells and so on, the hierarchical model cannot be correct. If so, alternative models proposed include (1) The marrow has a number of distinct stem cells to explain plasticity. Which predominates depends on various external stimuli. (2)

There are the equivalent of embryonic stem cells in bone marrow, which allows differentiation into all lineages. (3) Haematopoietic stem cells differentiate into the expected blood cells until there is a stress injury of some external stimulus, which allows some of these stem cells to form other lineages. (4) The final mechanism is dedifferentiation followed by redifferentiation based on the environmental stimulus to which the adult stem cell is being exposed. The latter is the likely mechanism to explain Dolly the sheep's origin from an adult differentiated mammary gland cell.

Potential uses of adult stem cells are comparable to what has been proposed for embryonic stem cells:

1. Research—understanding dedifferentiation and redifferentiation: The complex control of cellular differentiation is not understood, and so the option to have a model (the adult stem cell) to explore the molecular and cellular controls would provide invaluable basic scientific knowledge, as well as possible therapeutic options to induce cells to change their primary differentiation pathway.
2. Therapeutic: Some evidence is already available from mouse models that adult neurogenic stem cells can be used in the treatment of Parkinson's disease. Like

the embryonic stem cells, this would also open up the options for stem cells to be used in discovering new drugs as well as testing drugs for toxicity. Stem cell therapy for ischaemic heart disease is another focus of clinical research. As better treatments are reducing the acute mortality associated with coronary artery disease, the number of survivors developing congestive cardiac failure is increasing. Conventional drug regimens are not effective with end-stage heart failure, and there are inadequate numbers of cardiac transplants available. Stem cells have been tried, and are showing some promise (Box 6.4). Telomerase activity in adult stem cells is less compared to the level in immortalised embryonic stem cells (see Table 6.13). However, genetic manipulation provides the opportunity to add the telomerase gene to adult stem cells prior to their use. In theory, this would extend the life of these stem cells.

CLONING

In molecular medicine, "cloning" is a frequently used word with many different meanings. DNA can be cloned, cells can be cloned, monozygotic twins are examples of clones. Dolly the sheep proved that whole animals could be cloned. There are now unsubstantiated claims that human clones have been produced. In the context of cellular therapy, cloning refers to asexual reproduction brought about by the process known as somatic cell nuclear transfer (SCNT). In 1997, Dolly the sheep provided the proof that DNA from a differentiated tissue cell (mammary gland cell) could be taken and reprogrammed to produce a cloned animal copy. The process involved the removal of the nucleus (i.e., DNA) from the mammary gland cell. An egg from a donor sheep was next enucleated, and the nucleus from the mammary gland cell inserted into that egg, which was returned to a surrogate mother by the usual *in vitro* fertilisation techniques (Figure 6.9).

The genetic composition of the clone in this case is virtually identical to the mother, and there is no paternal contribution. "Virtually identical" is used because the recipient enucleated egg still has within its cytoplasm mitochondrial DNA, which contributes about 1% of the total DNA in the clone. This process is called SCNT, and although it has produced Dolly the sheep as well as a host of other animals, it is very inefficient and error prone; for example, it took 277 attempts to get Dolly and many of these produced malformed fetuses. It is assumed that the inefficiency of the procedure partially reflects the development of abnormalities. As indicated earlier, Dolly's genetic constitution excluded any paternal genes, and this is relevant because, as discussed in Chapters 4 and 7, regions of the genome are imprinted, and in these regions, either the maternal or the paternally inherited gene is critical to normal development. So it is not

> **Box 6.4 New therapies: Developing a cellular approach to the treatment of congestive cardiac failure.**
>
> Remodelling occurs post-myocardial infarction. This involves myocyte apoptosis, replacement of heart muscle with fibrous tissue, extension of the infarct because of inadequate blood supply, and eventual left ventricular dilatation. Cellular transplantation is one way to modify the remodelling process. For this, the following would be required: (1) cardiac myocyte stem cells to repopulate the damaged area and (2) stem cells to improve vascularisation of the infarct. Cardiac myocyte stem cells can come from embryonic stem cells, but there is also evidence that within the bone marrow there are mesenchymal-like cells that can be induced to differentiate into cardiomyocytes. Skeletal myoblasts have also been transplanted into damaged heart muscle, and although the cells that grew resembled skeletal rather than cardiac myocytes, there was evidence that contractility improved. To get around the hypoxia with subsequent extension of the infarct, peripheral blood-derived endothelial progenitor cells have been utilised to stimulate a more effective neovascularisation around the infarct. In the circumstances described, stem cells could be isolated from patients (i.e., autologous) or donors (i.e., allogeneic). A number of human clinical trials involving the injection of putative stem cells at the time of surgery or other interventions are presently under way to confirm earlier animal data or preliminary human studies which suggest that myocardial function can be improved with this form of therapy.

entirely surprising that SCNT is neither efficient nor reliable enough to be used for human reproductive cloning.

What has been described above is usually called **reproductive cloning**. In contrast is **therapeutic cloning**—another form of SCNT but the purpose now is to produce embryonic stem cells for research or therapy (Table 6.14). Two key issues remain unresolved. The first is the utility of embryonic stem cells compared to adult stem cells. This issue was discussed above. The second is the creation or use of embryos for research, and not fertility purposes. There is a division of opinion here, with some claiming that the latter is justified because of the downstream potential of embryonic stem cells. Others are less convinced of this argument particularly when balanced by the spectre of creating embryos and then destroying them for the purpose of research or perhaps the creation of "spare parts." This is not an easy debate, but a key priority will be to conduct high-quality and ethically based research to answer some of the issues raised about the various types of stem cells and their potential for treating human diseases.

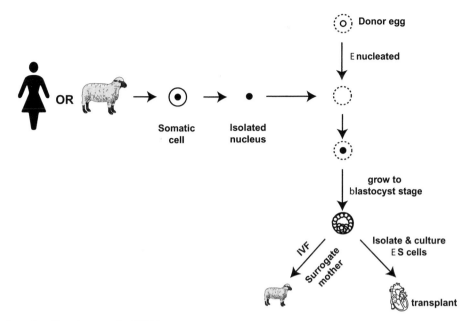

Fig. 6.9 Somatic cell nuclear transfer (SCNT).
The nucleus (•) from a somatic cell (mammary gland was used for the sheep Dolly) is isolated and then inserted into an enucleated egg. The egg with its new nuclear DNA complement is stimulated to divide and then reimplanted into a surrogate mother by IVF for reproductive cloning, i.e., the production of a live animal. Alternatively, the egg serves as a source of embryonic stem (ES) cells (therapeutic cloning) that might or might not have the same antigenic makeup as the somatic cell donor, depending on whether the donor cells are autologous or allogeneic. In theory, these cells could then be used for organ or tissue transplantation. Although somatic cell nuclear transfer produces a clone which is virtually identical to the donor, there are some differences because the enucleated donor egg still contains within its cytoplasm mitochondrial DNA (perhaps 1% of total DNA) that is genetically distinct to mitochondrial DNA in the donor. In this form of asexual reproduction, it is difficult to control for epigenetic changes (for example, imprinting, discussed in Chapters 4, 7), which might require genetic contributions from both the male and female germ cells.

Table 6.14 Therapeutic versus reproductive cloning

Property	Therapeutic cloning	Reproductive cloning
Alternative terminology	Cloning for biomedical research or experimental cloning.	Cloning to produce children or live birth cloning.
Purpose	Clone an early embryo from a patient with disease, then use the embryo to produce embryonic stem cells to treat the patient.	Infertile couples; death of a child.
Technical feasibility	Embryonic stem cell can be produced in this way.	Likely that in humans (as was found in animals) there is a significant risk of malformations or genetic abnormalities (related to failed epigenetic control of gene expression). Animals cloned in this way include Dolly and others although most died.
Acceptance	Another focus for debate with two polar views emerging. Some countries have made this type of cloning illegal.	Unacceptable scientifically (although some scientists have a dissenting view). Illegal in some countries.

XENOTRANSPLANTION

Xeno (Ξενο comes from the Greek "foreign or strange." Xenotransplantation describes the use of non-human organs or tissues for transplantation. The usual transplant involves an allograft (donor and recipient are the same species). The pressure to develop a technology such as xenotransplantation comes from the worrying trend that the demand for donors for kidney, liver, heart and lung allotransplants continues to grow while the donor rate is falling. No obvious solution is evident, although some countries have made some advance by using a "presumed consent" approach; i.e., failure to specifically opt out of being a donor at time of death allows organs to be removed from that person. However, it is unlikely that this approach will be universally acceptable. Stem cell technology may offer an alternative source of tissue, but this remains an unknown and at best a long-term goal. Therefore, an alternative source of organs or tissues for transplantation is those from other species.

The definition of what is a xenotransplant is problematic. There are potentially four different levels: (1) The most obvious example would be a solid organ such as a heart from a non-human primate or a pig. (2) The next level is the transplantation of cells; for example, pig pancreatic islet cells have been used in an attempt to treat diabetes in humans by xenotransplantation. (3) A third level is the use of animal extracorporeal organs or tissue to support a human until the latter's own tissues start to work. An example of this would be acute poisoning, which leads to temporary liver failure from which the patient will recover with time. The functions of the patient's liver can be taken over temporarily by perfusing the patient's blood through pig liver tissue. (4) Finally, and more controversial, is whether the presence of an animal-derived substance such as bovine serum albumin in a therapeutic product is a type of xenotransplant.

Despite some enthusiasm for whole organ xenotransplants in the 1980s–1990s, the present focus is on cellular xenotransplants for diabetes and conditions such as Parkinson's disease. Cells prepared from the donor animal are injected into the patient directly or encapsulated within a membrane. Although some preliminary results are promising, two major problems need to be resolved before xenotransplants can be more realistically assessed in clinical trials: (1) rejection and (2) risk of infection.

Rejection of xenotransplants involves both antibody and cellular responses to the foreign tissue. The best characterised is called hyperacute rejection and results from preformed antibodies to the animal tissue, leading to a rejection response within minutes of transplantation. At present, the most suitable animal for cellular xenotransplantation is the pig, and the basis for the hyperacute rejection in this animal is the presence of galactosyl α-(1,3)-galactosyl β-1,4-N-acetyl glucosaminyl groups (abbreviated to the gal epitope) on porcine tissue. One way to reduce rejection because of preformed antibodies in the human to the gal epitope is to make transgenic "humanised" pigs by taking out this gal epitope. Other genetic modifications of the pig are being tried to diminish the hyperacute rejection response.

The second obstacle to xenotransplantation is the unknown risk of horizontal transmission of pig endogenous retroviruses (abbreviated to PERV) and perhaps other infections to humans (see Chapter 8 for further discussion on animal-to-human infections or zoonoses). Apart from the animal husbandry approach to reducing infections in pigs (sterilisation of feed and water, avoidance of mammalian proteins in the feed and so on), there are breeding or perhaps genetic engineering options to reduce the endogenous load of PERV. Such "cleaner" pigs would be less likely to transmit PERV to the patient (and even the close contacts or community) as well as other porcine infections. The risk of animal-to-human infection has some experimental basis but is difficult to quantify in the real-time human situation. However, the transmission of bovine spongiform encephalopathy (see Chapter 8) from cattle to humans has sensitised regulators, scientists and the community to the potential for zoonotic infection. It will need continuing research and understanding if xenotransplants are to become acceptable to the wider community (both lay and scientific).

FUTURE

NOVEL PROTEIN EXPRESSION SYSTEMS

Pharmyard

The initial high costs in preparing genes and vectors and evaluating both efficacy and safety through a range of *in vitro* and *in vivo* parameters will be offset by long-term and reliable sources of therapeutic products. More efficient expression systems would reduce costs even further. Large transgenic animals (e.g., sheep, goats and pigs) are being investigated with this in mind. Recombinant drugs that are expressed in the milk or blood of these animals might provide a high yield source. Examples being evaluated include coagulation factor IX, α1 anti-antitrypsin (sheep milk) and tissue plasminogen activator (goat milk). The long-term developments in the

Table 6.15 Comparisons of various production systems for recombinant human (rh) drugs and vaccines (Ma *et al* 2003)

System	Cost	Production timescale	Scale-up capacity	Product quality	Post-translation modification	Risks of trans-gene	Storage costs
Bacteria	Low	Short	High	Low	None	Toxins	Some
Yeast	Medium	Medium	High	Medium	Some	Low	Some
CHO	High	Long	Very low	Very high	All	Viral, others	High
Transgenic animals	High	Very long	Low	Very high	All	Viral, others	High
Transgenic plants	Very low	Long	Very high	High	Some differences	Low (environmental risk not included)	Cheap

transgenic "pharmyard," as it has been called, are promising. Another interesting source for therapeutic products is plants (molecular pharming).

Pharming

It is claimed that transgenic plants used to produce recombinant therapeutic products (including vaccine) would involve lower costs, rapid scalability would be possible and human pathogens would not be a problem. Since plants have comparable (but not identical) mechanisms to synthesise proteins, post-translational modifications should be feasible. The option to take a tomato, banana or lettuce containing the drug of choice is obviously very appealing, and the results of trials currently under way will be awaited with considerable interest. This option will be particularly valuable for communities that cannot afford the expensive products being marketed using microbiologic fermentation or mammalian cell cultures. However, genetically modified (GM) food has already provoked considerable debate in various communities. The recombinant plants producing therapeutic products would also involve comparable risks in terms of inadvertent spread or incorporation into the food chain. The effect of the therapeutic products would also need to be modelled in terms of the ecology, particularly animals, insects and microorganisms that would be exposed. Nevertheless, a comparison of the various ways in which new recombinant approaches can be used to produce cheap drugs and vaccines suggests that molecular "pharming" needs serious consideration (Table 6.15).

TRENDS IN CELLULAR THERAPIES

DNA Vaccines

The hepatitis B virus has shown that the recombinant DNA approach to vaccine production can work in humans. The potential for innovative developments possible through recombinant DNA technology may enable difficulties associated with vaccination for other infections to be resolved. These are illustrated by reference to AIDS. The HIV's RNA is capable of integrating into host DNA, and so it can remain latent until it is activated. Thus, the use of a conventional live attenuated virus as a vaccine poses a potential risk if the integrated (latent) form were able to become activated and produce a mutated (wild-type) strain at a later stage (a similar problem was described earlier with the polio Sabin vaccine). The other standard approaches involving inactivated HIV-1 or subunit components are also unsatisfactory since Gp 120 (see Figure 8.2), an important antigenic surface envelope protein that enables the virus to attach to the lymphocyte's CD4 receptor, is subject to considerable antigenic variation. Finally, HIV may be transmitted by infected cells as well as in free virus form. Thus, intracellular virus may escape immune surveillance.

Gene Therapy

Apart from a substantial increase in the number of protocols for gene therapy, another significant development since the first edition of *Molecular Medicine* has been the change in emphasis from genetic disease to cancer and HIV (see Table 6.7). Monitoring by various regulatory bodies has ensured a safe and steady progression for gene therapy. However, a problem occurred in 1999 when a young 18-year-old male, Jesse Gelsinger, died as a direct result of gene therapy (Box 6.5, discussed also in Chapter 10). The consequences of his death were significant, including a reassessment of the regulatory procedures in the USA. With development of acute leukaemia in the two SCID-X1 children mentioned earlier, a further review of gene therapy was conducted, and again changes were made to the regulatory environment.

Future developments in gene therapy are likely to involve more *in vivo* strategies for gene transfer (see Figure 6.3). Therefore, better approaches that allow genes to be introduced and expressed in specific organs are needed.

The biggest challenge continues to be how to increase the efficiency of transduction or delivering sufficient numbers of genes to the required tissues. The long-term expression of the transgene will also require better harvesting of stem cells. Recent entries into RNA therapeutics are the double-stranded small RNA species that work through RNA interference (RNAi) (see Chapter 2). They have not yet reached the clinical trial stage but show promise *in vitro* and in animal studies to inhibit unwanted RNA species, for example, HIV. Like ribozymes, the RNAi molecules can be carried into cells using vectors such as retroviruses and adeno-associated viruses. Their specificity allows selective knocking out of aberrantly expressing genes in cancer or foreign genes in infections. Small RNA species can also fold into the 3D structure of proteins. HIV actually uses this approach to recruit viral and host proteins. These small RNAs have the potential to be used as decoys to inhibit unwanted proteins.

The adenosine deaminase deficiency and SCID-X1 examples described earlier illustrate gene therapy in which a normal gene is added to the patient's deficient cells. Hence, it represents addition rather than replacement therapy. Problems with addition therapy include random integration into the genome, leading to inefficient or inappropriate expression, and interference with the function of nearby normal genes. A way around this problem would be to replace defective genes by homologous recombination (Figure 6.10). *In vitro* studies using mammalian cells have shown that it is also possible to target by homologous recombination, although the frequency of this occurring is low, e.g., 1 in 100 to 1 in 100 000.

Two developments have made homologous recombination an achievable goal in mammals. First, PCR allows many cells to be screened for the appropriate recombination event. This is achieved by constructing DNA amplification primers that are able to detect the creation of a novel junction formed between target and incoming DNA. DNA inserted elsewhere in the genome is not detected because the junction fragment would not be present. Although homologous recombination is a rare event, it can be detected by PCR. The second development has been the availability of stem cells both adult and embryonic. A vector containing the gene of interest is transferred into these cells by physical means, e.g., microinjection. In the great majority of cases, random

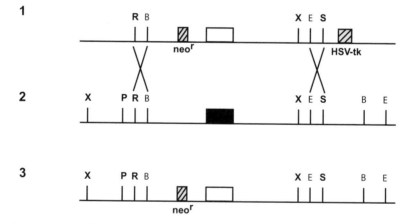

Fig. 6.10 Homologous recombination.
During gene transfer, the problem of random integration into the genome could be avoided if genes were able to be targeted to their correct position and then made to replace the defective gene. An incoming DNA segment is depicted in **(1)** and the correct place into which it needs to insert is **(2)**. The wild-type (normal) gene is shown as □ and the mutant as ■. R, B, X, E and S identify recognition sites for restriction enzymes. Identical restriction enzyme patterns for **(1)** and **(2)** show where the two DNA areas are the same; i.e., homologous recombination is possible at these loci. The incoming DNA segment has an additional two genes that will be used to distinguish cells which have undergone homologous recombination. The genes are neo^r (neomycin resistance) and HSV-tk (herpes simplex virus thymidine kinase gene). If there is **homologous recombination** at the corresponding DNA sequences marked by (X), the structure depicted in **(3)** will result; i.e., the mutant gene is replaced by the normal gene, which also brings with it neomycin resistance but not the HSV-tk gene. Another way in which **(1)** can insert into the genome is via **random integration**. In this case the whole segment of **(1)** will be acquired since random integration occurs through the ends of linearised DNA; i.e., **(1)** will simply link to the end of **(2)** or more likely another part of the genome, and so there will be both mutant gene, normal gene plus the neomycin and HSV-tk genes in tandem array. Cells that contain the neomycin resistance gene can be selected for by growing them in the presence of the drug G418, which will kill all other cells. The neo^r gene will select cells that have undergone either homologous recombination or random integration. The two options can be distinguished by a second drug (ganciclovir) that is cytotoxic to cells containing the HSV-tk gene; i.e., cells integrating the gene in a linear fashion will be destroyed. The result is selection for cells that have undergone homologous recombination. The cells integrating the appropriate gene are identified by PCR.

Jesse Gelsinger was an 18-year-old male with a mild form of OTC (ornithine transcarbamylase) deficiency, an X-linked genetic disorder. The defect in OTC involves an enzyme in the urea cycle leading to protein intolerance due to accumulation of ammonia in the body. Although not severely affected, this individual volunteered for a phase I gene therapy study (in a phase I study, safety rather than efficacy is the end point being measured) to correct the OTC deficiency. A number of individuals had been treated by this gene therapy approach before him, and on the basis of these treatments, there was some evidence that the viral vector being used (adenovirus) was associated with some adverse events including fever, thrombocytopenia and transaminitis. Nevertheless, the patient was treated with a relatively high dose, and he died from acute respiratory failure four days later. Subsequent review by the FDA identified a number of violations of the clinical trial rules. Shortcomings were also noted in the review process, as well as the regulatory protocol for notification of serious adverse events. This unfortunate event, while a complication of gene therapy, was potentially avoidable.

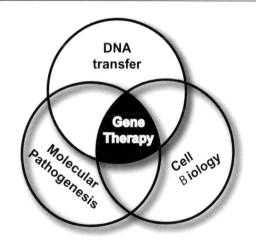

Fig. 6.11 Components of gene therapy.
The diagram illustrates the three basic ingredients for gene therapy. An understanding of molecular pathogenesis will allow the appropriate strategy to be designed, for example, to inhibit or replace a defect gene or mRNA product. Next is required an efficient DNA transfer, whether it be via a viral vector or some other means. Finally, knowledge of cell biology, in particular, the identification and manipulation of stem cells allows the introduced gene to have a long-standing effect.

insertion of DNA will occur. However, in a very few cells, the gene of interest will pair with its corresponding DNA sequence. Transferred and targeted DNA can then be exchanged by homologous recombination. Stem cells that have undergone targeting are selected by PCR and then cultured *in vitro* prior to injection back into the patient.

Stem Cells

In the past few years there has been a lively and at times emotive debate among scientists, politicians and the community on the relative merits of embryonic versus adult stem cells. Passionate beliefs have been expressed about the powers of stem cells to cure a wide range of human disorders. However, a lot more high-quality scientific evaluation is needed to determine what is fact, and this will be a challenge. In terms of molecular medicine, stem cells are key prerequisites to technologies like gene therapy. Without the stem cell becoming involved, a gene therapy approach will always be a transient effect limited by the half-life of the cell that has been genetically altered (Figure 6.11). Stem cells *per se* promise alternative therapeutic approaches that can be used alone, or a stem cell's genetic environment can be genetically manipulated to provide it with greater flexibility.

Cloning

While Dolly the sheep was a spectacular scientific achievement, her cloning has produced scientific, ethical and moral dilemmas with differing opinions within the community both lay and scientific. Surprisingly, some basic questions, such as whether Dolly prematurely aged (after all, she was derived from the cells of a six-year-old sheep), remain unanswered. As discussed in Chapter 10, the original discoverers of the technique that allowed Dolly to be produced have expressed dismay that this technology might be applied to human reproductive cloning because many scientific questions remain unanswered. This will be an important challenge for scientists and the community, with evidence-based decision-making countered by emotive issues such as infertility or a dying child or relative who might be given a second life through cloning.

Xenotransplantation

The future of xenotransplantation would seem more straightforward than cloning or stem cells. A number of countries have now established regulatory guidelines that will allow xenotransplantation to proceed subject to careful monitoring both short and long term. The risk for PERV is likely to remain unquantifiable until large enough studies are conducted and there is a longer follow-up period. Recent data are reassuring since they suggest that while PERV infection can occur *in vitro*, there is little evidence that this is accompanied by replication of PERV. Provided the risk/benefit analysis is carefully examined, xenotransplantation is likely to proceed, although it is difficult to predict its long-term success.

FURTHER READING

Recombinant DNA (rDNA) Derived Drugs

Hedenus M *et al.* Randomized, dose finding study of darbepoetin alfa in anaemic patients with lymphoproliferative malignancies. British Journal of Haematology 2002; 119:79–86 (*provides preliminary data on the EPO derivative in clinical trials*).

Keating CG, Noble S. Recombinant hepatitis B vaccine (Engerix B®). Drugs 2003; 63:1021–1051 (*provides a useful overview of one particular rhHBV vaccine but also general information*).

McClamrock HD. Recombinant gonadotropins. Clinical Obstetrics and Gynecology. 2003; 46:298–316 (*comprehensive summary of recombinant DNA–derived products for this hormone family*).

Samol J, Littlewood TJ. The efficacy of rHuEPO in cancer related anaemia. British Journal of Haematology 2003; 121:3–11 (*succinct overview of rhEPO in various haematologic malignancies*).

Sylvester RK. Clinical applications of colony stimulating factors: a historical perspective. American Journal of Health-System Pharmacy 2002; 59 (suppl 2): S6–S12 (*provides a pharmacist's perspective of the haematopoietic growth factors and their impact*).

http://www.asheducationbook.org/ (*this is the web address for the American Society of Hematology education section. Provides concise reviews on many topics, including haemophilia by PM Manucci. Unfortunately, stops at year 2002*).

Cellular Therapies

Aguilar LK, Aguilar-Cordova E. Evolution of a gene therapy clinical trial: from bench to bedside and back. Journal of Neuro-Oncology 2003; 65:307–315 (*nice overview of the steps involved in developing a clinical study in human gene therapy*).

Balicki D, Beutler E. Gene therapy of human disease. Medicine 2002; 81:69–86 (*useful overview of gene therapy particularly from the description of transfection methods and a summary of some gene therapy trials*).

Donnelly J, Berry K, Ulmer JB. Technical and regulatory hurdles for DNA vaccines. International Journal for Parasitology 2003; 33:457–467 (*excellent discussion of DNA vaccines*).

Dooldeniya MD, Warrens AN. Xenotransplantation: where are we today? Journal of the Royal Society of Medicine 2003; 96:111–117 (*concise overview of the xenotransplantation situation particularly in relation to rejection and risk*).

Hacein-Bey-Abina S *et al.* Sustained correction of X linked severe combined immunodeficiency by ex vivo gene therapy. New England Journal of Medicine 2002; 346:1185–1193 (*the first report of a gene therapy success in SCID*).

Hacein-Bey-Abina S *et al.* LMO2 associated clonal T cell proliferation in two patients after gene therapy for SCID-X1. Science 2003; 302:415–419 (*a summary of the insertional mutagenesis event leading to acute leukaemia after gene therapy*).

Hassink RJ *et al.* Stem cell therapy for ischemic heart disease. Trends in Molecular Medicine 2003; 9:436–441 (*brief, succinct overview of stem cells for the treatment of heart disease*).

Knowles LP. A regulatory patchwork—human ES cell research oversight. Nature Biotechnology 2004; 22:157–163 (*an informative overview of the various countries and the stances taken about human ES cell research*).

Korbling M, Estrov Z. Adult stem cells for tissue repair—a new therapeutic concept? New England Journal of Medicine 2003; 349:570–582 (*excellent overview of adult stem cells*).

McCormack MP, Rabbitts TH. Activation of the T cell oncogene LMO2 after gene therapy for X linked severe combined immunodeficiency. New England Journal of Medicine 2004; 350:913–922 (*an in-depth discussion of leukaemia developed in the 2 SCID-X1 children treated with gene therapy*).

Manno CS. The promise of third generation recombinant therapy and gene therapy. Seminars in Hematology 2003; 40 (suppl 3):23–28 (*provides a summary of both gene therapy and the most recent recombinant factor VIII products in haemophilia*).

Prentice DA. Stem cells and cloning. 2003. Benjamin Cummings, San Francisco. ISBN 0-8053-4864-6 (*small monograph but excellent overview of stem cells and cloning*).

Puthenveetil G *et al.* Successful correction of the human β thalassemia major phenotype using a lentiviral vector. Blood 2004; 104:3445–3453 (*summarises a promising early success for gene therapy in thalassaemia*).

Sullenger BR, Gilboa E. Emerging clinical applications of RNA. Nature 2002; 418:252–258 (*brief and comprehensive overview of RNA options for therapy*).

Urban E, Noe CR. Structural modifications of antisense oligonucleotides. Il Farmaco 2003; 58:243–258 (*provides both a simple and a more complex overview of oligonucleotides including the two applications described in the text*).

Weissman IL. Stem cells—scientific, medical and political issues. New England Journal of Medicine 2002; 346:1576–1579 (*provocative presentation of some issues*).

http://www.asheducationbook.org/ (*this is the web address for the American Society of Hematology education section. Provides concise reviews on many topics, including Stem cells by CM Verfaillie, MF Pera and PM Lansdorp. Unfortunately, stops at year 2002*).

http://www.newscientist.com/hottopics/ (*Internet address for New Scientist providing brief comments on topical subjects including cloning, gene therapy*).

http://www.wiley.co.uk/genmed/clinical/ (*Gene Therapy Clinical Trial Site—lists more than 636 gene therapy studies undertaken worldwide*).

http://www4.od.nih.gov/oba/rac/clinicaltrial.htm (*Internet address for the NIH's Recombinant DNA Advisory Committee—*

provides access to the list of gene therapy studies approved in the USA).

Future

Ma J K-C, Drake PMW, Christou P. The production of recombinant pharmaceutical proteins in plants. Nature Reviews Genetics 2003; 4:794–805 (*an interesting and provocative paper describing the potential use of transgenic plants for drug and vaccine production*).

Stevenson M. Therapeutic potential of RNA interference. New England Journal of Medicine 2004; 351:1772–1777 (*excellent overview*).

7

REPRODUCTION AND DEVELOPMENT

REPRODUCTION

The 1980s were the beginning of a new era in reproductive technology. Developments included the availability of *in vitro* fertilisation, fetal blood and tissue sampling and ultrasound examination. By the 1990s, the fetus could be visualised with impressive clarity using modern ultrasound equipment. Biopsy of fetal tissue became a routine procedure with chorionic villus sampling (CVS). Recombinant DNA techniques enabled fetal-specific DNA to be characterised. The last two developments made it possible to undertake many types of **prenatal diagnoses**. The diagnostic options continue to expand with the availability of the polymerase chain reaction (PCR) and, more recently, fluorescence *in situ* hybridisation (FISH). DNA technology has entered the traditional biochemically based discipline of **newborn screening**. The scope for DNA testing in the newborn has correspondingly increased. **Fetal therapy** has started. Blood transfusions and surgical corrections of some defects in the fetus are possible. Attempts at *in utero* gene therapy are conceivable options for the future. The neonate as a recipient of bone marrow has been explored using cord blood as a source of stem cells. Following on from the many changes described above have emerged **fetal medicine units**, which are primarily responsible for the care of the fetus *in utero*.

INFERTILITY

Infertility is defined as the inability to conceive after 12 months of regular unprotected intercourse. It affects approximately one in six couples wanting to start a family. The aetiology of infertility is complex with environmental, medical, surgical and genetic contributions interacting (Table 7.1). The application of cytogenetics and molecular genetics techniques to the study of infertility has enabled a number of chromosomal and gene mutations to be identified as causative factors (Table 7.2). However, these are rare explanations for a relatively common condition.

In males, the finding of bilateral congenital absence of the vas deferens is now firmly associated with mutations in the cystic fibrosis gene (*CFTR*—see Chapter 3 and Figure 3.12). Although these males as a group are more likely to have *CFTR* mutations, individuals are not so informative because they may be only heterozygotes for

Table 7.1   Causes of infertility

Cause	Examples
Lifestyles, events	Increased age, exercise.
Medical reasons	Hormonal, immunological or psychological issues; obesity, infectious diseases, blockage (surgical or other causes).
Genetic	Difficult to quantitate because likely to be associated with other factors. Genetic causes range from single gene defects to chromosomal abnormalities (Table 7.2).
Unknown	The cause of infertility in about 20% of couples is never explained.

Table 7.2   Genetic causes of infertility

Genetic defect	Examples
Chromosomal abnormalities	Translocations, inversions and other abnormalities. Aneuploidies—Klinefelter's syndrome (47,XXY); 47,XYY males; Turner's syndrome (46,X); and Down's syndrome (Trisomy 21). Trisomies in the fetus may explain a number of pregnancies that do not progress. Sperm from infertile males is more likely to have chromosomal abnormalities.
Complex syndromes	Prader-Willi and Angelman's syndromes demonstrate complex phenotypes including infertility (see Chapter 4). Prader-Willi syndrome has associated obesity that may contribute to the infertility. The cause of infertility is unknown, but it may be related to the primary imprinting defect.
Single gene disorders in females	Mutations in various genes from the gonadotropin releasing hormone receptor to the FSH and LH receptors and various steroidogenic pathways can lead to infertility.
Single gene disorders in males	Mutations in the cystic fibrosis gene (*CFTR*) are associated with bilateral absence of the vas deferens. *SOX9* and *SRY* mutations are associated with infertility. Mutations in the androgen receptor lead to androgen insensitivity syndrome.

Box 7.1   Molecular genetics of puberty (Seminara *et al* 2003).

Fertility in the female results from the cyclic and pulsatile release of GnRH, FSH and LH. Although the same hormonal pathways are involved in the onset of puberty, the complex interactions of factors initiating puberty have remained elusive. One approach to investigate this phase in development is to study, at the molecular level, "mistakes of nature" delaying puberty and, from this, identify the relevant genes. This approach is well illustrated in a family with infertility and consanguinity (three first cousin marriages). The family was thought to have idiopathic hypogonadotropic hypogonadism, i.e., absent spontaneous sexual maturation associated with low-normal range gonadotropins. Using the traditional positional cloning and linkage analysis, the site for the apparent genetic defect (which, because of consanguinity, suggests an autosomal recessive trait) was shown to be chromosome 19p13.3. In this region of 1 Mb was the gene *GPR54*, which is a G protein–coupled receptor gene and so a likely candidate in perturbing hormonal signals. G proteins are a family of membrane proteins that are only activated once they bind guanosine triphosphate (GTP). The activated G proteins then activate an amplifier enzyme on the inner surface of the membrane, leading to the release of second messengers such as cyclic AMP. Affected members in this family all had homozygous missense mutations involving the *GPR54* gene. This showed that the gene mutation is causative and confirmed the autosomal recessive inheritance. Definitive proof came from the study of *Grp54*-deficient mice that demonstrated the same phenotype as the family, i.e., hypogonadotropic hypogonadism. Therefore, *GRP54* is one genetic factor in the initiation of puberty. Its exact mode of action is now being studied.

---

a known mutation and so difficult to distinguish from an asymptomatic carrier. The second *CFTR* mutation in these individuals is unlikely to be detected with current technology, which focuses more towards the severe end of the cystic fibrosis spectrum. This second mutation must have a very mild effect; otherwise, cystic fibrosis would have developed. In contrast to a physical barrier to fertility demonstrated by *CFTR* mutations, females with single gene mutations affecting their reproductive capacity are more likely to have defects in their hypothalamic-pituitary-ovarian axis. For women, fertility (as well as

the onset of puberty—Box 7.1) is regulated by cyclic hormonal pulses starting with gonadotropin releasing hormone (GnRH) from the hypothalamus. This then induces the production of follicle stimulating hormone (FHS) and luteinising hormone (LH) from the pituitary. The latter two stimulate the ovary to release oocytes. The majority of cycling defects in the female become apparent during puberty, and most start at the level of the ovary.

ASSISTED REPRODUCTIVE TECHNOLOGIES (ART)

Assisted reproductive technologies (ART) have proven to be very successful in treating infertile couples. Since the birth of the first IVF (*in vitro* fertilisation) baby in 1978, the techniques have evolved to include ICSI (intracytoplasmic sperm injection) for infertile males. In addition, the embryo can be manipulated and cultured *in vitro* as well as biopsied for preimplantation genetic diagnosis (PGD). More than one million children have been born as a result of IVF. Until recently, it was believed that children born through ART developed normally and had the same frequency of malformations as the general population. However, reports now emerging suggest that this may need to be reviewed. They include (1) In one study, the risk for major birth defects was doubled following ART. (2) In another study, the risk for low birth weight in singleton pregnancies was increased by 2.6 times following ART. (3) In a few case reports, a higher risk for Angelman's syndrome and Beckwith-Wiedemann's syndrome was suggested following ART, perhaps more so after ICSI. If these observations are confirmed, it will be necessary to determine whether these adverse consequences are secondary to what is causing the infertility and/or the ART itself, particularly the use of ICSI, a technique that can select immature sperms. A third explanation for adverse consequences is an epigenetic defect, particularly imprinting, since it is noteworthy that both the Angelman's and Beckwith-Wiedemann's syndromes are examples of imprinted genetic disorders involving the expression of maternal genes (see Table 4.7).

Epigenetic Changes during Reproduction

The importance of epigenetic inheritance in normal human development and the aberrations in this complex form of inheritance leading to development of cancers were discussed in Chapter 4. Mention was also made in Chapter 6 that reproductive cloning by SCNT was a very inefficient process, with the success rate only about 1 in 100 that the embryo will develop into an adult. One explanation for this is that the reproductive cloning technique bypasses the normal epigenetic pathways in the developing embryo.

There are about 50 imprinted human genes, i.e., they are differentially expressed according to their parent of origin in either the oocyte or spermatozoon. Epigenetic modifications described in Chapters 2 and 4 such as DNA methylation or changes in the histones are thought to explain how copies of either the maternal or paternal alleles are switched off. Three critical periods during gametogenesis and embryo development require epigenetic modifications: (1) **Erasure:** During early development of the germ cells, the imprint is erased as shown by the primordial germ cells lacking methylation. (2) **Acquisition:** Thereafter, as the sperm and the oocyte mature, the methylation pattern is re-established for all genes (imprinted and non-imprinted). In sperm, the methylation of genes occurs very early on, so IVF procedures are unlikely to interfere with it. In comparison, the methylation of genes in the oocyte is slightly more delayed and, in theory, the oocyte is more susceptible to ART interfering with methylation, particularly if procedures such as

in vitro maturation are used. (3) **Maintenance:** At the time of fertilisation, the non-expressing male or female copy of the imprinted genes will continue to remain methylated. All other genes gradually lose their methylation. Therefore, (3) is another period during development that could lead to malformations if ART interferes with the maintenance of methylation (for imprinted genes) or the loss of methylation (for all other genes). Freezing of sperm or the use of cryoprotecting agents might interfere with methylation or chromatin structure in humans although it does not appear to do this in mice. Superovulation, with its exposure to gonadotrophins leading to the release of ova that might not have completed the methylation step, and extensive culture for *in vitro* maturation are procedures that could disturb the maintenance of an imprint.

Imprinting and Development

Genomic imprinting is found in eutherian mammals. Various theories have been proposed why some imprinted genes are preferentially expressed by the paternal alleles and others by the maternal alleles, but none is proven. Imprinted genes fall into two major functional groups: (1) those involved in fetal development (in general, these are more often maternally expressing genes) and (2) genes with a predominant effect on placental function (usually paternally expressing genes). The maternal versus paternal effect is demonstrated by two tumours. Ovarian teratomas are considered to have arisen from a single germ cell. They are composed of tissues with all three germ layers (ectoderm, mesoderm and endoderm) found in the fetus. All benign forms usually have a normal 46,XX (female) chromosomal complement. In contrast, hydatidiform moles affect the placenta and are invariably 46,XY. They are thought to have arisen from fertilisation of an empty ovum by a sperm. The haploid sperm chromosome complement is then duplicated. Thus, all chromosomes are of paternal origin.

In view of what has been described in relation to ART and the recent discussions centred around SCNT and reproductive cloning, a lot of interest has now shifted to IVF to determine if this produces any perturbations in the epigenetic patterns. At the very least, a change in the epigenetic imprint might explain the low birth weights observed. More seriously, subtle changes in the epigenetic pattern may not be discernable for some time into the future (and may not be apparent until the children born from IVF have children of their own). Because of the relatively limited numbers of IVF babies, small but potentially important effects will be difficult to detect for some time. Nevertheless, an interesting question has been raised, and prospective long-term follow-up and molecular analysis will be essential to address any concerns.

DEVELOPMENT

HOMEOBOX (HOX) GENES

An important advance made possible through molecular medicine has been the identification of genes involved in human development. From a better understanding of the normal molecular processes, we are obtaining some insight into malformations and their underlying mechanisms. The success of this work has depended on basic biological research utilising animal models such as the fruit fly *Drosophila melanogaster*, the mouse and, more recently, the zebrafish. The interspecies conservation of the important developmental genes has meant that the equivalent ones in humans are able to be identified and characterised. The significance of the animal model in understanding what happens in the human was acknowledged with the 1995 Nobel Prize for Physiology or Medicine (see Chapter 1). In both vertebrates and invertebrates, families of genes (such as *HOX*, *PAX* and *SOX*) and a number of other genes are shown to play key roles.

Mutations in the body form of *Drosophila*, which caused a part of the body to be replaced by a structure normally found elsewhere, were shown in the early 1980s to involve genes called *HOX* (abbreviation comes from h̲o̲m̲e̲o̲b̲o̲x̲). In *Drosophila*, the physical arrangement of these genes is identical to the order in which they are expressed along the anteroposterior axis of the embryo during development; i.e., the more 5' a gene, the more posterior is it expressed in the developing body. The *HOX* genes are called *HOM-C* in invertebrates.

A conserved DNA sequence is found in all *HOX* genes. It is called the **homeobox** (also known as homeodomain) and is about 180 bp in size. The 60 amino acid encoded by the homeobox has DNA-binding properties, and so the homeoproteins are transcription factors. Thus, this class of genes can regulate the expression of many other genes. Comparative DNA analyses have shown that homeobox genes evolved from common ancestral genes and their subsequent divergence reflects the morphological complexity of the organism in which they are found. For example, insects and *Drosophila* have a single cluster of the *Hox* genes; in vertebrates the number is four. In humans, the *HOX* gene clusters (*HOX-A* to *HOX-D*) are found on chromosomes 7p, 17q, 12q and 2q, respectively. An amazing observation is that in all species the genes remain aligned in the same relative order as in *Drosophila*. The conservation between the genes is so high that vertebrate genes can replace their invertebrate counterparts in transgenic *Drosophila* embryos. Thus, *HOX* genes encode a family of transcription factors of fundamental importance for body pattern during development involving the central nervous system, axial skeleton and limbs, gastrointestinal tract, and external genitalia. Humans, like most vertebrates have 39 *HOX* genes arranged in four clusters (Figure 7.1). These genes have evolved from a single ancestral gene by tandem duplication and then diverged, producing the *HOX* cluster. From this, multiple clusters were formed.

To understand the role of the homeoboxes in development, *Drosophila* or animals with mutations that occur spontaneously or are created by recombinant DNA means are studied. This approach has provided some important evidence to suggest that *HOX* is the mammalian equivalent of the *HOM-C* genes since structural deformities (of the head and neck) result. Despite the identification of these highly conserved genes, the search for natural mutants has been less fruitful. This may reflect the fact that mutations in *HOX* are expressed only as an abnormal phenotype when both alleles are inactivated (in contrast to the dominant effect from *PAX* genes—discussed further below). It is also possible that the paralogous genes from the various clusters can compensate for each other (Box 7.2). Paralogous genes are genes in the same species that are so similar in their nucleotide sequences that they are assumed to have originated from a single ancestral gene. Although the *HOX* genes (39 in total) are the best studied of homeobox genes, many other homeobox genes (about 200) are also found in clusters (such as PAX, described next) or dispersed throughout the genome.

OTHER GENES

Paired-box (PAX) Genes

Another conserved DNA sequence is present in mice and other species as divergent as worms and humans. This is called the *paired* box. The relevant genes are known as *PAX* (p̲aired b̲o̲x̲). In the human, nine of these genes are dispersed over many chromosomes. A 128 amino acid DNA binding domain in *PAX* is conserved in mammals and *Drosophila*. Like the homeobox, this sequence has the properties of a DNA transcription factor. Some of the *PAX* genes also contain homeobox domains. A number of natural mutants involving *PAX* produce clinical problems (Table 7.3). The finding of a relatively large number of developmental disorders associated with *PAX* contrasts with the *HOX* genes, and is explained by the dominant nature of the *PAX* mutations, which produce abnormalities if one of the two alleles is mutated. Like *HOX*, abnormal function in *PAX* (such as that occurring in association with a chromosomal translocation) can also lead to tumour formation.

An interesting developmental disorder in the mouse is called *Splotch (Sp)*. This disorder affects neural crest–derived components leading to the development of

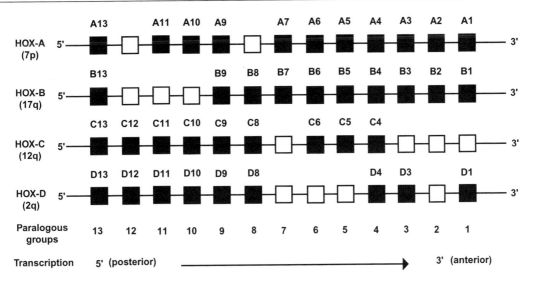

Fig. 7.1 HOX gene clusters.
In vertebrates (including humans), 39 HOX genes are organised into four clusters (on chromosomes 7p14, 17q21, 12q13 and 2q31). The genes can be vertically aligned into 13 paralogous groups determined by the homeobox DNA sequence homology. Genes in the same species that are so similar in their nucleotide sequences that they are assumed to have originated from a single ancestral gene are paralogs. The numbering of each gene in each cluster is based on their DNA sequence similarity and relative position to each other. Functional genes are represented by ■. In general, paralogous HOX genes (for example, HOX-A7 and HOX-B7) are more similar to each other than adjacent genes on the same HOX cluster (i.e., HOX-B7 and HOX-B6). All genes are transcribed in the same direction (→). However, the 3′ genes (head of body) are expressed before the 5′ genes (tail of body). In vertebrates, the 5′ HOX-A and HOX-D genes are involved in limb development.

Table 7.3 PAX genes in mammals (Chi and Epstein 2002)

Gene[a]	Organ/tissue involved	Chromosome	Human disease
PAX1	Skeleton, thymus	20p	No human disease described. Pax1-deficient mice have skeletal defects.
PAX9	Skeleton, cranial-facial, tooth	14q	Abnormal tooth development.
PAX2	CNS (central nervous system), kidney	10q	Renal coloboma syndrome.
PAX5	CNS, B lymphocytes	9p	Lymphoma.
PAX8	CNS, kidney, thyroid	2q	Thyroid dysplasia; thyroid follicular carcinoma.
PAX4	Pancreas	7q	No human disease described. Pax4-deficient mice have abnormalities in pancreas development.
PAX6	CNS, eye, pancreas	11p	Aniridia, cataracts.
PAX3	CNS, neural crest, skeletal muscle	2q	Waardenburg's syndrome, rhabdomyosarcoma.
PAX7	CNS, cranio-facial, skeletal muscle	1p	Rhabdomyosarcoma.

[a] All PAX genes have a paired domain. In addition, PAX 3, 4, 6 and 7 have a homeodomain.

Box 7.2   Development of the human body.

Despite considerable diversity, most animal bodies have basic similarities consisting of bilateral symmetry and a head-to-tail axis. Not surprisingly, for such an important process, genes involved in development have been well conserved during evolution. The *HOX* gene family plays a key role in the head-to-tail axis development. Mutations in two *HOX* genes (*HOX-D13* and *HOX-A13*) produce clear structural malformations affecting development of the hands and, to a lesser extent, the genitals (for example, hypospadias). It is interesting that one mutation involving *HOX-D13* is produced by a DNA triplet repeat (much like that in Huntington's disease—Chapter 3) although in this case the repeat is a poly alanine or (GCG,GCA,GCT,GCC)$_n$. This repeat is unusual because it has the four different codons for alanine. These mutations are thought to exert a dominant-negative effect; i.e., the abnormal protein from the one mutated allele interferes with the function of the remaining (normal) protein. Unlike in Huntington's disease, other mutations within the *HOX-D13* gene have been described. Not surprisingly, in view of their key role in development, homeobox (*HOX*) genes can also be associated with cancer. In some leukaemias, chromosomal translocations can lead to fusion proteins (like the *BCR-ABL* in chronic myeloid leukaemia—Chapter 4) with leukaemogenic potential. One *HOX* gene implicated in this way in myeloid leukaemia is *HOXA9*.

SOX Genes

Sox proteins constitute a large family of transcription factors characterised by a DNA binding HMG (high mobility group) domain of about 79 amino acids. This domain is highly conserved and was first found in the mammalian testis-determining factor *SRY*. Hence, the name *SOX* derives from <u>S</u>ry-type HMG b<u>ox</u>. The HMG domain has an interesting effect on DNA. First, it binds; then it distorts the DNA's shape, and by so doing it allows genes in the DNA to be expressed.

SOX genes comprise about 20 genes in 7–10 groups. These genes are involved in a diverse range of developmental and differentiation activities. However, it remains to be determined how many developmental abnormalities are associated with malfunction of the *SOX* genes since only a few examples have been identified to date. They mostly involve rare syndromes affecting skeletal or sexual development. This may be explained by redundancy in the various *SOX* genes.

SEX DETERMINATION

Male and female development in humans is genetically determined. A number of genes have now been shown to be necessary for early gonadal development. Based on mice gene knockout studies, these genes include *Emx2*, *Lhx9*, *Lim1*, *Sf1* and *Wt1* (Figure 7.2). This cluster of genes leads to the development of the primitive gonad that will differentiate into the testis or the ovary. The molecular events involved in the testis-determining pathway are better understood than what happens with the ovary, and so the former will be described. The gene that triggers the cascade leading to the development of the testis is *SRY* (<u>s</u>ex determining <u>r</u>egion of the <u>Y</u>). This gene is found on chromosome Yp11.3, and is intronless. Like *SOX* genes described above, *SRY* has a conserved DNA binding domain of 79 amino acids (HMG box). Thus, *SRY* is likely to function as a transcriptional regulator similar to the *SOX* genes.

SRY was shown to be the testis determining factor (TDF) in 1990. Confirmatory evidence for this includes (1) About 15% of 46,XY females (i.e., sex reversals) have mutations in *SRY*, predominantly in the HMG box. (2) XX mice have their sex reversed if the *Sry* gene is added as a transgene. However, since only a very small proportion of 46,XY sex reversals have *SRY* mutations, other genes must be involved in testis determination. Two (*SOX9* and *FGF9*—fibroblast growth factor 9) also play a key role, and it is likely that there are others.

The X and Y chromosomes evolved from autosomes about 300 million years ago. These two chromosomes are critical for sex determination and are structurally very different, with the X a more typical chromosomal, whereas the Y is atypical in many respects (Table 7.4). The testis determining factor on the Y chromosome is the *SRY* gene.

neural tube defects, specifically spina bifida in homozygous *Sp/Sp* embryos. Heterozygous-affected mice have white patches on a dark background coat, which result from defective neural crest–derived melanocytes. It has now been shown that *Splotch* is the result of mutations in the mouse's *pax-3* locus. The corresponding genetic locus in the human is *PAX3*. Mutations affecting this gene have been identified in the human autosomal dominant disorder called Waardenburg's syndrome. In this condition there is deafness and pigmentary disturbance. The two tissues involved are both of neural crest origin. Thus, a genetic defect has highlighted the role that a conserved gene plays in development of the neural crest in humans. The *Splotch* animal also provides a model to investigate normal neurological development, including the abnormality that produces spina bifida. This model has been used to study the mechanism underlying the protective effect of periconceptional folic acid against the development of neural tube defects. Surprisingly, the *Splotch* mouse is not protected by prophylactic folic acid, suggesting different mechanisms may be involved.

Table 7.4 Some differences between the X and Y chromosomes (Lahn *et al* 2001; Marshall Graves 2002; Hawley 2003)

Feature	X chromosome	Y chromosome
Size, genes	165 Mb; ~1500 genes (functions include housekeeping and specialised function in both sexes).	50 Mb; ~200 genes (many of these involved in sex differentiation and spermatogenesis).
Special features	Undergoes dosage compensation (X inactivation) to equalise the same number of genes in males (XY) and females (XX).	About 50% of the chromosome is heterochromatic (i.e., region containing repetitive DNA). Y is a gene-poor chromosome, and it does not recombine with other chromosomes (except at at the two ends where it recombines with the X chromosome). The region specific to the Y is NRY (non-recombining Y).
Pseudoautosomal regions (PAR)	Has corresponding PARs with Y.	5% Y sequence (at ends of chromosome) is PAR1 and PAR2 (Figure 7.3). This recombines with X chromosome, and so these genes are shared with X.
Evolution	Two parts to X: the ancient part is conserved in all three types of mammals. Most of Xp is found only in placental mammals and is a more recent addition. This section, which shares genes with the Y chromosome, is not X inactivated (gene dosage here would be the same in males and females).	Similar two parts in Y: the ancient part is conserved in all three types of mammals. There is a recent addition that is found on the Y only of the placental mammals.

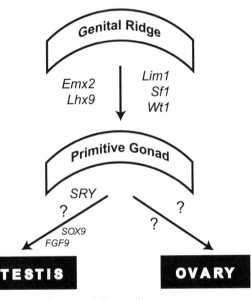

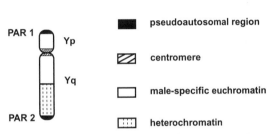

Fig. 7.3 Human Y chromosome.
Yp—short arm; Yq—long arm; PAR—pseudoautosomal region. The Y chromosome is small, gene-poor and has a lot of repetitive DNA. It also has two small pseudoautosomal regions at the ends (PAR1, PAR2). They recombine with genes mostly on the short arm of the X chromosome during meiosis. X inactivation involving the pseudoautosomal genes does not occur on the X chromosome because gene dose in males and females will be the same (unlike the majority of genes on the X chromosome). Most of the Y chromosome does not recombine and consists largely of repetitive DNA in the form of heterochromatin (see also Table 7.4).

Fig. 7.2 Development of the gonads.
From work in the mouse, a number of genes involved in development of the primitive gonad have been identified. However, little is known how it then matures into the ovary. Development of the testis is better understood; it relies on a number of genes, with one, *SRY*, being found only in mammals. Hence, there are other, more primitive genes to be found to explain maleness apart from contributions coming from *SOX9* and *FGF9*.

However, this gene is found only in mammals compared to, for example, *SOX9*, which is also present in birds and reptiles. Hence, *SRY* is only a "recent" evolutionary addition to the Y chromosome, and this is also shown by its position on the Y chromosome; i.e., the segment that is not ancient (Figure 7.3). Comparisons with different species confirm that other, more ancient genes are likely to be involved in determining maleness.

The atypical nature of the Y chromosome has led to amusing titles for scientific descriptions of this sex chromosome, including "The Rise and Fall of *SRY*," an article by J Marshall Graves in which she predicts loss of the Y in 5–10 million years, countered by a male's perception in "The Human Y Chromosome: Rumors of Its Death Have Been Greatly Exaggerated." Notwithstanding the speculation of the Y chromosome's past and future, it is critical for male sex determination, and as knowledge becomes available from sequencing and functional analysis of this chromosome, some new interesting twists in terms of its function and origin are sure to arise.

PRENATAL DIAGNOSIS

Detection of genetic disorders by DNA testing in adults is relatively simple since a number of tissues (blood sample, hair follicle, mouth wash) provide a source of DNA. In contrast, access to the fetus because of size and location is more difficult. Fetal tissues that can be used for prenatal diagnosis include fetal blood, amniocytes, chorionic villus sample and, rarely these days, a specific biopsy, e.g., liver (Table 7.5).

Chorionic Villus Sampling (CVS)

Chorion frondosum is tissue surrounding the developing embryo. It is fetal in origin and will eventually become the placental site. It can be biopsied by the technique of CVS first described in the late 1960s and now a routine procedure for prenatal diagnosis. CVS has been used in many centres throughout the world and, despite some concern about possible side effects including miscarriage and damage to the fetus, it has proven to be reliable and safe in the hands of the experienced operator. A CVS is an excellent source of DNA although it is essential that any contaminating maternal tissue is removed. This is relevant because PCR will be used in DNA testing.

CVS can be undertaken transcervically or transabdominally. The latter is the preferred method in most centres as it is technically easier. A number of trials have looked at the safety issues comparing amniocentesis and CVS. The conclusions from a Cochrane review in 2003 were (1) Second trimester amniocentesis is safer, in terms of pregnancy loss and spontaneous miscarriage, than transcervical CVS and early amniocentesis. (2) For earlier diagnosis, transabdominal CVS is preferred to early amniocentesis or transcervical CVS. (3) If transabdominal CVS is not possible because of technical difficulties, the options should be first trimester transcervical CVS or second trimester amniocentesis.

DNA TESTING

The indications for prenatal diagnosis can vary in different communities. In general, they involve the risk of serious genetic disorders in the fetus. Adult onset genetic defects that are not immediately life threatening (e.g., adult polycystic kidney disease) or for which effective forms of treatment are available (such as phenylketonuria) are contentious indications for prenatal testing (see Chapter 10). Sexing of the fetus (for example, at preimplantation genetic diagnosis) for family planning purposes is another controversial issue discussed in Chapter 10. Prenatal diagnosis, even if termination of pregnancy is not being considered as an option, is useful to allay anxiety in a couple or to allow the couple to plan for the birth of an affected fetus.

Scenarios involving DNA prenatal diagnosis include:

- DNA amplification by PCR to detect gene mutations or DNA polymorphisms in genetic disorders.
- Fluorescence *in situ* hybridisation (FISH) to identify chromosomal abnormalities.
- Fetal sexing in pregnancies at risk for X-linked disorders.

Table 7.5   Source of fetal tissue for prenatal diagnosis

Source	Logistics	Indications, complications
Fetal blood	Undertaken in second trimester (18–20 weeks). Ultrasound-guided umbilical vein puncture (cordocentesis) gives pure fetal sample in experienced hands. Rarely done these days because of technical difficulty, late diagnosis and availability of alternative approaches.	Source of DNA, detection of infectious agents.
Amniocentesis	Amniocytes are fetal cells shed into the amniotic fluid. They can be accessed from about 15 weeks until the end of pregnancy. Technically easy, but a disadvantage is the second trimester prenatal diagnosis. Can also be cultured to provide a substantial source of tissue. Attempts at early (first trimester) amniocentesis are associated with more side effects.	Used as a source of DNA or for chromosome analysis or enzyme analysis (metabolic disorders).
Chorionic villus sample (CVS)	The method of choice for prenatal testing as undertaken in the first trimester (10–11 weeks). Requires experienced operator. Transabdominal CVS also can be performed in the second and third trimesters, but there is more risk to the fetus.	DNA testing or chromosome analysis. Enzymes can be measured.

Table 7.6 Intrauterine infections detectable by prenatal DNA testing

Infection	Fetal consequences	Conventional test	DNA test
Cytomegalovirus	Mental retardation and deafness	Culture or serology	PCR of amniotic fluid
Varicella	Mental retardation and various effects on skin, eye, gastrointestinal, genitourinary tracts, growth retardation	Immunofluorescence, culture	PCR of amniotic fluid
Rubella	Numerous, affecting eye, heart, neurologic and immunologic systems	Viral culture (serology unreliable)	PCR of amniotic fluid
Toxoplasmosis	Chorioretinitis, hepatitis, hydrocephalus and pneumonitis	Culture amniotic fluid or fetal blood; serology less reliable	PCR of amniotic fluid

- Monitoring a risk pregnancy for immune haemolytic anaemia secondary to Rh (rhesus) incompatibility. The fetus (with an RhD negative mother and RhD positive father) is DNA tested. If the fetus is RhD negative, the couple can be reassured that all is well, and no further follow-up is necessary. An RhD-positive fetus requires careful monitoring, and intrauterine blood transfusions if necessary.

Intrauterine Infections

Maternal infections that can be transmitted to the fetus include bacteria, viruses and protozoans. Maternal infection is often mild or non-specific and can easily be missed. Effects on the fetus or neonate are variable, ranging from nothing to a fatal disorder. Thus, the screening of pregnant women for infections that can produce fetal abnormalities is a complex issue. A number of factors need to be considered: (1) the prevalence of the infection in each population, (2) the risk to the fetus or neonate, (3) the availability of treatment, (4) the sensitivity and specificity of screening tests, (5) access to the screening tests and (6) availability of counselling.

Some infections that can be transmitted vertically from the mother to her fetus and cause damage to the latter are rubella, syphilis, hepatitis B virus, cytomegalovirus, *Toxoplasma gondii*, human immunodeficiency virus (HIV), herpes simplex virus, group B streptococcus, *Chlamydia trachomatis* and *Neisseria gonorrhoeae*. In many communities the first three are routinely screened for during pregnancy. Whether it is appropriate to screen for others will depend on local circumstances such as the types of infections and the available facilities.

During pregnancy, acute infections in the mother that might involve any of the above pathogens require careful investigation. Conventional microbiological detection systems, described in Chapter 8, may or may not lead to a definitive diagnosis. The fetus introduces an additional complexity since infection in the mother, even if proven definitively, does not necessarily mean that the fetus will also be infected. Some examples illustrating the poten-

tial use of recombinant DNA tests in dealing with infections acquired during pregnancy are found in Table 7.6.

Maternal Screening

In many communities, it is recommended that pregnant women of a certain age group (for example, 35) have cytogenetic screening by amniocentesis or CVS. These tests are directed predominantly at detecting Down's syndrome in the fetus, although other cytogenetic abnormalities will also be found. Problems with this approach include the costs involved and the fact that most cases of Down's syndrome occur in women who are younger than 35.

Alternative screening strategies for detecting fetal abnormalities such as Down's syndrome and neural tube defects include

(1) Maternal serum testing. A number of biochemical markers can be measured in the maternal blood, e.g., α-fetoprotein, human chorionic gonadotropin, unconjugated oestriol and inhibin A. Results based on these studies identify pregnancies at risk for Down's syndrome and neural tube defects such as spina bifida. Cytogenetic analysis or ultrasound will confirm the screening tests. Maternal serum screening is usually undertaken in the second trimester.
(2) Fetal ultrasound. In the hands of experienced operators, ultrasound can detect subtle fetal structural anomalies. The risk for Down's syndrome can also be estimated from measurement of the nuchal thickness. Ultrasound is possible in the first trimester.

The above methods of screening have a detection rate of about 65–75% with 5% false positives. However, studies are now emerging to suggest that combinations of maternal serum screening and ultrasound might be undertaken in the first trimester with superior detection rates. The results from these studies are awaited to determine the optimal approach to screening for fetal abnormalities in pregnancy.

PREIMPLANTATION GENETIC DIAGNOSIS (PGD)

Preimplantation genetic diagnosis, or PGD as it is usually known, is a recent addition to the prenatal diagnosis options. It is particularly valuable in the scenario in which termination of pregnancy is not acceptable, or a couple with a history of infertility do not want to risk the potential miscarriage or loss of pregnancy that might follow a procedure such as CVS. PGD has developed as a result of improvements in ART such as embryo manipulation and culture, and the availability of diagnostic tests like PCR and FISH.

The technical demands of PGD prevent this from being a routine or inexpensive procedure, and so its availability is limited. In addition, potential and social issues related to this technology have made it controversial (discussed further below). Finally, as indicated earlier, there is some concern about the long-term safety of IVF, and so the use of PGD for trivial reasons would not be medically acceptable on a risk/benefit analysis. Figure 7.4 provides a summary of the procedure. Eggs obtained at IVF are grown in culture and fertilised. At either day 3 (cleavage embryo biopsy) or days 5–6 (blastocyst biopsy), embryonic material can be removed and examined by DNA (PCR) or FISH. Embryos in which genetic abnormalities are excluded can then be implanted into the mother by IVF.

Indications for PGD

(1) Sex determination—in the circumstance in which the couple are at risk of an X-linked disorder, and it might not be possible to detect a mutation. Female embryos would avoid this genetic disease. (2) Mendelian genetic traits—they include common conditions such as cystic fibrosis, thalassaemia. This is no different from prenatal diagnosis, except for the technical demands of embryo manipulation and culture, and the challenge of testing a very small amount of DNA. (3) Structural chromosomal abnormalities—FISH is used to screen a risk pregnancy. (4) Aneuploid testing—again FISH is used, and this indication for PGD includes a risk pregnancy as well the selection of euploid embryos to enhance the fertilisation success after IVF.

The first PGD was undertaken in 1992, and there is now experience in more than 1000 cases. The pregnancy rate following PGD is between 20–25%. A concern in relation to this procedure is the relatively high incidence of multiple pregnancies. There have been five misdiagnoses, a number which is surprisingly low in view of the amount of material that the laboratory has to study. Many centres recommend chorionic villus sampling to confirm the PGD result.

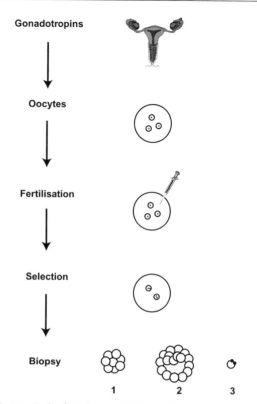

Fig. 7.4 Preimplantation genetic diagnosis.
Following exposure to gonadotropins, oocytes mature in the patient, and they are obtained under transvaginal ultrasound-guided aspiration. The oocytes are transferred to a culture medium (the diagram depicts three oocytes being cultured *in vitro*) and are subsequently fertilised. ICSI as a method of fertilisation is helpful if preimplantation genetic diagnosis is to be undertaken because then supernumerary sperms will not be present to interfere with the subsequent PCR. The successfully fertilised eggs are identified (two pronuclei) and separated from the others. These embryos are cultured further. By day 3, the cleaving embryo has reached the 8–16 cell stage, and a biopsy can be taken by removing 1–2 cells (at this stage all cells are totipotential). Although the amount of DNA is limited, this is the preferred method of biopsy. The alternative is to wait until days 5–6 when the blastocyst is formed (about 300 cells). Now a larger number of cells can be removed, and so there is more DNA. The downside of this is that the embryo has been in culture longer. A third method that is not used much is to biopsy a polar body from the mature oocyte. The advantage is that this DNA is extra-nuclear, and so the embryo is not damaged. The disadvantage is the DNA is only maternal in origin, and so information provided is limited. Following preimplantation genetic diagnosis (FISH or DNA), only normal embryos are reimplanted back into the mother.

Regulation of PGD

Regulation of PGD varies in different countries. In the UK, it is highly regulated with legislation requiring approval by the Human Fertilisation and Embryology Authority before a centre is licensed to undertake PGD. In contrast, there is no federal regulation in the USA covering PGD. The matter of regulation is mentioned because, as technologies improve, PGD will become more readily available although it remains costly (US$6000–10 500). Because it avoids termination of pregnancy, there is always the risk that PGD will be used to produce the perfect baby or the baby of choice.

Already in some countries, sex determination by PGD is advertised as a way of "family balancing" for socioeconomic reasons or sex selection for cultural reasons (discussed further in Chapter 10). PGD has been used controversially to select an embryo that would be genetically compatible with an affected sibling, and so it would at the time of birth provide a source of cord blood stem cells for transplantation. PGD is a valuable adjunct in the prenatal testing options although more work is needed to understand its possible consequences to fetal development. Better techniques are still needed for detecting DNA and chromosomal abnormalities.

NEWBORN SCREENING

Newborn screening programs are available in many communities. They are directed at diseases that satisfy the following criteria: (1) The disease occurs at reasonable frequency and is serious in nature. (2) Effective treatment is available. (3) There is an advantage in early treatment rather than waiting until clinical features develop. (4) Suitable laboratory tests are available; e.g., a screening test with appropriate **sensitivity** (the proportion of affected neonates with an abnormal screening test) and **specificity** (the proportion of all normal neonates with a normal screening test) and a confirmatory test can be undertaken if required (see Table 2.10 for further discussion on sensitivity and specificity). (5) The program has a clear cost-benefit.

The first example of this type of mass screening occurred in the 1960s and involved the testing of newborns' urine for phenylketonuria (see Table 2.9 for a summary of the various types of screening approaches). The scope for newborn screening increased with the availability of blood spots. They were obtained from heel pricks, with the blood being spotted onto filter papers. Newborn blood spots were initially tested for phenylketonuria using the Guthrie bacterial inhibition assay. The blood spots, or "Guthrie cards" as they are sometimes called, can be utilised to test for a number of diseases.

What newborn diseases are screened for will depend on at-risk conditions present in particular communities as well as political and community views of screening. In the USA, all states and the District of Columbia provide newborn screening for phenylketonuria and hypothyroidism. These conditions represent universal indications for newborn screening because treatment is relatively simple and cheap, compared to a missed or late diagnosis that leads to various serious consequences including mental retardation. However, apart from these two conditions, the targets for testing vary considerably from state to state in the USA, and from country to country. Table 7.7 lists diseases that would be useful to screen in newborns.

SCREENING METHODS

Approaches used in newborn screening extend from the original 1960s' bacterial inhibition assay (Guthrie test) to the modern tandem mass spectrometry. The latter has emerged because of proteomic initiatives and involves the use of two mass spectrometers in tandem. The first analyser separates molecules (amino acids, fatty acids or proteins) on the basis of size and charge. The separated molecules are fragmented further and passed through a second mass analyser, and this allows better separation based on size and charge. The profile generated is then analysed by computer to identify various compounds. The tandem mass spectrometry approach avoids the time-consuming use of liquid or gas chromatography. It can give results in minutes and assay a larger number of analytes simultaneously. Tandem mass spectrometry is a new development in newborn screening, allowing the testing for many more diseases. However, this will need to be matched with the appropriate data showing the benefits of early diagnosis. Another consideration is the current lack of uniformity between and within many countries, with different regions adopting their own specific newborn screening program priorities for various reasons mentioned above. This is presently a source of consternation but has the potential to become a bigger issue as a methodology such as tandem mass spectrometry expands greatly the range of disorders detectable in newborn screening.

DNA testing has not played a major part in the newborn screening program, apart from its addition to cystic fibrosis testing discussed in more detail below. Nevertheless, the applications of PCR on newborn blood spots are many. Therefore, decisions that involve the

Table 7.7   Some disorders that are included in newborn screening programs

Disorder	Clinical implications	Screening approaches[a]
Cystic fibrosis	Early diagnosis improves clinical outcomes, particularly malnutrition and long-term growth.	Biochemical test supplemented with DNA testing (see text for further discussion).
Galactosaemia	Potentially fatal if not recognised and treated by galactose free diet.	Biochemical test
Congenital adrenal hyperplasia	Potential for life-threatening salt-wasting syndrome, hypoglycaemia and incorrect sex assignment.	Biochemical test
Haemoglobinopathies	Potentially fatal complications (infection) in sickle cell disease can be avoided by early treatment (if diagnosis is known).	Biochemical test but could be developed into a DNA-based test
Homocystinuria	Developmental abnormalities affecting eye or skeleton. Potential for cardiovascular complications including thrombosis.	Guthrie bacterial inhibition assay
Biotinidase deficiency	Serious neurologic complications including epilepsy, developmental and hearing problems and acute metabolic problems that can be fatal. Preventable by early treatment with biotin.	Biochemical test
Maple syrup urine disease	Severe acidosis early in life can be fatal. Early treatment not as effective as in other conditions.	Guthrie bacterial inhibition assay
Fatty acid metabolism disorders	Wide-ranging group of metabolic disorders associated with clinical features of lethargy, failure to thrive through to coma and death. Treatments may not be available, but at least an early diagnosis is made.	Tandem mass spectrometry

[a] Included in the term "biochemical" is tandem mass spectrometry, which is likely to be useful for a number of the disorders listed above (see text).

testing of DNA from blood spots need to be taken with care and foresight (see Chapter 10). Factors to be considered when determining the types of screening tests undertaken in newborns are (1) cost effectiveness (clinical consequences, treatment options, frequency of disease, cost of testing), (2) community knowledge and acceptance, (3) availability of counselling and follow-up facilities, and (4) whether there are sufficient common mutations to make PCR a worthwhile approach. The last is limiting in many genetic conditions but will be less so as new technologies (e.g., DNA chips—Chapter 5) are developed. Another issue relevant to the blood spot and PCR is the potential misuse that can occur. This is the subject of much debate at present, with some laboratories specifically discarding the blood spots after a limited time to avoid future inappropriate use (for example, testing of the blood spot for legal purposes).

CYSTIC FIBROSIS

The cystic fibrosis newborn screening program is a useful model to discuss in greater detail. It utilises the biochemically based immunoreactive trypsin assay and has proven its effectiveness by the number of affected newborns detected. There are also some data to suggest that early detection of affected children will improve clinical outcomes. The availability of newborn screening has

reduced parental consternation since children with cystic fibrosis are likely to undergo repeated hospital and medical contacts before the diagnosis is firmly established. However, a problem with the immunoreactive trypsin screening assay is its low positive predictive value (the proportion of neonates with abnormal screening tests who are affected). Therefore, a large number of false positives will result. In these circumstances ~1% of infants need to be recalled, retested and if again positive, the diagnosis is confirmed with a sweat test. This is a time-consuming process. Apart from economic considerations, recall will provoke anxiety in the families concerned (Figure 7.5).

The false positive problem has now been solved by combining the immunoreactive trypsin with DNA testing for the ΔF508 mutation in the original blood spot. The finding of a homozygote for ΔF508 will confirm the diagnosis of cystic fibrosis without the necessity for further tests. A heterozygote for the ΔF508 mutation will either be a carrier of cystic fibrosis or have a second (unknown) mutation associated with the disease. The two will be differentiated by a sweat test. The exclusion of ΔF508 (which will be the usual outcome of DNA testing) will make it more likely that the immunoreactive trypsin result was a false positive, particularly in those communities where the ΔF508 mutation is the predominant mutation. Using this strategy, some newborns with cystic

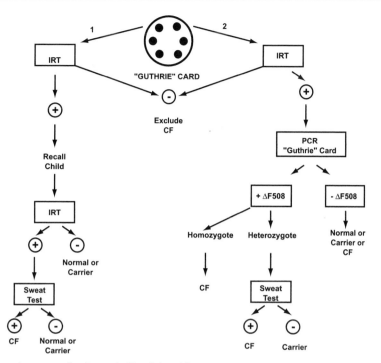

Fig. 7.5 Improving newborn screening for cystic fibrosis by adding a DNA component.
Strategy 1 (left) is the traditional biochemical approach to newborn screening for cystic fibrosis. Its main disadvantage is the number of infants recalled for sweat tests after the screening IRT (immunoreactive trypsin), many of whom will turn out to be negative. **Strategy 2** (right) incorporates a DNA step. This reduces the number of false positives (and so anxiety in the parents) because now there is an additional (DNA) step after the screening IRT. One disadvantage of this approach is that it will occasionally miss a child with cystic fibrosis caused by mutations other than ΔF508. How often this occurs will depend on the population involved (i.e., the frequency of cystic fibrosis caused by ΔF508 in that population). However, the small number of infants missed by the newborn screen incorporating the DNA test is considered to be justified because false positives are now virtually eliminated. Carriers of cystic fibrosis that are not due to ΔF508 will also be missed, but this is not a problem.

fibrosis will be missed since both their cystic fibrosis mutations will be non-ΔF508, but these will be few. Apart from the inadvertent identification of newborns as cystic fibrosis carriers, the combination of the traditional biochemical screen followed by confirmation using DNA analysis is now considered to be a useful development in newborn screening. The potential consequences of detecting asymptomatic carriers for cystic fibrosis are discussed further in Chapter 10.

FUTURE

NON-INVASIVE PRENATAL DIAGNOSIS

Fetal Cells in Maternal Circulation

A non-invasive source of fetal tissue is required as the next development in prenatal diagnosis. An early attempt at this involved the identification and isolation of the occasional fetal cells present in the maternal circulation during pregnancy. These cells included syncytiotrophoblasts, lymphocytes and erythroblasts (nucleated red blood cells). Monoclonal antibodies were described and reported to be specific for fetal syncytiotrophoblast (the outer layer of the chorion frondosum). These cells were then isolated by flow cytometry or antibody-coated beads. Although very few cells were obtainable, PCR could be utilised to overcome this particular problem. Lymphocytes became the second source of fetal cells. They are also few in numbers. However, on the basis of different maternal and paternal HLA antigen types, it would be possible to utilise antibodies to select fetal-specific lymphocytes, i.e., cells that carried a paternal HLA antigen type not found in the mother. These cells could then be separated from maternal cells as described above. A third and potentially very promising source of fetal tissue was the erythroblast. However, a major problem with this cell was the lack of antibodies that could identify unique determinants, i.e., present on fetal but not maternal cells.

Many technical problems made fetal cells a difficult target for non-invasive prenatal diagnosis. For example, how specific were the antibodies for fetal cells? What was the likelihood of maternal contamination being present in the sample? This was a particular issue as PCR would be needed. Finally, in terms of lymphocytes, there was the concern that some of these long-lived cells from an earlier pregnancy could interfere with results from the current pregnancy.

Free Fetal DNA in Maternal Circulation

The focus has now moved from fetal cells to free fetal DNA in the maternal circulation when it became evident that a significant quantity of this DNA was present in pregnant women. In 1997 Lo and colleagues showed that it was possible to amplify by PCR a Y chromosome–specific DNA fragment from maternal blood. It was also shown that this diagnosis could be made by the seventh week of gestation, and so an early prenatal test was feasible. Subsequently, it was demonstrated that the fetal Rhesus blood group could be identified this way because, if the mother was Rh D negative, the finding of Rh D positive DNA in the maternal circulation could only come from the fetus (excluding laboratory mistakes, contamination and so on).

Autosomal dominant genetic disorders involving an affected father and a normal mother would be particularly suitable for fetal DNA testing. A negative result would mean a normal fetus (or the PCR did not work, or maternal DNA had been tested). The PCR error could be controlled for by simultaneously using the PCR to assay for a neutral DNA marker. Maternal DNA contamination could be eliminated by confirming that the amplified DNA also demonstrated another DNA marker found only in the father. A positive result could only mean an affected fetus since the mother did not have this mutation. The same approach would be more limited in the autosomal recessive disorders because the mother's own DNA had the mutation. Therefore, the only unequivocal result would be a normal one. An abnormal result would always be difficult to interpret since it could have arisen from the testing of free fetal DNA or maternal DNA or a combination of both.

Free fetal DNA in the maternal circulation is present from the seventh week onwards. Using quantitative PCR, a recent observation has shown that the amount of fetal DNA varies with higher levels present in risk pregnan-cies such as aneuploidy in the fetus, or complications such as preeclampsia in the mother. Thus, studies are presently under way to determine whether free fetal DNA can be used as a prognostic indicator of a risk pregnancy. More information is needed about the source of this DNA. Can it persist from one pregnancy to another? In terms of a non-invasive form of prenatal diagnosis, this approach remains very promising.

FETAL THERAPY

A disappointment in molecular medicine has been the slow advances that have been made in fetal therapy. Although the availability and scope for prenatal diagnosis have improved considerably because of molecular diagnostics, the choices following prenatal diagnosis remain very limited if the baby is affected. It would be hoped that this is only an interim measure so that, in future, an adverse prenatal test is only the start of a fetal therapy option.

In utero fetal therapy is available in a very limited number of situations. Haemolytic disease of the newborn, usually secondary to rhesus immunisation (Rh disease), can be successfully treated by intrauterine blood transfusions (cordocentesis) until the fetus is considered to have reached an age where the risk of further transfusion is greater than the complications associated with prematurity. The ability to obtain pure fetal blood samples by cordocentesis means that the clinical progress of an affected fetus can be monitored more accurately through serial haemoglobin estimations rather than the less precise bilirubin levels in amniotic fluid.

In utero cellular therapies (allogeneic transplants, gene therapy, stem cell therapy and xenotransplantation) are potentially of benefit because the intrauterine fetal environment is ideal (i.e., highly proliferative, immunologically more tolerant and the number of cells required is relatively small). Prenatal correction might be used to minimise end-organ damage that would develop once the fetus was born. Fetal tissues for transplantation might also provide a better source of stem cells into which could be inserted normal genes. The underdeveloped immunological system in the fetus would be useful in situations in which transplantation from a genetically dissimilar donor will induce both graft rejection and graft versus host disease. Implicit in the scenarios described above is an early detection system (i.e., DNA analysis) for the underlying genetic defect. Despite the obvious advantages of *in utero* fetal cellular therapy, the risks of manipulating the fetus remain major limitations.

In utero surgery to correct abnormalities such as lung lesions, obstructive uropathy, abdominal wall defects, sacrococcygeal teratoma, congenital hydrocephalus and diaphragmatic hernia, neural tube defects and a number of other conditions is available in very few centres. Fetal surgery is still in its early days, so it remains an experimental form of treatment. One of the important problems to overcome before fetal surgery becomes a realistic option is the control of premature labour.

Treating the Neonate

The placenta, which would normally be discarded following childbirth, is now proving to be an important source of haematopoietic stem cells. Cord blood is rich in CD34$^+$ cells (see Chapter 6) and so provides an alternative to bone marrow for transplantation. There is also some preliminary evidence that cord blood is (1) less immunogenic, thereby reducing the risk of transplant rejection, and (2) immunologically less active, and so the frequency of the second problem related to transplantation (graft versus host disease) is also reduced.

Cord blood cell transplantations in the neonate have proven to be effective in the treatment of some genetic disorders, e.g., β thalassaemia. Therefore, the next step from this was to take cord blood at birth from a newborn with the genetic immunodeficiency disorder ADA (adenosine deaminase deficiency) and transduce it *in vitro* with a normal gene. The cord blood was then returned to the newborn within a few days. Early results suggest that there has been long-term expression from the transduced cells; i.e., DNA has been transferred into stem cells.

An HLA-identical match for bone marrow transplantation is difficult to find even with a number of potential siblings as donors. An alternative in this situation is cord blood, which is easy to obtain and store. It also provides a broad ethnic representation, which can be a problem with conventional sources for bone marrow or organ donation. The potential advantage of cord blood for transplantation has led to cord blood banks being established. A disadvantage of cord blood as a source of marrow for transplantation is the small volume obtainable, so that this form of treatment is predominantly used in children. However, the volume problem might be overcome by using recombinant DNA–derived growth factors to increase the number of stem cells circulating in the cord blood.

PREECLAMPSIA

Another future challenge for molecular medicine in the area of reproduction and development is a better understanding of diseases of pregnancy. The example to be used is preeclampsia, a hypertensive disorder in pregnancy associated with considerable morbidity and mortality both to the mother and the fetus (Box 7.3). Preeclampsia is found in about 6% of all pregnancies, with an estimated mortality of 15%–20% in developed countries. Despite the importance of this disorder, and the interest shown in understanding it, little progress has

Box 7.3 Preeclampsia.

Two challenges in antenatal care are prematurity and preeclampsia. Management of these conditions has not made significant inroads into improvements in maternal and fetal outcomes because so little is known about aetiology, and therefore treatment is essentially symptomatic. Preeclampsia is a hypertensive disorder of pregnancy (predominantly the first pregnancy). Its clinical features include hypertension after 20 weeks of pregnancy, which resolves soon after the baby is delivered. In addition to hypertension, there is significant proteinuria. The clinical spectrum is quite variable, ranging from transient hypertension in the latter part of pregnancy to a life-threatening disorder. Complications associated with preeclampsia include acute renal failure, *abruptio placentae*, cerebral haemorrhage, disseminated intravascular coagulation, liver disease, convulsions and shock. Complications in the fetus include hypoxia and growth retardation. Preeclampsia accounts for about 25% of low birth-weight babies, partly the result of the disease directly, and also secondary to earlier obstetrical delivery to relieve the complications in the mother. In the United Kingdom, it is estimated that 500 babies die each year as a result of this complication in pregnancy. A problem with management of preeclampsia is the inability to predict accurately who is at risk—hence, the importance of improving our understanding of this disorder so that earlier and more accurate detection and specific treatment directed to the cause can be initiated.

been made. Hence, an alternative approach is to use molecular strategies such as gene discovery.

Like the multifactorial conditions described in Chapter 4, preeclampsia demonstrates a number of similar features including (1) evidence for some genetic inheritance but no clear mendelian pattern, (2) evidence for paternal effects (perhaps these are the equivalent of the environment-gene interactions) and (3) a difficulty in establishing an unequivocal phenotype. On the other hand, an unusual feature not found in multifactorial disorders is the observation that preeclampsia is primarily a disease of first pregnancies.

The evidence for a genetic component to preeclampsia comes from the usual observations including greater risk if there is a positive family history in female relatives and greater concordance in MZ twins. However, the concordance in twins is not perfectly consistent with other factors involved in aetiology. Although preeclampsia is a disease of the first pregnancy, women who have had this problem and then change partners are again at risk if they become pregnant. The effect of the male partner in the aetiology of preeclampsia is also seen with a higher risk for preeclampsia in women with complete hydatidiform

Table 7.8   Some candidate genes proposed as genetic factors in preeclampsia (from Wilson *et al* 2002).

System	Genes	Mode of action
Blood pressure	Renin-angiotensin–related genes have been associated but confirmation equivocal.	Via effect on a key manifestation of preeclampsia, i.e., hypertension.
Vascular injury/remodelling	(1) Nitric oxide synthase. (2) Prothrombin, factor V Leiden, methylenetetrahydrofolate reductase.	(1) Impaired vasodilation, inhibition platelet aggregation (i.e., thrombosis). (2) Thrombotic predisposition.
Endothelial cell health	(1) Superoxide dismutase. (2) Lipoprotein lipase, apolipoprotein E.	(1) Impaired protection from superoxides. (2) Endothelial damage.
Immune abnormality	HLA genes.	HLA genes are relevant through impairing the normal suppression of the immune system during pregnancy.
Placentation	Genes in mice shown to influence various stages of placentation. The homologues in humans would be good candidates.	A key pathogenic feature of preeclampsia is poor placentation.
Other	(1) *IGF2*. (2) Mt DNA	(1) Insulin growth factor II plays a key role in human growth, including the fetus. (2) Impaired energy production would compromise placental development.

moles (these tissues are paternal in origin—discussed earlier under Epigenetics and Imprinting).

Traditional linkage studies using a positional cloning approach and candidate gene studies by positional cloning or case control associations have all been used to identify genes associated with preeclampsia. Many genes have been implicated, but like all other examples of complex genetic traits, the results are difficult to reproduce, and functional confirmation of the associations is often lacking. So the current thinking is that many genes are implicated, but none has shown unequivocal evidence for a role in the aetiology of preeclampsia (Table 7.8). Future developments in bioinformatics and statistical analyses of complex traits are now awaited to sort out a mass of information that must contain the answers being sought.

FURTHER READING

Reproduction

Aldashi EY, Hennebold JD. Single gene mutations resulting in reproductive dysfunction in women. New England Journal of Medicine 1999; 340:709–718 (*summary of genes associated with infertility in women*).

Lucifero D, Chaillet RJ, Trasler JM. Potential significance of genomic imprinting defects for reproduction and assisted reproductive technology. Human Reproduction Update 2004; 10:3–18 (*nice review of ART epigenetic inheritance mechanisms that could be perturbed*).

Seminara SB *et al*. The *GPR54* gene as a regulator of puberty. New England Journal of Medicine 2003; 349:1614–1627 (*original report illustrating how molecular analysis can translate into relevant clinical knowledge*).

Shah K, Sivapalan G, Gibbons N, Tempest H, Griffin DK. The genetic basis of infertility. Reproduction 2003; 126:13–25 (*broad overview of genetic causes of infertility*).

Development

Chi N, Epstein JA. Getting your Pax straight: Pax proteins in development and disease. Trends in Genetics 2002; 18:41–47 (*nice overview of PAX in health and disease*).

Gefrides LA, Bennett GD, Finnell RH. Effects of folate supplementation on the risk of spontaneous and induced neural tube defects in Splotch mice. Teratology 2002; 65:63–69 (*describes experiments testing the effect of folic acid in the Splotch mouse model*).

Goodman FR. Limb malformation and the human HOX genes. American Journal of Medical Genetics 2002; 112:256–265 (*describes mutations in HOX and their effects predominantly on limb development*).

Goodman FR. Congenital abnormalities of body patterning: embryology revisited. Lancet 2003; 362:651–662 (*gives overview of HOX and other genes involved in development and congenital abnormalities*).

Hawley RS. The human Y chromosome: rumors of its death have been greatly exaggerated. Cell 2003; 113:825–828 (*identifies potential new functions of the Y chromosome*).

Lahn BT, Pearson NM, Jegalian K. The human Y chromosome, in the light of evolution. Nature Reviews 2001; 2:207–216 (*comprehensive overview of the Y chromosome, its genes, evolution and association with disease*).

MacLaughlin DT, Donahoe PK. Sex determination and differentiation. New England Journal of Medicine 2004; 350:367–378 (*describes a large number of genes and mutations associated with sex determination and developmental problems*).

Marshall Graves JA. The rise and fall of *SRY*. Trends in Genetics 2002; 18:259–264 (*nice review illustrating how the Y chromosome and its genes have evolved and where it might be going*).

Wilson M, Koopman P. Matching SOX: partner proteins and co-factors of the SOX family of transcriptional regulators. Current Opinion in Genetics and Development 2002; 12:441–446 (*gives overview of SOX genes*).

Prenatal Diagnosis

Braude P, Pickering S, Flinter F, Ogilvie CM. Preimplantation genetic diagnosis. Nature Reviews Genetics 2002; 3:941–953 (*comprehensive scientific, medical and social overview of this subject*).

Bui T-H, Blennow E, Nordenskjold M. Prenatal diagnosis: molecular genetics and cytogenetics. Best Practice and Research Clinical Obstetrics and Gynaecology 2002; 16:629–643 (*particularly good overview of FISH*).

d'Ercole C et al. Prenatal screening: invasive diagnostic approaches. Child's Nervous System 2003; 19:444–447 (*brief but excellent summary of the three approaches used for prenatal testing*).

Flinter FA. Preimplantation genetic diagnosis. British Medical Journal 2001; 322:1008–1009 (*brief but useful summary of the UK situation with PGD*).

Gerber S, Hohlfeld P. Screening for infectious diseases. Child's Nervous System 2003; 19:429–432 (*useful summary of important intrauterine infections including approaches in detection*).

Wapner R et al. First trimester screening for trisomies 21 and 18. New England Journal of Medicine 2003; 349:1405–1413 (*multicentre study assessing first versus second trimester maternal screening*).

Newborn Screening

Carreiro-Lewandowski E. Newborn screening: an overview. Clinical Laboratory Science 2002; 15:229–238 (*comprehensive overview of the methods and indications for newborn screening*).

Khoury MJ, McCabe LL, McCabe ERB. Population screening in the age of genomic medicine. New England Journal of Medicine 2003; 348:50–58 (*general overview of population testing including the newborn program*).

Future

Flake AW. Genetic therapies for the fetus. Clinical Obstetrics and Gynecology 2002; 45:684–696 (*reviews some options in fetal therapy*).

Chan AKC, Chiu RWK, Lo YMD. Cell free nucleic acids in plasma, serum and urine: a new tool in molecular diagnosis. Annals Clinical Biochemistry 2003; 40:122–130 (*review from one of the leaders in this area and describes the value of free DNA in the blood in pregnancy and other situations*).

Hedrick HL et al. History of fetal diagnosis and therapy: Children's Hospital of Philadelphia experience. Fetal Diagnosis and Therapy 2003; 18:65–82 (*a summary of fetal therapy carried out in this hospital provides some insight into what is achievable surgically in the fetus*).

Wilson ML, Goodwin TM, Pan VL, Ingles SA. Molecular epidemiology of preeclampsia. Obstetrical and Gynecological Survey 2002; 58:39–66 (*very comprehensive overview of the potential genetic factors involved in preeclampsia*).

http://www.apec.org.uk (*The web site for the UK's Action on Preeclampsia charity set up to provide information and improve care in preeclampsia*).

8

INFECTIOUS DISEASES

DIAGNOSTICS

OVERVIEW

Phenotypic-based Tests

The traditional or **phenotypic** methods for the detection of pathogens are well tried, relatively cheap and available in most hospital and private pathology laboratories (Table 8.1). However, their utility needs to be considered in the context that (1) Direct visualisation or culture is not always feasible. (2) These techniques can be time consuming and technically difficult. (3) Phenotypic variation can occur during a pathogen's life cycle; e.g., eggs, larvae and adult forms of a species may alter depending on the stage of development, the associated host or vector and whether the organism is free-living. Thus, antibodies or isoenzyme techniques for detection may become limiting and dependent on the stages in the life cycle. (4) Host immune responses can be delayed, or conversely, they remain persistent even after resolution of a previous infection. (5) Cross-reacting antibodies acquired from natural infections or vaccination can produce false positive results.

Phenotypic methods can be used to discriminate between isolates, genera and species. These approaches are less effective when it comes to distinguishing differences within species. More recently, the emergence of new pathogens, or the concern about bioterrorism, has brought an added urgency to the development of more efficient and rapid methods to detect pathogens and predict their potential virulence.

Genotypic-based Tests

Many of the traditional diagnostic tests are now being complemented or replaced by **genotypic** analysis, i.e., detection of DNA or RNA fragment sizes or specific sequences through nucleic acid hybridisation techniques, or increasingly today DNA amplification by the polymerase chain reaction (PCR). An expression frequently found in microbiologic DNA testing is **NAT** (nucleic acid amplification technique). NAT refers to a number of techniques apart from PCR that can be used to amplify DNA, e.g., ligase chain reaction. For convenience, the description "PCR" will be used interchangeably with "NAT" to indicate broadly based amplification of DNA. To synthesise or design a DNA probe or primer, it is necessary to sequence an organism's DNA to identify species-specific regions (see Table A.2, which provides more information about probes and primers). Today, whole genomes from many pathogens have been sequenced, and this information is available on databases. From this sequence (1) DNA primers can be

Table 8.1   Traditional diagnostic approaches for detecting pathogens

Method	Comments
Staining and visual identification under the naked eye or microscope (light or electron)	Characteristics such as colony size, shape and colour give a clue to the underlying organism. Using light microscopy and staining (e.g., Gram stain) is the most rapid approach to identifying many bacteria. Viruses are visible with the electron microscope.
Culture and growth	Organisms display culture and growth characteristics; e.g., *M. tuberculosis* is a slow grower. Organisms can be further distinguished by their growth in particular media; e.g., chocolate agar is needed for fastidious growers such as *H. influenzae*.
Biochemical characteristics	The presence of catalase (degrades hydrogen peroxide) is used to differentiate many gram positive organisms.
Immunological, i.e., recognition of antigenic determinants related to an organism and the host's specific immune responses to it (production of antibodies)	An antiserum can be used to agglutinate *Salmonella* and *Shigella* species. Antibodies produced in the patient can be used to identify both active and past infections.

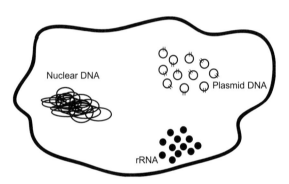

Fig. 8.1   Potential targets for nucleic acid (DNA or RNA) probes in pathogen detection.
Nuclear DNA or RNA and rRNA (ribosomal RNA) are ubiquitous and stable, providing useful targets for a wide range of hybridisation or PCR-based assays. Plasmid DNA can be present in multiple copies, although they can be lost or undergo rearrangements. Therefore, plasmid DNA is not a dependable target for testing but can provide information on infectivity or antibiotic resistance. Nuclear and rRNA probes provide wide-ranging genotypic assays extending across the entire taxonomic spectrum from Family to Genus to Species and Strain.

synthesised for PCR. (2) DNA probes can be isolated or synthesised (oligonucleotide probes) for nucleic acid hybridisation techniques.

Ribotyping: A useful target for a probe is repetitive DNA, an example of which is ribosomal RNA (rRNA). rRNA and the rRNA genes in chromosomal DNA provide naturally derived amplified products that enhance their hybridisation potential. In these circumstances, radio-labelled DNA probes can be avoided since the signal-to-noise ratio is increased because of the amplified target sequence (Figure 8.1). This method of differentiating between bacteria is called ribotyping. For ribotyping, DNA is broken into fragments using restriction

endonucleases. DNA probes that are specific for rRNA genes, then allow various fragments to be identified and compared. This approach is very similar to DNA mapping or Southern blotting in eukaryotic DNA (see the Appendix), and the bacterial hybridisation patterns are comparable to eukaryote DNA polymorphisms.

Pulsed field gel electrophoresis (PFGE): This type of DNA electrophoresis is capable of separating very large DNA fragments. These fragments result from digesting the organism's genome with a restriction endonuclease enzyme such as *Not*I that cuts DNA very infrequently (see Table A.1). The pattern of the fragments generated in this way is sufficiently discriminatory to allow bacterial strains to be distinguished. Read more about PFGE under *Legionella*.

DNA hybridisation: 16S rRNA and 12S rRNA contain highly conserved regions common to all bacteria as well as highly variable regions unique to particular species. Therefore, multifunctional DNA or RNA probes can be prepared since they can be generic for all bacteria, or specific for different species or strains.

PCR

It is reasonable to say that whatever can be done by the various DNA probes or hybridisation techniques just described can also be accomplished, and probably better, with PCR. Thus, there are genus-specific and species-specific PCR-based assays and many other strategies appropriate for a range of activities involving microorganisms in the environment, food industry, research and clinical practice. Ultimately, the most information that can be obtained about a microorganism's DNA is through sequencing of its genome. This is now increasingly possible as a consequence of the Human Genome Project and the sequencing of many model organisms (see Table 8.2). The spread of epidemics or hospital-acquired (nosocomial) infections will be followed and characterised with more accuracy by the identification of unique DNA fingerprints for individual pathogens. The value of the DNA approach to emergent infections was recently illustrated by SARS (severe acute respiratory syndrome—discussed in more detail below).

Real-time PCR is starting to make a significant impact on molecular microbiology. This technique is particularly valuable because the risk for contamination is reduced (real-time PCR is conducted in a closed system—see Chapter 2, Appendix), Results from real-time PCR are quickly available. Semi-automation is possible, which is essential when dealing with common infections for which the number of referred specimens would be high. Related to real-time PCR is quantitative PCR (Q-PCR). This technique is essential for accurate viral quantitation (HIV, HCV).

Test Utility

A number of parameters are used when assessing the value of a diagnostic test. **Sensitivity** refers to the percentage of true positives that will be detected as being positive by the test—or the probability that the test will be positive when the disease is present. **Specificity** refers to the percentage of true negatives that will be detected as being negative by the test—or the probability that the test will be negative when the disease is not present (Table 8.3). The above parameters are largely a product of the test and do not alter according to the population tested. However, they do not always directly demonstrate a test's usefulness. This is better assessed by the predictive value of positive and the predictive value of negative results coming from the test.

Positive predictive value (PPV) refers to the percentage of all positive test results that are truly positive (or the probability that the disease is present when the test is positive). PPV is influenced by the specificity but also by the prevalence of what is being tested for in that population. The higher the prevalence, the higher will be the PPV since there will be less chance of false positive results. The **negative predictive value (NPV)** refers to the percentage of all negative test results that are truly negative (or the probability that the disease is absent when the test is negative). NPV is influenced by test sensitivity as well as prevalence. A high NPV will be easier to obtain if the infection has a low prevalence in a population. **Tests with high PPVs** are required for conditions in which a false diagnosis will have significant consequences; e.g., treatment regimens are potentially toxic; there are medical or psychological stigmata associated with a pos-

Table 8.2 Some pathogens that have had their genomes sequenced, with the focus here on those likely to have major pathogenic effects[a]

Grouping	Examples
CDC category A[b]	*Yersinia pestis* (plague), *Bacillus anthracis* (anthrax)
CDC category B[c]	*Brucella melitensis* (brucellosis), *Coxiella burnetii* (Q fever), *Salmonella enterica* (typhoid fever), *Vibrio cholerae* (cholera), *Clostridium perfringens* (gas gangrene)
Others	*Borrelia burgdorferi* (Lyme disease), *Mycobacterium leprae* and *tuberculosis* (leprosy and TB), *Neisseria meningitidis* (meningitis), *Helicobacter pylori* (ulcers), *Treponema pallidum* (syphilis), various strains of *Chlamydia*, *Mycoplasma*, *Staphylococcus*, *Streptococcus*

[a] By December 2004, about 188 bacterial genomes had been sequenced. See this list as well as all other published genome projects (239 in total) on the Gold™, Genomes OnLine Database (http://www.genomesonline.org/). [b] CDC—Centers for Disease Control category A are organisms posing a major threat to national security particularly in relationship to high mortality, infectiveness and public health impact. [c] CDC category B is the next level because these organisms are easy to disseminate and cause morbidity, but not to the same degree.

Table 8.3 Sensitivity and specificity in an infectious disease scenario (180 individuals in the population—50 are infected and 130 are not infected, Strohl *et al* 2001)
(See Table 2.10 for more information on this topic.)

Description of test	Number infected (from total 50) who test positive	Number not infected (from total 130) who test positive	Comments
High sensitivity High specificity	50	0	All infected people detected and no false positives recorded.
High sensitivity Low specificity	50	60	All infected people detected, but also false positives are recorded.
Low sensitivity High specificity	30	0	No false positives, but missing some infected people (i.e., getting some false negatives).
Low sensitivity Low specificity	20	70	Not all infected people are detected (i.e., some false negatives), and some detected are falsely positive.

Table 8.4   Some examples of DNA diagnostics used for infectious disease detection and monitoring (Louie *et al* 2000; Gilbert 2002)

Organism	Traditional test	DNA test	Comments
C. trachomatis	Fastidious grower; antigen detection not sensitive or specific enough.	PCR very accurate with sensitivities and specificities in the 90–100% range.	Can be detected in a number of samples and can be multiplexed as a PCR test to detect other organisms in STD, e.g., *N. gonorrhoeae.*
M. tuberculosis	Fastidious grower, but culture still plays an important role in overall management and remains the gold standard. Only way in which to measure drug sensitivity.	PCR demonstrates high sensitivity and specificity (if specimen has acid fast-staining bacteria detectable).	Particularly useful when rapid diagnosis is essential, e.g., TB meningitis.
Herpes simplex virus	Difficult to detect with conventional assays in scenarios such as encephalitis.	PCR can detect virus in CSF.	Avoids need for brain biopsy.
Bordetella pertussis	Detection through swab culture possible.	PCR-based method more rapid and easier.	Early detection possible.
HIV, HCV	Cannot quantitate with conventional assays.	PCR quantitation now available in commercial kits.	Drug treatment in AIDS can be monitored by CD4$^+$ cell counts and viral loads.
S. aureus (methicillin), *Enteroccoci* spp. (vancomycin), *M. tuberculosis* (various drugs), HSV (acyclovir), CMV (ganciclovir)	Culture for sensitivity testing (when possible) takes a few days.	PCR-based assays targeting various antibiotic or drug-resistance genes can be completed in a few hours.	As more information becomes available about the genomes of pathogens, it will be possible to test for various DNA markers to predict the pathogen's effect.

itive test as might occur in the case of sexually transmitted diseases. **Tests with high NPVs** are required when it is essential that positives are not missed, e.g., blood screening tests, treatable infections with fatal outcomes if missed.

INFECTIOUS DISEASE DNA LABORATORY

To date, DNA testing in microbiology has been directed predominantly to the detection of organisms that are difficult to culture *in vitro,* or for various reasons (delayed transport, low titre pathogens, prior treatment with antibiotics) growth is unlikely. Infections in which there is a mix of pathogens might also be usefully approached through DNA analysis. Apart from the straightforward diagnostic applications, DNA microbiological testing has been used to detect antimicrobial resistance (methicillin-resistant *S. aureus* and vancomycin-resistant *Enterococci* spp) or toxigenic forms of *E. coli*. More recently, the availability of DNA technology to quantitate HCV and HIV has been useful in planning and monitoring treatment. Table 8.4 provides some scenarios involving DNA testing.

Unlike molecular genetic (DNA) testing, commercially produced kits have made an early entry into molecular microbiology, and they allow a wider range of laborato-

ries to participate in this technology. A very important boost to this area has been the Human Genome Project (Chapter 1). The successful completion of genomic sequencing for a number of microorganisms was both a catalyst for the human sequencing endeavour as well as the characterisation of genomes from additional model organisms. The next developments with a major impact on molecular microbiology will be microarrays (see Chapter 5, Appendix) and nanotechnology (Chapter 5). The latter will enhance the potential for point of care testing.

In contrast to the genetics DNA laboratory, the microbiology DNA laboratory faces greater technical challenges because infected samples invariably have little of the pathogen's DNA to function as a template for PCR, and the limited amount of DNA that is available will be mixed with DNA from many other organisms that are present. Because of the small target for PCR, it is not surprising that contamination from other PCR products (other samples or previous amplifications) becomes a major source of error. Another inherent problem in the microbiology setting is the potential for inhibitors in the various samples tested to interfere with DNA amplification. Finally, interpretation of some microbiological DNA results is difficult because the test shows only the presence of pathogen DNA. Whether this indicates active infection or residual DNA after the pathogen has died cannot be determined. On the other hand, the finding of mRNA (i.e., cDNA by PCR) is more likely to indicate an active infection. If one then adds that DNA tests are relatively expensive compared to the traditional approaches, it is not surprising that many of the optimistic predictions about molecular microbiology have been slow to materialise. However, there is a steadily growing acceptance of DNA diagnostics. The threat of bioterror or emerging global infections such as was feared with SARS will best be resolved if there is a DNA or RNA component as part of the overall containment strategy.

HUMAN IMMUNODEFICIENCY VIRUS (HIV)

In December 2003, an estimated 34–46 million people were infected with HIV, and more than 20 million had died from AIDS. Included in this number are 2.1 million children under 15 years of age, and more than 2 million HIV-infected women give birth each year (http://www.who.int/whr). All regions of the world have AIDS, including Africa, Asia, Latin America, North America, Europe, the Caribbean, and Oceania. Three quarters of cases are found in sub-Saharan Africa. A very worrying trend is occurring in India, with estimates of the HIV infected somewhere between 2.2 to 7.6 million people.

There are two viral types (HIV-1, HIV-2). HIV-1 was first isolated in 1984 (Figure 8.2). A year later, a similar virus (HIV-2) was found in Paris from West African patients with AIDS but seronegative for HIV-1. HIV-2 is predominantly found in a number of West African countries, including Senegal, Ivory Coast, Gambia, Ghana and Nigeria. It has spread to other countries that have historical or other links to the African sources; i.e., France, Portugal, Angola and Mozambique. However, with the ease of air travel, sporadic cases can appear in any country. Overall, HIV-2 spreads less rapidly and takes longer to

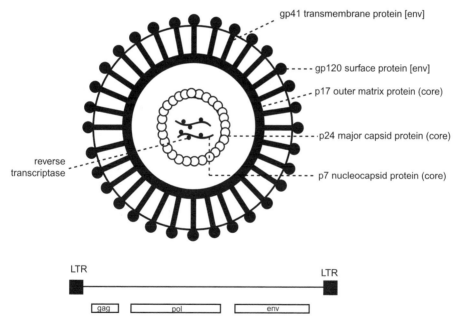

Fig. 8.2   Structure of HIV-1.
(Top) A schematic representation of the HIV genome (family Retroviridae, genus Lentivirus) shows that it comprises two identical strands of RNA within a **core** of viral proteins. Key viral enzymes are reverse transcriptase (viral DNA synthesis), integrase (integration into host cell chromosome) and protease (viral assembly). The core is surrounded by a protective envelope (env) derived from viral glycoprotein and membrane from the previous host cell. Components of the envelope are a transmembrane protein, gp41, and a surface protein, gp120. Abbreviations: gp41—glycoprotein molecular weight of 41 000; p24—protein molecular weight 24 000. **(Bottom)** A number of genes make up HIV. **Gag** (group specific antigens) codes for capsid, matrix and nucleocapsid proteins. **Pol** codes for reverse transcriptase, protease, integrase and ribonuclease. **Env** (envelope) codes for a glycoprotein precursor, gp160, that is cleaved to give envelope structural proteins gp120 and gp41. Although an RNA virus, HIV is able to produce DNA through its reverse transcriptase enzyme once it infects a cell. The viral DNA then integrates (via integrase) into double-stranded host DNA to produce the proviral form. Not shown are six regulatory genes (*rev, tat, nef, vpr, vpu* and *vif*) that give HIV a more complex structure compared to the animal retroviruses. LTR—long terminal repeats of the retrovirus.

develop than HIV-1. The lower viral load associated with HIV-2 may also explain why it is less likely to be transmitted from mothers to babies.

HIV-1 and HIV-2 are retroviruses (so called because they transfer the order of information from RNA to DNA) belonging to the lentivirus genus, which predominantly affects the nervous and immune systems. Transmission occurs by one of three routes: (1) sexually—both male to male and male to female, (2) blood or blood products, including HIV-contaminated needles and (3) perinatally—during birth or from breast feeding. There is presently no effective vaccine against AIDS. Perinatal transmission can be reduced by treating the mother with an antiretroviral agent, avoiding breast feeding and early weaning. Prevention is also possible through screening blood (discussed below) and the practice of safe sex. Treatment options have dramatically improved the outlook for AIDS. However, less than 10% of those requiring treatment in developing countries have access to antiretroviral drugs (Box 8.1).

The urgency and potential implications of the AIDS pandemic have resulted in many research programs directed to better diagnostic tests, more effective therapeutic regimens and a greater understanding of the viral biology. The identification of HIV viruses has required extensive changes to be made in blood screening protocols (see below). In all these areas, rDNA technology is playing a major role. In the context of the present theme of laboratory detection, molecular medicine has an important function to play since opportunistic infections are the chief cause of death in AIDS. All who have AIDS will develop infections at some time in their illness. The pathogens involved include those that normally do not produce overt disease such as *Candida albicans*, and the more exotic organisms that will not usually be seen in clinical practice, e.g., JC virus, which gives rise to a demyelinating neurological disorder called progressive multifocal leukoencephalopathy. In these circumstances, standard microbiological detection methods often prove ineffective.

Screening and DNA Testing for HIV

Many commercial ELISA (enzyme linked immunosorbent assay) kits detect antibodies to HIV-1 and HIV-2 (Figure 8.3). They are used to screen those who are at risk and blood donors. The assays are well standardised with good sensitivities and specificities. In HIV infection, screening tests must have the highest sensitivity so that positives will not be missed. False positives are then excluded by using a second assay with a high specificity, e.g., a western blot.

HIV-1 testing by DNA analysis can be directed towards the *gag*, *env* or *pol* genes of the virus (see Figure 8.2). Laboratory techniques and detection strategies (algorithms) utilising these sequences have been constructed

---

Box 8.1 Treatment of HIV/AIDS.

AIDS represents the clinical phase of HIV infection that by now is well established. Therefore, effective treatment must be started early and take into consideration a feature of this infection—the rapid development of mutant viruses. This occurs because (1) HIV infection is associated with significant viraemia. (2) The viral population is very heterogeneous due to an error-prone reverse transcriptase (unlike DNA polymerase, which makes few mistakes in the replication of DNA). Therefore, mutations are constantly being produced in the HIV genes, and mutant viruses will form. This, combined with the high viral titres present, facilitates the development of resistance. The key to success with AIDS treatment has been HAART (highly active antiretroviral therapy), which became available in 1996 and involves the use of multiple drugs directed to different viral pathways.

Class of antitretroviral drug	Mechanism of action	Examples
Nucleoside and nucleotide analogues	Act as DNA chain terminators, thereby inhibiting the reverse transcription of viral RNA into DNA	Zidovudine (AZT), lamivudine
Non-nucleoside reverse transcriptase inhibitors	Bind and inhibit viral reverse transcriptase	Nevirapine
Protease inhibitors	Block viral protease required for making the inner core of viral particles	Nelfinavir
Entry inhibitors	Block virions from penetrating target cells	Enfuvirtide

Combination drug therapies involve the use of three drugs having a synergistic rather than additive effect on reducing the viral load, and so they delay the development of drug resistance. Various combinations are used, for example, two nucleoside and nucleotide analogues with the third drug a non-nucleoside reverse transcriptase inhibitor or a protease inhibitor. More recently, a fourth class of drug with a completely new mechanism of action has been developed. This involves inhibiting the entry of HIV-1 into CD4+ lymphocytes. The major questions now requiring resolution include when to start HAART and what are the optimal drug combinations (more than 20 drugs are approved by the FDA for treating HIV infection). Despite the use of HAART, HIV re-emerges once therapy is stopped, indicating a virostatic rather than virocidal effect. Better drugs are still needed, but more importantly, a vaccine that will allow the prevention of HIV-AIDS.

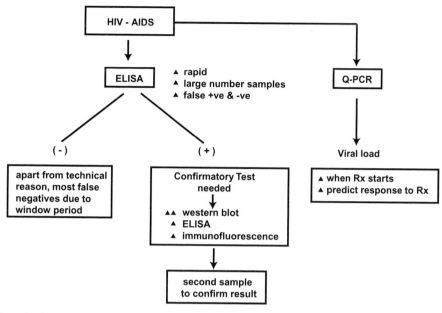

Fig. 8.3   Diagnosis of HIV/AIDS infection.
The ELISA test is the mainstay for routine HIV detection. It is rapid and able to test many samples. It has the disadvantage of both false positives and false negatives, and so confirmation by a western blot (or second ELISA or an immunofluorescence test) is essential. Apart from technical and clerical mistakes leading to a false negative result, the most common reason for this type of error is the window period since it takes between 4–12 or more weeks after infection for antibodies to form. A second independent sample is then tested to reduce the potential for technical or clerical problems before a definitive diagnosis is made. Quantitative PCR (Q-PCR) is used to assess viral load, which helps to decide when to start therapy and can be used to monitor response to therapy.

to optimise sensitivity and specificity and minimise false positive or false negative results. It has been estimated that one HIV-1 proviral copy can be detected against a background of 10^5 mononuclear cells in peripheral blood. Virus isolation is the only other method with comparable sensitivity although various problems with this test exclude it as a routine assay.

Variations of the PCR technique including nested PCR (see Chapter 2, Appendix) can be used to reduce the false positive rate, which is due, in most cases, to contamination of samples or reagents. On the other hand, false negative results can reflect the presence of inhibitors in the specimen being tested or chemical substances such as anticoagulants. Another source of a false negative result with HIV occurs because of the considerable diversity found in the viral DNA sequence so that a mismatch between the usual primer pairs inhibits the subsequent PCR step. This difficulty can be avoided by use of primer pair combinations derived from conserved segments of the viral genome. For the present, serological testing for HIV-1 and HIV-2 remains the preferred screening method for at-risk individuals, including blood donors, because it is fast, relatively cheap and able to be automated (see also blood screening below).

TRANSFUSION-RELATED INFECTIONS

A number of the important recent viruses have reached notoriety through their association with blood transfusions. Previously, blood transfusion services based their donor and blood screening programs on detecting antibodies or antigens in the donor or blood supply. However, this approach has been shown to be inadequate, and an important new addition to the screening protocols is the use of PCR to identify viral DNA or RNA. The advantages of a NAT assay include (1) higher sensitivity and (2) greater reliability during the window period, which is the time between a blood donor becoming infectious and donor screening tests becoming positive; i.e., seroconversion has occurred.

PCR-based assays for screening blood donations are used to test (1) pools of donations, for example, 16–24 donations simultaneously, and (2) individual donations. The former is more rapid and cheaper, but the very rare case of HIV and hepatitis C virus (HCV) being missed means the testing of individual donations is the method of choice. Screening by NAT has a number of potential drawbacks, including (1) It involves more sophisticated technology (therefore, trained staff are needed, and tests

Table 8.5 Some success stories with screening in blood transfusion services (Goodnough 2003)

Virus	Screening	Outcomes[a]
HIV	Donor and blood with full-range tests including NAT. First blood-related infections in 1982.	In the USA, there were 714 HIV cases in the year before testing was implemented. That number fell to about five cases per year in the next five years after NAT testing started in 1999. Risk of blood transfusion–acquired HIV in the post-NAT era is 1 : 1 900 000.
HCV	Donor and blood with full-range tests including NAT.	There are serious consequences of blood-borne HCV infection. However, following testing protocols, HCV is a rare complication. Risk of blood transfusion–acquired HCV in the post-NAT era is 1 : 1 600 000.
HBV	Donor and blood with third-generation serologic assays for hepatitis B surface antigen.	Risk of blood transfusion–acquired HBV is still relatively high (1 in 220 000) although only about 10% of post-transfusion hepatitis is due to HBV. Unlike HCV, risk of chronic liver disease is less with HBV.
CJD, vCJD	No useful test available.	In the USA, deferral protocols have been in place since 2000; i.e., individuals living in certain European countries during defined times are excluded. No direct evidence that blood can transmit CJD or vCJD, although the infection is transmitted with tissues such as cornea and pituitary extracts. In the UK, plasma for fractionation is obtained from the USA.
West Nile	Most useful is NAT for viral RNA. Serologic testing of donors is not sensitive.	In 2003, 23 cases were reported—resulting in the development in a very short time (2003) of a PCR-based test used to screen all donors. Also included in the donor questionnaire is history of fever with headaches in the week before donation (yes answer leads to deferral).

[a] It is worthwhile comparing infectious risks of blood transfusions with incompatible blood transfusions with their associated 10% mortality. The FDA reports over twice the frequency for incompatible blood transfusion compared to all infections, and in the UK, adverse events related to incompatible blood transfusions were 10 times higher than for infectious diseases.

are invariably more expensive). (2) Assays are technically demanding, including the strict requirement to prevent contamination. (3) Results from NATs take longer than those from serologic ELISAs (enzyme linked immuno-sorbent assays). However, in response to a number of cases of contaminated blood, many blood transfusion services have opted for NAT testing.

Blood transfusion services screen blood and donors for a range of infectious agents, e.g., hepatitis B virus (HBV), HCV, HIV-1, HIV-2 (and in some regions human T-cell lymphotropic virus types I and II [HTLV-I, HTLV-II]), syphilis, cytomegalovirus (CMV) and, more recently, West Nile Virus in the USA. Some screening tests are undertaken only in selected circumstances; e.g., CMV-free blood is required for an immunosuppressed patient, fetus or neonate.

The risks of blood-borne viral infections have dramatically decreased with the implementation of donor questionnaires that allow self-exclusion (particularly important for infections that are not routinely tested for), donor and blood testing with ELISA and PCR-based techniques. The latter's sensitivity is several magnitudes higher than culture or ELISA methods. To exclude the potential for slow virus transmission (i.e., CJD and vCJD)

through blood, transfusion services have deferred the use of donors who have lived in the UK over certain time periods (Table 8.5). Other reasons for deferral include fever with headaches the week before donation (possibility of West Nile Virus and others) or travel to risk regions (for example, the possibility of malaria).

Apart from HIV, another success with blood transfusion screening has been HCV. Since the single-stranded ~10 kb RNA sequence of the HCV was reported in 1989, it has been possible to use an enzyme immunoassay to detect antibodies against a number of viral antigens in patient testing or blood donor screening. However, the first-generation immunoassays gave rise to many false positives. This problem with serological testing was overcome by the availability of more sophisticated second- and third-generation enzyme immunoassays. Similar to what happens with HIV, the resolution of false positive results is undertaken by an additional assay such as an immunoblot that incorporates multiple viral-specific antigens. Viral antigens in many of the modern tests have been prepared by using rDNA expression systems such as *E. coli* or yeast (see Chapter 6). Indeterminate or equivocal results can be further evaluated with PCR.

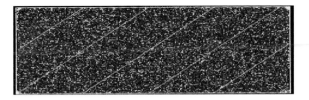

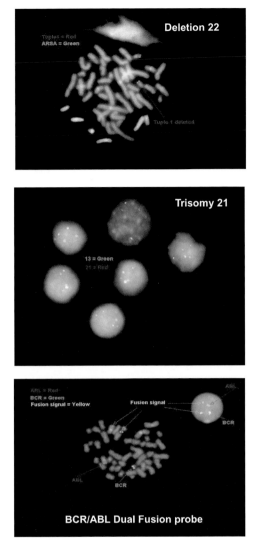

Deletion 22

Tuplet = Red
ARSA = Green

Tuplet deleted

Trisomy 21

13 = Green
21 = Red

ABL = Red
BCR = Green
Fusion signal = Yellow

Fusion signal

ABL

BCR

ABL

BCR

BCR/ABL Dual Fusion probe

Fig. 2.11 FISH.
Top: The metaphase FISH of a deleted chromosome 22 shows a chromosome spread. Chromosome 22 is identified using two DNA probes (coloured red and green). One of the two chromosomes 22 is abnormal; i.e., there is only the red probe because the area of chromosome 22 detected by the green probe has been deleted. **Middle:** This photo shows an interphase FISH, which has several advantages over metaphase FISH. It can score a larger number of cells, thus increasing the chance of detecting a chromosomal rearrangement, particularly in low-level mosaic states. Interphase FISH is more rapid, which is desirable in prenatal diagnosis. This particular picture illustrates trisomy 21 (Down's syndrome) because it has an additional signal with the red (chromosome 21) DNA probe. **Bottom:** This composite photo shows a metaphase and interphase FISH for the *BCR-ABL* fusion gene found in chronic myeloid leukaemia. The red and green DNA probes for the *ABL* and *BCR* genes will show up as a yellow signal when both are present in the translocated rearrangement (read more about chronic myeloid leukaemia and the *BCR-ABL* rearrangement in Chapter 4). *FISH provided by Dr Michael Buckley, Molecular & Cytogenetics Unit, South Eastern Area Laboratory Services, Sydney.*

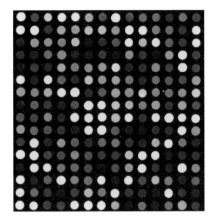

Fig. 5.2 A gene microarray. *(reproduced with the permission of Dr C Roberts and Dr S Friend, Rosetta Inpharmatics LLC, Seattle, USA).*
Top: This is a 1-inch × 3-inch glass slide onto which has been imprinted 25 000 genes in the form of oligonucleotides. The oligonucleotides are 60 mers (i.e., 60 bases in size), and they are imprinted *in situ* onto the glass slide using a conventional inkjet printer head. The longer oligonucleotides used for hybridisation (as discussed previously, a 20 mer oligonucleotide is generally sufficient to detect a specific region of the genome) provides this particular microarray with a technical advantage because, when constructing the array, any mistakes or errors in the inkjet printing will not be as critical as would be the case with a 20 mer array. On the periphery of the slide and diagonally are grid lines. They are used for alignment when a scanner reads the array. The scanner would be faced with the picture shown, i.e., 25 000 spots with various colours and intensities, including green, red, yellow (both red and green have hybridised), black (neither red nor green has hybridised). **Bottom:** This is a small section of the microarray magnified to show the individual spots and the colours generated. At this stage, the scanner needs to identify the various colours as well as their intensities, and these data need to be stored. Bioinformatics-based tools are needed for this aspect of the microarray, as well as the next step, which is not as technically demanding as the actual microarray but equally challenging because the vast amount of data needs to be interpreted in terms of potential biologic significance.

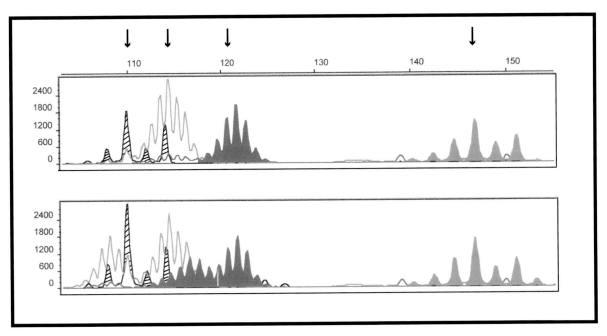

Fig. 3.8 Microsatellite instability detectable by DNA testing.
The ↓ indicate the position of four DNA markers that are used to detect microsatellite instability (from left to right they are D5S346 (stippled—black), BAT26 (open—green), BAT25 (filled—blue) and D17S250 (filled—green). The upper series shows the normal patterns for these markers; i.e., there is no microsatellite instability. The lower series comes from a tumour that has microsatellite instability. This is seen in BAT26 and BAT25 since the profile now shows that at these two loci a number of additional fragments are generated following PCR; i.e., there has been slippage reflecting the microsatellite instability.
Electrophoresis pattern courtesy of Le Huong, Department of Molecular & Clinical Genetics, Royal Prince Alfred Hospital, Sydney.

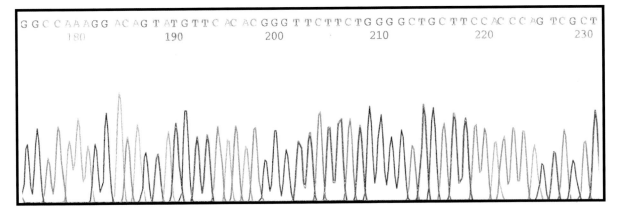

Fig. A.8 Trace of an automated sequencing run.
The automated DNA sequencer has revolutionised molecular medicine since it enables relatively cheap sequencing of long segments of DNA and readouts are automated. Each ddNTP is distinctly labelled with four different fluorochromes, and because dye terminator sequencing is used, the four ddNTPs are electrophoresed in the one gel lane or capillary run. The DNA sequence is written above each peak and has been determined by the sequencing software. If N appears in the DNA sequence, it means the automated sequencer software cannot decide which base is correct and visual inspection is necessary.

PATHOGENESIS

The pathogenesis of many infections, particularly viral ones, has been deduced from experimental strategies based on light and electron microscopy, cell culture and immunoassay. To these research tools can now be added nucleic acid (DNA, RNA) probes for *in situ* hybridisation or NAT. Advantages provided by DNA techniques include the ability to detect latent (non-replicating) viruses and to localise their genomes to nuclear or cytoplasmic regions within cells. Tissue integrity remains preserved during *in situ* hybridisation, and so histological evaluation can also be undertaken. Nucleic acid probe techniques or NAT can be manipulated to enable a broad spectrum of serotypes to be detectable. This is particularly valuable in those emerging infections where the underlying serotypes are unknown. More recently, the most powerful application of DNA techniques is the ability to sequence whole genomes, and so identify the pathogen, and from its genomic sequence (1) predict its role in disease pathogenesis and (2) identify regions of the genome suitable for NAT-based detection methods.

HOST RESISTANCE

The host's response to an infection involves a complex mix of genetic and environmental factors. In humans, evidence for a genetic component contributing to the outcome of an infectious disease comes from the observation that some ethnic groups are more resistant to infections, whereas others appear to have an increased susceptibility; e.g., resistance to malaria in black Africans and susceptibility to chronic carrier state for HBV in Chinese. Not all exposed to HIV-1 get infected, and those who do progress to AIDS show different responses. For example, there can be a rapid progression within 1 to 5 years, or individuals can demonstrate a more benign progression extending up to 20 years. Because host resistance is likely to represent multiple genetic effects—i.e., QTLs (quantitative trait loci), which interact with the environment—they are difficult to detect and so have been sought by the usual molecular genetic approaches, particularly association (case control) studies and the investigation of candidate genes (see Chapter 4 for further discussion of this strategy and QTLs).

HIV-1 Infection

HIV-1 targets the macrophages, monocytes and T lymphocytes that carry the CD4 cell surface protein. HIV-1 first attaches to CD4 and then the chemokine receptor CCR5. Once this occurs, the HIV gp41 protein can penetrate the cell membrane, enabling HIV to infect the cell. In some circumstances, particularly as the disease progresses, a mutation in the *env* gene enables a move away

from the CCR5 receptor to the CXCR4 receptor. Therefore, two host-related genetic factors can influence the infectivity of HIV-1 in humans. They are (1) chemokine receptor genes and (2) major histocompatibility complex (MHC).

CCR5 is a chemokine and functions as a co-receptor required for HIV to enter cells. CCR5 is highly polymorphic, with variants found in different ethnic groups. One important variant involves a 32 bp deletion in the coding region leading to a truncated protein. It is now known that homozygotes for CCR5 Δ32 (approximately 1% of Caucasians) have significant resistance to HIV infection because the virus has lost one of its entry points into the cell. Heterozygotes do not appear to have enhanced resistance to HIV infection although progression to clinical AIDS is delayed, and viral loads are lower. Other genetic variants related to CCR5 have been reported to influence HIV infection and progression (Table 8.6).

The second and probably the more important of the genetic factors influencing HIV infections came from an observation of apparently HIV-resistant female sex

Table 8.6 Genetic variants in chemokine receptors influencing HIV infectivity (O'Brien and Nelson 2004)[a,b]

Variant	Mechanism
CCR5	Δ32 defect decreasing the amount of this receptor. Homozygotes for this have a strong resistance to HIV infection (although this is not absolute). Progression to AIDS is slowed in heterozygotes.
CCR5	Apart from the 32 bp defect, mutations in the promoter have also been associated with increased gene function, thereby accelerating the development of AIDS.
RANTES (*CCR5* ligand)	Mutations in the promoter can affect both susceptibility as well as HIV progression, depending on how RANTES expression is affected.
CCR2	The variant V64I appears to delay progression of HIV perhaps by impairing the transition from the CCR5 receptor to an alternative one (CXCR4).
CXCL12	Variant in the 3' end appears to impair HIV entry into the cell perhaps through the CCR5 to CXCR4 transition.

[a] Unlike the protective effect of the CCR5 Δ32 defect, there is some controversy about the role played by the other genetic variations because they have all been identified through genetic association studies. [b] Two other important genes that influence progression from HIV to AIDS are *IL10* and *IFNγ*. These genes produce cytokines that inhibit HIV replication. Hence, genetic variants that lead to overexpression of these genes will favour a milder disease, and the converse occurs with genetic variants that are associated with reduced expression.

Table 8.7 The role of the MHC in controlling infections[a]

HLA class	Cells expressing this class	Peptides recognised	Immune response generated
Class I (HLA A, B, C)	All nucleated cells	Those derived from intracellular pathogens, e.g., viruses	CD8[+] T cells initiate cytotoxic cell response.
Class II (HLA DP, DQ, DR)	Expressed only on antigen-presenting cells, i.e., macrophages, dendritic cells and B lymphocytes	From extracellular and intravesicular pathogens, e.g., bacteria, fungi and some phases of viral infections	CD4[+] T cells lead to cytokine production and promote B cell antibody production.

[a] The MHC (in humans, it is usually called HLA) locus codes for cell surface proteins are important for immune surveillance since T cells do not respond to foreign peptides from a pathogen unless they are coated (or presented) with MHC proteins.

Table 8.8 Association between HLA type and susceptibility to infection (Carrington and O'Brien 2003)

Infection	HLA type	Mechanism
HIV	Homozygosity for any type	Faster progression of HIV because a narrower range of HIV peptides is presented for cytotoxic T cell response.
HIV	B27	Slower progression of HIV appearing to reflect a super efficient binding of p24 gag HIV epitope by those positive for HLA B27.
HIV	B57	Slower progression of HIV perhaps because B57 molecules target a wider range of HIV peptides involving gag and reverse transcriptase.
HIV	B35	Faster progression of HIV secondary to failure of B35 alleles to bind HIV peptides due to single amino acid change. Carriers have a more rapid progression, while homozygous individuals develop AIDS within half the median time it takes those who are B35 negative.
Malaria	B53	Protection from severe infection in West Africa.
Myco bacteria	DR2	Increased susceptibility documented in some populations (Asia, India) but not others.
HBV	DRB1*1301 and 1302 (subtypes of HLA DR13)	Protection against chronic HBV through enhanced viral clearance.
HCV	DQB1*0301	Enhanced clearance of HCV (Europeans and American blacks); however, in American whites associated with viral persistence.
	DRB1*0101	Contradictory results in relation to association with self-limiting HCV infection or enhanced clearance.

workers in Nairobi. These women were shown to have a particular HLA type (HLA A2, HLA A28), and it has been proposed that this enhanced their resistance through a better presentation of HIV antigens to T lymphocytes. Therefore, the virus was more rapidly cleared by the immune system. Subsequently, there have been many associations described between HLA types and increased resistance or greater susceptibility to HIV infection and its progression to AIDS (Table 8.7).

Another observation involving HLA is that mothers and children who are mismatched at the MHC are less likely to have perinatal transmission because immune surveillance is enhanced. However, the HLA protective effect is not observed if HIV is acquired through breast feeding. Other HLA associations in AIDS (as well as other infections) have been reported, including polymorphisms of interleukins IL4 and IL10 (Table 8.8). Like association studies in genetic disorders, the reports have produced

many correlations between genetic markers and response to infections although they often remain controversial. The confusion that can be found with association studies is well demonstrated by the HCV example in Table 8.8.

Malaria

Each year there are more than 500 million cases of falciparum malaria in Africa, and around 1–2 million deaths will result. A resurgence in malaria reflects socio-economic factors (poverty, overcrowding), emergence of mosquitoes that are resistant to insecticides and development of drug resistance. The two common forms of malaria (*P. falciparum* and *P. vivax*) produce severe anaemia, and in addition, *P. falciparum* is associated with cerebral malaria, respiratory and metabolic complications. This differing spectrum is partly explained by *P. falciparum* being able to invade a large proportion of red blood cells compared to *P. vivax,* which invades only the reticulocytes. Another explanation is the mode of entry of these parasites into red blood cells, with *P. falciparum* having a number of different invasion pathways compared to *P. vivax,* which is able to enter only red blood cells that carry the Duffy blood group.

Host factors providing some protection from malaria have been identified. They include single gene effects seen in haemoglobinopathies such as sickle cell anaemia or thalassaemia, and the Duffy negative blood group. The former are explained by suboptimal red blood cell environments that do not provide a good milieu for the parasite. The Duffy negative blood group has a more powerful protective effect since it is the receptor by which *P. vivax* enters the red blood cell. Thus, *P. vivax* is not seen in West Africa because the populations are Duffy negative.

Hepatitis B

Although vaccination for HBV has made a very significant reduction in new cases, worldwide there still remain about 350–500 million infected individuals. Long-term complications with this DNA virus occur in about 15% of those infected and include chronic hepatitis, cirrhosis and hepatocellular carcinoma. Why some go on to develop a chronic carrier state is not known, with the course of the illness after infection very variable. This has been explained on the basis that it involves the viral genotype and the host's immune responses; i.e., following acute infection, a vigorous T cell response against the various viral-specific components (core, envelope and polymerase) limits the infection. In contrast, those with a poor cellular immune response are more likely to develop a chronic carrier state. In this scenario, the MHC will play a key role since class II molecules from the HLA complex are critical for T cells to function. Evidence for an MHC-related effect in hepatitis B infection is presented in Table 8.8.

Animal Models

Polygenic traits contributing to pathogen susceptibility or resistance in humans would be difficult to separate into their individual components because of the genetic variability in the human genome. However, this can be overcome by using different strains of inbred animals, particularly mice. Since chromosomal segments in the mouse can be traced to their homologous or syntenic regions in humans, it becomes possible to identify a genetic locus or gene in a mouse and then go back to the corresponding region in the human genome to find the equivalent gene.

Forward genetic screens (phenotype to genotype) result from the observation that a particular mouse strain (or one created by, for example, a chemical mutagen such

as ENU—Chapter 5) has a phenotype that alters the animal's expected response to an infection. An example would be a mouse strain shown to have resistance to *Leishmania donovani* infection. From this observation, a gene was isolated by conventional positional cloning. The gene was called *Nramp1* (now renamed *Slc11a1*). Mice with mutations in this gene (particularly G169D) are resistant to three intracellular pathogens (*Mycobacterium bovis,* *Salmonella typhimurium* and *Leishmania donovani*). This gene is likely to play a key role in infection since it is exclusively expressed in macrophages, key cells in the control of the three intracellular pathogens just mentioned.

A reverse genetic screen (genotype to phenotype) is exemplified by the transgenic knockout mouse when the particular phenotype associated with a gene can be confirmed by inactivating that gene. One interesting knockout involves the removal of the TNFα (tumour necrosis factor) gene, which leads to a mouse that is very susceptible to the pathogens *M. tuberculosis* and *L. monocytogenes.* A similar susceptibility is seen if the mouse's interferon-γ gene is knocked out. Both these genes play key roles in the immune response.

Another strategy allowing a mouse model to be used for an infection that is found exclusively in humans is to "humanise" the mouse for the particular host-tropism that explains the species specificity. An example of this is the polio virus that does not normally infect mice. However, transgenic mice that express the human poliovirus receptor CD155 develop polio if infected intracerebrally, but not if they are infected orally. In addition to creating a suitable animal model, this experiment also indicates that the oral phase of infection with poliovirus is likely to involve another receptor.

DRUG RESISTANCE

Antimicrobial Drugs

Resistance to antibiotics has always been a concern. However, new antibiotics in the 1950s–60s temporarily addressed the problem, and this false sense of security was reinforced in the next two decades with newer and more powerful antibiotics. Today, the increasing trend in antibiotic resistance is alarming, and is not being matched by new drug development. This led to the World Health Organization in 2001 developing a plan to address what has become a global problem.

Resistance to antibiotics occurs by three major mechanisms: (1) reduced antibiotic uptake by a cell or increased efflux from the cell, (2) modification or inactivation of the antibiotic and (3) altering the target for the antibiotic. The development of resistance may be an intrinsic feature of an organism, or it may result from mutations or the acquisition of resistance genes. Transfer of resistance genes can occur via plasmids or transposons. The latter represent mobile DNA elements that can move between plasmids or between plasmids and chromosomes, thus having the potential to disseminate widely (and stably) drug resistance genes.

The genes involved in antibiotic resistance are being characterised, enabling a better understanding of how this develops and how to detect it quickly by DNA-based approaches, and finally, identifying strategies by which the resistance pathways can be bypassed. Apart from the inappropriate use of antibiotics in medicine, agriculture and animal feeds (about 50% of antibiotics are used in the beef and poultry industry!) contributing to the increasing resistance, other factors involved include travel and tourism.

Pneumococci are important pathogens in the community leading to central nervous system infections, pneumonia and otitis media in children. Resistance to macrolide antibiotics (e.g., erythromycin) in some countries is reaching alarming proportions; for example, 65% of *Streptococcus pneumoniae* are resistant to erythromycin in parts of Asia. Two major mechanisms explain resistance in this organism: (1) Bacteria carrying the *ermB* gene can methylate a specific amino acid on the ribosome (the site of action for macrolides) and so interfere with binding of the antibiotic to this region. (2) Bacteria carrying the *mefA* gene (an efflux pump gene) can export out the antibiotic and so protect the ribosome.

Multidrug resistance (MDR) occurs with both antibiotics and anti-cancer agents, and refers to the development of resistance to a range of drugs that may be unrelated or differ widely in their structure or target. Common pathways are seen with MDR. One of these involves the P-glycoprotein (*MDR1*) gene, which is a plasma membrane–spanning multidrug transporter protein. One of the most important mechanisms for MDR is the ability to extrude out the antibiotic (or anti-cancer drug) from the cell (Figure 8.4). Antibiotic resistance to many gram negative bacteria (*Pseudomonas aeruginosa*, *Acinetobacter* spp, and the Enterobacteriaceae) occurs through an efflux-mediated mechanism.

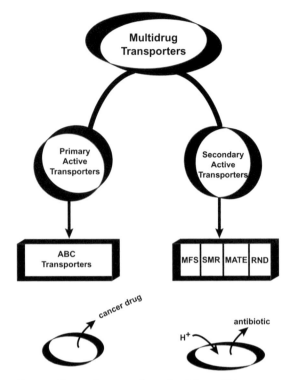

Fig. 8.4 Drug resistance pathways involving multidrug transporters.
Two major classes can be described. The first is the ABC (ATP Binding Cassette) transporters that utilise ATP to transport drugs out of the cell. This class is mostly found in eukaryotic cells and would be predominantly involved in resistance developing to cancer drugs, although it is also implicated in drug resistance for a wide range of pathogens. The second includes a number of families (MFS, SMR, MATE and RND). They move drugs out of the cell at the same time as they move in protons or sodium. Members of this class are found predominantly in prokaryotes and would be important in antibiotic resistance.

Mycobacterial Infection

Rifampicin occupies a pivotal place in the World Health Organization's multidrug resistance program for tuberculosis (TB) and leprosy. Despite this, about 3.2% of new TB cases are caused by multidrug-resistant strains. The development of resistance has major therapeutic and public health implications. Since *M. leprae* grows slowly, it is essential to confirm the development of drug resistance by alternative means to the traditional culture approach. Similarly, to avoid the use of what will be ineffective drugs (a cost consideration as well as a way to optimise antimicrobial therapy), the Centers for Disease Control recommends that resistance patterns for TB should be reported within 28 days of the specimen being

received. However, this is a challenge because of the slow-growing *M. tuberculosis*. Therefore, an alternative approach is to use molecular diagnostic techniques to identify resistance genes in the mycobacteria.

Until recently, the use of various drug combinations selected from isoniazid, rifampicin, ethambutol, streptomycin and pyrazinamide as well as improved living standards was instrumental in producing a declining incidence for tuberculosis. However, this is now changing. The number of new cases is increasing. The traditional dogma that active cases usually arise from infections acquired years earlier is no longer applicable since DNA studies have shown that approximately one third of active cases in some cities are the result of person-to-person transmission. Contributing factors to

Table 8.9 Drug resistance to mycobacterial infection (de Viedma 2003)[a]

Drug	Mechanism for resistance
Rifampicin	Involves the *rpoB* gene encoding the β subunit of the DNA-dependent RNA polymerase. Mutations in this gene are found in a specific 81 bp region.
Isoniazid	Involves at least four different genetic loci: *katG*, *inhA*, *ahpC* and *oxyR* genes, which affect different metabolic pathways. One particular mutation (*katG*315) has a frequent association with high-level resistance.
Ethambutol	Mutations are found in the *emb*CAB gene involved in metabolic pathways.
Streptomycin	Mutations are found in *rrs* (coding for 16S RNA) or *rpsL* (codes for the ribosomal protein 12S).

[a] Despite the many resistance-related mutations being identified, some resistant strains of *M. tuberculosis* do not have mutations in these genes.

Table 8.10 Drug resistance to malaria (Le Bras and Durand 2003)

Drug	Mechanism for resistance
Chloroquine	Impaired uptake of chloroquine by the parasite's digestive vacuole due to mutations in the *Pfcrt* gene is responsible for resistance to this drug. Mutations in genes *Pfmdr1* and *Pfcg2* may also be important.
Antifolates (pyrimethamine and proguanil)	Mutations in *Pfdhfr* gene (dihydrofolate reductase) which is the target of the antifolates. Point mutations interfere with inactivation of the dihydrofolate reductase.
Sulfonamides and sulfones	Mutations in *Pfdhps* gene (dihydropteroate synthase) which is the target of these drugs. Mutations interfere with inactivation of dihydropteroate synthase.
Sulfadoxine-pyrimethamine	Through selection of *dhfr* and *dhps* mutants.
Amino alcohols (quinine, mefloquine and halofantrine)	Unknown.

the increase in tuberculosis and the finding of multidrug-resistant strains are HIV infection, intravenous drug use and the decline in living standards resulting from political changes or war. At the molecular level, the genes associated with drug resistance are being defined (Table 8.9). To deal with these challenges, modern global strategies directed to tuberculosis (and leprosy) will require laboratory facilities that utilise state-of-the-art technology to provide health professionals with accurate information in an efficient and timely fashion.

Malaria

During the latter half of the twentieth century, chloroquine was the drug of choice for treating falciparum malaria. However, the spread of drug-resistant *Plasmodium falciparum,* and to a lesser extent *Plasmodium vivax,* has made the prophylaxis and treatment of malaria a global public health issue. Chloroquine-resistant *P. falciparum* is present in every major region where malaria is endemic. Air travel means that drug-resistant strains can appear in any city. Resistance is now spreading since the 1950s from a limited number of foci in South East Asia and South America. The epidemiology is consistent with multigenic effects, i.e., rare events occurring initially in South East Asia or South America and then spreading slowly worldwide. In contrast, resistance to pyrimethamine and proguanil has arisen independently in many different regions where these drugs have been used, consistent with a single gene being involved.

The molecular basis for resistance in malaria is slowly being unravelled, and many genes are now implicated (Table 8.10). As the molecular defects become more comprehensively characterised, it will be possible to predict the drug resistance pattern for various geographic

regions and so optimise drug selection. It might also be feasible to use other drugs or genetic-based therapies to bypass the resistance defect in the parasites, allowing the continuing use of what are cost-effective and relatively safe drugs such as chloroquine. Another positive development in the therapy of malaria is the availability of the DNA sequence from the genomes of the important human and mouse malaria parasites. This development will enable new targets to be found, i.e., genes for metabolic pathways that differ between the parasite and humans.

HIV

Monitoring response to AIDS drug treatment can be undertaken by measuring the CD4[+] cell count, as well as the assessment of viral load by Q-PCR. In the USA, about half the patients are infected with viruses that have some resistance to drugs. The mutations causing resistance, as well as their likely modes of action, are well understood for HIV. They include many single base changes and the occasional insertions of a few amino acids. Cross resistance, once it occurs, is restricted to drugs within the same class although all classes of drugs can develop resistance, which tends to occur slowly over a period of time (Box 8.1 describes the various drugs). Resistance is the result of many mutations accumulating within HIV, although one exception is the M184V mutation in reverse transcriptase, which alone produces complete resistance to lamivudine. When plasma viral loads rise despite therapy, a change in regimen is needed. Because of the cross-resistance within classes, what drug to substitute is not an easy decision. Assays to detect HIV resistance include genotypic (detecting RNA mutations) as well as phenotypic (*in vitro* viral cultures).

VIRULENCE FACTORS

Streptococcal Virulence Factors

Group A streptococcus (GAS) is responsible for a wide range of human infections. Virulence factors in this organism produce a broad spectrum of toxic effects from antiphagocytic, cellular adherence, internalisation, invasion and systemic toxicity. An important toxin associated with GAS is now better understood through analysis of its gene. This is the β haemolysin effect, which was difficult to study without DNA technology. The β haemolysin effect in GAS is due to a powerful cytotoxin called streptolysin S, which has a very broad toxic effect on cells and is non-immunogenic. The gene for streptolysin S has now been isolated and is called *sag*. It comprises 9 different contiguous genes *sagA* through to *sagI*. The corresponding gene for β haemolysin in group B streptococci is *cyl* and involves 11 contiguous genes.

Helicobacter Pylori *Virulence Factors*

H. pylori is an important gram negative slow-growing bacterium that colonises the gastric epithelium and can lead to chronic gastritis, ulceration and cancer. Infection is strongly associated with socioeconomic status, with prevalences >80% in developing countries and much lower ones (20–50%) in affluent regions. Disease outcome depends on a number of factors including bacterial genotype. Infection with *H. pylori* can be treated with antibiotics. However, because it is so common in some communities, a vaccine approach for prevention would be more economical and effective. To identify potential targets for vaccines in this organism, the various virulence factors have been identified. They include urease, vacuolating toxin, a cytotoxin and an adherence factor. The genome of this bacterium was sequenced in 1997, and from this it was possible to understand better the bacterium's acid tolerance and genes involved in pathogenesis. A number of the virulence-producing genes have been characterised. They include *vacA*—exotoxin-producing vacuoles in epithelial cells associated with greater risk for ulcers; *cag* PAI region associated with the induction of a more intense host inflammatory response leading to more severe disease in Western populations. The significance of *cag* PAI in the developing world is less obvious because almost all strains are *cag* positive.

CANCER

DNA and RNA viruses cause tumours in humans. Oncogenic viruses include hepatitis B (DNA), hepatitis C (RNA)—hepatocellular carcinoma, and papillomavirus (DNA)—cervical cancer. The Epstein-Barr virus (EBV), a DNA virus, is associated with the development of both lymphoid and epithelial tumours.

Hepatitis B Virus (HBV)

The hepatitis B virus is a hepatotropic DNA virus. Its replication cycle within the liver nucleus leads to the formation of mature virions via reverse transcriptase, and they are exported from the cell. Alternatively, the newly formed viral DNA can be recycled back into the nucleus for conversion to a plasmid form that generates further genomic transcripts and so maintains a constant intranuclear pool of templates for transcription. The HBV replication cycle within the hepatocyte is not directly responsible for liver damage, consistent with the observation that many HBV carriers are asymptomatic with minimal liver damage. It has also been observed that those with immune defects tend to be chronic carriers of HBV but usually in association with mild liver damage. Therefore, it has been proposed that liver damage following chronic HBV infection is predominantly the result of the host's immune response, in particular: (1) cytotoxic T cells, natural killer T cells and (2) inflammatory products, including tumour necrosis factor (TNF), free radicals and proteases.

Most primary HBV infections in adults are self-limited (compare this with hepatitis C virus discussed under Future). About 5% of primary infections in adults continue and lead to persistent infection. It is estimated that about 350 million people worldwide are HBV carriers. About 20% of chronic carriers go on to develop the serious complication of cirrhosis. Liver transplantation in HBV will be associated with a recurrence of the infection in more than 80% of patients unless preventative measures are undertaken to reduce the risk for reinfection. They include the use of hepatitis B immune globulin and an antiviral agent such as lamivudine that is also used for treating AIDS (see Box 8.1). Another serious consequence of chronic HBV infection is hepatocellular carcinoma, with carriers being 100 times more likely to develop this than non-carriers. Those who are HBeAg positive carriers are particularly at risk, which is not surprising since this serologic marker is associated with high HBV titres. Despite the HBV genome being sequenced and characterised, it is disappointing that even today there is incomplete understanding how this DNA virus leads to the development of hepatocellular carcinoma.

Papillomavirus

Papillomaviruses are DNA viruses that exhibit species specificity and induce hyperplastic epithelial lesions as a result of infection. Cervical cancer is the second most common cancer in women worldwide, and there is now unequivocal evidence implicating certain types of human papillomavirus (HPV) as the primary cause of cervical cancer and its precursor, cervical intraepithelial neoplasia. HPV is acquired mainly through sexual activity. More than 80 types of HPV have been identified, and about 40 can infect the genital tract.

Cervical cancer begins as a preinvasive neoplastic change in cells (histologically described as CIN—cervical intraepithelial neoplasia). This may regress, remain unchanged or progress to an invasive malignancy. Human papillomavirus has been implicated in the various stages of cervical cancer with different types correlating with the histological findings. Human papillomavirus has more recently been implicated in anal cancer (particularly in those who are HIV positive), penile, laryngeal and oral cancers. Human papillomavirus–related skin cancers can be found in the immunosuppressed.

Human papillomaviruses induce cellular transformation and may also interact with other viruses, oncogenes or carcinogens to bring about neoplastic changes. The virus is not the sole factor in the progression to invasive cervical cancer (Figure 8.5). The E6 and E7 oncoproteins code for proteins essential for viral replication in HPV types 16 and 18. E6 binds to and inactivates *TP53*, thereby disrupting a key cell cycle checkpoint (see Figure 4.13). The E7 oncoprotein binds to and inactivates retinoblastoma gene products, leading to further uncontrolled cell cycle progression.

The various types of human papillomaviruses can be distinguished on the basis of their DNA sequences. Comparisons between histological findings and viral types have enabled classification into three groups—low, intermediate and high risk (Table 8.11). Commercial kits based on PCR assays are now available to identify most of the high-risk HPVs. This test is considered to be very sensitive. However, as mentioned previously, a DNA test indicates only the presence of viral DNA. It cannot predict whether the infection is active or transient. Hence, the suggestion that HPV DNA testing would replace the traditional Papanicolaou smear is incorrect, particularly in young women who are more likely to have a transient HPV infection. However, used in conjunction with the Papanicolaou smear, the papillomavirus DNA screen will be helpful if there is a suspicious result, or this combination might become the preferred method for screening women over 30 years of age. In this way the combined tests will increase the overall sensitivity of screening, and so increase the interval between testing.

Epstein-Barr Virus (EBV)

EBV is a ubiquitous human herpes virus that infects most adults. After being infected, the individual remains an EBV carrier for life. In developing countries, primary EBV infection is asymptomatic and occurs early in life. In developed countries, the primary infection is delayed until adolescence or adult life, when it manifests as infectious mononucleosis. An *in vitro* characteristic of EBV is its ability to immortalise lymphocytes, an application that

Table 8.11 Classification of HPV types and their risk for cervical cancer (Munoz *et al* 2003)

High-risk HPV (probable high risk)	Indeterminate-risk HPV	Low-risk HPV
Types 16, 18, 31, 33, 35, 39, 45, 51, 52, 56, 58, 59, 68, 73, 82, (26, 53, 66)	Types 34, 57, 83	Types 6, 11, 40, 42, 43, 44, 54, 61, 70, 72, 81, CP6108

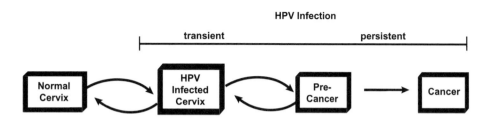

HPV Infection

transient   persistent

Normal Cervix → HPV Infected Cervix → Pre-Cancer → Cancer

Fig. 8.5 Human papillomavirus (HPV) in cervical infection.
Similar to what has been described for colon cancer, the development of cervical cancer is a staged process predominantly in response to infection with HPV. Infection can be transient (and so reversal of the HPV effects on cervical epithelium is possible), but once HPV infection is persistent, it can lead to pre-cancerous lesions that can progress to cervical cancer. Other factors associated with the development of cervical cancer include early sexual activity (<16 years of age), >4 sexual partners, and a history of genital warts. Other independent risk factors that might interact with the HPV include HIV infection, immunosuppression and cigarette smoking.

is very useful in the research laboratory to provide a permanent supply of a particular cell line (or an unlimited source of DNA). In the immortalised lymphocyte, EBV does not replicate but remains as multiple extrachromosomal copies of the circular viral DNA genome in the cell. Latent infection with EBV is accompanied by the expression of a set of viral latent genes. These genes and their respective proteins are well characterised and include (1) six nuclear antigens: EBNAs 1, 2, 3A, 3B, 3C and LP; (2) three latent membrane proteins: LMPs 1, 2A and 2B and (3) EBERs 1, 2—small, non-polyadenylated non-coding RNAs.

Inappropriate expression of EBV latent genes leads to tumour development. Tumours in which EBV is found include Burkitt's lymphoma, post-transplant B cell lymphoma, Hodgkin's disease and nasopharyngeal carcinoma. In each, various latent genes and proteins will be detected. Two important ones (these are always present in the laboratory-transformed lymphoblastoid cell lines) are EBNA2 and LMP1. The former is a transcriptional regulator; hence, inappropriate expression of this gene will have far-reaching effects on cell proliferation. LMP1 in experimental studies behaves as an oncogene. Nasopharyngeal carcinoma is a tumour that is predominantly found in China and South East Asia. The latency genes EBNA1, EBERs and LMP1 are frequently found in this tumour. A recent study used Q-PCR to quantitate EBV-specific DNA (in the form of EBNA3 and LMP1) in the plasma and tumours from nearly 100 patients with nasopharyngeal carcinoma. This study showed that the amount of EBV DNA in the plasma correlated closely with the stage of the cancer with the most DNA found in tumours that had metastasised. The authors (Lin *et al* 2004) suggested that monitoring for EBV DNA in the plasma was an effective way to predict outcome to treatment.

EPIDEMIOLOGY

Conventional typing of pathogens based on their phenotypes (phage susceptibility; biochemistry, antigen profiles; antibiotic resistance; immune response; fimbriation; etc.) is not always successful in epidemiological studies. A changing spectrum of infectious agents, particularly in the immunocompromised host and in hospital outbreaks, has meant that newer epidemiological approaches are required to complement or replace the more traditional methods. Five strategies based on characterisation of pathogen-derived DNA or RNA are useful in these circumstances: (1) nucleic acid hybridisation, (2) plasmid identification, (3) chromosomal DNA banding patterns, (4) PCR and (5) sequencing.

ANTIGENIC VARIATION

Influenza A

The three RNA influenza viruses (A, B, C) are distinguished by their internal group-specific ribonucleoprotein. Only influenza A and B are medically significant since epidemics or pandemics have not occurred with influenza C. Influenza A has the potential to produce pandemics because it infects other species apart from humans (for example, birds, pigs and horses). In contrast is influenza B, which infects only humans, and so its antigenic structure does not become sufficiently different to cause pandemics. Influenza A viral envelope has two important antigenic glycoproteins—haemagglutinin (HA—composed of 15 different types) and neuraminidase (NA—9 different types) (Figure 8.6). Although the envelope antigens are capable of producing many different combinations (as seen in water birds), a smaller number are found in humans (Table 8.12). To date, only a limited number have been implicated in human-to-human spread (H1N1, H2N2, H3N2, H1N2, H5N1, H9N2 and H7N7).

As the influenza A virus passes through its hosts, the most important of which in terms of global spread are the water birds, it undergoes genetic changes that are of two types: (1) **Genetic (antigenic) drift**—new viral strains frequently result from the accumulation of point mutations in the surface glycoproteins (HA and NA). These mutations are occurring continuously over a period of time. The new strains produced are antigenic variants that nevertheless remain related to viruses from preceding epidemics. Thus, they can avoid immune surveillance in those who have developed immunity, and so outbreaks occur. Because some immunity is still present, only epidemics result. (2) **Genetic (antigenic) shift**—this results from an abrupt major change in antigenicity of the HA protein or the HA and NA protein combination. Either this is the result of a novel HA alone, or in combination with a novel NA. This virus is antigenically distinct from previous ones and has not arisen from them by mutation. It is likely that the sources of the new viral gene are water birds that have a large reservoir of different HA and NA genes. They then get into humans via other animals such as pigs or directly from chickens such as occurred with the H5N1 virus discussed below. The new viral subtype is the precursor to a pandemic because populations have never been exposed to it before (Figure 8.7).

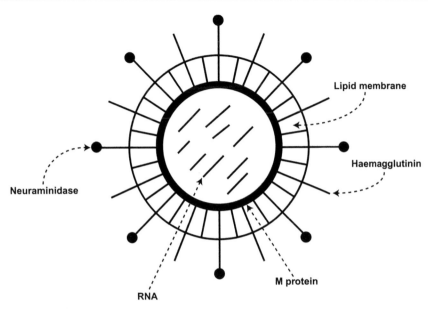

Fig. 8.6 Structure of the influenza virus.
This RNA virus has two key surface glycoproteins: (1) haemagglutinin (HA), which facilitates the entry of virus into host cells through attachment to sialic acid receptors, and (2) neuraminidase (NA), which is involved in the release of progeny virions from infected cells. The HA is the major determinant against which are directed neutralising antibodies and so also the target for influenza vaccines. In contrast, the NA is an important target for antiviral agents.

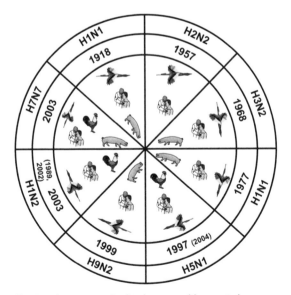

Fig. 8.7 Important animal-to-human and human-to-human influenza outbreaks.
Since the 1918 pandemic, a number of important outbreaks have been recorded (subtypes and dates are given as well as hosts involved). The hosts include water birds, chickens, pigs and humans. A worrying trend is the increasing number of new subtypes in humans, as well as an expanding animal involvement since 1997.

Table 8.12 Species range and types of the influenza A glycoproteins

Animal	Haemagglutinin	Neuraminidase
Water birds[a]	H1–H15	N1–N9
Humans[b]	H1–H3, H5, H7, H9	N1, N2, N7
Pigs	H1, H3	N1, N2
Horses	H3, H7	N7, N8

[a]Wild birds do not usually develop an illness following infection with influenza A. However, domestic birds (chickens, turkeys) can get severe infections, with mortality in some cases (such as the H5N1 subtype) nearing 100%. [b]Subtypes that currently infect humans are H1N1, H1N2 and H3N2.

Avian Influenza (Avian Flu, Bird Flu)

A contemporary worldwide threat to health is the large avian flu epidemic (influenza A—H5N1) presently infecting chickens across many South East Asian countries and mainland China. In 1997, the first cases of human infection from exposure to sick birds or their droppings were reported in Hong Kong. Eighteen patients were admitted to hospital and six died. Fortunately, the culling of more than a million chickens controlled this particular outbreak. Since then, there have been other human cases, with the most recent reported in Vietnam and Thailand.

By late 2004, 44 humans had contracted the avian flu, and 32 of them had died (73% mortality if the infected person was ill enough to be admitted). The "Spanish" (H1N1) flu epidemic in 1918 killed over 20 million people. However, mortality associated with this infection was estimated to be 2.5%, and so considerably less than what is being seen with H5N1 avian flu. It is possible that there have been milder cases of the H5N1 flu that have gone unrecognised, but this worrying comparison as well as the broadening host range for the H5N1 virus (particularly the involvement of the pig) remain important global concerns.

In contrast to the common human influenza virus (H3N2), which is highly contagious but rarely lethal, the avian flu in chickens is a particularly virulent type that can kill rapidly and that causes widespread organ damage. On the other hand, it is not easily transmitted from birds to humans or human to humans. So far only two probable cases of human to human transmission have occurred. However, swapping of genetic material should an individual be co-infected with both might produce a hybrid H5 (avian flu) N2 (human flu) virus with devastating effects. Another way for this mixing to occur (called reassortment) would involve an animal such as a pig being simultaneously infected with both the human influenza virus and the bird flu virus. This mixing might produce a new virulent virus if it had human genes (allowing human-to-human spread) and lethal genes from the bird virus. DNA sequencing of the viral genome from various outbreaks has shown: (1) Viruses from 1997 and 2003 Hong Kong infections have mutated; i.e., their DNA sequences are different. (2) Viruses from the most recent Vietnam and Thailand outbreak show that the virus is resistant to two of the four antiviral agents available. (3) Fortunately, the genes remain of bird origin; i.e., the virus has not acquired genes from human influenza virus, the precursor to a pandemic.

Other outbreaks of bird flu recently reported show subtypes H7N3, H5N2, H7N2, H2N2. These outbreaks have involved poultry in various parts of Canada and the USA. The H7N3 infection resulted in two humans getting a mild flu-like illness following exposure to sick poultry. One person died in New York in late 2003 from an H7N2-type influenza A infection although details of how this occurred are not known. The culling of many birds has controlled the risk of avian flu to humans, but the virus continues to re-emerge in poultry in several Asian countries, with a suggestion that H5N1 might be increasing in pathogenicity and becoming more widespread. Close surveillance by the WHO, the CDC and other bodies continues. Another concern is that clinical features and travel history of individuals who might have SARS (discussed below) and flu virus H5N1 are similar. Hence, vigilance by health professionals and airport monitoring continues, particularly in those with a travel history and respiratory symptoms suggestive of these

infections. In this unpredictable environment, the value of rapid DNA-based diagnostics is obvious.

Vaccines

Characterisation of the different viral types and their mode of development depends on the analysis of the HA and NA amino acids, and more recently, the sequencing of the single-stranded RNA genome of this virus. For this purpose, the WHO has a number of collaborating centres that sequence (by RT-PCR) various viral strains, particularly the HA and NA genes, to monitor the changes as they occur. The ability to identify the functional components of the viral ribonucleoprotein complex and then produce these by rDNA means will provide a potential source of antigens for immunisation programs.

Another contribution from molecular medicine has come in developing a vaccine to the potentially serious avian flu virus. Normally, influenza vaccines are prepared in chicken embryos, but since H5N1 is very lethal to chickens, it cannot be grown in this way. An additional step is necessary using recombinant DNA means to take out the H5N1 genes required for the vaccine and place them in a less toxic laboratory virus that will grow in the chicken embryos.

TAXONOMY

Echinococcus

Worldwide, the tapeworm *Echinococcus granulosus* is an important public health problem leading to the development of cysts in many body organs (hydatid cyst disease). An interesting feature of this parasite is the phenotypic heterogeneity observed in the various isolates, including those obtained from different intermediate hosts such as sheep, cattle, goats and camels. The parasite's variability has made it more difficult to understand its life cycle, to distinguish pathogenic from non-pathogenic types and to determine the best approaches to treating animals that are carriers. However, these issues are now less problematic, with DNA studies providing a more objective way to distinguish the different types, including the potential to distinguish the parasites in their intermediate hosts, making it possible to determine which parasitic strains are more likely to be infective for humans.

The DNA-based taxonomic classification of *Echinococcus* has identified nine different genotypes (G1–G9). Some of these genotypes are found in particular animal intermediate hosts; for example, G1 is the sheep form; G2, Tasmanian sheep; G6, camel; G5, cattle; and so on. The horse strain (G4) is not infective to humans. Another strain (G8) is associated with *E. granulosus* infections in Alaska and was not considered to be highly pathogenic until DNA analysis of a severely affected individual showed that this was not the case (Box 8.2).

Anopheles

Anopheles are insects that are medically important because of their association with malaria, filariasis and arbovirus infections. There are nearly 500 recognised species of the *Anopheles* mosquito. However, only a small number are vectors for human diseases. The importance of being able to identify them is illustrated by a mosquito eradication program in Vietnam that failed because the wrong mosquito was targeted. This error was discovered by DNA typing, and it occurred because the female forms of different strains in that particular geographic region were difficult to distinguish, and so a non-vector form of mosquito was mistaken for one involved in the spread of malaria.

Taxonomic classification of these mosquitoes relies on the traditional morphology, to which has been added chromosome analysis, and DNA markers. Recently, the publication of the complete genome sequence for *Anopheles gambiae* has opened up a new source of information that will allow even better characterisation of the *Anopheles* mosquito. There is also the potential for microarray studies of gene expression to understand better the mosquito's metabolic pathways and so expand the options for therapeutic targets.

Fungi

Fungi are eukaryotes that belong to a separate kingdom from plants and animals. They are ubiquitous in the environment and, in addition, have many commercial applications. In humans, infections caused by fungi (mycoses) can be superficial (e.g., skin, hair, nails) or deep (e.g., pneumonia, systemic infection, septicaemia). At particular risk are the immunocompromised for whom mycoses can be life-threatening. Treatment options in fungal diseases are limited. Until recently, there were few medically important fungi. Today, many new infectious opportunistic fungi are emerging.

The large diversity in fungal morphology, their ecological habitats and the wide spectrum of clinical consequences have complicated diagnostic approaches and our understanding of infections caused by these organisms. An important objective in studying fungi is to deduce evolutionary comparisons and, from them, determine relatedness. Fungal phylogenetics relies on parameters described earlier such as morphology, biochemistry and staining characteristics. In addition, life cycles (e.g., morphological appearances, particularly of the sexual reproductive structures) provide additional sources of information for comparison. However, traditional typing methods are not always helpful or sensitive enough and even life cycles can be uninformative; e.g., the presence of asexual forms of *Coccidioides immitis* has made classification particularly difficult.

An early breakthrough in fungal taxonomy based on DNA typing was the finding that *Pneumocystis carinii*, a very common and serious infection in AIDS as well as in other immunocompromised patients, is a fungus and not a protozoan as considered initially. Culture of human-derived *P. carinii* was always difficult. Its life cycle and metabolic processes were poorly understood. However, based on rRNA and DNA sequence comparisons, this organism is now considered a fungus. Today, there are new opportunities in fungal research as a result of the number of organisms that have had their genomes sequenced. This started with the *S. cerevisiae* yeast genome project that was completed in 1996, and there are many others in progress.

Genomics approaches are now very much part of basic research into fungi, and so microarrays (both genomic and protein) are being applied to study important pathogens. By cataloguing the gene profile of these organisms, it is expected that virulence pathways can be identified and then targeted by specific therapies. In an

Table 8.13 Water supply and nosocomial infections (Merlani and Francioli 2003)

Organism	Traditional sources of infection	New source of infection from DNA analysis
Pseudomonas aeruginosa	In intensive care units, *P. aeruginosa* infections can originate from the endogenous gut flora of the patients.	Genotyping of the organisms showed in some studies that the taps in the patients' rooms were the source of infection in more than a third of cases.
Mycobacteria spp.	Non-TB forms of mycobacteria can cause a wide range of infections including abscesses from the use of catheters or endoscopes, the surgical sites or sites related to dialysis.	Non-sterile ice and the inappropriate uses of tap water (rinsing of instruments) have been shown by DNA typing to be another source of these infections.
Fungi	Aspergillus is a life-threatening infection in immunosuppressed patients and is thought to be caused by fungal particles in the air.	In a three-year prospective study, aspergillus was detected in a hospital's water system. Higher titres were found in bathrooms. Genotyping one isolate of *Aspergillus fumigatus* from an infected patient showed that this fungus was the same as isolates recovered from the shower wall in the patient's room.

epidemiologic sense, an understanding of virulence pathways facilitates the investigation of disease outbreaks, particularly nosocomial ones, since it allows a mix of organisms that are likely to be present to be accurately typed and, from this, determine which strains or subtypes are responsible for the infection.

NOSOCOMIAL INFECTIONS

Nosocomial (hospital acquired) infections include bloodstream infections, pneumonia, ventilator-associated pneumonia, intra-abdominal sepsis, wound infections and infections in immunocompromised patients. Pathogens causing these infections are often resistant to many antimicrobial agents. Suboptimal treatment within the hospital environment only exacerbates the problem of drug resistance. In one study of intensive care patients, the effect of bloodstream (bacteraemia and septicaemia) nosocomial infections increased the stay in intensive care units as well as the associated financial costs.

It is important in nosocomial infections to have methods to type pathogens to enable their source(s) and mode of spread to be identified. A composite of DNA polymorphisms will provide a unique DNA pattern for an individual person. DNA fingerprinting, as it is popularly known, is now well established in forensic practice (see Chapter 9). The DNA/RNA profile of infectious agents is similarly being exploited for diagnostic or epidemiological purposes. DNA markers can be selected from a conserved region of the microbial genome if less discrimination is required. On the other hand, choosing a DNA probe from a highly variable (i.e., polymorphic) region of the genome will allow discrimination between closely related organisms. The ultimate in DNA fingerprinting is now available in infectious diseases, and this is the use of DNA sequencing including the option to sequence the pathogen's entire genome.

Legionella

Legionella spp. are ubiquitous bacteria present in domestic and industrial water systems, tanks and other sources of pooled or collected water. About 45 different species of the genus *Legionella* are described. It is now considered to be a relatively common pathogen involving both community acquired as well as nosocomial legionellosis. Infection with this organism is easily overlooked because traditional diagnostic tests are not sensitive enough, particularly with species other than the common *L. pneumophila*. Although air conditioners and cooling towers are recognised sources of this organism, more recently the finding of *Legionella* spp. in drinking water provides another explanation for sporadic or larger outbreaks.

Legionella spp. are found in the water supply of many hospitals. In one study, 17 of 20 hospitals tested were positive, and genotyping showed that each hospital had its own different *Legionella* spp. It is considered that water-borne *Legionella* causes infection when contaminated aerosols are inhaled or there is aspiration of contaminated water. The link between hospital water supplies and various other types of nosocomial infections is slowly growing as DNA typing provides the evidence for similarity between the infecting organisms and those found in the water supply (Table 8.13).

A proportion of cases with Legionnaire's disease is considered to be nosocomial in origin. Both sporadic cases and outbreaks of this respiratory disorder have been reported. The organism (e.g., *L. pneumophila*) is difficult to grow, and so it is frequently necessary to obtain bronchial aspirates or lung biopsies for direct detection. Serological testing is the mainstay of diagnosis although a number of DNA-derived tests are now being developed. These tests can utilise DNA hybridisation with probes that are often rRNA-specific. Another useful rDNA approach to fingerprinting bacterial genomes such as *Legionella* spp. is pulsed field gel electrophoresis (PFGE).

The technique allows large segments of DNA to be separated by using restriction enzymes such as *Not*I and *Sfi*I that digest DNA very infrequently. The upper limit for resolution in conventional DNA gel electrophoresis is approximately 30 kb, whereas for pulsed field gel electrophoresis it is approaching 10 Mb (10×10^3 kb). Pulsed field gel electrophoresis has been particularly attractive in microbiological work because the genomes of infectious agents are relatively small. For example, total DNA from *Legionella* spp. can be cleaved into a limited number of fragments (5–10 with *Not*I, depending on the strain), and these fragments or fingerprints can be directly visualised from an ethidium bromide stained gel, thereby avoiding the requirement for a DNA probe and a hybridisation step. However, the technique is very demanding. Not surprisingly, PCR-based approaches are increasingly becoming the preferred method for DNA fingerprinting. Typing of *Legionella* spp. for epidemiological purposes follows similar strategies.

Methicillin Resistant Staphylococcus Aureus (MRSA)

The first isolate of methicillin resistant *S. aureus* was reported in 1961, one year after methicillin became available. This was followed by the emergence of vancomycin-resistant MRSA in 1996. MRSA developed when *S. aureus* acquired a large segment of DNA (called SCC*mec*) that integrated into the *S. aureus* chromosome. This segment of DNA looks like an antibiotic "resistance cassette" because it contains genes for β lactam resistance and genes for resistance against non–β-lactam antibiotics. Vancomycin resistance is thought to have occurred with the emergence of a strain of *S. aureus* that had a thickened cell wall, thereby preventing the vancomycin from getting into the pathogen.

Antibiotic-resistant *S. aureus* is an important and frequent cause of nosocomial infections in hospitals. When this occurs, infection control protocols need to identify the source(s) of the methicillin-resistant *S. aureus* and then determine whether there has been a breakdown in infection control practices so that more effective preventative measures can be implemented in future. For the above to occur, the different bacterial subtypes need to be distinguished. Phage typing has been the traditional tool in epidemiological studies involving *S. aureus*. However, this method has poor reproducibility, it is not able to type all isolates and it is impractical for most laboratories since it requires a large number of phage stocks to be maintained. Phage typing is now being replaced by DNA-based tests. Like the example of *Legionella*, MRSA typing can involve a wide range of techniques from PFGE to PCR. Although these techniques are often more discriminatory than the traditional approaches, they now need to be standardised to allow better inter-laboratory comparisons to be made.

EMERGING INFECTIONS

Emerging infections describe infections that are newly identified or whose incidence in humans has significantly increased in the past 20 years. Many factors contribute to the emerging infections, including (1) changes in human behaviour, (2) globalisation (i.e., increased travel, tourism), (3) technologic advances, economic development and changes in the environment and (4) lapses in public health measures including those resulting from poverty and war. Very few of the emerging infections represent novel pathogens. Most are the result of a change in the epidemiology or virulence of a pathogen, or secondary to microbial adaptation. Important sources of emerging infections are animals.

ZOONOSES

In the past, the emergence of infectious agents reflected changed patterns of human movements that disrupted traditional geographical boundaries. For example, yellow fever is thought to have emerged in the New World as a result of the African slave trade, which brought the mosquito *Aedes aegypti* in ships' water containers. More recently, *Aedes albopictus*, a potential vector for dengue virus, has become established in the USA following its conveyance from South East Asia in old car tyres. With this, the threat of dengue in the North American continent has become real. Most emergent viruses are zoonotic; i.e., they are acquired from animals, which are reservoirs for infection. Thus, completely new strains are less likely than the appearance of a virus following a change in animal reservoirs. This is particularly relevant to the modern world where the consequences of easy migration, deforestation, agricultural practices, dam building and urbanisation are making, and will continue to have, a major impact on the ecology of animals. At the same time, humans have increasingly populated rural areas and are pursuing more outdoor recreational activities. There is also a growing trend for keeping exotic animals as household pets. Changes in global climate may also contribute directly, through their effects on vegetation, insect and rodent populations.

Table 8.14 lists a number of zoonoses that have become established as new infectious diseases or are emerging as problems for the future. Some of them are newly acquired in the West, whereas others remain endemic to specific countries. In many cases, molecular

Table 8.14 Examples of zoonoses and new human infections

Pathogen	Clinical problems	Epidemiology	Emergence	Role of DNA technology
West Nile Virus RNA Flavivirus related to Yellow fever; Japanese encephalitis	Febrile illness complicated by meningoencephalitis, weakness and paralysis.	Transmitted by mosquitoes; blood or organ donation; pregnancy, lactation; infected needles or laboratory specimens. The virus is maintained by a bird-mosquito-bird cycle.	Isolated in 1937 from Uganda and found in many parts of the world. Appeared in the USA in 1999 and has rapidly spread across North America.	NAT is used to screen blood donors and is potentially useful for immunosuppressed individuals who cannot mount an antibody response for serologic testing.
Monkeypox virus DNA virus related to smallpox	Self-limited febrile illness with vesiculo-pustular eruptions. Can be confused with more serious illnesses.	Primary animal reservoir is the rat.	Recognised in 1958 and remained localised to Africa. In 2003, outbreak in Midwest USA shown to be monkeypox. Traced back to rats imported from Africa to which native prairie dogs were exposed and became infected and then infected humans. Appears to be contained.	DNA characterisation helped in identifying this virus as monkeypox.
Ebola RNA filovirus	Haemorrhagic fever in humans (mortality 20–100%)	Example of increased human-to-animal contact in tropical forest. Animal host remains unknown (bats are suspected).	First isolated in 1976 from Sudan and Zaire. Since then, sporadic outbreaks have occurred. Remains isolated to Africa.	PCR-based assays for rapid and sensitive diagnostic tests described.
Lassa fever RNA virus	Haemorrhagic fever with 20% having severe multisystem disease.	Rats infect humans (by contact or if eaten). Virus excreted in human urine/semen for months after infection.	Endemic in West Africa since 1950s.	Nil
Hantavirus RNA virus	Haemorrhagic fever with renal syndrome and pulmonary syndrome (severe disorders).	Infection occurs through exposure to rodent excreta.	Isolated in 1979 in Korea. Now established within the Eurasian continent. Outbreaks in the USA in 1993 and 1998 thought to be due to climatic changes, increasing vegetation and so an increase in the rodent population. Now established throughout North Central and South America.	The types of Hantaviruses in different locations show a close relationship to their rat host. But now shown by DNA analysis that hantaviruses can jump from one rat species to another. This will make it more difficult to know from studying the rat populations if a particular type is likely to be virulent.
Lyme disease Bacterial spirochaete *Borrelia burgdorferi*	Early non-specific malaise can be complicated by arthritis, neurologic and cardiac problems.	Tick (*Ixodes* spp.) transmitted disease. Mice, rodents and birds are the intermediate hosts.	First recognised in the USA in 1957; since then reported in many countries.	DNA characterisation useful for epidemiologic purposes and to explain different clinical features in various countries.
HIV RNA lentivirus	Serious acquired immunodeficiency disorder.	Cross-species transmission from non-human primates followed by human-to-human spread.	Evidence for the link between non-human primate and human disease includes (1) similar viral genomes, (2) prevalence in the natural host, (3) geographic co-location.	DNA technology has been helpful in all phases of this particular disease from diagnosis to prognosis (in terms of viral load determination and detection of viral resistance).

Table 8.14 *Continued*

Pathogen	Clinical problems	Epidemiology	Emergence	Role of DNA technology
BSE, CJD, vCJD Bovine spongiform encephalitis, Creutzfeldt-Jakob disease and variant CJD	Fatal spongiform encephalitis with a long incubation peiod.	BSE detected in mid-1980s in cattle, resulting from transmission of prions in meat and bone meat products used as animal feed in the UK.	vCJD, the human equivalent of BSE, identified from 1996 due to the ingestion of meat from infected cattle. Theoretical risk that it can be transmitted through blood, although not proven.	Infectious agent does not have nucleic acid and so not detectable by conventional means. DNA analysis has shown mutations in the *PRNP* gene (see also Figure 8.8 and Chapter 4).
PERV Pig endogenous retrovirus	Xenotransplants (especially from pigs) considered as a source of organs or cells because there are insufficient human donors.	Concern that endogenous pig retroviruses (PERV) might jump the species barrier particularly if immunosuppression is needed for the xenotransplant.	Theoretical risk based on *in vitro* evidence that PERV can infect human cell lines. Nevertheless, a concern holding back xenotransplantation.	DNA testing can detect PERV but does not necessarily indicate infectivity.

medicine offers little to current management including diagnosis, although much knowledge about these infections has come from analysis of the organisms' DNA or RNA. However, the potential for spread of any disease is high because of international travel or the mass dislocation of large populations following civil unrest. There is also an increasing awareness that a number of pathogens could be used for bioterrorism. Some of the zoonoses associated with a viral haemorrhagic clinical picture can be confused with other clinical infections including malaria, leptospirosis and *Neisseria meningitidis,* and in these potentially fatal conditions, a rapid screening test is essential. In terms of bioterrorism and the differential diagnosis of haemorrhagic fevers, PCR (NAT) based assays are presently the only options with the potential to allow rapid and sensitive diagnostic tests to be developed.

SARS

SARS (severe acute respiratory syndrome) attracted a lot of publicity and provoked fear in 2003. It has been described by some as the first pandemic of the twenty-first century since it involved over 8000 patients and caused more than 800 deaths in 30 countries on five continents (Peiris *et al* 2003). The economic impact of this infection in Hong Kong was estimated to be nearly US$6 billion. SARS also represents a contemporary paradigm that illustrates the value of molecular-based DNA approaches in dealing with a serious emerging infection. The journal *Science* acknowledged the impressive work undertaken to control SARS, citing it as the outstanding achievement of the year: "SARS: a pandemic prevented" (see Table 1.1). A chronology of events in the SARS story is given in Table 8.15.

Traditional approaches such as viral culture, electron microscopy and serology played a key role in character-

ising the SARS virus now shown to be a coronavirus (CoV). Nevertheless, the molecular approach enabled the following key information to be obtained in a very short time frame:

- Molecular typing of this virus from an outbreak in Taiwan confirmed that the virus was identical in its DNA sequence to a virus isolated from a Hong Kong outbreak; i.e., human-to-human spread had occurred through an individual who had been in Hong Kong. Similar comparisons were possible, linking various outbreaks through molecular epidemiology.
- Rapid whole genome sequencing of the viral RNA provided a firm base for the development of PCR-based diagnostic assays, a critical requirement for acute infections such as SARS-CoV.
- In searching for animal reservoirs, RT-PCR–based techniques were used. They enabled the SARS-CoV to be detected, as well as identified genetic differences between the human and animal virus.
- Future understanding of the virus, how it evolved, what strains were pathogenic and its mode of spread to populations could be addressed through DNA approaches. For this, the US National Institute of Allergy and Infectious Diseases released a genome chip containing the full 29 700 base pairs of the SARS viral genome.

The SARS outbreak ended just as quickly as it started. Only the occasional cases were reported in early 2004, and none was reported after the end of April that year. However, there remain many unanswered questions, including the inconsistent human-to-human transmission with documented cases of one infected passenger transmitting to a number of others on a particular flight, while on another flight no transmission occurred despite a number of passengers being ill with this virus. These inconsistencies were explained by proposing that some

Table 8.15 Chronology of events in the SARS story

Date	Event
November 2002	Atypical pneumonia cases in Guangdong Province, China.
January 2003	Outbreaks of pneumonia in same province.
February 11, 2003	WHO notified about outbreaks of pneumonia in Guangdong Province. At this stage concern that this represents influenza A (H5N1) cases identified in Hong Kong.
February 21, 2003	Resident from Guangdong Province checks into Hong Kong hotel. He becomes ill and is hospitalised. Subsequently shown that he has infected at least 17 other visitors/guests at the hotel (they will later travel to Vietnam, Singapore and Toronto to spread SARS locally).
February–March 2003	Contacts at Hong Kong hotel become ill, and some infect health care workers. Hotel contact dies in Toronto.
March 12–14, 2003	WHO issues global alert; clusters of atypical pneumonia are reported in Singapore and Toronto (later, it will become evident that these clusters are linked to the Hong Kong hotel).
March 15, 2003	The atypical pneumonia is called SARS; more than 150 new cases are reported; WHO issues a travel warning.
Mid-late March 2003	Novel coronavirus isolated from SARS patients.
May 12, 2003	The complete sequence of the coronavirus is announced; virus now called SARS-CoV.
June 2003	Virus related to SARS-CoV isolated from animals.
July 2003	Official end of the SARS outbreak in humans.
August 2003	The identification of SARS-CoV as the cause of an outbreak of mild respiratory infection in a Canadian nursing home is recanted. Error based on laboratory misdiagnosis.
September 2003	Laboratory-acquired SARS-CoV infection (mild) reported in Singapore.
October 2003	Concern expressed that the definitive animal reservoir for SARS-CoV remains unknown and so future outbreaks are possible. Virus detected in civet cats and a raccoon dog in a Chinese market. These are a likely source for the human infections, but the source for these animals remains unknown.
December 19, 2003	Breakthrough of the Year by *Science* magazine: SARS: A pandemic prevented.
January–April 2004	In January, two suspected cases of SARS-CoV confirmed in Guangdong. The last cluster of nine cases (one fatal) reported in China in April. No further cases of SARS described after April 29, 2004.

of those infected were "super-spreaders." Another intriguing observation with SARS is the relatively large numbers of health workers that have been infected; i.e., SARS transmission seems to be more likely within a hospital setting.

Although a number of emerging viruses have been described recently, SARS-CoV is particularly worrying because it is one of the few that has shown human-to-human transmission, although fortunately this seems to be relatively inefficient. For the future, a key issue will be early and prompt recognition of new cases to enable traditional public health measures (the reason for effective containment on this occasion, although the seasonal changes may also have contributed) to be implemented even faster. The WHO has also flagged the importance of laboratory containment when dealing with this virus. This became an issue when two of the nine persons infected in China in 2004 worked in a reference laboratory conducting research into SARS-CoV. A similar scenario was reported earlier from Singapore. The latter case was clearly documented on DNA sequencing of the virus to be due to a contaminated laboratory culture that the scientist had been working with three days before showing signs of the infection.

An issue that is still outstanding with SARS-CoV is how this virus is normally spread. Communicable respiratory infections traditionally infect others because of droplet spread or contact with contaminated surfaces. Some preliminary data are now being reported that SARS-CoV involves airborne transmission. If proven, this will be an interesting public health challenge requiring a better understanding of this mode of transmission, including its relationship to modern housing, travel and working environments. Detecting viral particles involved in airborne spread will not be easy, and for this PCR will become an indispensable tool for the public health professional.

BIOTERROR

After the 2001 anthrax bioterror cases in the USA, the world has been placed on alert. A new form of modern terrorism had been launched. In Chapter 9 reference is made to the US government's move to develop "microbial forensics" as a way in which to deal with bioterrorism or biocrimes involving the use of human or animal pathogens. DNA technology will be a particularly valuable asset in microbial forensics, allowing the rapid identification of various pathogens, and as part of the detective work, to find the source of the infection. As discussed further in Chapters 1 and 9, it is fortuitous that a component of the Human Genome Project was the sequencing of genomes from a number of pathogens. This information will be needed in combating bioterrorism.

Biological warfare or bioterrorism is possible through many ways, including the contamination of food or water

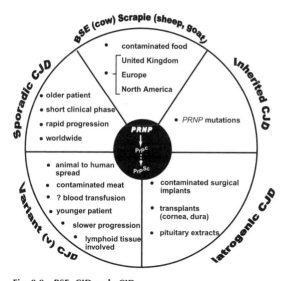

Fig. 8.8 BSE, CJD and vCJD.
BSE (mad cow disease) is caused by a prion, i.e., abnormal form of a protein (PrP) that can spread from one organism to another and interfere with the shape of the normal protein. The gene producing the PrP protein is called *PRNP*, and the two forms are PrP^c (normal protein; the c = cellular) and PrP^{Sc} (abnormal protein; the Sc = scrapie). In the UK, BSE resulted from the feeding of contaminated food to cattle. A similar disease is found in sheep and goats (called scrapie). In humans, the same disease is called CJD (Creutzfeldt-Jakob disease). There are a number of forms of CJD. The sporadic type is rare and occurs in the elderly. The inherited form is also rare, and it can be shown to have mutations in the *PRNP* gene. Iatrogenic CJD occurs from contaminated instruments or human products. Because the prion is not a traditional pathogen, sterilisation methods are not effective. Variant CJD (vCJD) is the result of direct animal-to-human spread (through contaminated beef products). vCJD has not been transmitted by blood, but because of the long incubation period possible (perhaps 30 or more years), this cannot be excluded, and some disturbing experimental data suggest that this is possible.

Box 8.3 Bioterrorism using anthrax (Griffith *et al* 2003).

Anthrax is caused by *B. anthracis*, a gram positive spore-forming organism. It is usually acquired by humans through exposure to infected animal products or contaminated dust. The major forms of anthrax are cutaneous (95% of cases, with a mortality of about 20%) and pulmonary (100% mortality if not treated before symptoms develop). In October and November following the September 11, 2001, World Trade Centre disaster, 7 cutaneous and 11 inhalational cases of anthrax occurred in Florida, New York City, New Jersey, Washington DC and Connecticut. They were examples of bioterrorism since all bacteria came from the one source based on DNA typing, which showed an identical genotype. The likely route of infection was intentionally contaminated mail. Apart from two cases, all others were mail or government workers, or employees at news media outlets. The two exceptions were a 61-year-old woman in New York City and a 94-year-old woman in Connecticut. Both died after developing inhalational anthrax. It was possible to detect a likely source of infection in the case of the Connecticut victim. The postal processing and distribution centre (PDCs) for the letter-related cases had been identified through the finding of contaminating bacteria. However, for this victim, no samples from her home or personal items tested positive for *B. anthracis*. It was known that she had received mail (including junk or bulk mail) from one of these PDCs. It was also subsequently shown that her local PDC (Southern Connecticut) had a heavily contaminated sorting machine. From this, it was assumed that *B. anthracis* identical to those involved in the other outbreaks had infected her through a letter directly contaminated in a PDC implicated in the earlier infections, or secondarily contaminated in the Southern Connecticut PDC.

---

supplies, or via infected animals including insects. A more serious form would involve the use of aerosols containing the pathogens because now there is the risk for infecting a very large number of people. The length of the incubation period is also a consideration since it would, to some degree, influence the number infected before containment or treatment was initiated. Biosensors are being developed to assist early detection. In some infections, DNA-based chip technology is being tried as a way in which microbes can be detected directly through their DNA sequence. This will ensure rapid detection and typing of pathogens.

Two pathogens of particular relevance to bioterror are anthrax and smallpox. DNA-based genotyping has already proven its value in the case of the 2001 US anthrax cases (Box 8.3). Promising reports now suggest that a pathogen such as anthrax could be detected within an hour by real-time PCR. Smallpox is another serious infection since routine vaccination was stopped in the USA in the early 1970s, leaving many unvaccinated and so vulnerable targets in this and other countries. In Asia, case fatality rates of around 30% were observed during epidemics, and there are no known treatments for smallpox. As part of its biologic warfare program, the former Soviet Union produced smallpox, anthrax and other pathogens, and it remains a concern that some could fall into the hands of potential terrorists.

FUTURE

Over the past century, there has been a steady decline in infectious diseases in affluent countries, and these diseases have become relatively minor problems when compared to the increasing role played by chronic and neoplastic diseases. During the 1980s this situation started to change with the onset of AIDS and its related infections such as tuberculosis. Other emerging infections have appeared and are important concerns for the future.

The hepatitis C virus is an example of a pathogen that is starting to impact on the health and well-being of many communities. Although it does not have the same emotive force as HIV, it is an important modern scourge that will need to have its spread controlled more effectively, and those infected will need access to antiviral agents (Box 8.4).

A key to controlling the spread of modern pathogens will be effective vaccines that are developed on the basis of the organism's genome sequence (some call this reverse vaccinology). This is illustrated by the group B meningococcus for which a vaccine could not be developed by traditional means. However, soon after the genome was sequenced, it was possible to identify many potential vaccine candidates that are now being tested in clinical trials. The genomic era has opened up new opportunities for the control of infectious diseases.

The SARS example also illustrates the value of genomics in public health medicine. In particular, the very rapid sequencing of the genome of what appeared to be a coronavirus-like pathogen enabled a definitive diagnosis to be made and potential vaccine targets to be identified, some of which are already in the animal phase of testing.

Developments in technology are addressing many of the potential problems related to DNA testing in microbiology. A challenge for the future will be a better understanding of what the finding of a DNA (or RNA) sequence means in terms of infectivity and pathogenesis. The traditional approach has many years of experience to help in interpreting a result such as a positive culture. The same experience is not available with DNA testing, but it will be necessary to do the hard work and determine the significance of various DNA tests. An example would be persistence of mycobacterial growth in culture despite treatment. This would suggest a compliance issue or the development of drug resistance. The finding of a persistently positive DNA test in this circumstance would be more difficult to interpret. The validation of positive DNA test results will be necessary for a wide range of infections; for example, the finding of CMV DNA by PCR in a patient's serum could mean active disease or latent infection. How these issues are resolved will be impor-

tant for the integration of molecular diagnostics into routine microbiological laboratory practice.

No doubt the potential for genomics will allow novel ways to detect infection with pathogens. One early model described involves the use of microarrays to detect gene profiles in the patient rather than attempts at detecting pathogen-specific nucleic acids. Using this approach, it was claimed that peripheral blood mononuclear cells exposed to *M. tuberculosis*, a protozoan (*Leishmania donovani*) and a worm (*Brugia malay*) demonstrated discrete expression profiles particular to the infectious agent.

Box 8.4 Hepatitis C virus (HCV).

Thirteen years after the term "non-A, non-B" hepatitis was first used, the HCV virus was found. A milestone in molecular virology came when HCV was cloned in 1988, years before it was visualised, cultured or even characterised. Based on its amino acid sequence and genomic organisation, HCV was classed in a separate genus in the viral family *Flaviviridae*. HCV is an RNA virus and has six genotypes (1–6). The main source of infection today is intravenous drug use. Transfusion-related HCV is rarely seen because of the effective screening methods in place. Unlike HBV, mother-to-infant and sexual transmission is rare. Nearly 170 million people worldwide are infected with HCV, which is now the most common reason for liver transplantation in the USA. Unlike HBV, which infrequently leads to chronic hepatitis, about 70% of those infected with HCV develop this complication. Like HBV, the hepatic damage caused by HCV is thought to be secondary to an immune response rather than a direct viral effect. In addition, age, male sex, alcohol consumption, coexistent HIV infection, low $CD4^+$ count and metabolic disorders such as diabetes predispose to the development of liver cirrhosis and subsequently hepatocellular cancer. Like HIV, excellent serologic or DNA-based assays are available for accurate diagnosis. Quantitative RNA assays and genotyping enable predictions to be made of how an individual will respond to antiviral therapy. Treatments using interferon alpha and ribavirin are effective, particularly with HCV genotypes 2 and 3 (88% response rates compared to 48% for the other four genotypes). However, because of inadequate detection (or the money to pay for the above drugs), there is considerable concern that what is being seen at present is only the tip of the iceberg since the virus has a long incubation period, and many who are carriers remain asymptomatic until complications develop.

FURTHER READING

Diagnostics

Domig KJ, Mayer HK, Kneifel W. Methods used for the isolation, enumeration, characterisation and identification of *Enterococcus* spp. 2. Pheno- and genotypic criteria. International Journal of Food Microbiology 2003; 88: 165–188 (*provides overview of the different phenotypic and genotypic assays used in various microbiological applications*).

Allain J-P. Genomic screening for blood-borne viruses in transfusion settings. Clinical Laboratory Haematology 2000; 22: 1–10 (*comprehensive overview of NAT technology as well as effects it has had on HIV, HCV and HBV*).

Clavel F, Hance AJ. HIV drug resistance. New England Journal of Medicine 2004; 350:1023–1035 (*comprehensive review of this topic*).

Forster SM. Diagnosing HIV infection. Clinical Medicine 2003; 3:203–205 (*succinct summary of diagnostic tests for HIV AIDS*).

Fraser CM. A genomics based approach to biodefence preparedness. Nature Reviews Genetics 2004; 5:23–33 (*lists all the pathogens that have been sequenced*).

Gilbert GL. Molecular diagnostics in infectious diseases and public health microbiology: cottage industry to postgenomics. Trends in Molecular Medicine 2002; 8:280–287 (*in-depth analysis of infectious disease diagnostics*).

Goodnough LT. Risks of blood transfusion. Critical Care Medicine 2003; 31:S678–S686 (*excellent overview of all risks associated with blood transfusion*).

Louie M, Louie L, Simor AE. The role of DNA amplification technology in the diagnosis of infectious diseases. Canadian Medical Association Journal 2000; 163:301–309 (*useful overview of what DNA diagnostics can achieve*).

Pealer LN et al. Transmission of West Nile virus through blood transfusion in the United States in 2002. New England Journal of Medicine 2003; 349:1236–1245 (*describes the West Nile virus and transmission through blood*).

Strohl WA, Rouse H, Fisher BD (Harvey RA, Champe PA series editors). Lippincott's Illustrated Reviews: Microbiology. 2001. Lippincott Williams & Wilkins, Philadelphia (*well-presented overview of many topics in infectious disease*).

http://www.aidsmap.com (*UK-based charity web site providing a vast array of data on HIV-AIDS*).

http://www.who.int/whr (*World Health Organization 2004 Report on AIDS*).

Pathogenesis

Buer J, Balling R. Mice, microbes and models of infection. Nature Reviews Genetics 2003; 4:195–205 (*summarises the utility of mouse models in infectious diseases*).

Carrington M, O'Brien SJ. The influence of HLA genotype on AIDS. Annual Review of Medicine 2003; 54:535–551 (*summary of various HLA associations with infections*).

Chinen J, Shearer WT. Molecular virology and immunology of HIV infection. Journal of Allergy and Clinical Immunology 2002; 110:189–198 (*provides relevant overview of HIV and AIDS*).

de Viedma DG. Rapid detection of resistance to *Mycobacterium tuberculosis*: a review discussing molecular approaches Clinical Microbiology and Infection 2003; 9:349–359. (*useful

summary of resistance genes with a comprehensive overview of the molecular diagnostics technologies*).

Ganem D, Prince AM. Hepatitis B virus infection—natural history and clinical consequences. New England Journal of Medicine 2004; 350:1118–1129 (*useful review of the hepatitis B virus*).

Lage H. ABC transporters: implications on drug resistance from microorganisms to human cancers. International Journal of Antimicrobial Agents 2003; 22:188–199 (*reviews the multidrug transporters*).

Le Bras J, Durand R. The mechanisms of resistance to antimalarial drugs in *Plasmodium falciparum*. Fundamental & Clinical Pharmacology 2003; 17:147–153 (*comprehensive report on molecular basis for drug resistance*).

Lin J-C et al. Quantification of plasma Epstein Barr virus DNA in patients with advanced nasopharyngeal carcinoma. New England Journal of Medicine 2004; 350:2461–2470 (*shows how DNA quantification can help to plan treatment for this tumour*).

Low DE. The era of antimicrobial resistance—implications for the clinical laboratory. Clinical Microbiology and Infection 2002; 8 (suppl 3):9–20 (*describes antibiotic resistance in Streptococcus pneumoniae*).

Mansky LM. HIV mutagenesis and the evolution of antiretroviral drug resistance. Drug Resistance Updates 2002; 5:219–223 (*summarises how drug resistance evolves in HIV*).

Miller LH, Baruch DI, Marsh K, Doumbo OK. The pathogenic basis of malaria. Nature 2002; 415:673–679 (*useful review*).

Moss SF, Sood S. *Helicobacter pylori*. Current Opinion in Infectious Diseases 2003; 16:445–451 (*well-presented overview of this pathogen*).

Munoz N et al. Epidemiologic classification of human papillomavirus types associated with cervical cancer. New England Journal of Medicine 2003; 348:518–527 (*lists the various HPV types and their risk for cervical cancer*).

Nizet V. Streptococcal β hemolysins: genetics and role in disease pathogenesis. Trends in Microbiology 2002; 10:575–580 (*comprehensive description of genes involved in the β haemolysin effect*).

O'Brien SJ, Nelson GW. Human genes that limit AIDS. Nature Genetics 2004; 36:565–574 (*succinct summary of genetic variants and their possible effects on HIV infection*).

Segal S, Hill AVS. Genetic susceptibility to infectious disease. Trends in Microbiology 2003; 11:445–448 (*succinct overview of the association and linkage study to identify genetic susceptibility factors in infectious diseases*).

Young LS, Murray PG. Epstein Barr virus and oncogenesis: from latent genes to tumours. Oncogene 2003; 22:5108–5121 (*useful summary of EBV and pathogenesis of cancer*).

Waggoner SE. Cervical cancer. Lancet 2003; 361:2217–2225 (*summary of cervical cancer from epidemiology to treatment*).

http://www.who.int/emc/amrpdfs/WHO_Global_Strategy_English.pdf (*The WHO's global strategy for containment of antimicrobial resistance*).

Epidemiology

Hien TT, de Jong M, Farrar J. Avian influenza—a challenge to global health care structures. New England Journal of Medicine 2004; 351:2363–2365 (*brief summary; for regular updates see the CDC web site: http://www.cdc.gov/flu/avian/index.htm*).

Hiramatsu K, Cui L, Kuroda M, Ito T. The emergence and evolution of methicillin resistant *Staphylococcus aureus*. Trends in Microbiology 2001; 9:486–493 (*comprehensive description of the genetic factors for MRSA*).

Krzywinski J, Besansky NJ. Molecular systematics of Anopheles: from subgenera to subpopulations. Annual Review of Entomology 2003; 48:111–139 (*comprehensive review of taxonomy for this important mosquito*).

Lorentz MC. Genomic approaches to fungal pathogenicity. Current Opinion in Microbiology 2002; 5:372–378 (*describes genomic approaches to studying fungi*).

Merlani GM, Francioli P. Established and emerging waterborne nosocomial infections. Current Opinion in Infectious Diseases 2003; 16:343–347 (*interesting summary of nosocomial infections associated with drinking water*).

McManus D *et al*. Molecular genetic characterization of an unusually severe case of hydatid disease in Alaska caused by the cervid strain of *Echinococcus granulosus*. American Journal of Tropical Medicine and Hygiene 2002; 67:296–298 (*brief report on how DNA typing can be used to understand disease heterogeneity*).

Nicholson KG, Wood JM, Zambon M. Influenza. Lancet 2003; 362:1733–1745 (*comprehensive overview of influenza*).

Roig J, Sabria M, Pedro-Botet ML. *Legionella* spp.: community acquired and nosocomial infections. Current Opinion in Infectious Diseases 2003; 16:145–151 (*useful summary of Legionella*).

Webby RJ, Webster RG. Are we ready for pandemic influenza? Science 2003; 302:1519–1522 (*specifically looks at relevant issues and pandemics*).

Emerging Infections

Drosten C, Kummerer BM, Schmitz H, Gunther S. Molecular diagnostics of viral hemorrhagic fevers. Antiviral Research 2003; 57:61–87 (*overview of the type of PCR tests that might be used for viral hemorrhagic fevers*).

Griffith KS *et al*. Bioterrorism-related inhalational anthrax in an elderly woman, Connecticut, 2001. Emerging Infectious Diseases 2003; 9:681–688 (*provides some insight into the investigative approaches that would be required in the case of bioterror*).

Irani DN, Johnson RT. Diagnosis and prevention of bovine spongiform encephalopathy and variant Creutzfeldt Jakob disease. Annual Review Medicine 2003; 54:305–319 (*comprehensive review of CJD*).

Lim PL *et al*. Laboratory acquired severe acute respiratory syndrome. New England Journal of Medicine 2004; 350:1740–1745 (*brief account of how a case of SARS resulted from a laboratory source of infection*).

Ludwig B, Kraus FB, Allwinn R, Doerr HW, Preiser W. Viral zoonoses—a threat under control? Intervirology 2003; 46:71–78 (*excellent summary of zoonoses well worth reading*).

Makino S, Cheun H. Application of the real time PCR for the detection of airborne microbial pathogens in reference to the anthrax spores. Journal of Microbiological Methods 2003; 53:141–147 (*illustrates how real-time PCR could be used for rapid DNA detection of pathogens*).

Pealer LN *et al*. Transmission of West Nile virus through blood transfusion in the United States in 2002. New England Journal of Medicine 2003; 349:1236–1245 (*interesting discussion on the implications of WNV and blood transfusion*).

Peiris JSM, Yuen KY, Osterhaus ADME, Stohr K. The severe acute respiratory syndrome. New England Journal of Medicine 2003; 349:2431–3441 (*excellent overview of the SARS story*).

Petersen LR, Marfin AA, Gubler DJ. West Nile Virus. JAMA 2003;290:524–528 (*comprehensive clinical, epidemiologic and historic overview of WNV*).

Reed KD *et al*. The detection of monkeypox in humans in the western hemisphere. New England Journal of Medicine 2004; 350:342–350 (*useful overview of how this infection was quickly identified in the USA*).

Roy CJ, Milton DK. Airborne transmission of communicable infection—the elusive pathway. New England Journal of Medicine 2004; 350:1710–1712 (*brief editorial on SARS and a possible mode for its spread*).

Future

Lin B, Vahey MT, Thach D, Stenger DA, Pancrazio JJ. Biological threat detection via host gene expression profiling. Clinical Chemistry 2003; 49:1045–1049 (*interesting proposal using host rather than pathogen microarray gene profile for rapid detection of infection*).

Poynard T, Yuen M-F, Ratziu V, Lai CL. Viral hepatitis C. Lancet 2003; 362:2095–2100 (*summary of hepatitis C*).

Gold™, Genomes OnLine Database (http://www.genomeson line.org/ (*this Internet site lists all genome projects that have been completed*).

9

FORENSIC MEDICINE AND SCIENCE

HISTORICAL DEVELOPMENTS

FINGERPRINTING AND PROTEINS

The traditional (dermatoglyphic) fingerprints made their appearance in the 1890s and were adopted by various courts over the next few decades (Table 9.1). This adoption was followed by the use of protein polymorphisms to compare crime scene samples (blood, semen, tissue) with blood samples taken from the accused. Genetic differences detected through protein polymorphisms have been used in forensic laboratories since the late 1960s. Initially, protein markers were based on the ABO blood groups. Subsequently, other blood groups, serum proteins, red blood cell enzymes and then histocompatibility (HLA) antigens have been typed. These markers complemented the fingerprints and, in some cases, became the primary forensic evidence in the courts.

A major disadvantage of protein polymorphisms is their limited degree of variability (known as polymorphism). Thus, the finding of commonly occurring protein polymorphisms in two samples (e.g., samples from the

crime scene and the accused) would be of doubtful value if the probability was sufficiently high that they could represent chance events. For protein markers, the probability that coincidence could explain genetic identity is in the vicinity of 1 in 100 to 1 in 1000. Therefore, the legal emphasis with protein polymorphisms focused predominantly on **exclusion** (i.e., samples from a crime scene and the accused had different protein polymorphisms) rather than **positive identification** (i.e., samples from a crime scene and the accused had the same protein polymorphisms).

Other problems inherent in protein analysis included (1) The relatively large amount of tissue required for testing. (2) The ease with which proteins degrade. These considerations are particularly relevant to the crime scene where the ideal laboratory conditions will not be found, and tissue available for analysis will, more often than not, be limited in amount and quality. (3) Finally, evidence derived from protein markers was unlikely to be helpful or even available for a crime committed in the

221

Table 9.1 Some key developments in DNA-based forensic evidence

Year	Event	Role played by DNA
1890s	Recognition of the traditional (dermatoglyphic) fingerprint as a unique identifier of an individual	Traditional fingerprints gradually accepted by the courts without the rigour now imposed on DNA fingerprints. Nearly 100 years later, the DNA fingerprint arrives.
1985	Immigration authorities deny entry of a Ghanaian child into the UK	One of the first non-criminal uses of DNA evidence which confirmed that the Ghanaian boy was related to a woman with UK residency, and it was likely that she was his mother, and not his aunt. On the basis of this, the child was allowed entry into the UK.
1987	Youth arrested for two rape-murders committed in 1983 and 1986	DNA excludes the individual as a suspect but links the two crimes. Subsequently, a new suspect is convicted.
1989	New York Supreme Court—*State of New York versus Castro* involving a double murder	One of the earliest cases in which the courts questioned the validity of DNA evidence.
Mid-1990s	Move to PCR-based STR (short tandem repeat) analysis	This represented a key technologic advance in DNA fingerprinting.
1994	"DNA fingerprinting dispute laid to rest" (Nature 1994; 371:735)	Two key players in a prominent dispute about the value of DNA fingerprinting publish a joint paper indicating that concerns about the scientific basis for DNA fingerprinting are now resolved.
1995	OJ Simpson declared innocent in a case involving a double murder	This case emphasised the importance of crime scene investigation and chain of custody of samples for DNA testing.
1995	DNA national forensic database established in UK	Early example of how a national database can help solve a number of crimes. On the other hand, it also illustrated the potential civil liberty issues that will arise.
1996	US National Institute of Justice commissioned report on DNA evidence in the courts	Study identified that 28 individuals convicted of serious crimes (some on death row) were subsequently exonerated because of DNA evidence.
2001	Anthrax bioterror threat	Potential for bioterrorism becomes real and expedites the development of DNA-based response systems.

past because protein, unlike DNA, does not last or cannot be stored for long periods.

DNA

In 1978, the first human DNA polymorphism related to the β globin gene was used to identify a genetic disease. In 1980, it was reported that restriction fragment length polymorphisms (RFLPs)—i.e., small variations in DNA detected with restriction endonucleases (see Chapters 1, 2)—were dispersed throughout the entire human genome. More complex and so potentially more informative DNA polymorphisms were described in 1985. They were called **minisatellites**. DNA rather than the protein polymorphisms now opened up the potential for a more sophisticated approach to tissue comparisons. In the forensic scenario, this would lead to an exclusion or even a positive identification of an accused since the chance that a match between DNA markers taken from the crime scene and the accused being coincidental were highly unlikely, with probabilities between 1 in 10^5 to 1 in 10^6

achievable (contrast this to the 1 in 100 to 1 in 1000 given earlier for protein-based markers). Subsequently, the availability of PCR and the finding of another DNA polymorphism (the **microsatellite**) propelled DNA fingerprinting into becoming powerful evidentiary material for the courts.

Not surprisingly, it soon became possible in British and North American courts of law for DNA evidence to be used in civil and criminal cases (see Table 9.1). Today, courts are familiar with DNA evidence and allow it for criminal trials and paternity disputes. DNA evidence has also been shown to be invaluable in excluding an accused, as well as identifying a likely, suspect. In a USA National Institute of Justice review of DNA evidence in the courts, it was shown that at least 28 individuals accused of serious crimes (including crimes for which there was the death penalty) were subsequently freed because of DNA evidence proving their innocence!

An appealing aspect of the DNA fingerprint lay in the robustness of DNA, so that samples from a crime scene,

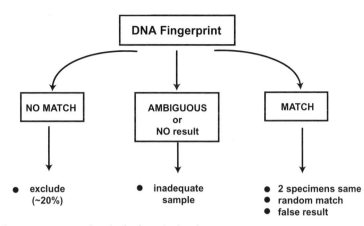

Fig. 9.1 Outcomes from a DNA comparison in the forensic situation.
DNA results comprise one piece of evidence in the forensic case. A number of outcomes are possible: **No match:** If the test is done properly, this is a powerful reason for excluding an individual as a suspect. **Ambiguous or no result:** The crime scene is not always an optimal source of material for DNA. In this circumstance, it is important that the laboratory does not waste police or court time with evidence that will not be acceptable. **Match:** The finding of a match is only the beginning. This result can be interpreted in a number of ways, and it is the function of the laboratory, expert witnesses and the court to determine which option is the most likely in this particular case, including (1) DNA samples from the crime scene and the suspect are the same; i.e., the suspect has powerful evidence implicating him or her. (2) DNA samples from the crime scene and the suspect are the same by chance. A mathematical value needs to be attached to distinguish which of (1) and (2) is the more likely. (3) The result is false and could come from a number of errors, including collection or processing of the sample; misinterpretation or incorrect reporting of the laboratory results; and even criminal intent on the part of the police, laboratory staff or the victim. Some possibilities for (3) will be excluded if the chain of custody is firmly established.

even long after the crime was committed, remained admissible as evidence. Another advantage lay in the intrinsic variability of DNA (i.e., its polymorphism content) so that exclusion was not necessarily the only option available. Thus, it became possible to aim for a unique DNA profile for each individual similar to the traditional (or dermatoglyphic) fingerprint (Figure 9.1).

In the late 1980s–early 1990s, DNA fingerprinting passed through an unsettled period with the courts. This period started with a pretrial hearing for a double murder case involving the *State of New York versus Castro*, at which time DNA evidence was first seriously questioned by a number of leading scientists. This case led to the demonstration of suboptimal laboratory practices as well as doubtful interpretations of the statistical significance of DNA polymorphic data. Evidence based on DNA studies in this case was thereby deemed inadmissible (although the accused later confessed). Subsequently, a number of other cases had to be withdrawn by the prosecution because DNA data comprised an important component of the evidence. Cases already decided were appealed. The scientific controversies about DNA fingerprinting continued into the 1990s, particularly in respect to the significance of identical matches. Public interest reached extraordinary levels with the trial of OJ Simpson in the USA in 1995 when, despite very strong DNA evidence presented by the prosecution, the accused was

acquitted. This case highlighted that DNA evidence is only as good as the laboratory that produces it, and equally important, the way in which police and forensic experts handle the evidence to ensure that the chain of custody cannot be challenged.

The early adoption of forensic DNA technology, particularly by commercial companies, produced a specific set of problems. They related to patented DNA polymorphisms that could not be used by competitors. Comparisons were not possible, and so the standard of quality assurance was severely compromised. The way in which the likelihood of DNA matches was calculated was another controversial issue following some absurd claims made about matches. Particular concern was expressed when the accused came from a minority ethnic group. For example, in the *State of New York versus Castro* example, the laboratory reported that a DNA match between a blood stain found on the accused and blood from the victim had a 1 in 10^8 probability of occurring by chance alone. However, the comparisons to derive the chance association were considered invalid for a number of reasons, including the fact that they had not been made against an ethnic group to which the accused belonged, i.e., Hispanic.

The problems described have now generally been resolved with input from government and the involved laboratories. In a very short time, DNA technology has

---

Box 9.1   CODIS and validation of STRs for forensic purposes.

When the FBI developed its DNA database in the early 1990s, an important component to this work was the selection and validation of STRs (microsatellites) that would be used for such a database. Thirteen STRs are now validated for CODIS. All but two are on different chromosomes. None is in a coding region (thereby avoiding potential ethical issues such as the finding of a genetic abnormality in an individual without that person's consent, and the possibility that there could be some selective effect on the STR if a coding DNA polymorphism was used). In addition, a 14th DNA marker is available involving the amelogenin gene (located on both the X and Y chromosomes) and so useful to sex a sample of DNA. Using the CODIS 13 STRs, Chakraborty *et al* (1999) estimated that a random match with these markers would occur by chance 1 in 1 trillion times! Today, many countries now use kits containing various combinations of these STRs numbering from 9 to 13. Recently, the value of the amelogenin marker has been questioned since in some rare instances, a small deletion on the X chromosome takes away the amelogenin on that chromosome, and these DNA samples would then be falsely typed as female (see also Figures 9.3 and 9.4 for further discussion on the STRs and the calculation of probabilities).

---

had a major impact on the judicial system, and this impact is extraordinary given the slow pace with which the legal system usually moves. In 1995, it became legal in the UK for law enforcement agencies to take DNA in the form of hair roots or buccal swabs (i.e., non-intimate samples) from those convicted of serious crimes. The aim was to establish a national DNA database from such offenders. This approach was successful, and by 1999 the database had >700 000 DNA profiles, and from these, a number of crimes were able to be solved. In 1990, the FBI set about creating a DNA database of forensic DNA profiles (CODIS—Combined DNA Index System). Today, CODIS is a national system assisting criminal investigations at the local, state and federal levels.

Initially, there were two databases: (1) convicted offenders and (2) forensic crime scene samples from unknown offenders. However, the utility of CODIS prompted the US Government to expand this database to include two additional ones for (1) relatives of missing persons and (2) unidentified human remains. CODIS has also had an indirect benefit on DNA fingerprints by validating a set of STRs that could be used with greater confidence (Box 9.1). Despite the success stories associated with the various police DNA databases, the ethical and privacy issues emerging from such databases are considerable and will be discussed further in Chapter 10.

REPETITIVE DNA

SATELLITE REPEATS

Only 1–2% of the DNA making up the 3.3×10^9 base pairs of the human haploid genome codes for genes. About 50% of human DNA is composed of repetitive DNA sequences that appear to have no function. The remainder of the human DNA is non-coding and non-repetitive, and also appears to have no function. Unfortunately, the term **junk** has been used to describe DNA that is non-coding and not related to genes. Today, this name is inappropriate since the distribution of repetitive DNA is non-random, and there remains some interspecies homology. Of particular relevance to this chapter is that a proposed role for repetitive DNA is as a hot spot for recombination. This is appealing as an explanation since the repeat sequences have no apparent coding (exon) function, and so there would be less evolutionary pressure for conservation. A greater degree of mutational activity would thus be possible at these loci. In an evolutionary sense, this would be useful to develop new genes (see Chapters 1 and 2 for further discussion on the role of non-coding DNA).

Irrespective of the functional role for repetitive DNA, it forms the basis of the DNA fingerprint. Repetitive DNA can be divided into two major classes: the tandem repetitive sequences (known as **satellite** DNA) and the **interspersed** repeats. The term "satellite" is used to describe DNA sequences that comprise short head-to-tail tandem repeats incorporating specific motifs. They make up one third of DNA repeats and are exemplified by the minisatellites, microsatellites and macrosatellites. A summary of the satellite DNA repeats is given in Table 9.2, and they are illustrated in Figure 9.2. The interspersed repeats are not described further since they have never been applied to the forensic scenario.

MICROSATELLITES

DNA polymorphisms now used in the forensic situation are the microsatellites; they are also called **STR**s for short tandem repeats or **SSR**s for simple sequence repeats (see Figure 9.2). These single locus VNTRs consist of tandemly repeated simple nucleotide units of about 2–6 base pairs.

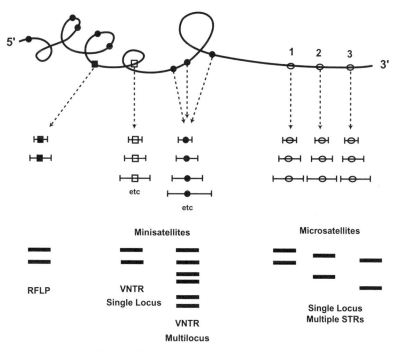

Fig. 9.2 The various types of polymorphisms used in DNA testing.
■ Indicates an RFLP present at a single locus, and producing two polymorphic bands (large and small) of fixed size. The number of combinations generated by this biallelic polymorphism is limited to large/large, small/small and large/small. □ Two polymorphic bands are also obtained for the single locus VNTR minisatellite, but these polymorphisms are more informative because there is greater variability between the sizes present for each of the two bands and so there is more chance that individuals will have different profiles. Combining a number of different single locus VNTRs produces an even more characteristic set of markers per individual. ● is the most informative of all polymorphisms because the multilocus VNTR pattern is a composite of many VNTRs that are scattered throughout the genome. A complex DNA profile (fingerprint) results. ○ represent microsatellites. Each is a separate locus producing a different profile, like the VNTR single locus. However, with PCR, it is easy to type a number of microsatellites to build up a complex DNA profile, which provides good discriminatory power between two samples.

The best described are the dinucleotide repeats involving bases such as adenine and cytosine (AC)$_n$, where n (the number of repeats present) varies from 10–60. For forensic purposes, tri and tetra repeats are preferred—for example (AACT)$_n$—because they produce more technically satisfactory results. Each STR identifies one unique segment of the genome. Microsatellites, because of their potential hypervariability, are more informative than the biallelic RFLP system, but less than the minisatellites. Nevertheless, the microsatellites can be assayed by PCR, and their value or informativeness is increased by measuring a number of them and adding the information obtained from each marker.

Today, it can sometimes be embarrassing to view an old minisatellite DNA profile that was tendered in evidence 10–15 years ago. Depending on the standard of the DNA testing laboratory, the quality of the Southern blot can be pretty ordinary, and band shifts are often present. Band shifts refer to different mobilities

for the same DNA fragment. They occur because of imperfections in gels resulting in non-uniform electrophoresis. However, these technical issues are now a thing of the past as a result of DNA profiler kits being produced commercially. These kits are validated and quality controlled, and the same DNA polymorphisms can be sought in different laboratories, making it easier to implement internal and external quality control measures and to compare data (see Box 9.1). Results for evidentiary specimens can be rechecked at a future time or by an independent laboratory because PCR is used. Previously, this was not an option with the minisatellites because most of the DNA from the crime scene would have been used in the initial Southern blot test. PCR also allows automation to be used, and so the sizing of the DNA fragments is no longer done visually but with proprietary software, thereby increasing the accuracy and reproducibility of the DNA fingerprint (Figure 9.3).

Table 9.2 Some examples of repetitive DNA in the human genome

Type of satellite	Features	Utility in forensic analysis
Minisatellite	Comprise core DNA repeats, e.g., $(AGAGGTGGGCAGGTGG)_{29}$. They are relatively large (1–30 kb) and are either single or multilocus. A combination of 4–6 of these markers would enable a reasonable DNA fingerprint to be established, including the potential to obtain a close-to-unique profile for an individual. Repeats called VNTRs (variable number of tandem repeats).	These markers were the original DNA fingerprints used until the early 1990s. The limitations of these DNA markers included the requirement for a large amount of DNA from the crime scene, the tedious and demanding Southern blotting technique and, finally, the band patterns obtained were difficult to interpret, particularly when there were band shifts[a].
Microsatellite	This repeat is <1 kb and comprises a simple sequence repeat (abbreviated to SSR) or short tandem repeats (STRs) such as $(XX)_n$, $(XXX)_n$ etc., where X are different nucleotide bases—e.g., $(AC)_n$.	SSRs are presently the ideal DNA repeat for forensic work because they are detectable by PCR (so only a very small amount of DNA is needed from a crime scene). By assaying a number of SSRs (e.g., 13), it is possible to sum the results and so obtain a powerful discriminator to compare crime scene and accused DNA. SSRs can be automated, thereby avoiding technical issues such as band shifts.
Macrosatellite	Although repeat units are small in size (hundreds of base pairs), they are repeated many thousands of times, making for a large DNA repeat that is not useful for forensic analysis.	These repeats are mostly found in the centromeres and telomeres of chromosomes.

[a] Band shifts were a major problem in the early forensic laboratories and resulted from the use of minisatellites and the type of electrophoresis undertaken with lack of automation, making it particularly difficult to establish standards in and between laboratories.

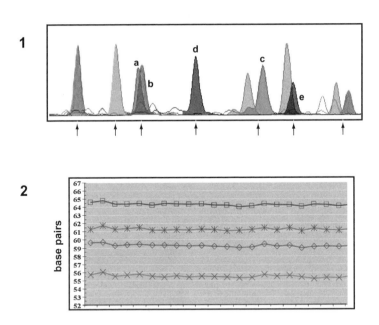

Time (6 months)

Fig. 9.3 Automated measurements of DNA band sizes.
To size a DNA fragment obtained by PCR, computer-based automated fragment sizing can now be used. This approach ensures both accuracy and reproducibility. **(1)** Each ↑ shows a DNA polymorphism (SNP) detectable by PCR. In total, seven different SNPs can be multiplexed, and so detected with the one PCR. Next, PCR products are electrophoresed and distinguished from each other on the basis of fragment size. When there is overlap in sizes (for example, fragments a and b), the primers for PCR are labelled in different colours, and so fragments with approximately the same size remain distinguishable. Each fragment has the potential to have two different alleles; for example, a and c, d and e are alleles for the same two fragments. These alleles are separable on the basis of size and colour, thereby enhancing accuracy of the electrophoresis. **(2)** This graph illustrates regular QA measurements over a period of six months for four DNA fragments in the size range 55 to 65 bp. The graph confirms the reproducibility of the DNA electrophoresis with very little drift over this period of time.

OTHER REPETITIVE DNA

MtDNA

Apart from nuclear DNA, another target for the forensic laboratory is mitochondrial DNA (mtDNA), which is highly variable in a region known as the D loop. This variability predominantly results from single base changes and some length polymorphisms. Advantages of mitochondrial DNA in the forensic scenario include (1) The exclusive maternal origin facilitates the determination of family relationships. (2) Thousands of copies are present in each cell (in comparison, there are only two copies of nuclear DNA per cell). Therefore, even smaller crime scene samples can be tested. (3) Hair is frequently found at the scene of a crime, but there are many potential sources for this tissue, e.g., victim, accused, police, bystanders, etc. Individual hairs must be studied. Nuclear DNA (present as two copies per cell) is extractable from hair roots but not the shafts, and so hair has limitations in forensic DNA testing. However, since multiple copies of mitochondrial DNA are present in the shafts, it becomes possible with PCR to type individual hairs without the necessity for roots to be present. Other useful sources of mtDNA are bones and teeth. These are robust specimens from which DNA can be extracted even though many years may have lapsed since the crime.

Disadvantages of mtDNA testing in forensics include the ease with which contamination occurs compared with nuclear DNA testing. Therefore, technical demands are greater. Another concern about mtDNA is heteroplasmy, i.e., the presence of one or more mtDNA types in an individual. It has been shown, for example, that hairs from an individual might demonstrate a different mtDNA profile because of heteroplasmy (see Chapter 4 for further discussion on mtDNA).

SNPs

The recently described single nucleotide polymorphism (SNP) provides an alternative to the microsatellite for DNA forensic fingerprinting. Indeed, the SNP has the potential to produce an even more unique profile of an individual than the microsatellite because of the numbers of these polymorphisms in the human genome (in the millions). However, costs for SNP analysis remain high, and because of their biallelic nature (in contrast, the microsatellites have a larger number of alleles and so better discriminatory power), it has been estimated that somewhere in the vicinity of 25–45 SNPs would be required to match the power presently available through the standard 13 SSR profile kit (Chakraborty et al 1999). However, SNP costs will only come down, and in the longer term, it is highly likely that the forensic laboratory of the future will provide even better profiles of DNA

from crime scenes by using a large number of SNPs as the basis for the genetic DNA fingerprint.

Y Chromosome STRs

The use of the Y chromosome for forensic analysis is only a fairly recent development because the appropriate STRs had to be found. For example, the X and Y chromosomes share some DNA sequence, and this shared DNA would not be an appropriate place for an STR that was to be Y chromosome specific (see Chapter 7 for more information on the X and Y chromosomes and their evolution). There are now a number of validated Y chromosome STRs, and they are particularly useful in sexual assault crimes because the DNA from female (victim) cells in these cases does not interfere with examination of the male-derived tissue (blood or semen). The alternative, when there are mixed tissues, is to differentially extract DNA from sperm and the female victim's own tissues. However, in some cases differential DNA extraction is not possible, and now the Y chromosome STRs would be useful. Another application for Y chromosome STRs is in mass disasters or missing person identification when DNA from only male relatives was available. The downside of the Y chromosome STRs is their reduced power of discrimination (only one band will be present because there is only one Y chromosome), and so a number of Y-STRs will be required.

DNA AMPLIFICATION (PCR)

Four properties of DNA amplification by PCR make it ideal for the forensic situation: (1) Minute amounts of evidentiary material left behind at the scene of the crime will provide enough template for DNA analysis. (2) Degraded DNA can still be amplified, since only a small segment of DNA is required for primers to bind in PCR. (3) It remains possible to retest the sample in another laboratory or at some future date. (4) Automation is available, leading to less chance of contamination and more accuracy in fragment sizing.

Balancing the above are a number of potential problems of particular relevance to forensics including (1) PCR-based errors. Does exposure to the environment with its consequent DNA-damaging effects lead to errors in PCR? Experience would now suggest that this does not occur. In other words, if amplification is possible after environmental exposure, then the end product is free of artefacts resulting from damaged DNA (because PCR amplifies very small fragments, and so even damaged DNA remains a suitable template for the PCR to work). There is always the potential for errors to occur through misincorporation by the *Taq* polymerase enzyme or differential amplification of DNA sequences. These errors are not necessarily a problem with PCR in forensic practice since the test can usually be repeated, and modern

Taq enzymes are less likely to produce a misincorporation (see Chapter 2, Appendix for further discussion of PCR). (2) Contamination. The effect of contaminating sources of DNA on PCR has already been mentioned in relation to genetic disorders and the detection of pathogens. Contamination occurs in the ideal laboratory despite strict care and the highest standards. Sources of contamination include the laboratory scientists, equipment or, more frequently, other amplified products. Contamination becomes an even more significant issue in the poorly controlled crime scene, with the environment itself a potential source of many different DNAs.

FORENSIC ANALYSIS

LABORATORY

Crime Scene

At the scene of a crime, there may be stains (e.g., blood or semen on the victim's clothing) or other tissues (e.g., skin, hair under the victim's fingernails) that require identification and characterisation. DNA from evidentiary samples is used to build the DNA profiles. They will then be compared to DNA patterns from the suspect(s) or evidence linked to the suspect(s). In the crime scene, it is important to realise that there are likely to be a number of DNA sources, e.g., the victim him/herself, third parties or the environment. The potentially complicating issues facing the forensic scientist are well illustrated by the evidentiary samples obtainable in a case of rape. In this situation, DNA can come from (1) the victim in the form of blood, body tissues or secretions (vaginal, anal or oral in origin) and bacteria; (2) one or more assailants; (3) semen from earlier voluntary intercourse; (4) animals or bacteria from the crime scene and (5) a possibility that the source of DNA was planted by a third party, including the victim or police.

DNA testing remains feasible in the complex circumstances just described. DNA originating from microorganisms or other animals does not usually cross-hybridise with human-specific DNA. DNA obtained from sperm is more robust when it comes to isolation procedures. Therefore, laboratory protocols can be designed to utilise this property and enhance the isolation of sperm DNA at the expense of DNA from other tissues. Thus, the level of the contaminants can be reduced or even excluded. The alternative described earlier is the use of Y chromosome–specific STRs. The problem of multiple human DNA sources can also be overcome. First, the blood or tissue from the victim is obtained to identify that person's DNA band patterns, which are then subtracted from the overall profile. The contribution from an innocent third party (e.g., sexual partner) can be treated in the same way. Next, DNA patterns from evidentiary samples are compared to those obtained from potential assailant(s). From these comparisons it might be possible to identify specifically the accused(s) as the source of the DNA stain or semen. Alternatively, blood from the victim may have spilled or splashed onto an assailant's clothing or the crime scene, e.g., a car. DNA isolated from the blood spots will subsequently provide important evidence connecting the victim with the individual wearing the clothing or the crime scene.

No matter how good the DNA fingerprint is, its value is ultimately dependent on a well-established chain of custody. This is critical to avoid the criticism that the police or others tampered with the evidence or planted false evidence. Anything less will invalidate what might be very persuasive DNA evidence.

Technology

The same DNA methods routinely used in the research or hospital diagnostic laboratories will be technically more challenging in the forensic situation. In the latter, DNA is frequently tested under suboptimal conditions. For example, the stability of DNA will depend to some extent on the way it has been maintained or stored. This cannot be controlled at the crime scene. The older the sample or the longer it has been exposed to the environment (particularly ultraviolet light, moisture, high temperature and microorganisms), the more degraded becomes the DNA. Artefacts or technical problems resulting from contamination or degradation have led to a substantial proportion of evidentiary samples being considered inconclusive on the basis of DNA tests.

An important goal of the Human Genome Project was technology development (see Chapter 1). There are now new sophisticated methods for DNA analysis such as fluorescein-labelled DNA primers with PCR, the use of capillaries for DNA electrophoresis and the sizing of DNA fragments with lasers and computer software. These methods have all contributed to the forensic laboratory becoming a highly sophisticated DNA analysis facility, thereby providing the highest quality DNA fingerprinting.

The problems created by incorrect application or interpretation of molecular techniques in forensic medicine have provided a useful example of how utilisation of this technology must be appropriate and scientifically based. Following on from the *State of New York versus Castro* case and various other examples of suboptimal laboratory protocols, more stringent legal requirements are now in place to ensure that laboratories involved in DNA typing are able to maintain the highest standards, including regular quality assurance and proficiency testing.

Quality Control

Quality control programs were particularly difficult to organise in the beginning since many of the commercial forensic laboratories utilised their own specific single locus VNTR probes protected by patents. Therefore, inter-laboratory comparisons were impossible. This became less of an issue as microsatellite markers and PCR technology were utilised. Today, various externally based testing programs involving the use of unknown samples give the courts an indication of a laboratory's performance in DNA typing. The availability of commercial kits adds another layer to quality control and allows inter-laboratory testing and results to be compared directly.

Another aspect of quality assurance reflects the types of evidentiary samples provided. For example, some specimens are fresh, whereas others have been exposed to the environment for variable periods of time. The different methods used to store the evidentiary sample once it had been collected might also be another variable in the DNA patterns obtained. Finally, there is the problem of contaminating DNA and the overall effect this will have on the test results. An experienced forensic laboratory will need to know how the above variables might influence the testing procedures and, if necessary, have the appropriate control samples or data should they be required.

Statistical Comparisons

One advantage of DNA testing is its ability to show identical patterns (genotypes) between two samples being compared, and then to take this one step further and demonstrate that the genotypes are likely to be unique; i.e., the probability of them being present in another individual is insignificant. In this respect, it is essential to know the frequency in the population of the various markers that make up the genotype. This forms the basis for the statistical calculation that the two specimens are likely to be derived from the same source (Figure 9.4). The product rule approach described makes a number of assumptions; e.g., there is random mating and the alleles for the multiple STR markers segregate independently of each other (i.e., there is no linkage disequilibrium).

For some time there was considerable debate about (1) random mating and the effects of linkage disequilibrium in ethnic or minority groups within a community and (2) whether a single allele present for one marker represents homozygosity for that marker, an additional null allele that has not been typed or two alleles that cannot be distinguished. These were some of the questions that had been asked by scientists and the courts. Although the debate had been vigorous, it ultimately enhanced the

quality of the science. A landmark article in *Nature* in 1994 reaffirmed the value of DNA fingerprinting, and stated that a number of the earlier controversies had now been resolved (see Table 9.1).

Today, there are now sufficient DNA databases with which to make direct comparisons even in sub-populations. For example, a large city in the USA might have STR allele frequencies for a variety of populations including Black, Caucasian, Hispanic, American Indian and Asians. Despite all the rhetoric about regional and ethnic factors and their potential effects on the frequencies of DNA markers, population geneticists now agree that the differences these make to the final calculations are minimal. For example, a probability of 10^{-7} might be reduced to 10^{-6} because of population differences. A tenfold or even hundredfold difference in probability should not in itself be sufficient to convict or acquit a defendant.

Exclusions

Evidence based on DNA testing will be used by the prosecution to confirm a link between the victim and the accused. On the other hand, DNA evidence could turn out to be more beneficial to the defence if DNA patterns excluded a match. An accused who is on trial because of evidence obtained from an eyewitness may request DNA testing as the only means by which his/her innocence can be proven (a not-infrequent scenario identified by the Institute of Justice report mentioned earlier). DNA fingerprinting will save time in police investigations since suspects can be quickly excluded by their DNA profiles. Despite being acquitted of a crime, an individual will suffer humiliation and possible stigmatisation following arrest and trial. This can be averted if DNA testing avoids the arrest in the first place. Two experienced forensic laboratories (the US Federal Bureau of Investigation and the British Home Office) have reported that DNA testing allowed suspects to be excluded in approximately 20–25% of cases.

EXPERT WITNESS

One important aspect of DNA fingerprinting is the way in which the results are presented to the judge and jurors, with the latter unlikely to have much knowledge of an already complex subject. For example, a claim by the prosecution that a DNA match between the accused and blood obtained from the victim's clothing represents a 1 in 10^6 chance of a random event is very persuasive evidence. However, an equally crucial component to this evidence is the requirement to explain to the court how the test was done and what its potential drawbacks are, including the methods used to assess the statistical probability of a random match. In this respect it is essential that expert witnesses have both a scientific and practical

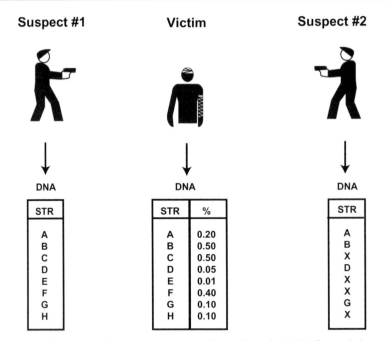

Fig. 9.4 Calculating the statistical chance of a random match using the product rule in DNA fingerprinting.
Following a violent crime, blood is found on the clothes of two suspects. DNA is prepared from the victim, and the profile is then compared to DNA from the blood on the suspects' clothing. The results of eight STRs (A–H) are given for DNA from the victim (centre). For simplicity, only one allele is used although each STR can have up to two alleles. The frequency (%) for each allele in the population is also given. Even without any calculations, suspect #2 can be excluded since, at four alleles (marked with an X), the DNA profile taken from the blood on his clothes has different fragment sizes than blood from the victim. In contrast, the DNA profile from the blood on the clothes from suspect #1 has the identical STR profile. The chance that this is a random event is calculated by multiplying together the frequencies of all the allelic components making up the profile (called the product rule). In this case, the chance that the DNA from the victim and the DNA from the blood on suspect #1 is a random event is 1 in 10^7. Using more STRs would increase the probability that the two samples were related although the statistical chance that this is a random match is already extremely low.

knowledge of the subject and that they can provide this information in a meaningful way. The challenge of imparting complex technologic information in an understandable format to the jury is particularly relevant to DNA evidence. In the OJ Simpson trial, the evidence from the prosecution's expert witness in DNA statistics was described by a local newspaper as "dry as sand and about as digestible" (Weir 1995). The dilemma surrounding the presentation of complex scientific DNA evidence remains a major challenge.

The legal profession itself has taken on the importance of DNA evidence in terms of self-education. In the USA, a body called EINSHAC (Einstein Institute for Science, Health and the Courts) specifically promotes the awareness and understanding of science-based evidence in the courts, including what are likely to be challenges in the future. The Federal Court of Australia has taken the challenge and looked at new ways in which complex technical evidence can be presented in a more conducive and rational atmosphere than might be possible under the adversarial system. Initiatives by this court include the

appointment of third-party experts that have been agreed to by both the defence and the prosecution, with the role of the expert being to facilitate the court's interpretation of data. Expert conferences have also been considered to allow the experts (without the adversarial environment) to reach an agreement or clearly define the limit of consensus and disagreement.

Proposed reforms to the way in which complex scientific evidence might be given are criticised within the legal profession because they are perceived as having the potential to introduce bias (that of the expert), and also junk science to the courts. These concerns are certainly valid. On the other hand, from a scientific viewpoint, it is frustrating to observe that molecular science can sometimes produce results in the forensic scenario that might be explicable following a reasoned debate. In the adversarial system, this can be difficult to achieve, and what should be credible evidence can be rejected. A good lawyer will understand the advantages and disadvantages of the technology and use this information to the client's benefit. This may be particularly an issue in the adversarial court system (judge and jury) where lawyers have the opportunity to influence the jury more than might be possible in the inquisitorial (judge alone) court system.

AUTOPSY

A final application of DNA analysis is the forensic autopsy to determine cause of death. This application is well illustrated by a case involving a 44-year-old psychiatrically disturbed woman who had attempted suicide on two occasions. She was found dead in the bath, and although the cause of death determined at autopsy was drowning, it was not clear whether this was a suicide or there were other explanations. DNA testing to look for mutations associated with sudden cardiac death showed that she had a mutation which produced the long QT syndrome. Therefore, in her case, a more likely explanation for the drowning was ventricular arrhythmia, a common complication in this genetic disorder, followed by drowning rather than suicide (see Table 3.2 for more information on the long QT syndrome).

HUMAN RELATIONSHIPS

TISSUE TYPING

The genes which comprise the MHC (major histocompatability complex or HLA locus) on the short arm of chromosome 6 are divided into three classes occupying a physical distance of approximately 3600 kb (Figure 4.2). The HLA antigens continue to remain an important set of markers for tissue typing, but they are no longer used for comparative studies of populations that are now undertaken with DNA-based markers. Traditionally, HLA typing involved serologic testing although very subtle differences between antigens shared by individuals at the HLA-D locus were detected by a functional assay known as the mixed lymphocyte reaction. HLA typing is required in most types of tissue transplantation since the closer the graft is to the donor in the immunological sense, the less likely graft rejection is to occur. Comprehensive tissue typing can be time consuming, and despite histocompatibility at various HLA loci, examples of rejection still occur, which may in part reflect hidden antigenic determinants. Today, kits allowing the identification of all HLA classes by hybridisation-based DNA assays are commercially available. They are routinely used in the tissue typing laboratories.

PATERNITY TESTING

DNA typing using STRs has also replaced the HLA polymorphisms in paternity testing. Two paternity testing scenarios can be considered: (1) straightforward trio cases—mother, child and alleged father and (2) more complex motherless cases, for example, child and alleged father. In paternity testing, the STRs provide high sensitivity (few false negatives) but low specificity (false positives can occur because unrelated people can share STR alleles). However, this is less of an issue in the trio case because apart from a random match, the problems of mistyping through silent (null) alleles or allele dropout (see Figure 2.8) are less relevant as all three individuals are being examined and silent alleles will be detected.

The chance of a random match will never be completely excluded, but the chance can be reduced further, if necessary, by using a larger number of STRs or other DNA polymorphisms such as mtDNA or Y chromosome–specific STRs.

Two-person (motherless) paternity testing cases are more of a dilemma because it is not known which alleles in the child have come from the mother. Therefore, assumptions need to be made, and more complex statistical analysis programs are required. With a missing parent, the risks from mistyping due to null alleles or allele dropout are now a real issue with the STRs. Similarly, determination of paternity in alleged incest is difficult to resolve even with the STRs. A brief commentary on paternity and kinship testing by Wenk *et al* (2003) suggests that it is unfortunate that STRs have replaced the more informative minisatellites in the difficult paternity cases, as these markers are potentially more useful. This is true, however, for the reasons mentioned in the Appendix, very few diagnostic laboratories use a Southern blot, and so the minisatellite is no longer a practical proposition. In the longer term, the SNP-based DNA fingerprint will provide the same or even a better level of discrimination than the minisatellites.

KINSHIP TESTING

In addition to assessing inter-family relationships for genetic disorders, the utility of DNA markers in legal assessments of relationships within families has become increasingly common. This is particularly so in immigration cases, and as illustrated in Table 9.1, this was one of the first legal applications of DNA fingerprinting. Of particular relevance is the distinguishing of uncle-nephew pairs; i.e., an individual wants to emigrate to a country and claims that a relative there (for example, an uncle) is the father. Because of this close relationship, STRs, even using a large battery such as nine markers, may not be discriminatory enough to prove conclusively whether the case is a father-son or uncle-nephew. As for the "motherless" paternity case, other DNA markers can be tried, but these particular circumstances remain problematic and will, in the longer term, require more sophisticated DNA fingerprints. SNPs might be the answer, but they would need to be tested in large numbers and presently such testing would be too costly.

The case of missing persons is another area in which DNA fingerprinting is being applied, although like the kinship examples illustrated above, the technology has still some way to go before it has the same impact that DNA fingerprinting has had in the criminal scenario.

TISSUE IDENTIFICATION

HUMAN REMAINS

DNA analysis has a role to play in the identification of human remains. For example, the availability of parental DNA samples might allow identification of a body when conventional means (physical appearances, dermatoglyphic fingerprints, dental charts) have been unsuccessful. Dissimilar DNA profiles will exclude a relationship, and similarities in DNA polymorphisms will confirm that the deceased is likely to be or is related. This approach was critical to identify dismembered bodies resulting from the crash of a SwissAir flight from the USA to Geneva in 1998 (Box 9.2). DNA fingerprints for military personnel are used in combat situations when identification of body parts may be required.

Teeth are important evidentiary material in forensic cases since they are more resistant to postmortem degradation as well as extremes of environmental conditions. Teeth are also easy to transport and serve as a good source of DNA. Comparisons of antemortem dental records with skeletal remains provide useful means to identify individuals—even in a mass grave. In affluent societies, dental records may be decisive in determining the identity of individual victims. However, in less affluent communities, and these are more likely to be involved in human rights abuses leading to mass murder, dental records are unlikely to be available. In this situation, the only option for identification is DNA analysis.

CANNABIS

An unusual but relevant application of tissue identification in forensic science has involved the plant *Cannabis sativa*, which is the source of cannabis (marijuana or hemp). Throughout history *Cannabis* has been used to produce hemp fibre for rope and fabric (plant stems), food and oil (plant seeds) and as an intoxicant (plant flowers and leaves). Two distinct types of *Cannabis* are grown depending on whether fibre or the intoxicant (marijuana) is required. Used legally, *Cannabis* is a multipurpose plant with considerable economic potential.

There is interest now in determining if DNA fingerprinting can provide a *Cannabis sativa* gene profile that will distinguish fibre and intoxicant plant varieties. Another forensic goal is to characterise genetically the plants so that illegally grown or seized material can be traced back to the original sources.

The work has followed the standard DNA fingerprint strategy with the identification of STRs in *Cannabis sativa* and then a comparison of the profiles in various plants (fibre versus intoxicant and plants from various geographic locations). Disappointingly, it has not been possible to use STRs to distinguish unequivocally the fibre and drug varieties, and this is consistent with *Cannabis* being a single species. Additional or better DNA markers than the STRs are now being sought. However, the STRs appear to be useful in discriminating between different groups or accessions of these plants, suggesting that this approach will have value in the courts to link various crops, thereby providing information whether seized plants are being produced locally or imported.

Box 9.2   DNA fingerprinting in the case of a mass disaster (Carmody 2003).

SwissAir flight 111 crashed on 2 September 1998 with the loss of 229 lives (215 passengers and 14 crew). Bodies were dismembered when the plane fell 4 km off the US coast at the beginning of its flight from New York to Geneva. The grim task ahead was to identify individuals, put together the human remains (there were 1277 crash scene samples) and perhaps determine cause of death, since a charred body might suggest where an explosion had occurred. Also found were various personal effects (toothbrushes, combs, hair brushes). These effects would be used to extract DNA for matching to family samples. More than 300 living family members gave their DNA for comparative analysis, and within 3.5 months, the forensic laboratory had unequivocally identified the 229 victims.

TERRORISM

The use of chemicals and infectious agents in war has been recognised for more than a thousand years. However, these agents as weapons of terrorism are a new development, with a well-documented case dating back only to 1984 in the USA (Table 9.3). The events of September 11, 2001, and the subsequent anthrax threat in the USA have highlighted inadequacies in how potential weapons of bioterror, particularly chemical and microbiologic, are dealt with in contemporary society. In this forensic scenario, the application of modern genomic and proteomic technologies will make important contributions. It is interesting to note that many reasons were given for the sequencing of model organisms in the Human Genome Project. However, at the time, a defence to bioterror was not a prominent justification, but it now turns out that this is likely to be a very important contribution from the Human Genome Project.

MICROBIAL FORENSICS

A new discipline called microbial forensics has been developed. It is defined as "a scientific discipline dedicated to analysing evidence from a bioterrorism act, biocrime, or inadvertent microorganism/toxin release for the purpose of identifying those responsible for the crime" (Budowle *et al* 2003). The US Government has now developed a plan that will allow future use of infectious agents for bioterror to be dealt with more effectively. Since part of the bioterror agenda is to inflict fear, panic and economic chaos in addition to the actual morbidity and mortality, the traditional public health approach to an infectious disease crisis is not adequate, and new strategies are needed. Refer to Chapter 8 for further discussion on bioterror, specifically on pathogens such as anthrax and smallpox.

Priorities that will need to be addressed in microbial forensics include (1) Development of rapid DNA-based (or proteomic-based) diagnostic strategies for infectious agents. A DNA chip containing information about infectious agents and/or a protein chip able to detect toxins will eventually comprise the front line to rapid diagnostics. (2) Understanding pathogenesis, including the host-pathogen interactions. Knowledge of the organisms' transcriptome will be invaluable in this goal. Fortunately, a number of the key bacterial pathogens have now been completely sequenced, and this information will provide an invaluable resource to move ahead in understanding where the genes are by using *in silico* methods, and from this what are the functions of these genes. (3) The final challenge will be how to treat or manage bioterror-

Table 9.3 History of biological warfare and bioterror (Fraser 2004)

Time	Agent	Effect
Greeks, Romans and Tartars	Bodies of humans and animals were used to poison drinking water or spread infections.	Plague outbreak in the fourteenth century attributable to the Tartars catapulting the bodies of plague victims over the walls into the city of Caffa.
Seventeenth and eighteenth centuries	British and French soldiers used smallpox via blankets.	Smallpox used to kill American Indians.
World War I	German plan to use glanders to infect horses (and then humans) in the USA. Various neurotoxic chemicals used.	Not implemented. Estimated to cause >1 million casualties.
World War II	Japanese use of anthrax, cholera and plague.	Used against the Chinese.
Cold War 1970s–80s	Accident in Soviet Union weapon's laboratory.	Outbreak of inhalational anthrax.
1984 USA	Religious cult in Oregon USA spread *Salmonella* to prevent voting in an election.	>750 cases of food poisoning with a delay of more than a year to determine the cause.
1980–88, Iran-Iraq war; 1987–88, Kurds in Iraq	Chemical warfare using mustard and other gases.	Difficult to confirm number of casualties, but one estimate is >10000 killed by chemical weapons.
1993–1995 Japan	Japanese cult releases sarin and also botulinum toxin and anthrax.	>5000 injured and 12 deaths due to Sarin in Tokyo subway; 7 deaths from Sarin in Matsumoto.
2001 USA	Anthrax dispersed by mail.	Five deaths. Crime remains unsolved.

related outbreaks. Vaccines have always played a key role in the control of infectious diseases, but they may not be enough in the bioterror scenario, with the infectious agent having the potential to be distributed widely and acutely or indolently and not recognised for some time. New therapeutic approaches (for example, gene therapy based) may be needed to deal with bioterror-sponsored outbreaks.

A sobering statistic is the US budget for bioterror which will be US$5.2 billion in 2004. This is 15 times that spent on bioterror in 2001. One use of this money is to fund two national biocontainment laboratories that will have the infrastructure and human resources to deal with future threats. An interesting comment made by L Richardson reflects how this initiative also has the potential to increase the risk for bioterror because at present a limitation with terrorists is their lack of expertise in dealing with biologic weapons. The concern lies with the possibility that terrorists might now be given the opportunity to acquire skills by recruiting from those who have been trained in the two new facilities.

Looking more positively into the future, the threat of bioterror has focused the politicians, lawmakers, scientists and the community, so it is likely that our knowledge and understanding of infectious agents will be dramatically increased in a very short time frame. One hopes this new knowledge will be put to good use in combating infections that are scourges in many communities throughout the world but, until recently, not a sufficient cause of concern to the countries with the money and the resources needed to tackle them.

An interesting debate has also developed around scientific research and the publishing of data that could inadvertently be used for bioterrorism. In particular, the availability of DNA sequences might provide information allowing terrorists to genetically engineer their organism to make it more virulent, or make detection more difficult. In this environment it is likely that research involving potential bioterror weapons will be monitored and censored by government because of its security risk. The US National Academies has produced the report "Biotechnology Research in an Age of Terrorism," which is available on the Internet. In this report, experiments of concern are identified such as those that would (1) demonstrate how to render a vaccine ineffective, (2) confer resistance to useful antibiotics and antiviral agents, (3) enhance the virulence of a pathogen or change a non-pathogen into a pathogen, (4) increase the transmissibility of a pathogen, (5) alter the host range for a pathogen, (6) enable the evasion of diagnostic tests and (7) enable the weaponisation of the agent.

Checks and balances as well as appropriate review processes are identified in the National Academies report to mitigate the risks while at the same time avoiding the potential that restrictions placed on sensitive work will prevent the sharing of information through traditional routes such as conferences and publications.

CHEMICAL WEAPONS

Genomics has less to offer when it comes to chemical weapons. However, the pressure to develop better and faster DNA sequencing platforms has led to new proteomic approaches such as MALDI-TOF mass spectrometry (see Chapter 5). Mass spectrometry has become a powerful tool for the analysis of chemicals that could be used for this type of terrorism. Although dealing with chemicals, genes may also be important in understanding how the chemicals are metabolised or exert their toxic effects. This is illustrated by a study that was conducted to determine if death following sarin poisoning in the Japanese subway attack in 1995 could be attributed

---

Box 9.3   Sarin bioterror: The molecular basis for this agent (Costa *et al* 2003).

Sarin is a highly toxic nerve agent first produced for chemical warfare in Germany in 1937. It works like an insecticide by inhibiting acetylcholinesterase, producing a neurotoxic effect with immediate death occurring because of respiratory failure. Sarin was used as a form of bioterror in Japan in 1994, 1995 and 1998. Soldiers in the Gulf War were exposed to low levels of sarin. In the 1998 Japanese subway attack, 10 people died. The effect of genetic variants on susceptibility to drugs was considered as one explanation for the deaths in Japan. Specifically, it is known that sarin is metabolised by paraoxonase-1 (PON-1—see Chapter 5 for more discussion) and that one particular genetic variant of PON-1 (arginine 192) is associated with reduced sarin hydrolysing activity. In addition, this genetic variant is more common in the Japanese (35% are homozygous for arginine 192) compared to other ethnic backgrounds (9% of Northern Europeans are homozygous for arginine 192). Therefore, the question was asked whether the deaths in the Japanese victims were partly explained by a genetic predisposition. A comparison of the arginine 192 DNA profiles of the 10 victims showed a similar distribution to the normal Japanese population. This confirmed that death was due to high toxicity of sarin rather than race-dependent genetic variation. Another observation to emerge from molecular analysis involving sarin is that treatment with PON-1 can reduce the severity of insecticide exposure in mouse models. However, it is unlikely that the genetic variants found in association with PON-1 exert a powerful enough effect to be useful in this regard, unless a genetically engineered form of PON-1 can be made that has a significantly enhanced sarin catalytic potential.

to an individual's genetic ability to metabolise this neu-rotoxin (Box 9.3). Although the study produced a nega-tive result, it illustrates an additional genetic-based

approach to understanding chemical weapons, and so the development of antidotes or counter-terrorism strategies.

FUTURE

The courts and scientists demonstrate an interesting dichotomy when it comes to DNA fingerprinting. The courts are conservative and prefer to deal with prece-dents. On the other hand, scientists are constantly striving to develop new techniques, and in the field of molecular medicine, the changes seem to occur on a daily basis. Some adjustment, perhaps by both parties, will be required to ensure that this does not interfere with the long-term development and utilisation of DNA fingerprinting.

An important ethical consideration is privacy. This issue will have already involved a limited number of individuals and their families who have undergone DNA testing for the diagnosis of genetic disorders. It will become a more widespread concern as the forensic DNA databases continue to expand and affect a larger number in the community. Similarly, the utilisation of genetic information by government or the courts will need a deli-cate balancing act to ensure that good is not outweighed by the harm that indiscriminate use of personal informa-tion can have for an individual and that person's family.

Just as has been described in Chapter 5, the genomics and proteomics era has now emerged with the comple-tion of the Human Genome Project. The information that is available in the databases will drive the development of new technologies. This development can only enhance and produce even more sophisticated DNA fingerprints. The potential power of SNP-based digital coding can be compared to the very familiar bar code seen on all super-market products. There are both good and bad outcomes for the digital coding of humans. It will take a wise society to extract only the good. Wisdom cannot come from ignorance, so education and familiarity with DNA fingerprinting are critical to the long-term ethical use of this technology.

FURTHER READING

Historical Developments

www.ncjrs.org/txtfiles/dnaevid.txt (Convicted by juries, exoner-ated by science: case studies in the use of DNA evidence to establish innocence after trial—*National Institute of Justice review of DNA evidence in the court*).

Repetitive DNA

Chakraborty R, Stivers DN, Su, B, Zhong Y, Budowle B. The utility of short tandem repeat loci beyond human identifica-tion: implications for development of new DNA typing systems. Electrophoresis 1999; 20:1682–1696 (*provides an*

overview of the standard SSR profile used today and com-pares with SNPs).

Carey L, Mitnik L. Trends in DNA forensic analysis. Elec-trophoresis 2002; 23:1386–1397 (*comprehensive and well-written summary of various technologies and polymorphisms used in forensics*).

Forensic Analysis

Lander ES, Budowle B. DNA fingerprinting dispute laid to rest. Nature 1994; 371:735–738 (*historical perspective describing some of the scientific concerns about DNA fingerprinting*).

Lunetta P, Levo A, Mannikko A, Penttila A, Sajantila A. Death in bathtub revisited with molecular genetics: a victim with suicidal traits and a LQTS gene mutation. Forensic Science International 2002; 130:122–124 (*forensic case showing value of DNA testing even postmortem*).

Weir BS. DNA statistics in the Simpson matter. Nature Genetics 1995; 11:365–368 (*interesting scientific overview of this well-known trial*).

http://www.einshac.org/ Internet address for EINSHAC—Einstein Institute for Science, Health and the Courts (*EINSHAC is a US-based organisation with a mission to make science accessible to the instruments of justice*).

Human Relationships

Wenk RE, Chiafari FA, Gorlin J, Polesky HF. Better tools are needed for parentage and kinship studies. Transfusion 2003; 43:979–981 (*brief but succinct summary of family relation-ship testing with DNA fingerprints*).

Tissue Identification

Carmody G 2003. Identification of victims using DNA from rel-atives: the Canadian experience. XIX International Congress of Genetics 2003. Abstract 6H, p55.

Gilmore S, Peakall R, Robertson J. Short tandem repeat (STR) DNA markers are hypervariable and informative in *Cannabis sativa*: implications for forensic investigations. Forensic Science International 2003; 131:65–74 (*an interesting but unusual application for DNA fingerprinting*).

Terrorism

Budowle B *et al*. Building microbial forensics as a response to bioterrorism. Science 2003; 301:1852–1853 (*describes the US Government's strategy for microbial forensics*).

Costa LG, Cole TB, Jarvik GP, Furlong CE. Functional genomics of the paraoxonase (*PON1*) polymorphisms: effects on pesti-cide sensitivity, cardiovascular disease, and drug metabolism. Annual Reviews of Medicine 2003: 54:371–392 (*summarises the various functions of paraoxonase and the associated molecular variants*).

Fraser CM. A genomics-based approach to biodefence pre-paredness. Nature Reviews Genetics 2004; 5:23–33 (*an excellent summary of how genomics will provide critical information about infectious agents used in bioterror*).

Richardson L. Buying biosafety—is the price right? New England Journal of Medicine 2004; 350:2121–2123 (*brief perspective on issues related to two new biocontainment laboratories for the USA*).

Stenger DA, Andreadis JD, Vora GJ, Pancrazio JJ. Potential applications of DNA microarrays in biodefense-related diagnostics. Current Opinion in Biotechnology 2002; 13:208–212 (*looks at future technology that could prove valuable in microbial forensics*).

http://www.niaid.nih.gov/biodefense/ (*Internet address for the NIH's biodefence research strategy*).

http://www4.nationalacademies.org/news.nsf/isbn/0309089778 ?OpenDocument (*provides a summary of the US National Academies report "Biotechnology Research in an Age of Terrorism"*).

http://www:dhs.gov/dhspublic/ (the US's Department of Homeland Security giving details on biosecurity).

10

ETHICAL, LEGAL AND SOCIAL IMPLICATIONS

INTRODUCTION

OVERVIEW

The practice of medicine has the welfare of others as its focus. The fundamental principles of good medical care are also the basic principles of ethics, as demonstrated by (1) **Integrity** as a guiding value for researchers and practitioners with a commitment to honest, ethical conduct in the search for knowledge or delivery of care. (2) **Respect** for persons in their need, their welfare, autonomy and rights to confidentiality and privacy. (3) **Beneficence** in the responsibility of the practitioner and the researcher to do good and minimise harm (non-maleficence), and in time to provide benefit. Application of the principle of beneficence is also intended to ensure that research and new developments are socially worthwhile. (4) **Justice** requires equity in the provision of care, and accountability in the use of public funds for research and the distribution of benefits.

During the Nuremberg trials, the judges issued a 10-point statement that became known as the Nuremberg Code. For medical research the code required that (1) The subject give voluntary consent, and that any risk involved should never be greater than the likely benefit. (2) The research be justifiable and socially worthwhile. (3) The investigator has appropriate knowledge and skills, and is aware of what has been done previously in that field. The World Medical Association embraced the Code, and in 1964 expanded it to the *Declaration of Helsinki*, which made recommendations to guide physicians in their conduct of medical research on humans. They were intended to improve diagnostic and therapeutic procedures, which of necessity would sometimes be combined with professional care. The original *Declaration of Helsinki*, although reviewed and amended by a number of World Medical Assemblies (the last being in 2004), remains essentially the same. Various countries have, in addition, developed their own sets of guidelines based on similar principles; for example, in Australia there is a *National Statement on Ethical Conduct in Research Involving Humans*.

Because of the significance of genes and their genetic information, developments in molecular technology have attracted the attention of the public and the media, and have been monitored or reviewed in a number of enquiries. Input into these deliberations from laypersons has enabled more broad-based assessments of protocols and options. These activities have led in the UK to the establishment of the Human Genetics Commission. The purpose of this body is to provide prompt, centralised responses to expected new developments in genetic technology. These responses give government and the public some degree of assurance that a committee of experts and community representatives is available to guide future decision-making in areas of contention. US President GW Bush created a *President's Council on Bioethics* in 2001 to advise the President on ethical issues related to advances in biomedical science and technology. Topics to be addressed by the Council include cloning, stem cells, gene patents, biotechnology and public policy. In Australia, the Federal Government commissioned a far-reaching enquiry relating to human genetic information. This has resulted in an extensive report *Essentially Yours: The Protection of Human Genetic Information in Australia*.

To date, there have been few instances of unethical behaviour by physicians or scientists in the field of clinical genetics or genetic engineering. Established and accepted principles for the conduct of research using DNA have been followed. Nevertheless, careful monitoring is needed to ensure that the applications of molecular medicine continue to remain medically, scientifically and ethically sound. Molecular medicine is particularly vulnerable because there are financial rewards to be gained from intellectual property.

The risk of unethical practice is less in a community that is well informed and actively involved in decision-making. The community also has a responsibility. Lobbying is one route by which politicians can be made aware of priorities in health care. The lobbying must be balanced to ensure that facts are not overtaken by personal perceptions and agendas that have the potential to divert energies and resources to technologies that are unlikely to deliver what they promise. An example of this was seen in the recent stem cell debate when facts and emotion were difficult to distinguish at times. Training in ethics for health professionals and scientists is a necessity, and the lack of such education in many curricula needs to be addressed since situations with ethical, legal and social implications are likely to arise, particularly in the field of molecular medicine.

As has been shown consistently in molecular medicine, the advances in this field will be rapid and substantial. The likely innovative developments need to be balanced by their potential ethical and social implications. Professional knowledge and expertise carry with it the responsibility for the careful weighing of risk versus benefit for the individual (or the community as a whole). Therefore, while the degree of benefit may not always be predictable, at least there should be no harm to emerge. The predictable risk should never outweigh a possible benefit. A number of issues that have potential ethical and social implications are summarised in this chapter. They illustrate the present situation as well as make predictions about the future. As indicated in Chapter 1, the Human Genome Project has been an outstanding technological success. It has also captivated public attention and stimulated community and scientific debate about the ethical, legal and social implications (ELSI) of genomics research.

INFORMED CONSENT

The process of obtaining informed consent for a variety of medical procedures involving routine clinical activities or clinical research is well established. Molecular medicine must fall within this framework, but it is made more difficult by the rapid changes that occur. This complicates the issue of informed consent because the health professional must provide factual information so that appropriate decision-making can follow. Our understanding of genes and genetic disorders has the potential to change, so what was contemporary knowledge at the time of the consent process may be viewed in a different light some time later. Such an example occurred in Huntington's disease DNA predictive testing. In the early 1990s, the normal reference range for the expanded triplet repeat had not been established, and there was some inter-laboratory variation in the way the triplet repeats were counted (see below, as well as Chapter 3 for further discussion about Huntington's disease DNA testing). Some laboratories reported an abnormal number of triplet repeats as 36 or more; others used greater than 37 or greater than 38 repeats. Today, there is unanimous agreement that 40 or more repeats are abnormal and 100% predictive for the development of Huntington's disease. During this transition period, the consent process had the potential to be flawed because of the uncertainty with repeat numbers. In some cases, patients had to be recalled years later to revise what had been discussed during earlier counselling visits.

The consent can also be complicated by real or potential conflicts of interest. This is illustrated by the Jesse Gelsinger case summarised in Box 6.5. Here, clinical investigators and their University had a financial interest in the company producing a gene therapy vector that ultimately caused the death of a subject. This conflict of interest may or may not have been fully explained to the patient prior to the clinical trial, but it remains a contentious issue. However, gene therapy trials are expensive and so often require a sponsor. A sponsor in the context of a clinical trial is the individual, company or organisation that takes responsibility for running and/or financing the trial. The sponsor will often be the company that manufactures the product to be studied in the trial. Therefore, conflicts of interest may be inevitable, but they become a particular issue when the investigator is also involved with the company.

CONFIDENTIALITY AND PRIVACY

Doctor-patient confidentiality is well established. The health professional's duty of care to patients is protected by law. Many countries now have specific privacy laws enacted to protect the individual. However, the concept of privacy may need to be reconsidered within molecular medicine because any DNA-based knowledge has the potential to affect the health of others, since in terms of genetic disorders, the same DNA is shared by other family members. Whether one is dealing with the straightforward single gene defect, such as cystic fibrosis, or the more complex multifactorial disorder such as heart disease, the genetic elements are shared. What risk this constitutes can sometimes be accurately determined, but in other cases all that is certain is that there is an increased (or decreased) but difficult to quantify risk. In this circumstance, an interesting but important dilemma will be to define the boundaries for the physician's duty of care (see further discussion under Medico-legal Challenges). This issue will become particularly relevant with direct-to-public DNA genetic testing.

DNA TESTING

DNA has an intrinsic versatility that makes it ideal for genetic testing, as described in Chapter 2. On the other hand, this versatility can lead to unnecessary or inadequate testing, which may cause more harm than good.

PREDICTIVE (PRESYMPTOMATIC) DNA TESTING

Although rare, Huntington's disease is a useful model illustrating the advantages and disadvantages of DNA testing. Until DNA tests became available in 1986, the individual who was at risk for Huntington's disease had to wait until the fourth decade or so before the first clinical features became apparent. This uncertainty was further compounded by first-hand experience of the inexorable clinical deterioration that occurs in this disorder because a parent or close relative would have been affected.

Huntington's disease was first localised to the short arm of chromosome 4 in 1983, although the gene itself was not isolated for another decade. However, even without the actual gene, an **indirect** approach to predictive (presymptomatic) diagnosis became available. Linkage analysis enabled the mutant phenotype to be predicted as part of a family study. From this, the *a priori* risk of 50% for an offspring of an affected individual could be lowered to 1–5% or increased to 95–99% depending on whether the individual had inherited a DNA marker that co-segregated with the normal phenotype or the Huntington's disease phenotype. The risk estimates given included a conservative 1–5% error rate, which reflected the recombination potential for the DNA markers in linkage with the putative Huntington's disease gene. With the isolation of the causative gene in 1993, DNA testing progressed to the next level, i.e., **direct** DNA predictive testing for Huntington's disease by looking for a mutation in the appropriate gene.

The positive side to DNA testing in Huntington's disease is that the risk for an individual can be assessed at any time during life. With this knowledge, informed decisions about lifestyle and other key issues such as reproduction can be made. With DNA testing, prenatal diagnosis becomes an option, whereas previously, at-risk couples often elected to have no children or adopt or undergo *in vitro* fertilisation (IVF) by donor insemination. In some cases, the replacement of uncertainty with a DNA diagnosis, whether it is favourable or not, enables the individual to adapt and lead a better quality life.

There are also negative aspects to DNA testing. For linkage analysis, it is essential to study a number of family members. Key individuals are particularly important since it is they who allow linkage between a clinical phenotype and a DNA marker to be established. In the context of Huntington's disease, a key individual is one who has unequivocal evidence for the disorder. Less helpful is the family member who is relatively young, and so a confident diagnosis of not affected is never a certainty (see Chapters 3, 4). Apart from the resource-intensive nature of linkage analysis, a number of ethical issues must be considered in undertaking a family study, particularly one that may need to be extended to include distant relatives. Confidentiality will invariably become a concern. Even the first step, which involves construction of a pedigree, is undertaken without the consent or knowledge of family members. Disclosure of the pedigree may lead to key individuals being coerced by others into supplying information or giving blood. This could reach the courts of law, with an uncooperative relative being threatened by legal action. At present, the example given above produces no major dilemmas since the unwilling participant would not be compelled by the courts to give blood or have his/her privacy infringed by the enforced disclosure of personal details. Nevertheless, a court might be convinced to rule on the grounds that good for the majority (the person wanting testing and his/her family) would outweigh concerns by one individual.

Another issue that can be contentious in Huntington's disease is the time to test for this disorder. DNA predictive testing is usually undertaken for an individual who is at 50% risk; i.e., a parent has Huntington's disease. However, in some circumstances, the risk remains at 25% because the at-risk parent does not want to be tested, examined neurologically or even acknowledge that he/she is at risk. A dilemma then occurs when the person at 25% risk is tested and shown to carry the Huntington's disease mutation. In this situation, the genetic status or risk of other family members can be changed (Figure 10.1). Ethical dilemmas also arise as a consequence of too much information becoming apparent following DNA testing. Some examples are given in Box 10.1.

PRENATAL DIAGNOSIS

The increasing scope described for DNA predictive testing is also seen in prenatal diagnosis. Prior to the availability of chorionic villus sampling, prenatal detection of genetic diseases was possible only during the second trimester of pregnancy. Delayed results meant a difficult termination of pregnancy should this be requested. Today, first trimester diagnosis by chorionic villus sampling makes termination of pregnancy less traumatic and increases the number of disorders detectable, since DNA can be isolated with relative ease from chorionic tissue. The above, as well as the increasing

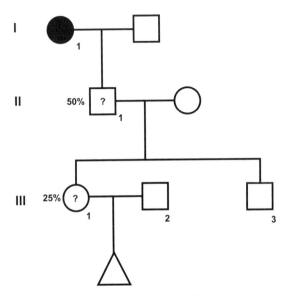

Fig. 10.1   DNA testing for Huntington's disease.
A female individual with Huntington's disease (●) is depicted as I-1. Her son (II-1) has a 50% *a priori* risk for Huntington's disease. Her granddaughter (III-1) has a 25% *a priori* risk. In this situation, DNA testing is not usually undertaken for the latter because her risk has not reached 50% (this would occur if her father has a positive DNA test, or he develops signs or symptoms of Huntington's disease). The dilemma arises if the granddaughter wishes to know whether she carries the mutation for Huntington's disease (this often becomes an issue if reproduction decisions are to be made or there is a pregnancy, as has occurred in this example), and the asymptomatic father refuses to be seen by a neurologist or undertake a DNA test or for whatever other reason does not want to be involved. Even without the father's participation, DNA testing in the granddaughter can be undertaken. A negative result in her does not alter the status quo. However, a positive result will mean her father must have Huntington's disease. It will also increase the risk for her brother (III-3) from 25% to 50%. Does she tell the brother this information? Will it be possible to prevent the information that the father is affected from reaching him? If either or both siblings are affected, they will no longer be able to seek help and support within the family framework because doing so would establish the father's genetic status.

Box 10.1   Case studies: Some ethical dilemmas in DNA predictive testing.

A possible finding in DNA linkage testing is non-paternity. Such an example is given in Figure 10.2. This finding can be put aside and not disclosed to comply with the principle of non-maleficence. However, what if the non-paternity then excludes the consultand or other family members from being at risk? Autonomy and beneficence in respect to the consultand must be balanced by confidentiality and privacy of a family member. Will the ethical dilemma be made easier if the latter is deceased but the spouse is alive? Another, less common example of too much information coming from direct gene testing involves monozygotic twins at risk for a genetic disorder. This would occur if one of the pair wanted to have DNA predictive testing but the other declined. However, because they are monozygous, a result from one would automatically apply to the other. The first individual has the right of autonomy; the second is entitled to privacy, including the right to know or not to know. Involving both is the principle of beneficence. In some DNA testing programs, this potential dilemma would be resolved by linking the testing to both or none; i.e., beneficence overrides.

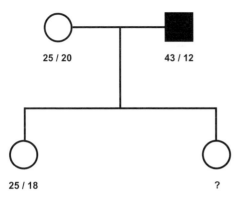

25 / 20   43 / 12

25 / 18   ?

Fig. 10.2   DNA test complicated by non-paternity.
A symptomatic male with Huntington's disease (■) has 43 and
12 triplet $(CAG)_n$ repeats (see Chapters 2 and 3 for further
discussion of triplet repeat expansion). The 43 repeats confirm
the diagnosis of Huntington's disease in him. His two
daughters are at 50% risk. One daughter is tested (the other
indicated by a ? does not want to know or be tested). The
daughter who is tested has $(CAG)_n$ repeats of 25 and 18.
Provided this is not a laboratory or blood collection error,
these results show (1) She does NOT have Huntington's
disease because neither of her $(CAG)_n$ triplets is 40 or more in
number. (2) The other reason she does not have Huntington's
disease is the likelihood that the father is not her biological
father. This is assumed because the 25 repeat in the daughter
has come from the mother, and her 18 has come from another
person since her putative father has repeats of 43 and 12.
Non-paternity would need to be confirmed by a panel of
DNA markers (see Chapter 9). Although the daughter who has
not been tested remains at 50% risk, the finding of non-
paternity in the sister adds an additional complexity. Because
families with genetic diseases often know a lot about their
disease (the Internet has made dissemination of knowledge
considerably easier for laypersons), patients and families not
infrequently ask the clinician to go over the actual DNA
results with them. In this particular case, such a request would
quickly show up the possibility of non-paternity and add an
additional complication to the family that is already under
stress.

**Box 10.2   Case study: Failure to diagnose
haemoglobin Bart's hydrops fetalis.**

The development of hydrops fetalis (i.e., congestive
cardiac failure in the fetus) can be caused by a
number of conditions, including infection, Rh
haemolytic disease and a form of α thalassaemia
known as haemoglobin (Hb) Bart's hydrops fetalis.
With the increasing use of ultrasound during pregnancy, the fetus can be monitored, and anomalies
detected. In one particular case, a fetus was demonstrated to be developing hydrops fetalis, but the
underlying cause was not adequately investigated,
even though the parents were from South East Asia,
an area in which Hb Bart's hydrops fetalis is common.
The fetus was transfused *in utero*. The child was born
alive, but unfortunately the basis for the illness was
shown belatedly to be Hb Bart's hydrops fetalis, a
condition that is not usually treated because it is fatal
in utero or the baby dies soon after it is born. A diagnosis of Hb Bart's hydrops fetalis in the affected fetus
could have been made very easily by looking for the
common deletion in the α globin gene known to
cause this problem. Apart from a very ill newborn
who will forever need regular blood transfusions and
iron chelation therapy to remove excess iron, this
case illustrates another ethical dilemma. Research
studies are being conducted in South Asian countries
where Hb Bart's hydrops fetalis is common. These
studies are looking at the consequences of intrauterine blood transfusion. To date, the results are preliminary, and although there have been a small number
of survivors, there remains some doubt about the
quality of life that these individuals are experiencing.
However, for counselling purposes, parents with an
at-risk pregnancy may now be faced with an additional choice—the possibility of intrauterine blood
transfusion, the efficacy of which remains in doubt.

availability and commercialisation of DNA diagnostic
kits, has meant that the decision whether or not to
undergo prenatal diagnosis is being faced by an increasing number of couples.

The ethical, cultural, religious and social implications
of terminating a pregnancy will not be discussed here as
they are generic to medicine and not specifically molecular medicine. These issues will influence a couple's final
decision in relation to prenatal DNA testing. Complicating the decision-making will be the knowledge that many
of the genetic disorders that were once untreatable (for
example, cystic fibrosis) are now able to be managed
better. In some cases, options such as bone marrow transplantation can lead to cures. Nevertheless, the medical
and psychosocial implications of having a child with a
genetic disorder are considerable, and appropriate

screening for early detection of risk couples, as well as
sufficient time for decision-making, is appropriate in
these circumstances. The dilemmas facing the couple can
be even more difficult if there is no first-hand experience
of the genetic disorder being tested, a scenario that is
increasingly the norm as more screening options are used
during pregnancy. It is also possible that failure to use a
DNA test can itself lead to an ethical dilemma (Box 10.2).

Sex selection by prenatal testing or other forms of
sexing (for example, preimplantation genetic diagnosis
[PGD] or sperm sorting) are practised in some communities. The standard sex ratio for newborn boys to girls
is 105 : 100. In some countries, this ratio is closer to
120 : 100 (data from the US President's Council on
Bioethics web site). Determining the fetal sex because
there is risk for an X-linked genetic disorder is accepted
in medical practice, but a DNA test to identify fetal sex

for family balancing or when one sex is preferred over the other is contentious for reasons including (1) Limited resources are being used to undertake non–medically-related DNA testing. (2) The test is a form of sex discrimination; i.e., one sex (usually male) is preferred over the other. (3) Long-term societal effects, if the trend leading to a predominance of males continues. (4) The slippery slope argument, i.e., sexing now, and when the genes for other traits (personality, performance and so on) are found, they will be requested.

Surprisingly, sex selection has kept a low profile in the community and media, In affluent countries, those who can afford the test can have the relatively expensive PGD or sperm sorting so they do not have to terminate a pregnancy. Those who cannot afford it must rely on chorion villus sampling or amniocentesis and a possible termination of pregnancy. Ethicists have defended sex selection for various reasons, including individual autonomy and the basic right of a couple to choose the sex of their baby. On the matter of sex selection, a working paper (US President's Council on Bioethics) designed to promote discussion but not representing official policy made three recommendations: (1) Do nothing, and hope that it will work out. (2) Engage the various medical governing bodies to encourage self-regulation. (3) Take legislative action. To date, this issue has not progressed further because no doubt the path ahead was going to be very difficult.

Somewhere between prenatal testing for a serious genetic disorder and prenatal testing for sex selection lies prenatal tests that are being undertaken for adult-onset genetic disorders. In these circumstances, the fetus, if affected, will not develop the genetic disorder for perhaps 20–40 years into the future, by which time new treatments or even cures are likely to be in place. These are problematic prenatal diagnoses, particularly for the laboratory staff, who often work under a degree of stress in this area. One hope for a solution acceptable to all is that the developments in fetal therapy come soon so that the ethical, social and legal issues related to prenatal testing are resolved by testing the fetus for abnormalities, and then following this with the appropriate treatment to correct the defect identified.

POPULATION SCREENING

Many types of screening programs could be implemented on the basis of DNA testing (see Table 2.9). The utility of DNA testing strategies, particularly the potential for PCR to test many samples quickly and relatively cheaply, has meant that widespread screening of the population becomes a practical consideration. As more genes are sequenced, the number of mutations identifiable by PCR will increase. Gene chips and related technologies will open up the options for widespread DNA testing.

Two early examples of selective screening programs targeted to at-risk populations illustrate the potential advantages and disadvantages of this type of testing. Tay Sachs disease is a fatal neurodegenerative disorder of childhood. It is inherited as an autosomal recessive trait. Since the early 1970s, individuals at risk for Tay Sachs have been screened and counselled. The incidence of Tay Sachs disease has been reduced without the societal problems that developed following the implementation of population screening for sickle cell disease, another autosomal recessive disorder. Sickle cell disease leads to considerable morbidity and mortality although the ultimate outcome is not entirely genetic in origin as environmental factors are important. The US-based sickle cell screening program, which was also started in the early 1970s, was targeted to the at-risk African American population. The initial version of this program produced more harm than good. Results led to a lowering of the self-image, overprotection by parents and discrimination. The discrimination came from employers, insurance companies, health insurers and potential spouses. Why did these two programs produce different outcomes?

One reason for success with Tay Sachs screening was the target group, which comprised individuals of Jewish origin who had better educational opportunities and social infrastructure. Another contrast between the two programs was the close community consultation undertaken prior to genetic testing for Tay Sachs. Because of the problems associated with sickle cell screening, changes were made to the program including the removal of legal compulsion to screen, and improved counselling and education facilities. These changes enabled more successful testing to be pursued. Experiences with these programs illustrated the necessity for counselling and public education to explain the significance of mass screening results as key ethical considerations in program design.

The modern screening dilemma is illustrated by cystic fibrosis. DNA amplification by PCR enables the common ΔF508 defect to be detected in peripheral blood or more accessible tissues such as cells present in a buccal swab. As summarised in Chapter 3, more than 1300 mutations produce cystic fibrosis although ΔF508 is the most common in those of northern European origin. Other mutations are much less frequent (see Table 3.8). The ethical and social issues raised include the use of a test that will not detect all those who are affected. For example, if only the ΔF508 mutation is sought, false negative results in couples from a population with a frequency for the ΔF508 mutation of 70% will be $0.51(1 - (0.7 \times 0.7))$; i.e., approximately half the couples will not be informative by this approach. The detection of the less common mutations (some of which are present only at a 1–2% frequency in the population) will add to the workload but not increase substantially the information to come from the screening program.

Debate continues about the value of mass population screening in contrast to testing individuals or at-risk families, i.e., selective screening. Even if laboratory facilities are available, major efforts directed towards genetic counselling and public education would be required to ensure that those tested fully understood the implications of the results. The financial resources to carry out a mass screening program would be enormous. In view of this, and the inability to detect all mutations with present technology, the majority of reviews and enquiries have recommended limited screening—perhaps of pregnant populations or selective screening of groups who are at higher risk than the general population.

Additional problems that need to be resolved before embarking on a widespread cystic fibrosis screening program include the uncertainty of the mutations in respect to disease severity. Thus, counselling in a number of instances will be difficult and incomplete. An example can be seen from the IVF programs which have shown that some cases of infertility in otherwise healthy males are due to cystic fibrosis. If, in this circumstance, the female partner carries the ΔF508 mutation, offspring conceived by IVF using sperm isolated from the infertile male have a 50% chance of being compound heterozygotes for ΔF508 and one of the mild defects carried by the infertile male. The phenotype that this produces will be difficult to determine since without IVF-based technology infertile males would not have transmitted that particular mutation.

Many of the newborn screening programs have now successfully incorporated DNA analysis for cystic fibrosis into their standard biochemical testing protocol (see Chapter 7). This has streamlined the laboratory's work and has reduced anxiety in families because the number of newborns required to be recalled for definitive testing is reduced. However, the advantages might need to be balanced with the growing cohort of children in the population who have been identified as carriers of the cystic fibrosis mutation. What effect, if any, this information will have on the way they develop should be an important subject for future research. Finally, in all examples of population screening, data storage would have to be secure to ensure confidentiality.

COUNSELLING

Individuals seeking advice because of a family history of genetic disease or a couple in the prenatal diagnosis situation are given the best information available at the time that will allow them to make their own decisions. However, the rapid advances in molecular genetics can both assist and complicate this type of counselling. Information gained from molecular genetics can make the counselling process more difficult because it provides some data but not enough for the couple to make a decision. For example, a question frequently asked is How

severely affected will be an offspring with a particular genetic disorder? In β thalassaemia, the molecular basis for the milder, non-transfusion-dependent form called "thalassaemia intermedia" has been determined in some cases. However, other factors are also involved, making population-based data less relevant to the individual case. Has knowledge of the molecular pathology of thalassaemia, arguably the most intensively studied of all Mendelian genetic defects, reached the stage that the co-inheritance of β thalassaemia with other genetic changes (e.g., an increase in fetal haemoglobin or coexisting α thalassaemia) will enable a confident prediction of severity to be made? Unfortunately, the answer is no.

Therapeutic options, particularly those related to molecular medicine, are also changing. For example, gene therapy in cystic fibrosis and the availability of an oral chelating agent in thalassaemia will become future treatment considerations. These treatments should improve the outlook for both disorders and need to be considered by prospective parents.

Criticism has been levelled at commercial DNA testing laboratories that make available a DNA result without considering whether the test is indicated, or there are the resources that would allow the test's implication to be understood by the patient or even the health professional. This problem could be resolved if the provision of DNA testing were clearly linked to counselling, but unless there is a national health system, this would be difficult to enforce. Particularly worrying is direct laboratory-to-patient DNA testing. A study by Williams-Jones (2003) identified three such examples following an Internet search. A related activity is the initiative by some companies to provide genetic (DNA) profiles for the purpose of **personalising health care**. In the UK and USA, companies offer cardiovascular, osteoporosis, ageing and other genetic screens. These screens are predominantly based on SNP markers that have been associated with an increased risk for disease. From the population-based genetic profile, the customer (the person is a true customer now, and not a patient in the traditional sense) can decide to make changes in lifestyle—for example, taking vitamins—in response to the genetic findings.

The era of personalised medicine has much to promise, and was discussed in some detail under Pharmacogenetics in Chapter 5. However, the concept can also be criticised because the scientific information provided is often incomplete, and it is doubtful that the public can make a full assessment of the genetic data when, in many cases, scientists themselves remain unclear how reliable is the SNP association with a particular disorder.

DISCRIMINATION AND STIGMATISATION

To the individual, DNA screening has the advantage of detecting potentially reversible problems early, or allowing preventative measures to be implemented prior to

Box 10.3 The case of the Burlington Northern Santa Fe (BNSF) Railroad.

Based on an unfortunate ignorance of genetics, the BNSF attempted to DNA test employees who had claimed compensation for carpal tunnel syndrome developed during work. The DNA test was undertaken because it was purported to detect a gene mutation that produced a hereditary neuropathy associated with pressure palsy. When this testing became public, the company had to back down and settle out of court in response to legal action from the employees. A number of inappropriate issues were identified in this case: (1) DNA tests had been undertaken without consent. (2) The test itself was of no value because there was little evidence that the DNA mutation would predict the development of carpal tunnel syndrome in the circumstances being investigated. (3) The use of the DNA test was illegal as the US Government in 1995 legislated to define an individual's genetic status as a disability in some circumstances. For an employer to use this information in employing or dismissing an employee would be discriminatory.

disease being established. In the workplace, there is the potential to know, through DNA testing, which industrial toxins are particularly noxious to an individual based on that person's genetic makeup. The disadvantage of the DNA test in these circumstances is the risk that this will lead to loss of insurance or employment (and so health insurance if this is linked to employment). An example used to illustrate how discrimination in the workplace has occurred because of genetic (DNA) testing is the Burlington Northern Santa Fe Railroad case (Box 10.3). However, this is not a particularly good case because it illustrates more a basic ignorance of genetics than a DNA test being used in the workplace with the potential to discriminate against employees. The fact that this example has been quoted by a number of sources might even be a positive message; i.e., this type of discrimination is not a major problem, and so there is time to ensure that inappropriate DNA testing in the workplace does not occur.

On the other hand, DNA testing in the workforce (included in this is the sporting arena) that is conducted appropriately might be beneficial and should not be totally excluded. An interesting case is the unconfirmed report that boxers with the apolipoprotein E genetic type known as E4 (*APOE4*—the same genetic predisposition marker in late-onset and sporadic Alzheimer's disease) are at greater risk of chronic traumatic brain injury. Where does this place the various boxing federations as well as sporting institutions or organisations that train boxers? Should *APOE4* DNA testing be required of

boxers? One could argue that this type of DNA test in the workplace or sporting arena might be justified if it identified a predisposition to serious brain damage from boxing. It is surprising that advice from some lawyers is that under discrimination and disability laws, it would be difficult to exclude *APOE4* positive individuals from being trained in boxing or prohibit them from taking part in matches! This example demonstrates how discrimination and disability laws may not function as they are meant in the DNA environment. There is little doubt there would be a medico-legal obligation on the part of the medical practitioner to warn a boxer that there might be a DNA genetic test that will provide a risk of long-term brain damage in boxing (assuming this association is eventually proven to be correct).

In late 2004, the German Government came under criticism for producing draft legislation that would allow employers to undertake DNA tests on applicants for certain jobs. The example given was DNA testing for vision disorders in bus drivers. This alarmed members of ethics committees because it represented the beginning of the "slippery slope" in their view. If legislation is enacted, the long term effects of this government initiative will be important to monitor.

Insurance companies are aware of the potential for DNA screening or testing. A number of reasons are given to justify the use of DNA testing, including the duty of the insurance company, which is a business, to the other policyholders and shareholders. A UK House of Commons committee was informed by representatives of the insurance industry that providing life insurance for the ~5% of individuals presently excluded because of pre-existing medical conditions would at least double the cost of current premiums.

Insurance companies base their policies on actuarial calculations related to the likelihood of death, loss or damage. In terms of life insurance, those who fall into high-risk groups are usually able to be insured although at a much higher premium. The medical examination, the medical history and, in some cases, a compulsory test for HIV are accepted as part of the prerequisites for obtaining life insurance. However, the potential to utilise the DNA or genetic makeup of an individual has opened a Pandora's box. Governments have responded in a number of ways, from proposing a form of universal insurance coverage that would have no preconditions to a complete ban of genetic testing when it comes to insurance matters. However, in most countries there remains some uncertainty about the way in which the insurance industry will use DNA testing, and so governments have adopted a wait-and-see approach. At present, many insurers follow a policy that a DNA test will not be specifically requested in an individual with a family history of genetic disease. However, if the DNA test has been undertaken, the insurance company will expect that the result is made known.

A number of key issues still need to be resolved. For example, will DNA testing comprise another component of the actuarial assessment or will it be used to deny insurance or make the loaded premium unattainable? Presently, a family history of Huntington's disease will place an individual who wants to take out life insurance in the at-risk category, and the premium will be modified accordingly. However, how would the insurance company react to the following two scenarios? (1) Prior to insurance being obtained, the at-risk individual voluntarily undertakes DNA testing and is shown to carry the mutant gene. (2) Following the insurance premium being negotiated, DNA testing identifies the mutant gene. It is presumed that in both the above hypothetical situations, the finding of a normal gene will place the person in the general population risk category.

TESTING CHILDREN

One important advantage of DNA testing is that it can be undertaken at any age. This can also lead to an ethical dilemma when healthy children are tested. In these circumstances, informed consent is given by their parents. Why are children being tested? An acceptable medical indication is a childhood disorder or an adult-onset disorder for which early intervention will improve the prognosis. Non-medical indications that have been identified include lifestyle planning decisions such as choosing the appropriate environment in which to bring up a child with a genetic disorder. Dubious indications include the options for planning of future educational, career or reproductive decisions by the child.

Ultimately, it is the parents acting as the child's guardians who will make the request for DNA testing. Whether this is to relieve anxiety on their part or a legitimate medical indication or a combination of both can be difficult to assess. Relief of parental anxiety must be balanced by the problems of DNA testing mentioned already, and the potential harm to a child whose disclosed genetic status may lead to an unnecessary change in the way the child is allowed to develop, as well as negative emotional consequences and damage to self-esteem. However, the effect of DNA testing, particularly predictive testing in childhood, may be less traumatic than previously thought. In fact, judging from the results of a recent study, children seem to tolerate these DNA predictive tests better than adults, at least for the short term (Box 10.4). More research is needed to determine the effects of DNA genetic testing in children.

To avoid difficult decisions in these circumstances, some DNA programs—for example, Huntington's disease—will not test individuals unless they themselves are able to give informed consent; i.e., they have attained the age of 16–18 years. The Huntington's disease example is relatively straightforward since this adult-onset disease has no known treatment, which can influ-

> **Box 10.4 Predictive genetic testing in children (Michie *et al* 2001).**
>
> The purpose of a study by Michie *et al* was to compare the emotional impact of DNA predictive testing on children and adults who had been tested for familial adenomatous polyposis (FAP), a genetic form of colon cancer discussed in more detail in Chapter 3. Measures such as depression, anxiety, optimism and self-esteem were sought. Follow-up was relatively short at 12 months, and so the results reflect shorter rather than long-term consequences. Surprisingly, it was shown that children (aged from 10 to 16 years) tolerated predictive testing for FAP mutations, perhaps even better than adults. Children receiving positive or negative results did not experience greater anxiety or depression than adults. Another surprise was the finding that there are actually very few studies that have attempted to determine the effect on a child who undertakes this type of DNA testing. Another observation requiring more research was the finding that it might not be necessary to subject everyone to the standard counselling (pre- and post-test) protocol, but it might be more efficient and less resource intensive to target individuals who tend to pessimism and are low in self-esteem, as they appear to be particularly at risk of responding negatively to this type of DNA testing.

ence its natural history. On the other hand, more debate has occurred in respect to familial hypertrophic cardiomyopathy, another autosomal dominant disorder that usually manifests in adult life (see Chapter 3 for further discussion on the molecular genetics of this disorder).

Views held in favour of testing children for familial hypertrophic cardiomyopathy include the relief of anxiety for parents and the child, if the underlying mutation is excluded, or the test ensures that those identified as carrying the mutation are more effectively followed and encouraged not to participate in competitive sports, which are associated with sudden death. Some medical surveillance measures and surgical interventions could prevent complications, particularly ventricular arrhythmias. The opponents to this view point out that no medical intervention has yet been shown to alter the natural history of the disorder, and the harm coming from early knowledge that the child carries a defective gene could have a negative effect on that individual's subsequent development. In this debate it is interesting that the proponents for DNA testing are more likely to be the clinicians (e.g., cardiologists, paediatricians) who deal directly with affected individuals, whereas the opponents for testing are health professionals specialising in genetics. Informed debate and objective research are now required to enable a consensus to be achieved.

Table 10.1 **Points to consider when DNA tests are undertaken in children or adolescents[a]**

Issues	Benefits	Harm-causing
Medical	Improved prevention or therapy. More effective surveillance. Improved diagnosis and prognosis.	Failure to adequately implement the benefits.
Psychosocial	Reduction of uncertainty and anxiety. More time for adjustment. More time for planning in a range of activities.	Altering self-image. Increased anxiety and guilt. Altered expectations for a range of normal activities. Distorted perception of the child. Discrimination at school, employment or insurance. The finding of non-paternity, adoption.
Reproductive	Informed family decisions by the parents	

[a] Policy statement from American Society of Human Genetics and the American College of Medical Genetics.

Guidelines are helpful in complex cases, but flexibility needs to be maintained to deal with those situations in which DNA testing will be beneficial to the child. In the context of a research protocol, the testing of children can occur, but due consideration will need to be made since the process is more complex than that involving the mature adult who can consider the implications for him/herself. A series of conclusions and recommendations relating to the issues of DNA testing in children and adolescents may be found in a 1995 policy statement from the American Society of Human Genetics and the American College of Medical Genetics (Table 10.1).

RESEARCH

PURSUIT OF KNOWLEDGE

The pursuit of knowledge for its intrinsic value is becoming more difficult in today's outcome-driven priority-based research. Investigators are frequently encouraged to strive for a marketable end-product that can be patented, the outcomes measured, and results obtained within a designated time frame. While shrinking resources are in part the reason for this type of rationalisation, the trend has the long-term potential to stifle creative or innovative work. Another consideration is the increasing government influence in the type of research undertaken. Since governments often provide a significant proportion of the research funds, there is some justification in their demand for greater input into the types of work being undertaken. National priorities or specific targets can be identified, and funding is used to drive research in these directions. The comments made above have particular relevance to molecular medicine, which is very much technology based, and so requires considerable infrastructure and does not necessarily produce results within a short time frame. In this environment, publicly funded researchers may be disadvantaged compared to their counterparts in industry because the latter can take a longer term view of outcomes, with the increased costs of this strategy eventually passed on to the community.

The molecular medicine era is one of the most exciting in the modern history of medicine. Many developments have occurred at a time when there have been significant shifts in the way in which research is being undertaken. The free-thinking or hypothesis-driven research is giving way in some circumstances to goal-driven activities that have attracted funding because of perceived commercial benefits. The Human Genome Project illustrates another approach in which a mammoth undertaking will be resolved by technological blitzkrieg. For some time purists did not perceive genome mapping as anything other than data collection because it did not follow the traditional avenues in research. However, there is little doubt that information coming from the technology-driven Human Genome Project has produced very useful and practical knowledge and has opened up directions for the more traditional type of research. Times are changing rapidly. The effects that these shifts in strategies will have on future developments in molecular medicine will take time to assess.

PATENTS AND INTELLECTUAL PROPERTY

A patent is an intellectual property right acquired by the inventor of a new, inventive and useful product or process. The raison d'être for a patent is to encourage an inventor to place an invention in the public domain in exchange for certain rights for a limited period (usually 20 years). The patent allows the individual to (1) stop others from exploiting the invention during the life of the patent, (2) exploit the patent and (3) licence the patent to others. Criteria required before a patent is granted include (1) It is a novel or new invention. (2) An inventive (non-obvious) or innovative step is involved. (3) Usefulness (utility) must be demonstrated. Within the above framework, there are differences in interpretation. For example, the UK's Nuffield Council on Bioethics has claimed that the inventiveness step on patents for DNA sequences is easier to satisfy in the USA. In contrast, it is said that the utility component of a patent is more strictly observed in the USA.

In some countries, notably those in the European Union, ethical considerations may exclude a patent from being filed. In contrast, countries such as the USA, Canada and Australia make no provision for ethical and social considerations in the patent process. Countries are also bound by international treaties on patents such as the World Trade Organization (TRIPS) agreement, and they restrict what a country can do with a patent. For example, compulsory licensing can be evoked by government in certain circumstances, but the TRIPS agreement requires appropriate compensation to be paid. The monopoly granted by a patent also overrides anticompetitive legislation, and this is particularly concerning for DNA testing because it can mean that restricted licensing has the potential to impact negatively on quality assurance. In the case of the Myriad Genetics patent with *BRC1*, the company's original position was that all DNA testing for breast cancer would be conducted in its USA-based laboratory.

Some consider patents as a necessary evil because without them the huge costs involved in developing new drugs or therapies would be a barrier to progress. A number of the very important recent gene discoveries have been made by private companies. The medical and scientific community is now strongly encouraged to derive benefits from the patenting of important discoveries. Patents have become important criteria for the investigator's productivity and competitiveness in the peer review process for research funding. Many governments establish a direct link between improvements in health through generation of wealth in biotechnology.

Until the molecular era, patents were obtained for therapeutic substances or technologies produced by processes that were more easily identified as being inventive. Now genes, gene sequences and their products are the subjects of patents. For example, a research group at the US National Institutes for Health (NIH) caused an uproar by filing patents for more than 6000 anonymous human brain–derived DNA sequences in 1991. These "genes" were isolated from a brain cDNA library, and their uniqueness demonstrated by sequencing a segment of the cDNA and showing on DNA database searches that the sequences were not present in the databases. Thus, they represented unique DNA segments called ESTs (expressed sequence tags) that, since they come

from a cDNA library, were likely to be segments of genes with, as yet, unknown function. Despite this, an attempt was made to patent these unknown "genes." Groups on both sides of the Atlantic were drawn into the NIH controversy, which was only defused when the patent applications were withdrawn.

The dilemma arising from the patenting of a DNA sequence reflects the philosophy behind a patent, i.e., a novel idea or invention that has some utility. Since the above cDNA clones had no known function, their utility was difficult to assess. At the time, the view held by many in the community, both lay and scientific, was that genes, humans and components of humans are not **inventions** but **discoveries**, and so inappropriate objects for patents. Many are still of this view, but the fact remains that the complete human genome has been patented. Both academic institutions and commercial companies continue to pursue the option to patent DNA segments in anticipation that this will prove to be financially rewarding some time in the future. Alternatively, the value of a patent might lie in the ability to block competitors. As well as the basic philosophy on the appropriateness of patenting the human genome, another concern about these types of patents is their broad claims. For example, the Myriad Genetics *BRCA1* patent claims rights over any use of the DNA sequence, including DNA genetic testing or any therapies that might emerge from knowledge of the DNA sequence.

Negotiations between patent holders and Directors of DNA testing laboratories, individual researchers and various organisations are often conducted under a commercial-in-confidence agreement, which means that discussions are secret. In this circumstance, individuals (and even organisations such as a University) are disadvantaged because of the non-disclosure, and there are the very high costs required to litigate. Whilst the proponents of patents point out that they can be challenged in the courts, it is instructive to review the Roche company's web site (http://www.roche-diagnostics.com/ba_rmd/pcr_litigation_chronology.html), which summarises ongoing litigation with Promega, another company, over the patent covering PCR (and no one disputes that PCR is a true discovery, unlike the patenting of a stretch of human DNA sequence). The reader will find the graphical summary of the chronology particularly informative. It leaves little doubt that very few would have the resources or the ability to take on a major company to challenge a patent. In the most recent court ruling of May 2004 (and there have been many in this long-lasting saga, which started in late 1993), both Roche and Promega are claiming victory although Roche will appeal one aspect of the ruling.

There is growing evidence that the patents issue has had a direct and negative effect on a number of DNA diagnostic services. Some have had to stop a particular DNA test, whereas others have not developed tests

---

Box 10.5   Patents and provision of clinical genetic DNA testing (Cho *et al* 2003).

A survey of 201 US-based laboratory directors enquired whether patents had interfered with their provision of a DNA testing service. Of the respondents, 25% indicated that they had to stop a DNA test, and 53% stated that they had not developed a new test because of the patent issue. In all, 12 genetic DNA tests could not be performed, and these tests were common ones in many laboratories. The 12 included *BRCA1*, *BRCA2*, *HFE* (haemochromatosis), FAP (familial adenomatous polyposis), Huntington's disease and Factor V Leiden (for predisposition to thrombosis). Overall, 22 patents covered the performance of these 12 tests. Perhaps one of the more irritating findings in this survey was that 13 of the 22 patents were based on research that had been funded by the US Government.

Box 10.6   Protecting public access to gene sequence information.

Four examples regarding the protection of public access to gene sequence information are given: (1) The SNP Consortium Ltd (an amalgamation of industry, academic centres and charity) was formed specifically to develop a high-quality single nucleotide polymorphism (SNP) map. This information would be available at no cost, it would not be patented and the data would be placed in the public domain. The SNP Consortium's aims were both altruistic and pragmatic—the latter being a defensive move to stop other companies from acquiring exclusive intellectual property rights to SNPs (read more about SNPs in Chapters 2 and 5). (2) Another example to suggest that the patent system needs reviewing in the area of genomics is the filing of patents for the SARS viral sequence by two not-for-profit public organisations (the US Centres for Disease Control and Prevention and Canada's British Columbia Cancer Agency). The justification given by these organisations was that this was the only way to ensure that the viral sequence is freely available to all because even publishing the viral sequence will not prevent others from filing patents. The CDC and the BCCA have publicly declared that this will now allow them to licence gratis any work that is for public good. (3) Cancer Research UK, a charitable organisation, obtained a Europe-wide patent on the breast cancer gene *BRCA2*. The purpose was to stop US-based Myriad Genetics from utilising its own *BRCA2* patent and so restricting access by European workers to DNA research and diagnostic testing on *BRCA2*. Cancer Research UK indicated that it would allow publicly owned laboratories to use information based on this gene sequence for no charge. (4) PUBPAT (Public Patent Foundation) was formed to represent the community against wrongly used patents and unsound patent policy (see http://wwwpubpat.org). Appeals or issues raised by PUBPAT have resulted in the US Patent and Trademark office re-examining or rejecting potentially lucrative patents involving industry and one university.

---

because of patent issues (Box 10.5). In this milieu, it is critical that government takes a greater role directly or indirectly in the negotiations to ensure that they are fair, and in the case of a dispute, there is the option through government of taking the patent holder to court.

In May 2004, in response to appeals from scientific institutes, the European Patent Office revoked Myriad's patent for breast and ovarian DNA testing that had been granted in 2001. This decision was based on the grounds that the monopoly would jeopardise the development of research and identification of new tests, and on review, the European Patent Office concluded that the discovery was not novel and errors had been made by the company. This decision which can be appealed is likely to be an important precedent for the future.

Another controversy surrounding patents involves Genetic Technologies Ltd. This Australian company claims a patent covering all non-coding DNA sequences, i.e., more than 98% of the human genome, as well as genomes in other organisms including plants. This has profound implications for DNA diagnosis, as well as research, because we are now dealing with the great majority of total DNA. For example, there will be few DNA primers required for PCR and DNA mutation testing that do incorporate knowledge of non-coding DNA in their design. Litigation has already started, and the results of this court case will be awaited by many.

There are alternatives to litigation in challenging a patent. One involves the taking out of a defensive patent (Box 10.6). Some legal experts have indicated that the patent system in respect to genomic research is healthy because it has been possible to protect the public interest by defensive patents, and there is always the potential for national compulsory licensing (although this is

rarely invoked). However, the necessity for these options only confirms that the present system cannot be sustained without having a long-term detrimental effect on the community and the advancement of molecular medicine. The costs alone in taking out and then defending a defensive patent are considerable and available to few organisations.

As shown in Table 10.2, patents are a legal minefield, with various options and strategies available to press home an advantage or, alternatively, ensure the competition is disadvantaged. Since money is increasingly being equated with medical discoveries, the obvious

Table 10.2 Patent variations, issues and dilemmas[a]

Patent issue	Explanation
Dependent patent	Patent on an invention that cannot be exploited without encroaching on an earlier patent (dominant patent).
Blocking patent	Patent used to inhibit developments by others.
Defensive patent	Patent taken out to prevent others from patenting (see Box 10.6)
Patent thicket	Multiplicity of overlapping patents making it difficult for others to navigate through this web to develop their own new technology.
Royalty stacking	Multiplicity of overlapping patents leading to the need to pay multiple licence fees.
Reach-through claims	Claims made by patent holders to future intellectual property in new products that might result from the use of a patented invention. These claims have the potential to restrict the licensee's rights to future inventions that might emerge.
Patent pools	One mechanism used to deal with multiplicity of patents. It involves a cooperative arrangement allowing the owners of several patents required for some product to licence or assign rights at a single price.
Licencing	Means by which patented technology is legally transferred to others for certain uses and under certain conditions. Unlike laws against anti-competitive practice, a patent is anti-competitive, and licensing (with very rare exceptions) is decided solely by the patent holder. Exclusive licensing has the potential to increase the value of the patent but diminish the value of the product at least in terms of clinical care.

[a] From Australian Law Reform Commission 2004 report on Gene Patenting and Human Health (http://www.alrc.gov.au/).

question to follow is Why are subjects of research studies not also sharing in the spoils? Previously, these subjects would have been satisfied with an altruistic motive for their participation. In particular, a number of cogent arguments have been made that, without the very large pedigrees required for positional cloning, genes such as *BRCA1* would not have been discovered.

GENETIC REGISTERS, DATABASES AND DNA BANKS

Repositories

The keeping of registers is an important component of clinical genetics. Registers come in various forms from local lists of genetic diseases to national registers. The compilation of a pedigree is one form of a register. As

indicated previously, many individuals identified on the pedigree are not aware of its existence or have given permission for their inclusion. In fact, to acquire informed consent in these circumstances is both difficult and in itself an intrusion of an individual's privacy.

A further extension of the genetic register is the availability, in a central database, of a list of names or identities of individuals who have a particular type of genetic disorder or have participated in a clinical trial. The WHO is actively considering the feasibility of an international clinical trials register. The significance of this in providing information for health planning or to assist other family members is balanced by the potential for unauthorised disclosure of such data. The privacy issue is particularly significant when third parties (e.g., employers, insurance companies or the courts of law) may gain access to this information. In these circumstances, the ethical principles of autonomy and confidentiality outweigh the considerations that third parties may derive benefit or harm from such information. However, this can be interpreted in different ways, and a court of law may consider the reverse holds in a particular situation.

A different database involves the storage of DNA fingerprints. As indicated in Chapter 9, DNA can provide a unique profile for an individual. These fingerprints are now being deposited in centralised police databases, although the indications for taking them and the length of time they are stored vary in different jurisdictions. However, unlike the traditional fingerprint (which would not be universally available for all individuals in a community), there will be alternative sources of DNA for most, if not all, members of a community. These sources would have been obtained for specific purposes (e.g., Guthrie spots during newborn screening or inadvertently through routine blood counts or typing specimens from blood donors) but could now be used (or misused) for other purposes.

DNA Storage

As indicated previously, DNA is ideally placed to be stored (banked) for some future use. Guidelines for formal DNA banks have been established to cover various aspects of this activity (see Chapter 5). It is important to emphasise that the word **depositor** rather than **donor** is preferred because the individual giving the sample should maintain ownership and is not usually acting as a donor in the broadest sense.

Storage of human tissue (a source of DNA) or DNA itself can be undertaken for a number of reasons. For example, as part of the routine laboratory practice, residual tissue sections, blood samples or blood spots present on Guthrie cards obtained from a neonatal screening program can be kept for a period of time. There may even be medico-legal reasons for long-term storage. Tissue storage has produced additional ethical and legal considerations since archival samples are able to be tested using DNA. An example of this was the suggestion that the long-standing dilemma whether US President Abraham Lincoln had Marfan's syndrome could be resolved through a PCR study of tissue fragments and blood spots taken from clothing worn at the time of his assassination. This study became possible when fibrillin, the gene for Marfan's syndrome, was found. Opinions about the ethical aspects of such a study were divided. One group considered that there were no legal and ethical issues involved. Another questioned the ethics of this proposal.

Guidelines and agreements define the rights of the individual who has had DNA banked, but less clear can be the status of DNA or tissue that has been kept as part of the laboratory's routine activities. In both circumstances, dilemmas arise when material has been stored for one purpose, but then another DNA test becomes available and the stored material would be helpful in defining the genetic status of other family members. A topical example is the Guthrie card. In many communities, newborn screening is mandatory. Thus, newborn blood is spotted onto Guthrie cards, and these cards are used to screen for a number of medical conditions. Could blood taken from an infant for the purpose of neonatal screening be used to test for other genetic diseases? Could DNA from the Guthrie card be used to establish the identity of a body that would otherwise not be identifiable? The good that can come to the individual or the community needs to be balanced with the person's right to privacy. This will not always be easy.

deCODE

A large-scale DNA bank and centralised database that has provoked controversy and discussion is the one from Iceland, a relatively closed community with well-documented genealogies. This facility was established by the Icelandic government and has been licensed to a private company called deCODE. The purpose of this resource is to facilitate gene discovery. The Icelandic repository of DNA and information involves large numbers (potentially the whole population of Iceland) and includes three repositories: (1) a database of established family relationships in the forms of genealogies, (2) a database of phenotypes taken directly from the medical records and (3) a DNA collection to provide the genotypic data.

The first model for this resource provoked considerable debate because there was no consent required to include medical information (genealogies were already in the public domain). The consent issue remains the subject of controversy that has been only partially addressed with the addition of an opt-out system; i.e., access to medical record data is automatic unless the individual specifically says no. Safeguards in relationship to privacy and confidentiality were included in new legislation. Nevertheless, there are still many unanswered questions about this type of repository, including Who has access, what type of research will be undertaken, what are the commercial implications for the work and how secure are the data?

A comparable resource (UK Population Biomedical Collection) is being set up in the UK. The aim is to focus more on interactions between genes and environment, and so the resource is expected to have a large data set (up to 500 000 people) and will require long-term studies of an epidemiologic nature linking DNA data with medical records and family histories obtained through family physicians.

GENETIC AND CELLULAR THERAPIES

GENE THERAPY AND XENOTRANSPLANTATION

The issue of germline versus somatic cell therapy has already been raised in Chapter 6. The former is prohibited. This reflects the inevitable consequence of germline manipulation; i.e., any changes that result (good or bad) will be transmitted through the germ cells to offspring in subsequent generations. On the other hand, somatic cell gene therapy is considered similar to other accepted medical interventions, for example, manipulations required for bone marrow or organ transplantation. Ultimately, it is the risk-to-benefit ratio that determines whether somatic cell therapy is medically and ethically acceptable in individual cases. Thus, a potentially life-threatening disorder for which there is no effective treatment or what is available is associated with significant complications is the first prerequisite. Once the somatic cell therapy approach can be justified as technically and therapeutically acceptable, it is allowed to proceed after protocols are reviewed by the appropriate monitoring bodies, and a follow-up process is set into place.

Somatic cell gene therapy, particularly that involving retroviral vectors, has been associated with only one direct complication (acute leukaemia) in two children with SCID (severe combined immunodeficiency—see Chapter 6 for more details). However, many protocols have been utilised in situations in which the patient's clinical state was terminal, and death, as a consequence of the underlying disorder, soon followed. The long-term effects of gene therapy need continued monitoring; for example: (1) What is the risk in humans that retroviruses can produce cancer? (2) Is there any leakiness from the somatic cells so that the introduced foreign gene can reach the germline? (3) How beneficial will gene therapy turn out to be?

The use of a xenotransplant including one that has been genetically engineered to make it more human-like, and so reduce the risk for rejection, brings with it the additional concern that viruses or other infectious agents could cross the species barrier. This type of gene therapy must be considered within appropriate ethical and regulatory frameworks since the risk is not only to the patient, but to the wider community.

STEM CELL THERAPY

The ethical, moral and social issues concerning termination of pregnancy have, with recent technological developments, become relevant to research on or manipulation of the human embryo. A key question is What constitutes an individual, and when during development does an individual become a distinct entity? There are two differing views: (1) An individual becomes a discrete entity at the time of fertilisation. In this case, embryo research would be forbidden since it would have the potential to interfere with the individual's right to life. (2) The alternative opinion is that an individual cannot exist until about 14 days after conception, by which time the primitive streak is being formed. Prior to that time, it is possible for the embryo to split and so produce identical twins; i.e., it is not a distinct entity. After considerable debate it is difficult to see a consensus emerging from these divergent views.

Other issues to consider include public opinion and concern how this technology might be used, particularly the potential for human cloning. In terms of ELSI, it is worth noting that the unduly complex terminology has complicated understanding and discussion, particularly the concepts of reproductive cloning and therapeutic cloning. The US President's Council on Bioethics has attempted to simplify what are confusing descriptions by suggesting Cloning-for–Biomedical-Research (therapeutic cloning) and Cloning-to-Produce-Children (reproductive cloning). This distinction has some merit and conveys

to the public (and health professionals) more meaningful information.

Various constraints are placed on embryo research; for example, research is permitted only in the case of spare embryos obtained at IVF that would otherwise be destroyed. There is a prohibition on transfer of embryos to the uterus of any species once they have been genetically manipulated. Reproductive cloning is prohibited. In the recent and, at times, divisive debate about the merits of embryonic stem cells versus adult stem cells, a compromise adopted by some governments was to agree to use existing embryo cell lines (they would have been destroyed eventually), but not allow the creation of new ones specifically for the purpose of stem cell research.

Governments' responses to the stem cell debate have, not surprisingly, been mixed. This is exemplified by the European Union where some countries have adopted a liberal view, whereas others have strict laws prohibiting any experimentation on the embryo. There is also the added problem of differing views on the use of IVF embryos that would otherwise be discarded versus embryos that are specifically created by therapeutic cloning (Cloning-for-Biomedical-Research). The regulatory anomalies arising in the stem cell area are well described in the paper by Orive *et al* (2003).

Cloning-to-Produce-Children (reproductive cloning) is technically difficult and very likely to be associated with deformities. A response to the sensational announcement by two fertility experts that they were about to clone a human is found in the 30 March 2001 issue of *Science*. In his letter, I Wilmut, one of the scientists involved in the creation of the cloned sheep Dolly, described the inefficiency of reproductive cloning, as well as the many abnormalities found in animals created in this way. An explanation for these defects is provided in terms of epigenetic changes that are important for DNA function, but which may not be adequately controlled in the cloning process (see Chapters 2, 7 for more discussion on epigenetics and development). Dr Wilmut concluded with a comment that failures with human reproductive cloning are inevitable, but in addition, there will be collateral damage flowing to therapeutic cloning as the community recoils from an inappropriate use of recombinant DNA technology.

GOVERNMENT

ALLOCATION OF RESOURCES

The rapid advances in medical knowledge are starting to produce ethical dilemmas even in the most affluent of societies. Health and welfare policies in communities are highly variable, ranging from complete coverage of each

individual by the State to a user-pay private enterprise system. Advocates for the different systems consider theirs to be the best, and perceived deficiencies are resolved by the politically expedient term of "fine tuning." Hence, many of the systems have developed chronic problems centred on the conflicting demands yet limited resources, and the public's perception of what rights are basic to life within that community.

Information about the human genome, the expanding DNA testing opportunities and alternative therapeutic regimens through gene and cell therapies will put further pressure on the health system. The exponential growth in knowledge of the single gene disorders has been impressive but is still in its infancy for the more complex polygenic and multifactorial conditions. The issues of what are priorities will need to be addressed. Since many of the present-day systems are already stretched in terms of resources, it will be interesting to see what priority molecular medicine is given by governments. In some cases, the pendulum might swing too far the other way, and important everyday problems (e.g., the aged, the disabled, the mentally ill, HIV-AIDS and other infectious diseases) might be further disadvantaged because health dollars are diverted to molecular medicine by the promises that it will provide a panacea for all problems, and a reservoir for attracting money through its biotechnology potential. Only informed debate will ensure that the allocation of resources for the many health priorities is appropriate and fair.

Consumer pressure and lobbying will increasingly play an important role in the progress of molecular medicine, and so it is essential that the consumer and the community remain supportive (see also Future below). Already there exists a strong lobby group that has well-established negative views about gene technology. In some ways this has been beneficial to the development of molecular medicine because it has ensured that interested parties from either extreme have not been able to coerce politicians into promoting unsavoury aspects of gene technology. However, it is essential that the public continue to be educated through exposure to information that is both factual and understandable. The more sensational media headlines (e.g., "*Infidelity—it may be in your genes*") have been less helpful in this respect.

REGULATION

There are many forms of regulation. The extremes are self-regulation, which is popular with government when it comes to many types of industry, to the passage of laws with the appropriate penalties clearly identified for non-compliance. In between, there are the options to set up guidelines and advisory bodies that make recommendations to institutions, organisations or government. These advisory bodies have varying degrees of power. In terms of molecular medicine, the power of advisory bodies can include the withdrawal of research funding that is restricted to the recipients, and there is always the use of moral or peer pressure. If all else fails, the advisory bodies still retain considerable clout since there are the medico-legal options that would be difficult to defend if

the recommendations of a reputable advisory body had not been followed.

For molecular medicine, self-regulation is a less attractive option because of the emotive issues placed on DNA and genes. Self-regulation does not prevent irresponsible statements, such as the comments about the inevitability of human cloning. Self-regulation has also taken a battering recently with a number of large corporations shown to have acted unethically or dishonestly. So, the regulatory model for molecular medicine must be a mix of advisory guidelines and the passage of appropriate laws. The community feels safer if there is a law because the ground rules are well established and the penalties clearly identified. However, laws are not always helpful in a field such as molecular medicine because changes occur so quickly. In this situation, the law can easily address one particular issue, but then an unforeseen or new development occurs and the law needs to be changed or reviewed. This is not an easy task.

The formal monitoring processes involving gene therapy in the USA are a model that illustrates how a mix of statutory regulation and advisory guidelines can be applied in molecular medicine. Apart from the Institutional Review Board (IRB), the key regulatory body is the USA's Food and Drug Administration (FDA). A second body is the Recombinant DNA Advisory Committee (RAC). The FDA's role is in the development of safe and effective biological products from their initial investigational phase to commercial production. The RAC is a technical committee of the NIH and considers all gene therapy proposals that originate from NIH-funded projects or institutes. Both the FDA and the RAC are critical for the effective implementation and evaluation of gene therapy in the USA, and indirectly in other countries. The FDA has a primary role in ensuring the safety and quality of the therapeutic product, while the RAC focuses on the scientific, safety and ethical values of the clinical trial. Public scrutiny of gene therapy proposals to the RAC is possible, and this provides a forum for open public debate about social and scientific issues related to DNA technology. This dual arrangement has worked well, although some changes had to be made after the death of Jesse Gelsinger (see Box 6.5) because the regulatory process had itself produced some confusion on the reporting of serious adverse events.

At the local level, a number of committees are in place to monitor activities related to genetic engineering. In particular, with gene therapy these are the Institutional Biosafety Committee (IBC, which predominantly deals with the laboratory issues), the Institutional Review Board or Institutional Ethics Committee (IRB or IEC, which predominantly deals with the human ethical issues) and the Animal Welfare Committee (AWC). From time to time, ad hoc committees are formed to review and examine topical or sensitive matters. Membership of monitoring committees is usually designed to ensure that there is an appropriate mix of professionals, consumers, ethicists and other interested parties. The work done by these committees, which is often voluntary and very time consuming, has been a key factor in the smooth progress of molecular medicine to date. Because of the increasing administrative demands of regulatory issues, there has been some debate about the greater use of national or more central committees to deal with particular issues rather than the institutional-based committees. This is not meant to replace the latter, which must continue to play a key role in assessing local conditions as well as ongoing monitoring, but to supplement their expertise particularly with rapidly evolving technologies. There is also some sense in reducing the administrative demands for multiple reporting.

QUALITY ASSURANCE

The rapid advances in molecular medicine and the negative effects they can have on laboratory practices were well illustrated by suboptimal standards practised in some forensic laboratories (Chapter 9). An important ethical issue (and a legal one) in laboratory and clinical practice includes the obligation to ensure high standards. To this end, quality assurance programs have been developed. On the other hand, the many rapid discoveries in molecular technology, particularly in the area of DNA diagnostic testing, can overtake the implementation of formal quality assurance programs. Deficiencies in these circumstances may not reflect a primary reluctance by the laboratory to participate in such a program but may result from the intense pressure to start a new diagnostic test because a probe or DNA sequence is available. Thus, expansion to increase the quantity of testing can take priority over the quality of the results. Ultimately, a balance is essential to ensure that data provided by a laboratory are of the highest standard possible given the resources available.

In general, the increasing use of commercial DNA diagnostic kits will lead to improved quality assurance, but this will come at a cost. Because of costs, many laboratories produce their own in-house kits, which over the years have proven to be very effective and relatively cheap. However, there is now a move by the regulators to require that these kits are also validated in terms of how they are produced and what they purport to test. This increasing regulation would seem reasonable and can only lead to higher standards. However, this will need to be balanced by the increasing burden of regulation on laboratory directors so that the effects of patents and regulatory demands will inevitably mean that fewer DNA testing laboratories will compete. The winner in this environment is likely to be the commercial sector, which is often more efficient for various reasons, including the large numbers of samples being tested. However, these laboratories are often selective and provide only tests that are commercially attractive.

MEDICO-LEGAL CHALLENGES

Health professionals are expected to work under best practice guidelines, and there is the threat of medico-legal action if these practices are not followed. This is a reasonable arrangement but can become distorted, depending on community expectations and the types of medical or legal systems. In some cases, medical practice becomes defensive based rather than evidence based. Unexpected legal decisions add further uncertainties (Box 10.7). In the case illustrated, the dilemma involves the requirement for confidentiality between patient-doctor versus the duty to warn others in the family. In this particular example, the appeals court ruled

---

Box 10.7   Legal Case: Duty of Care—*Safer v. Pack.* Superior Court of New Jersey, Appellate Division. 677 A.2d 1188 (1996).

Robert Batkin was treated by Dr George Pack for colon cancer. Treatment included colectomy. Mr Batkin died in 1964 (his daughter Donna Safer was aged 10 at the time), and Dr Pack died in 1969. In 1990 (26 years after her father's death), Donna Safer was diagnosed to have colon cancer with metastases. She was treated with surgery and chemotherapy. In 1991, it was discovered that Robert Batkin had died from familial adenomatous polyposis (FAP), an autosomal dominant genetic condition, which meant that his daughter Donna was at 50% risk of developing FAP. A complaint was filed in 1992 against the Estate of Dr Pack alleging a violation of duty (professional negligence) for not warning Donna, who could then have been followed, and with prophylactic colectomy, she would have avoided the development of colon cancer. The first court hearing dismissed the case, with the judge ruling that the doctor had no legal duty to warn the child of a patient of a genetic risk. This ruling was based on (1) A genetic disease was different from an infectious disease because there is no potential risk to the community at large. (2) Another court case involving a familial medullary thyroid carcinoma came to the conclusion that a physician owed the patient's child no duty to warn. However, on appeal, this judgement was overturned with the appeals court ruling that Dr Pack should have breached patient-doctor confidentiality with Robert Batkin and warned Donna Safer. Comment: The traditional legal process, which is often based on precedent, and can also produce inconsistencies in interpretation would be particular concerns in molecular medicine, a field in which changes can be expected to occur quickly.

that confidentiality should be breached to warn a daughter who was at risk of a serious genetic disorder in her father. Still to be determined legally is the physician's responsibility in terms of duty of care; for example, does it stop with the immediate first-degree relatives or more distant ones? This issue is particularly relevant to molecular medicine since even without the courts' varied interpretations, uncertainties based on rapidly changing knowledge are likely. Too much uncertainty in molecular medicine cannot be healthy for its development.

Another future challenge is population screening. If we take cystic fibrosis as an example, many responsible bodies have indicated that screening for this disorder should not be universal for the reasons mentioned earlier. Other bodies or individuals have not agreed, and have pushed forwards the proposal that all should have this test. So, unless the community can reach some consensus (this would be a useful role for a central body such as the UK's Human Genetics Commission mentioned earlier), the threat of litigation will almost always win over what might have been reasonable clinical practice determined on the basis of available health dollars, counselling and laboratory resources, and other competing health priorities.

FUTURE

GENETICS TO GENOMICS EVOLUTION

All the ethical and social implications of DNA testing raised earlier have centred on single-gene disorders (genetics era). In these circumstances, diagnosis is relatively straightforward, and the resulting phenotypes are more or less predictable. However, in the not too distant future, the more complex multifactorial and somatic cell disorders will become increasingly recognisable at the DNA level (genomics era). Thus, coronary artery disease, neuropsychiatric disorders, dementias and cancers will have their underlying DNA components defined. On the positive side, there is the potential to understand more fully the DNA-to-DNA and the DNA-to-environment interactions that are necessary to produce the various clinical phenotypes. This will facilitate the implementation of more effective preventative and therapeutic measures.

On the negative side of the ledger will be the increasing gap that is emerging between what is known about many diseases and what can be done to treat them. There will also be the potential for misuse of this information by the individual, the State, industry or third parties. Taking multifactorial genetics one step further will be the DNA characterisation of genes involved in the individual's fundamental makeup, including physical features, behavioural and other components such as intelligence. This knowledge is presently a long way in the future, although reports of a "gay" gene, "ageing" genes and other sensational-sounding genetic discoveries keep appearing periodically. Fortunately, these are often one-off reports, but with time the genetic contributions to various forms of behaviour will be identifiable.

The education of health professionals and the public in the many far-reaching aspects of molecular medicine is a key priority for the future. Educational resources, particularly at the school and tertiary levels, are required to ensure that the full medical, ethical and social implications of new developments relevant to genetic engineering can be appreciated by the majority, and debate is not distorted by vocal minority groups. Informed community input involving ordinary human wisdom is required to play a role in the direction or application of research and clinical aspects of molecular medicine. A difficult to predict variable in the genetics-to-genomics evolution will be the effect of the media and the Internet. Self-regulation of the former is essential for free speech. In the genomics era, this will place increasing pressure on media to report in an ethical manner. The influence of the Internet is already substantive in genomics, and this will only grow. What constraints, if any, are possible on the Internet remain unknown.

Traditional medical practice is very much driven by the finding of a diagnosis, and then the steps to treat the pathology follow. The centrepiece is the label or diagnosis from which can flow various decisions. Molecular medicine complicates this through the potential to predict development of disease or genetic traits that might contribute to a disease development. Hence, a new concept of **predisease** needs to be considered by health professionals and third parties including employers and insurers. Predisease is simply an increased risk state. It does not mean the person has the disease or indeed will develop the disease. The person may also die from other unrelated causes before the predisease has time to develop.

Does the researcher who is studying the molecular basis of genetic disorders have specific responsibilities? Early studies of Huntington's disease suggested a single mutation for this disorder. This is now confirmed to be correct, and so linkage analysis, apart from the risks of non-paternity or recombination, was an appropriate way in which to undertake predictive testing for 10 years before the actual gene was found. However, a different story occurred with adult polycystic kidney disease, which involves three distinct loci with the second (APKD2) only becoming apparent three years after DNA testing for this disorder had started. Therefore, the researcher has an obligation to report findings as accurately as possible so that locus heterogeneity can be anticipated.

Related to the above is the question When does the research phase of gene discovery move to the routine clinical diagnostic phase? This can also be exemplified by contrasting *BRCA1*, *BRCA2* (breast cancer) and *APC* (bowel cancer) genes, and their implications for cancer development. Unlike *APC*, which is an accepted DNA test for familial adenomatous polyposis (see Chapter 3), there remain a number of uncertainties about the *BRCA1* and *BRCA2* genes. Policy statements from learned professional bodies have been produced, to indicate that until further knowledge becomes available, DNA testing in this circumstance is still within the realm of research or should be very strictly controlled in a clinical genetics specialty service.

Pressure from peers or consumer groups can be another reason why a DNA test is prematurely developed from what should still be a research study. In these circumstances, the researcher has the important task of ensuring that data (comprising laboratory, counselling and educational components) are unambiguous in terms of the test's status so that patients and families can make realistic decisions. Planning should also consider how patients and families will be followed in the not unlikely circumstance that new knowledge becomes available, and this changes the significance of the DNA test.

HEALTH, INDUSTRY AND THE COMMUNITY

Just as patents have provided the catalyst to encourage pharmaceutical and biotechnology companies to invest heavily in the development of new drugs, so has the patent on genes and DNA sequences provided the incentive for companies to utilise their considerable resources for gene discovery ventures directed to genes associated with common or high-profile diseases. However, the legality of various patents related to gene sequences is starting to be tested in the courts of law, and it will be interesting to see how the patents stand up to this level of scrutiny.

Another concern about the industry connection with molecular medicine must lie in the increasing control that industry is now exerting on DNA testing, particularly through the availability of DNA testing kits. As nanotechnology allows DNA testing to be miniaturised and so taken to the bedside or the consulting office, it will take away the control from the traditional DNA diagnostic laboratory and allow individual health professionals to provide this type of testing. Another option that is already under way is direct-to-patient DNA testing, with ordering through the Internet a likely scenario. This will be a challenge for the future as the potential good of molecular medicine in these circumstances is likely to be surpassed by the spectre of inappropriate genetic profiles, inadequate information, little or no counselling, and deficient follow-up of individuals and their families. Very much related to the issues just described is the decision-making process that will be needed to determine what are reasonable health-related DNA tests, and what DNA tests are predominantly looking at traits or other characteristics that have less relevance to health.

When it comes to health matters, the community is usually very positive about medical research. The health professionals involved in molecular medicine are generally well-informed and well-intentioned individuals who are there to progress the development of molecular medicine and so improve the health and well-being (and wealth) of the community. However, there is the risk that a community which is constantly exposed to sensational news items about genetic discoveries will become increasingly sceptical by the "genehype," and those in molecular medicine will lose support. Unfortunately, this matter is predominantly in the hands of the media, which complicates predictions about future developments. Nevertheless, it is relevant to note that scientists both in private and public organisations are encouraged to publicise their findings in the media, but more responsibility is needed. The reporter must satisfy an editor, and one way in which this is possible is through sensationalising a story. The ammunition for this can only come from the scientist.

An interesting observation has been made about patents and the community perception of stem cells (Caulfield 2003). The controversy about the use of embryonic stem cells remains, and many in the community will need to see that these cells live up to the many promises before they change their views. In the meantime, there is the risk that patents on stem cells (there are more than 2000 patents filed on human and non-human stem cells with more than 500 specifically to embryonic stem cells) will further alienate the community as the perception will only grow that human embryos are being used as a commodity for research and then patented for profit.

MOLECULAR MEDICINE TEAM

One of the significant changes in internal medicine in the past few decades has been the move to increasing specialisation. As part of this trend, the team approach to treatment of patients with particular disorders has developed. For example, the improved survival associated with cystic fibrosis has been attributed to modern therapies, as well as the comprehensive management that becomes possible in special cystic fibrosis clinics. In this environment, patients have improved access to medical and ancillary services, e.g., physiotherapy, genetic counselling.

Molecular Medicine Team

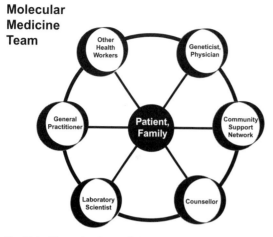

Fig. 10.3 The molecular medicine team.
A suggested team that could be assembled to deal with a molecular medicine–based clinical problem is illustrated. The relative contributions from the components that make up the outer circle would depend on the underlying disorder. The primary care physician (general practitioner) and the disease or community support network play a key role as the health professionals and community members who, in the longer term, will have the closest contact with patients and their families.

The demands being placed on genetic counsellors and clinical geneticists will continue, particularly as the multifactorial disorders come on line. There will be few of these health professionals who can provide comprehensive counselling on a wide range of issues including the clinical consequences, recent advances in molecular medicine and potential future developments. The complexities and rapid developments in molecular medicine require a team approach, since it is unlikely that any single individual will have the breadth and depth of knowledge required to understand all genetic disorders and then convey this information to the patient in ways that are meaningful. Improved ways of communicating with patients and families are needed. Hence, the concepts of e-Counselling and e-Consultations introduced in Chapter 5 need serious consideration.

Continuing professional education remains the cornerstone of modern medical practice. This is an even greater challenge for molecular medicine because the changes occurring are many and rapid. In this environment, specialisation is one way in which to ensure knowledge remains contemporary. Another way in which to maintain high-quality professional education is to work within a team, thereby utilising stimulation and feedback as incentives for learning. Implementation of the team approach, an example of which is given in Figure 10.3, will be critical to the long-term success of molecular medicine.

FURTHER READING

Introduction

Clayton EW. Ethical, legal and social implications of genomic medicine. New England Journal of Medicine 2003; 349:562–569 (*describes some contemporary issues in the genomics era, particularly discrimination and the duty of care*).

Smith L, Byers JF. Gene therapy in the post-Gelsinger era. JONA'S Healthcare Law, Ethics and Regulation 2002; 4:104–110 (*provides an overview of this death and the contributing ethical issues*).

http://www.wma.net/e/policy/b3.htm (*reference to the World Medical Association Declaration of Helsinki—Ethical Principles for Medical Research Involving Human Subjects*).

http://www.hgc.gov.uk/ (*United Kingdom's Human Genetic Commission—"an advisory body on how new developments in human genetics will impact on people and health care"*).

http://www.nhmrc.gov.au (*National Statement on Ethical Conduct in Research Involving Humans 1999*).

http://www.bioethics.gov/ (*US President's Council on Bioethics—web site provides interesting background papers on a number of topical ethic debates*).

http://www.alrc.gov.au/ (*Australian Law Reform Commission from which can be found "Essentially Yours: The Protection of Human Genetic Information in Australia"*).

DNA Testing

Evans JP, Skrzynia C, Burke W. The complexities of predictive genetic testing. British Medical Journal 2001; 322:1052–1056 (*provides a good overview of predictive genetic testing and its potential ethical issues*).

Gottlieb S. US employer agrees to stop genetic testing. British Medical Journal 2001; 322:449 (*news report of the Santa Fe railway company and DNA testing for carpal tunnel syndrome—see also the reference to Burke et al below*).

Haga SB, Khoury MJ, Burke W. Genomic profiling to promote a healthy lifestyle: not ready for prime time. Nature Genetics 2003; 34:347–350 (*illustrates negative aspects of personalised medicine provided through commercial DNA testing laboratories*).

Jordan B, Relkin N, Ravdin L, Jacobs AR, Bennett A, Gandy S. Apolipoprotein E epsilon 4 associated with chronic traumatic brain injury in boxing. The Journal of the American Medical Association 1997; 278:136–140 (*a preliminary association study suggesting that the APOE4 type can predict chronic brain damage in boxing*).

Lloyd FJ, Reyna VF, Whalen P. Accuracy and ambiguity in counselling patients about genetic risks. Archives Internal Medicine 2001; 161:2411–2413 (*looks at some of the complexities associated with genetic counselling*).

Michie S, Bobrow M, Marteau TM. Predictive genetic testing in children and adults: a study of emotional impact. Journal of Medical Genetics 2001; 38:519–526 (*this study seems to agree with other work which suggests that children may not respond so negatively to predictive DNA testing*).

Points to consider: ethical, legal, psychosocial implications of genetic testing in children and adolescents. American Journal of Human Genetics 1995; 57:1233–1241 (*policy statement from the American Society of Human Genetics and the American College of Medical Genetics*).

Williams-Jones B. Where there's a web, there's a way: commercial genetic testing and the Internet. Community Genetics 2003; 6:46–57 (*provides examples of direct company-to-patient DNA testing for adult-onset disorders*).

http://www.genovations.com/patient_overview.html (*Genovations™ company and reference to its Internet "Patients Guide to Genomics"*).

Research

Cho MK, Illangasekare S, Weaver MA, Leonard DGB, Merz JF. Effects of patents and licences on the provision of clinical genetic testing services. Journal of Molecular Diagnostics 2003; 5:3–8 (*reports on a survey to determine the effects of patents on the provision of clinical genetics and DNA testing services*).

Gold ER. SARS genome patent: symptom or disease? Lancet 2003; 361:2002–2003 (*informative summary of why the CDC patented the SARS genome to ensure this information would be preserved for public good*).

Kaye J, Martin P. Safeguards for research using large scale DNA collections. British Medical Journal 2000; 321:1146–1149 (*a brief summary of the Icelandic and proposed UK large-scale DNA collections for genetic research*).

http://www.nuffieldbioethics.org/patentingdna/index.asp (*comprehensive document on The Ethics of Patenting DNA by the Nuffield Council on Bioethics*).

http://www.hugo-international.org/hugo/patent2000.html (*patent policy from HUGO—Human Genome Organization*).

http://snp.cshl.org/ (*web site for the SNP consortium. Provides information on members as well as a catalogue of SNPs chromosome by chromosome*).

http://www.gtg.com.au/ (*web site for the Australian-based company GTG Ltd, involved in patenting non-coding DNA sequences*).

http://www.decode.com/ (*web pages for the Icelandic-based deCODE genetics*).

http://www.alrc.gov.au/ (*Australian Law Reform Commission from which can be found the 2004 report on Gene Patenting and Human Health*).

http://www.roche-diagnostics.com/ba_rmd/pcr_litigation_chronology.html (*Roche company web site summarising its ongoing dispute with Promega involving the PCR patent—worth reading*).

Genetic and Cellular Therapies

Jaenisch R, Wilmut I. Developmental biology: Don't clone humans! Science 2001; 291:2552 (*succinct and well-founded scientific reasons are given why reproductive cloning in humans would be dangerous*).

Kaji EH, Leiden JM. Gene and stem cell therapies. The Journal of the American Medical Association 2001; 285:545–550 (*a brief overview of the gene and stem cell area including perceived ethical issues that might arise*).

Orive G, Hernandez RM, Gascon AR, Igartua M, Pedraz JL. Controversies over stem cell research. Trends in Biotechnology 2003; 21:109–112 (*summarises the difficult political and regulatory issues interacting with ethical discussions on stem cells*).

Spink J, Geddes D. Gene therapy progress and prospects: bringing gene therapy into medical practice: the evolution of international ethics and the regulatory environment. Gene Therapy 2004; 11:1611–1616 (*provides an overview of different national ethical and regulatory issues in gene therapies*).

Government

http://www4.od.nih.gov/oba/rac/guidelines/guidelines.html (*NIH's Recombinant DNA Advisory Committee [RAC] providing guidelines for DNA and gene transfer research*).

http://www.fda.gov/cber/infosheets/genezn.htm (*a summary of the role of the FDA in human gene therapy*).

Future

Burke W, Pinsky LE, Press NA. Categorizing genetic tests to identify their ethical, legal and social implications. American Journal of Medical Genetics 2001; 106:233–240 (*illustrates how criteria can be developed to determine the value of a DNA test*).

Caulfield TA. From human genes to stem cells: new challenges for patent law? Trends in Biotechnology 2003; 21:101–103 (*views are expressed about the patenting of stem cells and how this might alienate public opinion*).

APPENDIX

MOLECULAR TECHNOLOGY

PREAMBLE

Developments in molecular medicine are generally technology driven. Hence, some knowledge of key techniques will help in understanding the full implications of the molecular medicine revolution. In previous editions of *Molecular Medicine*, a chapter was dedicated to technology. For this edition, technology is included as an Appendix to signify that an understanding is useful but not essential. Descriptions given in the Appendix provide a brief and simplified overview of techniques. References are available if a more comprehensive understanding of the subject is needed. Some techniques described in earlier editions of *Molecular Medicine* are no longer included. They are manual DNA sequencing, functional cloning, physical and genetic maps, PFGE (pulsed field gel electrophoresis), somatic cell hybrids, scanning methods (SSCP, CCM, DGGE) and gene decoding (replaced with functional genomics)

DNA AND RNA

SOURCES AND PREPARATION

As noted in Chapter 2, DNA is a particularly useful analyte because in terms of inherited genetic disorders, it is present in all nucleated cells and is the same at all times during development. For practical purposes DNA is usually isolated from blood. Non-intimate sources of DNA include buccal swabs (by scrapping away a few buccal cells) and hair follicles. A fresh or archived biopsy specimen is also a useful source of DNA. In terms of early prenatal testing, DNA from fetal tissue can be prepared from a chorionic villus sample.

DNA preparation is simple with commercial kits allowing a chemically safe technique. These kits contain enzymes that destroy the cell membranes, thereby releasing DNA and digesting proteins. Proteins and DNA are separated from each other by elution through a column or the use of magnetic beads. In contrast to DNA, RNA is more difficult to obtain and prepare. For genetic diagnosis, RNA must be isolated from diseased tissue, and so

Table A.1   Cleavage sites and properties of restriction enzymes[a]

Restriction enzyme	Bacterial source	Recognition DNA sequence	Properties
EcoRI[b]	Escherichia coli strain RYI	5'-G*AATTC-3'	Six base pair recognition sequence.
HpaII	Haemophilus parainfluenzae	5'-C*CGG-3'	This enzyme would cut more often because it recognises a four base pair sequence. Like EcoRI, it generates a sticky end because the cut is not symmetrical.
SmaI	Serratia marcescens	5'-CCC*GGG-3'	In contrast to HpaII, a blunt end is produced because the cut is in the centre of the six base pair recognition site.
NotI	Nocardia otitidis	5'-GC*GGCCGC-3'	This is a rare cutting enzyme because of the long recognition sequence. It would be useful for a technique such as pulsed field gel electrophoresis (PFGE[c]).

[a] Only a single DNA strand is shown for greater clarity. However, restriction enzymes recognise the above sequences in the two strands of DNA and then cut at the *. In the case of EcoRI, two fragments produced will be xxx G—and—AATTC xxx (where x represents the remaining DNA).
[b] Restriction enzymes are named after the bacteria from which they have been isolated. Thus, EcoRI comes from E. coli strain RI. [c] PFGE is a type of electrophoresis used to separate very large DNA fragments.

a specimen of peripheral blood may not always suffice. However, it has been shown that peripheral blood lymphocytes are "leaky" when it comes to RNA, and using a technique such as PCR, it is possible to detect many RNA species although the lymphocyte may not specifically produce that particular protein. In terms of preparation, RNA is very sensitive to degradation by RNAase enzymes that are ubiquitous. RNA preparation kits are commercially available, and they contain reagents to inactivate RNAase and allow RNA to be separated from DNA and protein.

In most clinical applications, DNA is preferred because it is easier to use. However, RNA is sometimes better because it provides additional information such as (1) How does a single base mutation interfere with gene function? Just looking at the DNA sequence might not always give the answer. In contrast, the RNA profile can demonstrate the presence of an abnormal mRNA species. From this, the single base change might suggest the underlying defect to be a splicing mutation. (2) Is there a mutation in a large gene? Using genomic DNA to look for a mutation means that both exons and introns will be included. Introns are often very large and so make analysis of the gene technically very demanding, and in addition, interpretation of mutations in introns is problematic. However, RNA contains only the coding sequences (i.e., exons) and so facilitates analysis of the gene for mutations including splicing errors since most of these errors occur within exon-intron boundaries. The significance of changes in the exon's DNA sequence is easier to determine (see Table 2.5).

Although some laboratories prefer DNA to be collected in specific tubes and anticoagulants, in practical terms most anticoagulants suffice. Occasionally, the lab-

oratory receives a clotted blood specimen. This is also acceptable but more difficult to work with, and hence the preference for anticoagulated blood. DNA is very hardy. Storage at room temperature is adequate during transport, although to avoid bacterial growth, it is preferable to use a container with a cooling block. In contrast, RNA readily degrades, and so collection, storage and transport are best dealt with by consulting the laboratory and using specific protocols.

RESTRICTION ENDONUCLEASES

Restriction endonucleases (also called restriction enzymes) have a unique property in relation to DNA. They recognise specific base sequences in the double-stranded DNA, usually four to six nucleotides in length, and cut the DNA at these sequences. Because a double-stranded structure is needed for restriction enzymes, they do not digest RNA (see also Chapters 1 and 2 for more discussion of these enzymes).

The restriction enzyme forms a key component to DNA mapping, described below. Restriction enzymes that recognise a four base pair sequence (for example, CCGG) will digest DNA frequently, whereas restriction enzymes that recognise a longer DNA sequence (for example, GCGGCCGC) will give larger fragments (Table A.1). Although DNA mapping is not often used today (it has been largely replaced by PCR), restriction enzymes are still useful because they can recognise changes in the DNA sequence brought about by mutations. This is sometimes called an RFLP test, i.e., restriction fragment length polymorphism test, which can be used to cut DNA fragments generated by PCR (Figure A.1).

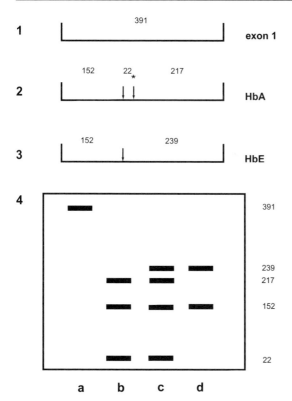

Fig. A.1   Using a restriction enzyme to detect a single base change in DNA (Glu26Lys) in the β globin gene producing the HbE defect.
The single base change that produces HbE is situated within the recognition site for the restriction enzyme *Hph*I. The normal DNA sequence is—GGT<u>GAGG</u>—while the change leading to HbE is—GGT<u>AAGG</u>—(GAG to AAG is glutamic acid to lysine; see Table 2.1). *Hph*I recognises the five base pair sequence GGTGA and cuts the DNA when this sequence is present. **(1)** Using PCR, a 391 bp DNA fragment that overlaps the HbE mutation is amplified. **(2)** In normal DNA (HbA), the presence of two *Hph*I recognition sites in this fragment (↓↓) produces three fragments of DNA 152, 22 and 217 bp in size. **(3)** The HbE mutation interrupts the second *Hph*I recognition site (*), and so when there is HbE, only two fragments are produced (152 and 239 bp). **(4)** This diagram illustrates the typical profile that would be seen after the restriction fragments are electrophoresed: a—uncut fragment, b—normal pattern, c—heterozygous HbE, d—homozygous HbE. The single base change described in this example is a DNA mutation. However, if the single base change does not cause disease or alter the coding sequence, it is called a DNA polymorphism. Detecting single base changes whether they are disease causing or DNA polymorphisms involves the same strategy. When DNA polymorphisms are detected in this way, they are called RFLPs (<u>r</u>estriction <u>f</u>ragment <u>l</u>ength <u>p</u>olymorphisms). Fragments depicted in **(4)** are not drawn to scale.

DNA MAPPING AND RNA CHARACTERISATION

The DNA mapping and RNA characterisation technique could have been left out, but they are historically important and still occasionally used in the diagnostic DNA laboratory. DNA mapping or gene mapping (a more appropriate name for this technique would be restriction enzyme DNA mapping) has also been called Southern blotting or Southern transfer after the person who first described it (E Southern). DNA mapping describes the construction of a restriction enzyme map for a particular segment of DNA, i.e., the pattern of fragments produced by cutting DNA with a number of these enzymes.

At any DNA locus, multiple sites will be recognised by restriction enzymes. The restriction map is represented by a composite of digestion fragments produced for that locus. Disruption of the normal map will indicate an alteration in DNA sequence. A change in only one of many restriction sites occurs when a discrete modification such as a point mutation affects a single nucleotide base (see Figure A.1). This may indicate a genetic disorder or a neutral change that has given rise to a DNA polymorphism. An alteration in more than one restriction site usually indicates a structural rearrangement such as a deletion has occurred (Figure A.2).

DNA fragments generated by restriction endonucleases are separated into their different sizes by electrophoresis in agarose gels. DNA fragments are then made single-stranded by treatment with sodium hydroxide. Single-stranded DNA fragments are transferred from agarose to a more robust medium such as a nylon membrane. The DNA transferred to the nylon membrane is stuck onto it by baking the membrane. The transfer step is called Southern blotting. DNA is now ready for hybridisation to a DNA probe, which used to be labelled with ^{32}P, but today is more likely to be labelled with a non-radioactive marker detectable by chemiluminescence (Table A.2).

The result will be binding of the single-stranded DNA probe (which provides the specificity for the DNA map) to its corresponding single-stranded DNA fragment. The annealing is detected by autoradiography. The upper limit of resolution for DNA mapping is approximately 30–40 kb. Differences in restriction fragments 100–200 base pairs in size can be detected by conventional DNA mapping, but generally when dealing with a small-sized fragment, PCR is preferred. As indicated earlier, gene mapping is rarely used in the diagnostic scenario because PCR has now taken over in the majority of cases. Gene mapping's main advantage is its ability to detect large, uncharacterised gene deletions, i.e., deletions without known break points as this information would be required to make DNA primers for PCR.

RNA is not "mapped" because restriction endonucleases digest double-stranded nucleic acid, not single-

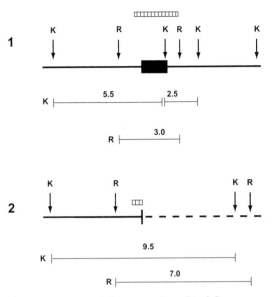

Fig. A.2   How a restriction enzyme is used to define a specific DNA locus by a gene map.
Restriction enzymes recognise a specific base sequence and cleave DNA at that sequence. A change in the DNA sequence will disrupt the site(s) of cleavage, altering the map of fragments. A small change, such as point mutation, may affect only one site; a larger change, such as a deletion, will affect more than one restriction enzyme cleavage site. A DNA probe is selected (hatched box) to hybridise against a gene or segment of DNA (■). A probe is necessary to locate which region of the genome is to be mapped. Two restriction endonucleases (designated K and R) are used to digest DNA in the vicinity of this gene or segment of DNA. **(1)** The restriction enzyme digests give 5.5 kb and 2.5 kb bands with enzyme K and a single 3.0 kb band with R. This combination of restriction fragments in association with the probe used provides a "gene map" for this particular locus. Note that to the right of the 2.5 kb fragment there is another potential fragment generated by K. However, since this fragment does not overlap the probe region (hatched box), it is not seen in the map. **(2)** Any rearrangement around the region picked up by the DNA probe, such as a deletion (------), will disrupt the restriction enzyme pattern since the novel DNA sequence found in association with the deletion will have different restriction endonuclease recognition sites; i.e., enzyme K now gives a 9.5 kb fragment, and enzyme R, a 7.0 kb fragment. Note: DNA fragments in a DNA map are large and usually measured in kilobases (kb), whereas PCR fragments are much smaller (measured in hundreds of base pairs—bp). This means a DNA map is not generally useful for measuring small deletions or point mutations, compared to PCR, which is a lot more versatile (see Figure 2.12).

stranded RNA. Characterisation of RNA is based on assessing size of the RNA species as well as determining its tissue specificity. The latter is a significant difference between DNA and RNA, requiring RNA to be prepared from transcriptionally active tissues. Electrophoresis of RNA is also undertaken in a denaturing gel to prevent secondary structures from forming during electrophoresis. Transfer of RNA from an agarose denaturing gel to nitrocellulose membranes is known as "northern" blotting. A northern blot (and there is also a "western" blot, which refers to protein separations) is usually written in lowercase because it does not refer to an individual's name like Southern blotting.

POLYMERASE CHAIN REACTION

PCR (polymerase chain reaction) and its variations are described in Chapter 2 (see also Figure 2.5). PCR is mentioned again in the Appendix to emphasise its importance in molecular medicine. A key step in amplifying a segment of DNA is the designing of appropriate oligonucleotide primers that will anneal to either side of the DNA region of interest. Software is now available to help in the designing process, and many companies will produce the primers. Primers about 16–20 bp in size suffice to give the PCR its specificity; i.e., it is unlikely that two primers of this size will bind to any other region of the genome. Two primers are needed, one for each side of the fragment to be amplified. The four nucleotides, *Taq* DNA polymerase along with magnesium and a suitable buffer complete the reaction mixture needed for PCR. All this is done in what is called a **DNA thermal cycler**, a machine with properties of a refrigerator and an oven to produce the various temperature changes 94°C → 55°C → 72°C, etc., during each cycle. Being an exponential process, PCR allows a defined segment of DNA to be amplified many millions of times in around 0.5–3 hours. Small volumes of DNA can be used (as exemplified by the forensic scenarios in Chapter 9).

A variant of PCR that is frequently used and merits understanding is **RT-PCR** (reverse transcriptase-PCR) (see Figure 2.2). This approach is necessary when a gene is too large for direct DNA mutation detection using genomic DNA. For example, the cystic fibrosis gene (*CFTR*—cystic fibrosis transmembrane conductance regulator) has 27 exons distributed over 250 kb of DNA. However, genetic information from these exons only extends over 6.5 kb (i.e., this is the size of the mRNA). The difference between 250 kb and 6.5 kb is represented by large intronic sequences. Although mutations in introns can lead to genetic disease, the majority of mutations will be found in exons or intron-exon boundaries, and very occasionally in the 5' or 3' flanking regions. Therefore, DNA mutation detection with genomic DNA would involve analysing 250 kb of DNA—a challenge that is beyond the routine DNA diagnostic laboratory. A shortcut is to analyse only the 6.5 kb of exonic DNA since this is the place where the great majority of mutations will be found. To do this requires mRNA from which is prepared cDNA by the technique of RT-PCR. Another shortcut in analysing large genes is dHPLC (discussed below under DNA Characterisation).

Table A.2 Various types of DNA probes[a]

DNA probe	Description	Applications
PCR primer[b]	16–20 bp synthetic single-stranded DNA sequence.	Used to bind to either side of a DNA fragment that is to be amplified by PCR.
cDNA probe	Larger fragment (up to kilobases in size) made from mRNA by RT-PCR and so more likely to bind to exon sequences.	Used in gene mapping to detect a target sequence and so determine the restriction enzyme cutting pattern around that sequence. Used also in hybridisation assays to detect mutations.
Genomic probe	This is also a larger fragment (up to kilobases in size) but can comprise exons, introns and non-coding sequences.	Used in gene mapping and hybridisation mutation detection as described above. Because it also incorporates non-coding DNA sequences (contains repetitive DNA), it may not be as specific as the cDNA probe.

[a] Probes can be labelled with different coloured dyes (or chemiluminescent material), and their annealing to target DNA sequences is detectable by sensitive laser cameras (or by autoradiography when the light is released onto an X-ray film). [b] The terms **probe** and **primer** can be confusing. A DNA probe refers to a fragment of DNA that is used in a hybridisation type reaction to detect its corresponding (i.e., complementary) fragment in DNA. In this way a gene or DNA fragment can be identified if the probe is labelled with a chemical such as fluorescein. A DNA primer is also a segment of DNA that binds to its complementary partner in the genome. However, a DNA primer is not used for detecting a specific DNA sequence *per se*, but for other purposes, particularly PCR. The DNA primer in PCR provides the specificity for binding to a particular region, which is then amplified.

An important recent development in PCR is **Q-PCR** (quantitative PCR—see Chapter 2) based on **real-time PCR**. Methodologies for quantitating the amount of DNA or RNA have been available for some time. With the Southern blot, densitometry was used to compare the density of the autoradiograph pattern between a test sample and control, and from this, an estimate was made of gene copy number or whether segments of a gene had been duplicated or deleted. In the past, Q-PCR was undertaken by quantitating the product at the end of PCR. However, this approach was less reliable than the Southern blot because PCR generates DNA copies in exponential fashion. In this circumstance, the presence of inhibitors or limiting reagents would have a significant impact on the PCR end product; i.e., quantitation was unreliable. This situation changed with the availability of real-time PCR.

Real-time PCR, as its name implies, allows the reliable detection and measurement of PCR products while they are being generated during each cycle. Real-time PCR platforms utilise fluorescent-based probes or other dyes that can intercalate within the DNA. Thus, the accumulation of DNA during each PCR cycle can be monitored as it is happening, i.e., in real time. Another advantage of real-time PCR is that the reactions are undertaken in a closed system; i.e., PCR products are analysed as the reaction proceeds and plotted out in the form of a graph (see Figure 2.7). This means that gels are not necessary, and since PCR products are not moved from one tube to another, there is less chance of contamination. Real-time PCR has many applications in molecular medicine, particularly quantitative and qualitative analysis of pathogens in molecular microbiology, and determining gene copy number in genetic diseases and cancers.

Apart from its ability to amplify DNA, another useful property of PCR is its versatility since it can be used to detect a wide range of DNA mutations from single base changes to large deletions and gene rearrangements. Microsatellite polymorphisms make a DNA fragment smaller or larger. These polymorphisms are detectable with PCR (discussed further in Chapter 2 and Figure 2.12). However, this versatility is also a drawback with PCR if a large deletion (or rearrangement) is not suspected. In this circumstance, the failure to utilise the appropriate primers will mean only the normal allele amplifies. This will lead to a misinterpretation of the final PCR pattern as being normal (Figure A.3, **1a**). The other problem with PCR is the potential for contaminating DNA to produce an erroneous pattern (Figure A.3, **2**).

ELECTROPHORESIS

Whatever is done to DNA after it is isolated (for example, cutting with restriction enzymes for a gene map or amplifying a specific fragment by PCR), the next step usually involves separation of DNA by electrophoresis. Since DNA is negatively charged, it is an anion and moves towards the positive electrode (the anode). Sizing of separated DNA can be undertaken by using standard size markers, and from this, it becomes possible to characterise the DNA or determine whether a mutation is present.

The separation of DNA has, until recently, been undertaken using slab gels made up of agarose, or for smaller DNA fragments, polyacrylamide is required. Many laboratories still use the traditional slab gel, and it remains the workhorse for post-PCR or post-gene map analysis. Slab gels and their inherent difficulties were tolerated in many laboratories but became a problem in forensic cases because in these circumstances the matching of DNA profiles required a very exact measurement of DNA fragment sizes (see Chapter 9). However, the making of a slab gel leads to some variability in texture, which can

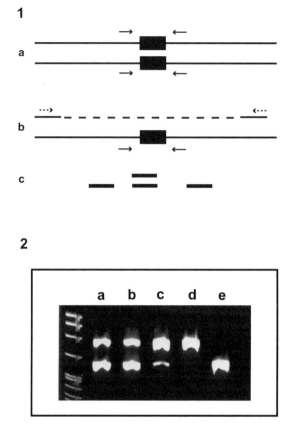

Fig. A.3 Intrinsic faults in PCR.
(1a) This figure depicts a gene (■) that will have two copies because there will be two chromosomes with that gene. PCR primers (→←) designed on each side of the gene will allow it to be amplified. **(1b)** One of the two genes and the adjacent primer binding sites are now completely deleted (------). Therefore, the primers will no longer work because they cannot bind to this deleted segment. Normally, this would not be a problem if a deletion was suspected because new primers (dashed arrows) could be designed to amplify across the new segment of DNA brought about by the deletion. **1c-left**: The PCR pattern for the normal **(a)** structure would be a single band. **1c-middle**: For the deleted segment **(b)**, there would be two bands because the second primer set (--> and <--) could be designed to amplify a different sized (larger) DNA fragment. However, if the deletion was not suspected, then the deletion would not amplify with the → ← set of primers, and the pattern obtained **(1c-right)** would incorrectly imply that there was only a normal fragment. **(2)** This figure shows ethidium bromide–stained PCR fragments across a deletion in the β globin gene with the lower band being normal and the upper band representing a deletion. **(2a)** and **(2b)** are a couple who are carriers for a Filipino-type β globin gene deletion (producing thalassaemia). **(2e)** is a normal control (lower band only), and **(2d)** is a homozygous affected individual with this deletion; i.e., only the top band is present. **(2c)** is the offspring of the couple and is a diagnostic problem because there is a lower band but with a greatly reduced intensity. This is an example of a contaminant (normal DNA) that gives **(2c)** the appearance of a heterozygous (carrier) individual although the individual is actually homozygous affected. The different band intensities would suggest a problem in the PCR, but contaminating DNA is not so easily detected in most circumstances.

influence the conductance of an electric charge through the gel. This can lead to inconsistency in fragment size calling. Slab gels are also time consuming to make and run, and it is difficult to automate them.

In many laboratories, slab gels have now given way to capillary gels. These gels are commercially available and involve a very fine capillary that is packed with various gels manufactured specifically for different conditions. Capillary gel electrophoresis has revolutionised gel electrophoresis because it is very fast and very reproducible and can be automated. Size measurement by capillary electrophoresis is now undertaken by computer software, which takes away another source of human error in DNA analysis.

QUALITY ASSURANCE AND QUALITY CONTROL

Molecular biology (and the more clinically relevant molecular genetics) laboratories are unusual compared to traditional laboratories because they utilise a diverse range of technologies (one need only look at the many different approaches available to detect single base DNA mutations). This fact, as well as the rapid changes that are occurring, makes it more difficult for the traditional QA/QC programs to be implemented. In many molecular genetics laboratories, in-house kits comprise the standard approach to DNA testing. Regulatory bodies are only now starting to tackle the huge logistic problem involving the validation of in-house DNA diagnostic kits. These comments are made to emphasise to the health professional that while developments in molecular medicine are impressive, they can still be associated with mundane errors such as sample mislabelling or sloppy laboratory practice. Hence, DNA diagnostics should be treated like any other laboratory-based test. The health professional ordering the test should be aware of its limitations. If in doubt, the DNA test should be repeated.

GENE DISCOVERY

POSITIONAL CLONING

Positional cloning refers to the discovery of new genes on the basis of chromosomal location rather than functional properties (Figure A.4 and also Chapter 3).

Chromosomal Location

The first step in positional cloning is to obtain, if possible, a clue as to the likely locus or chromosome involved. This information can come from (1) case reports describing the chance occurrence of the disease with another disease that has a known chromosomal location, (2) observations in which chromosomal rearrangements have been shown to occur in association with the disease of interest and (3) DNA linkage studies. An example is illustrated by the familial colon cancer called FAP (familial adenomatous polyposis). The clue that this disorder was associated with the long arm of chromosome 5 came from the chance observation of a deletion involving this chromosome in a family with FAP. Positional cloning was then started at the chromosome 5q locus and eventually led to the underlying FAP gene being found (see Chapter 3 for further discussion of FAP).

Candidate Genes

Another approach in positional cloning, if a disease locus has not been identified, is to look with DNA markers derived from candidate genes (i.e., likely genes) or even look directly for mutations in candidate genes. For example, a good candidate gene for a heart muscle disorder would be myosin, a component of muscle (see Gene Discovery—Familial Hypertrophic Cardiomyopathy in Chapter 3 for further discussion of candidate genes). Whether a locus is known or a candidate gene is available, ultimately it will be necessary to confirm that a gene is the causative one. This usually requires the demonstration of a DNA mutation. In some cases final confirmation of the mutation's significance comes from an animal model that is shown to carry the mutation in association with the appropriate clinical picture.

In Silico Positional Cloning

Positional cloning was a time-consuming and difficult strategy for gene discovery. It was demoralising if the gene could not be found. However, positional cloning has been simplified because the Human Genome Project has allowed all the genome DNA sequence (3.3×10^9 bp) to be accessible through the Internet. Thus, new genes can be sought by computer-based strategies, and human subjects are not necessarily needed for this aspect of positional cloning (Figure A.4). The in silico option for positional cloning is still difficult and time consuming because bioinformatics programs are not sophisticated enough to help with annotating a DNA sequence. However, this problem will be resolved in the near future, allowing the computer-based scientist to sift through DNA sequences predicting where likely genes are to be found and what they do.

ASSOCIATION STUDIES

Positional cloning was first mooted as a strategy for identifying new genes in the mid-1980s. However, successes came slowly, and by the early 1990s there was considerable scepticism about this approach to gene discovery. By the mid-1990s, a limited number of success stories in positional cloning were reported. Today, this has changed and few would question the value of positional cloning in gene discovery. Association (e.g., case control) studies are presently occupying the same ground as positional cloning in the early 1990s with much criticism because of poor reproducibility and the number of false associations that have been reported.

Association studies are presently one of the few approaches available for identifying genes in complex genetic disorders (see Chapter 4). Although case control

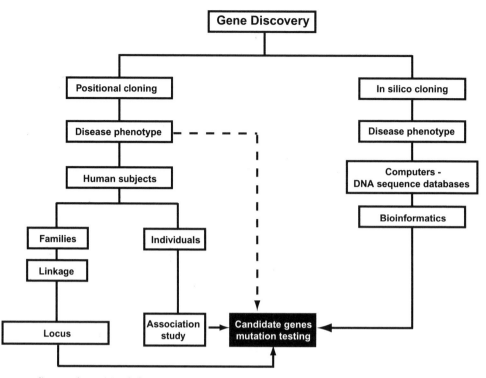

Fig. A.4 Gene discovery by positional cloning.
Positional cloning: The disease of interest is assessed through its phenotype (i.e., clinical picture), which should be straightforward. Knowledge of the genetic locus involved in the disease is next needed (this can be bypassed if a candidate gene can be identified ↓). To find a locus, human subjects with the disease are studied. If there are large families with affected individuals and if the phenotype is easy to define, linkage analysis becomes a key step in positional cloning because it allows a locus to be found. Families usually need to be large (multigenerational), and the phenotypes must be clearly identified so that in each family member DNA markers can be compared to the clinical phenotype until one or more DNA markers show consistent co-inheritance in every patient with the disease. Statistical programs calculate how closely inherited are the phenotypes and DNA markers, and on the basis of this calculation, the probability that this is a true linkage can be assessed. One measure of closeness is called the LOD score (LOD means $\log_{10}$ of the *odd*s). A LOD score of +3 is the minimum usually accepted for a positive linkage; i.e., the chance that the DNA marker and the clinical phenotype are inherited together is about 1 in 1000. On the other hand, A LOD score of −2 is enough evidence to exclude a locus from the particular disease. Once a locus is known to be involved in the disease, genes in that region are sought. Pre-Human Genome Project, this aspect of positional cloning could take years of work. Today, all the genome has been sequenced, so the investigator need only go to the database to look at DNA sequences in the region of the positive LOD score. If the investigator is lucky, DNA sequences will have already been annotated to identify the position of genes. If less fortunate, raw DNA sequences will need to be annotated so that likely genes can be predicted. If a large family is not available or the mode of inheritance is now known or the phenotype is more subtle, an alternative approach to linkage analysis can be followed. This is an association (case control) study and is discussed in more detail in the text. *In silico* **positional cloning,** i.e., positional cloning using computers: The disease phenotype in this approach is broader and can include, as well as the clinical picture, knowledge of a molecular pathway that might be abnormal; e.g., a known conduction defect will give an important clue to a likely candidate gene. The use of computer-based information and computer resources (rather than human subjects) to search for genes requires sophisticated bioinformatics software. The final step in both types of positional cloning is to prove that a gene is causing the disease. The most direct way to do this is sequencing the gene in a number of affected individuals and showing that they (and no normal controls) have a mutation that causes malfunction of the gene (see Table 2.5 for more information on how this is assessed). Sometimes it is also necessary to produce the same gene defect in something like a transgenic mouse to confirm that the mutation is causing a disease (see also Table 4.1).

studies in various forms have been around for some time, the availability of DNA polymorphisms dispersed through the genome (i.e., the SNPs—single nucleotide polymorphisms) has raised the stakes with the potential for these DNA markers to be used in comparing genetic profiles between disease and control populations to identify modifying and QTL-like genes in the multifactorial genetic disorders (see Chapter 4 for more information on complex diseases).

To date, association studies have produced many interesting hypotheses but few major discoveries. Although a lot of these studies have not been reproducible, it is worthwhile repeating that the study of complex genetic diseases is a real challenge because the gene effect is marginal, and often multiple genes are involved. In addition, there can be an unknown environmental contribution as a confounding variable. For the present, association studies remain controversial, but they are probably the way forward for the complex genetic disorders, examples of which are given in Chapter 4. A prediction for the future is that association studies will justify the effort when more SNPs can be analysed (presently their high cost is a major limitation), and there are better statistical analysis programs to make sense of the mass of data that can be generated.

CLONING DNA

Cloning DNA involves the production of unlimited pure quantities of the same DNA fragment using a vector and host system. The principles behind cloning are summarised in Figure A.5, and a summary of cloning vectors is given in Table A.3. The first step is to decide the size of DNA to be cloned. Thus, a limited segment of DNA needs to be cloned if the gene being sought is small (for example, the globin genes). On the other hand, trying to clone a large gene (for example, cystic fibrosis) or cloning a large DNA fragment in which there is a gene of interest will mean a different approach is necessary; i.e., the largest possible DNA segments will need to be cloned. Once size is known, the appropriate vector system required to carry the cloned DNA fragment can be selected. Random shearing by physical methods or restriction endonucleases will break DNA into fragments. The DNA of interest will be found within one or more of these fragments. The cloned gene is then confirmed usually by DNA sequencing.

The simplest cloning system involves a plasmid vector. In this situation, genomic DNA and plasmid are both digested with the same restriction endonuclease, creating, if possible, fragments with unpaired bases at each end (sticky ends). Mixing the two together in the presence of DNA ligase will allow ligation to occur so that each individual plasmid will have stitched to it one fragment of genomic DNA. Using electrical or calcium chloride shock, it is possible to get the plasmids plus their cloned inserts to be taken up (in a process called **transformation**) by *E. coli*. The host *E. coli* divides and is then plated out as a lawn on an agarose plate. If all goes well, there should be a broad representation of DNA from most of the genome (including the target DNA) inserted into the plasmids. Thus, a **library** of DNA fragments has been produced. The next step involves screening of the library to find the target DNA. For screening, a replica of the colonies on the agar plate is obtained by placing the plate against a nylon membrane. DNA that has been transferred to the nylon by this contact is then made single-stranded and can be screened for target DNA with the appropriate probe. When a positive colony is found, it is traced back to the relevant colony on the agar plate, which is then isolated and used as a source of the cloned DNA (Figure A.6). More complex levels of cloning are possible for larger DNA fragments. The vectors for this include bacteriophage and YACs—yeast artificial chromosomes (see Table A.3).

Table A.3   Different vectors used in DNA cloning

Vector	DNA size to be cloned	Features
Plasmid	<10 kb	Technically simple but inefficient.
Bacteriophage "phage"	~20 kb	Useful in terms of insert size and efficiency.
Cosmid	~40 kb	Allows for larger inserts to be cloned.
P1 phage	<100 kb	Somewhere between cosmids and YACs.
YAC (yeast artificial chromosome)	300–1000 kb	Large inserts possible but difficult to use. Particularly valuable in "walking" along a chromosome in positional cloning. Walking refers to the use of overlapping YAC clones (called contigs). In this way a large segment of DNA is characterised to find a gene.

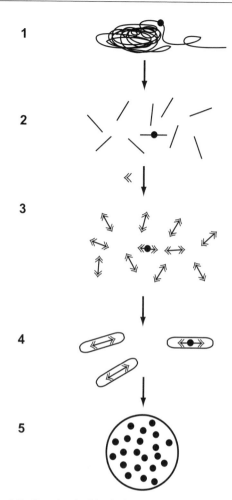

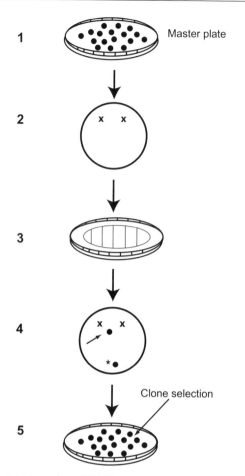

Fig. A.5 Steps involved in cloning DNA.
(1) Genomic DNA with target DNA depicted as (•). (2) DNA is randomly fragmented, by physical methods or restriction enzymes, into many pieces, which are then size-selected. (3) The correct-sized DNA is then joined, using the enzyme DNA ligase, to the DNA vector (≪) such as a plasmid or a phage. There will be many ligated DNA segments. One hopes the target gene will be included in this **library**. (4) Vector plus inserts are taken up into a host such as *E. coli* in a process known as transformation. (5) Vector, inserts and host are plated out onto an agarose plate. Here, each *E. coli* forms a discrete colony. The colonies can be screened to identify which one contains the DNA segment of interest (see Figure A.6 for a description of screening). The relevant colony can then be used to grow up large quantities of bacteria containing the DNA of interest. The latter is then purified from bacterial products.

Fig. A.6 Screening a DNA library.
(1) A master plate of agarose on which is growing a lawn of *E. coli*. The position of individual DNA clones will be seen as colonies (plasmid clones) or plaques (phage clones). (2) A nylon membrane that has the shape of the master plate is marked (× ×) for correct orientation relative to the master plate. (3) The nylon membrane is placed on the master plate. Portions of the colonies/plaques (containing DNA) will stick to the nylon membrane. DNA from these is made single-stranded and the gene of interest found by hydridising with a radiolabelled probe. (4) → identifies a positive clone/plaque on the membrane; ★ is a positive control. (5) Using the position of the positive clone relative to the × × markers, it becomes possible to go back to the original master plate and select the right colony/plaque.

DNA CHARACTERISATION

DNA SEQUENCING

The gold standard in DNA analysis (of normal DNA or to detect mutations in DNA) is sequencing of the individual nucleotide base pairs in the DNA. As a direct result of the technological developments from the Human Genome Project, DNA sequencing is now affordable. It is technically simple because of automation. Many central and commercial sequencing facilities will take a DNA sample in the morning and email the sequencing results later that day.

The methodology for sequencing DNA has evolved rapidly since the first descriptions in the mid-1970s by F Sanger, A Maxam and W Gilbert. Two procedures were initially developed. The first utilised a base-specific chemical modification followed by cleavage of modified DNA. This method is no longer used for routine DNA sequencing because the reagents are toxic, including the use of radioactive phosphorus, and better methods allow automation. Maxam and Gilbert sequencing, as it is sometimes called, is still valuable in specialised investigations including the interactions between DNA and proteins known as "footprinting."

The second approach to DNA sequencing, which still forms the basis for what is used today, is enzymatic primer extension. It is also called the dideoxy chain termination method. For this, single-stranded DNA is extended using DNA polymerase. The reaction mix also has ddNTP (dideoxy nucleotides ddGTP, ddCTP, ddATP and ddTTP). When ddCTP is incorporated into the extension, it produces a stop at each of the C nucleotides in the DNA. A similar stop occurs with the other ddNTPs (Figure A.7).

When first used, dideoxy sequencing required a pure template, and so DNA had to be cloned, which was a very tedious step. Today, an alternative to cloned DNA is DNA amplified by PCR. DNA sequencing can also be obtained as part of the actual PCR by cycle sequencing. The use of four different dyes for the ddNTPs (dye terminator sequencing) was a major improvement because instead of running each reaction down a separate lane in the gel as depicted in Figure A.7, it was possible to mix the ddNTPs and run this mixture down a single lane as the A, T, G and C stops could be detected through different colours (A—green, T—red, C—blue and G—black). A few years ago, manual sequencing was considered to be successful when it produced 200–400 base pairs of readable DNA sequence. Today, with automated sequencing, greater than 700 base pairs are obtained, or the run is considered suboptimal. The instruments in use rely on variations of fluorescence labelling, PCR and capillary gel electrophoresis (Figure A.8).

Although DNA sequencing is the gold standard and sophisticated software is used to assist in reading the sequence, it is still necessary to appreciate that both the software and the naked eye can make mistakes when reading DNA sequences. Very occasionally, mutations in DNA (especially single base changes producing a heterozygous defect) are missed by the sequencing methodology. Hence, it is usual to sequence both the sense and antisense DNA strand when looking for an unknown mutation.

DNA MUTATION DETECTION

Detecting point mutations or deletions or other rearrangements in DNA is a key technology that is used in all DNA diagnostic laboratories. Methods involved are many, and so it is difficult to provide a comprehensive overview of all techniques. This fact leads to a problem with quality assurance programs discussed earlier. Some of the common DNA mutation detection techniques are summarised in Table 2.7, and they are now described in more detail. In brief, DNA mutation detection techniques may be considered under six different categories: (1) sizing of DNA fragments, (2) hybridisation-based assays, (3) protein truncation tests, (4) quantitation of DNA, (5) DNA sequencing and (6) melting property between probe and target DNA (described under DNA Scanning—dHPLC).

Sizing of DNA Fragments

Sizing of DNA fragments is the most straightforward of the post-PCR techniques to detect changes in DNA. An example of this is the ΔF508 mutation in cystic fibrosis. This mutation involves a single amino acid deletion (phenyl alanine at position 508). Therefore, a PCR product obtained from a normal gene and a mutated one is distinguishable by a three base pair deletion in the latter (Figure A.9). Other variations of the sizing theme include restriction fragment length polymorphism analysis (see Figure A.1) and measurement of the number of CAG triplet repeats in a disorder such as Huntington's disease (see Figure 3.5).

Hybridisation Assays

Hybridisation assays are the most popular of the assays for detecting mutations; they rely on the specific binding of a single-stranded DNA segment or probe to its corresponding partner based on the complementary A to T and the G to C binding. By altering the hybridisation and washing conditions, it is possible to distinguish mutant from wild-type (normal) DNA fragments that may differ in sequence by a single base.

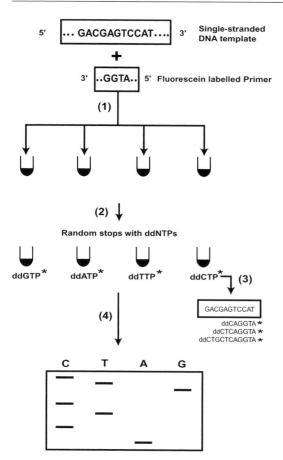

5′ ··· GACGAGTCCAT···· 3′   **Single-stranded DNA template**

+

3′ ··GGTA·· 5′   **Fluorescein labelled Primer**

(1)

(2) ↓

Random stops with ddNTPs

ddGTP*   ddATP*   ddTTP*   ddCTP*   **(3)**

GACGAGTCCAT

ddCAGGTA ★
ddCTCAGGTA ★
ddCTGCTCAGGTA ★

(4)

C   T   A   G

Fig. A.7   DNA sequencing with the dideoxy chain termination method.
Single-stranded DNA template and single-stranded primer labelled with fluorescein are prepared. The two are allowed to anneal and are then aliquoted into four tubes. **(1)** In the presence of DNA polymerase and a mixture of the four deoxynucleotides (dGTP, dATP, dTTP, dCTP), there is extension from the primer/template double-stranded site. **(2)** Random stops in the extension are then produced by adding to each tube one of the dideoxynucleotides (ddNTP). **(3)** As illustrated in one tube, ddCTP will produce random stops wherever there is a cytosine nucleotide, and so a number of double-stranded DNA products are formed of varying sizes. Note that the ddCTP-containing fragments are complementary to the original template sequence. The remaining three dideoxynucleotides will do likewise in their individual reactions. The result is a mixture containing variable lengths of extended DNA segments. **(4)** Each mixture is electrophoresed in the gel track corresponding to the dideoxy nucleotide added, e.g., G = ddGTP, etc. The DNA sequence is read from bottom to top. In this example the sequence reads: ACTCGTC, which represents DNA sequence 3′ to 5′ following annealing between ------GGTA------- (primer) and its complementary sequence in the template (------CCAT------). See also Figure A.8, which provides practice at reading an actual sequencing gel.

ASOs—allele specific oligonucleotides: Sometimes called "dot blots." This well-established technique utilises two oligonucleotide probes labelled with a non-radioactive dye that is detectable by colorimetric, chemiluminescent or fluorescent means. For example, to identify the single base change associated with the β globin gene codon 39 C → T mutation producing β thalassaemia, two probes are designed to hydridise to the C (wild-type) and T (mutant) DNA sequence as well as DNA on either side of these bases. The probe (a 19 mer, i.e., 19 bases in size) binding to wild-type sequence is 5′ CCTTGGACCCAGAGGTTCT 3′ and the probe binding to mutated DNA is the antisense sequence 5′ AGAACCTC TAGGTCCAAGG 3′. The single bases underlined in the middle of the probes represent the C (wild-type) to A (mutant) sequences. By altering the hybridisation conditions (temperature for annealing and then washing the filter to break apart any weak non-specific binding), it is possible to distinguish a fragment of DNA containing either the C or the A at the site of the mutation. Binding of both probes = heterozygous; binding of wild-type probe alone = normal; binding of mutant probe alone = homozygous affected (Figure A.10).

OLA—oligonucleotide ligation assay: This very useful technique relies on the hybridisation of two probes adjacent to each other. They are first hybridised to the DNA sequence of interest. A perfect match (i.e., hybridisation has occurred) is followed by a ligation between the two adjacent probes. This is illustrated using the same codon 39 β globin gene mutation described in the ASO technique (Figure A.11).

ARMS—amplification refractory mutation system: Like OLA, this hybridisation-based reaction is simple, reliable and does not involve the use of restriction enzymes or ASO-based hybridisation. Its disadvantage is that it can give false results if there are polymorphisms at the primer binding site. The principle behind ARMS is that DNA primers that are mismatched at the 3′ end will not function as primers in PCR (and no product will be generated). For ARMS, three primers are required and two reactions are undertaken. One primer is common in each of the two reaction tubes. The second primer is changed at the 3′ end so that it will bind only with the mutant sequence. The third primer is changed at the 3′ end so that it will bind only with the wild-type sequence. This is illustrated using the codon 39 β globin gene example (Figure A.12).

Protein Truncation Test (PTT)

The protein truncation test (PTT) is a fairly specialised approach to detecting single base changes that produce a frameshift (and so premature stop codon) or a single base change leading to a nonsense codon. In both cases, transcription of the mRNA containing the single base

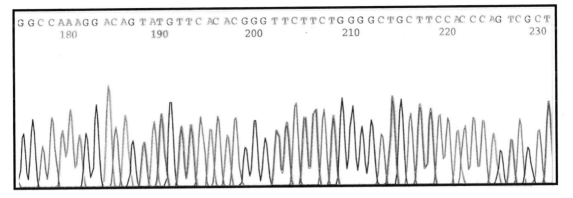

G G C C A A A G G AC AG T AT G T T C AC AC G G G T T C T T C T G G G G C T G C T T C C AC C C AG T C G C T
     180   190   200   210   220   230

Fig. A.8   Trace of an automated sequencing run.
The automated DNA sequencer has revolutionised molecular medicine since it enables relatively cheap sequencing of long segments of DNA and readouts are automated. Each ddNTP is distinctly labelled with four different fluorochromes, and because dye terminator sequencing is used, the four ddNTPs are electrophoresed in the one gel lane or capillary run. The DNA sequence is written above each peak and has been determined by the sequencing software. If N appears in the DNA sequence, it means the automated sequencer software cannot decide which base is correct and visual inspection is necessary. (see colour insert)

change will lead to a truncated protein because of the premature stop codon (Figure A.13).

Q-PCR

Previously, DNA quantitation was a tedious and inaccurate procedure. Today, the availability of real-time PCR allows accurate and reproducible quantitation of DNA or RNA (see Chapter 2 and Figure 2.7 for more information). Real-time PCR is faster than other types of PCR because the results are immediately printed in graph form (there is no need to take the final PCR product and assay it by gel or hybridisation onto a filter). Real-time PCR assays are also closed (i.e., PCR products are not taken out to be measured), and so the risk for PCR-based contamination is significantly reduced. Real-time PCR is particularly valuable for detecting large gene deletions, to determine gene dosage in oncology usually or to quantitate viral load.

DNA SCANNING

Previous editions of *Molecular Medicine* described a number of scanning methods including SSCP (single-stranded conformation polymorphism), DGGE (denaturing gradient gel electrophoresis) and CCM (chemical cleavage of mismatch), which allowed DNA to be scanned (**scanning** refers to a method that allows likely sites of mutations in DNA to be identified; the term **screening** is also used, but this can be confused with DNA screening to detect carriers of genetic diseases). Today, most scanning techniques have largely been replaced by <u>d</u>enaturing <u>h</u>igh <u>p</u>erformance <u>l</u>iquid <u>c</u>hro-

matography (dHPLC or DHPLC). By scanning, a large gene can be analysed to detect where DNA changes are present. These changes can be neutral polymorphisms or actual mutations. The scanning simply identifies where these changes occur, and then DNA sequencing is necessary to determine the type of change that has been detected.

dHPLC

Reference was made earlier in the section on PCR to the *CFTR* gene, which is very large in terms of its genomic size (250 kb). However, it has 27 exons, which span a much smaller area of 6.5 kb. One shortcut to analysing this large gene has already been described by using RT-PCR. Another approach is to take each of the 27 exons individually and scan them with dHPLC. Since most of the mutations will be found within the exons or exon-intron boundaries, dHPLC will identify which of the 27 exons are likely to have a DNA change, and so which exon should be sequenced.

dHPLC is an important new development in molecular medicine. It allows high-throughput semi-automated scanning with close to 100% sensitivity and specificity. Thus, it has replaced other methodologies such as SSCP, DGGE and CCM. The method relies on reduced melting temperatures for heteroduplexes compared to homoduplexes (Figure A.14). With dHPLC, DNA is partially denatured with heat. Heteroduplexes formed are separated from homoduplexes by ion pair reverse phase liquid chromatography on a special affinity column. Heteroduplexes containing a mismatch will elute before homoduplex molecules. An altered elution pattern gives the

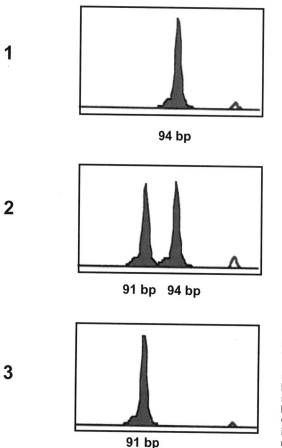

94 bp

91 bp   94 bp

91 bp

Fig. A.9   DNA mutation detection on the basis of DNA fragment size.
A combination of capillary electrophoresis and software-based fragment sizing means that PCR fragments can now be measured very accurately. **(1)** This figure shows a 94 bp fragment from a normal *CFTR* gene. This amplified product comes from exon 10 and detects the common cystic fibrosis mutation ΔF508. **(2)** This figure is a heterozygous carrier of the same mutation. The allele with the ΔF508 deletion involving 3 bp (now sized 91 bp) is easily distinguished from the wild-type allele. **(3)** A homozygous-affected individual with cystic fibrosis shows only the 91 bp allele.

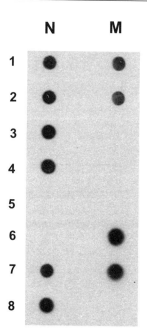

Fig. A.10   ASO—allele specific oligonucleotide.
ASO probes are synthesised so that one probe hybridises to the normal (wild-type) sequence and the second to the mutant DNA sequence. Amplified, single-stranded DNA is fixed onto nylon membranes in the form of dots (hence, the name dot blots). In the example given, DNA has been amplified in the region of the β globin gene located at the first exon-intron boundary. A mutation involving the first intron at position 5 (a G is changed to a C) interferes with the consensus sequence important for RNA splicing. Therefore, a thalassaemia will result from this mutation. N = normal probe (will detect G at this site), M = mutant probe (will detect C at this site). **(1,2)** are duplicate samples for unknown A; **(3,4)** are duplicate samples for unknown B. **(5)** is the no DNA control. **(6)** is the homozygous-affected (i.e., C) control. **(7)** is the heterozygous control (i.e., both the C and G hydridise), and **(8)** is the normal control (i.e., only the G will hybridise). From this ASO it is evident that unknown A is heterozygous for the IVS1–5 mutation, and unknown B does not have this defect. *ASO courtesy of Stuart Cole, Department of Molecular & Clinical Genetics, Royal Prince Alfred Hospital, Sydney.*

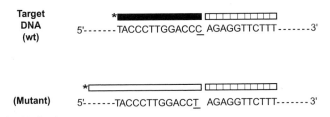

Fig. A.11 OLA—oligonucleotide ligation assay.
First, the region of interest (codon 39 in the β globin gene) is amplified using PCR. Three different DNA probes are needed for the post-PCR assay to detect the mutation: first, a common primer (represented by hatched bar); second, one specific for the normal sequence (black bar), and third, one that will hybridise to the mutant sequence (open bar). Each of the non-common primers is labelled with a different coloured dye indicated by *. In addition, the mutant primer is slightly larger than the normal one, and so the fragment generated will be both different in colour and in size. The three actual probes would be longer than what is depicted, but for convenience the probe lengths have been shortened. The common probe will bind to its complementary region in the β globin gene sequence. If the DNA is normal, only the normal primer will bind to the DNA, and so this primer + common probe are able to be ligated using the enzyme DNA ligase. The ligated product will be detectable following electrophoresis (by colour, chemiluminescence or fluorescence). Fluorescence is particularly valuable because multiple colours can be used, and so multiplexing is possible. If a mutant sequence is present, the mutant primer will bind, and it can be ligated to the common primer. Like ASOs, two results are possible: (1) a ligated product with both primers = heterozygous; (2) ligation only with one of the primers (i.e., either homozygous normal or homozygous mutant, depending on which primer has ligated).

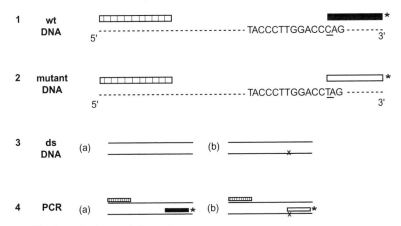

Fig. A.12 ARMS—amplification refractory mutation system.
The same codon 39 β globin gene example is given. (1,2) Three primers in ARMS are used in the PCR step (compared with three probes for the ligation step in OLA): a common primer (hatched bar) and two other primers mismatched at their important 3′ ends so that one primer binds only to normal sequence (black bar) and the other to the mutant sequence (open bar). In (3), the double-stranded DNA is shown with a mutation x in (b). In (4), PCR involves the denaturation into single-stranded DNA and then annealing of the primers. The common primer anneals in all cases, but because of the mismatches at the 3′ end, only the wild-type primer anneals to normal and the mutant primer to mutant DNA. These two primers are labelled with different fluorochromes (*) so they can be readily distinguished in the one reaction. Finding both colours would indicate heterozygote; only the mutant colour, homozygous mutant; only the wild-type colour, homozygous normal.

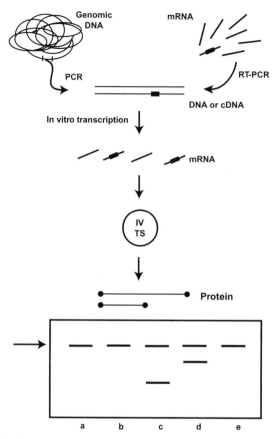

Fig. A.13 PTT—protein truncation test.
PTT is used if a point mutation leads to a premature stop codon so that one of the two alleles for a gene will produce a smaller peptide; i.e., it is truncated. A segment of the gene of interest is first amplified from mRNA by RT-PCR (see Figure 2.2). Alternatively, individual exons from genomic DNA are amplified by PCR. The ■ indicates the position of a premature stop codon on one DNA strand. cDNA or genomic DNA is then made into mRNA with an *in vitro* transcription system. Two RNA species will result since one will also contain the premature stop codon. The final step involves an *in vitro* translation system that allows protein to be produced from the RNA. Two distinct proteins will result, with the mutant one being smaller. The proteins are separated by electrophoresis, with the smaller (truncated) one having a faster mobility (→ indicates the normal protein species). All the steps described in PTT can be undertaken with commercial kits. Lanes a, b and e show normal samples, and the samples in lanes c and d have an additional band, suggesting truncated proteins of different sizes.

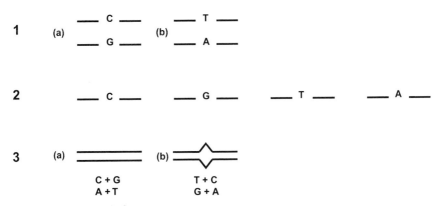

1 (a) ___ C ___   (b) ___ T ___
       ___ G ___   ___ A ___

2   ___ C ___   ___ G ___   ___ T ___   ___ A ___

3 (a) ========   (b) ===><===

           C + G   T + C
           A + T   G + A

Fig. A.14   Heteroduplexes and homoduplexes.
Using the same codon 39 β globin gene mutation example, **(1a)** and **(1b)** illustrate that in a heterozygote individual, there will be in one gene the C–G sequence on complementary DNA strands, and on the mutant gene, the complementary DNA strands will have T lining up with A. **(2)** If DNA is melted (through heat or denaturant chemicals), four single-stranded DNA species will be present as shown. **(3)** When DNA is allowed to anneal, the homoduplexes are represented by normal hybridisation involving C with G or A with T. However, heteroduplexes will also form; i.e., there will be an inappropriate hybridisation with some T + C and G + A joining. Because of the mismatch, the heteroduplexes are less stable and break apart more easily if heated or exposed to chemical denaturants.

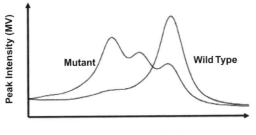

Fig. A.15   dHPLC elution profile.
The normal (wild-type) gene profile is shown. The mutant profile is distinctly different. The dHPLC profile simply indicates that the two alleles have migrated differently; i.e., there is a DNA polymorphism or a DNA mutation present. DNA sequencing will distinguish the two possibilities. *Profile courtesy of Dr Bing Yu, Department of Molecular & Clinical Genetics, Royal Prince Alfred Hospital, Sydney.*

clue that a DNA sequence change is present—but does not indicate what this change will be (Figure A.15). dHPLC has one drawback: This is the requirement for wild-type and mutant DNA species to be present to form a heteroduplex. Thus, autosomal recessive conditions are not suited to dHPLC detection because the underlying mutation involves a homozygous change. This can be overcome if the mixture containing the homozygous mutation is spiked with normal DNA.

LINKAGE ANALYSIS

Although described in Chapter 2, linkage is worth mentioning again because it remains a useful technique in both the diagnostic and research DNA laboratories. A key

to understanding DNA linkage analysis is the DNA polymorphism; the reader should be familiar with this variation in DNA. An example of a linkage study involving the *CFTR* gene is given in Figure A.16. This linkage study was necessary because it was not possible to find the mutation causing cystic fibrosis in one partner for a couple at risk of having children with cystic fibrosis.

In the cystic fibrosis example given, the diagnosis has been established indirectly (a direct diagnosis refers to mutation detection). However, the chance that recombination might cause an error needs to be considered in any linkage study. In this case, the risk for recombination is nearly negligible because the polymorphism is intragenic, i.e., within the gene. Another potential source of error is mislabelling of specimens, particularly in respect

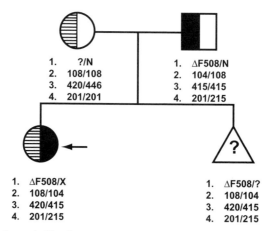

1.	?/N
2.	108/108
3.	420/446
4.	201/201

1.	ΔF508/N
2.	104/108
3.	415/415
4.	201/215

1.	ΔF508/X
2.	108/104
3.	420/415
4.	201/215

1.	ΔF508/?
2.	108/104
3.	420/415
4.	201/215

Fig. A.16   DNA linkage analysis for cystic fibrosis.
A couple has a child with cystic fibrosis (←). The father is a known heterozygote for ΔF508. The mutation in the mother cannot be found. The fetus during this pregnancy (Δ) has a chorionic villus sample taken for prenatal diagnosis. If the fetus does not have ΔF508, then it does not matter that the mother's mutation cannot be identified because the fetus will not have cystic fibrosis (the worst case scenario would be a carrier for cystic fibrosis). However, finding ΔF508 in the fetus would now increase the risk of cystic fibrosis to 50% rather than the *a priori* 25% risk. One way around this dilemma is to do linkage analysis using the known cystic fibrosis–affected child as the way in which to assign which DNA polymorphism goes with the cystic fibrosis defect. Three polymorphisms are used. They are intervening sequence (IVS)6, IVS8 and IVS17b. Because the three are found within the *CFTR* gene, recombination is not an issue with this particular test. The three polymorphisms are listed as **(2)–(4)** and the numbers refer to the size of the DNA fragment. Based on the polymorphisms found in the affected child and the parent from which they are inherited, it can be said that 108, 420 and 201 fragments go with the cystic fibrosis mutation coming from the mother. However, the 108 and the 201 results can be ignored because the mother is homozygous for them. The only useful polymorphism in this particular family is the 420 fragment since unequivocally this goes with the cystic fibrosis mutation in the mother. The fetus is shown to be ΔF508 positive, and in addition, the fetus has the 420 marker; i.e., the fetus has cystic fibrosis (see also Figure 3.11 for a more straightforward example of linkage analysis in cystic fibrosis).

to the relationship between various family members. Again, in this particular example, this is less likely because one of the parents is known to be a carrier for the ΔF508 mutation, and so this particular DNA sample can be confirmed to be correct. What might not be possible to exclude in linkage studies is non-paternity. This can produce a wrong result because an erroneous linkage becomes possible.

FUNCTIONAL GENOMICS

Another description for functional genomics is transcriptome analysis. This refers to the use of DNA to determine gene function (or gene malfunction in the presence of a DNA mutation).

IN VITRO *EXPRESSION*

The most basic expression vector is a plasmid that has an origin of replication, a selectable DNA marker to allow it to be detected and also provide a growth advantage, and a multiple cloning site into which the gene to be expressed is inserted. The vector with the inserted gene is then transfected into a host cell such as *E. coli* and allowed to replicate. Selection, usually via an antibiotic resistance gene, ensures that only *E. coli* with the plasmid insert will remain viable (Figure A.17). In the prokaryotic system, bacterial proteins are expressed, but eukaryotic-derived products will be degraded unless they are fused to a bacterial protein. Expressed protein is isolated, purified and able to be used as a therapeutic agent. More sophisticated *in vitro* expression systems utilise yeast, insect or mammalian-derived vectors. In the non-bacterial expression systems, protein products can undergo post-translational changes. They can be significant for biological activity (see Chapter 6 for a description on how expression vectors are utilised to produce drugs and vaccines).

In vitro expression systems are also used to test gene function, particularly that involving promoter regions. For example, the enzyme chloramphenicol acetyl transferase can be included in an expression plasmid. It is then possible to insert eukaryotic promoter sequences 5′ to the chloramphenicol acetyl transferase gene. The activity of these promoters can be tested by measuring how much chloramphenicol acetyl transferase is made. RNA transcripts can be produced from plasmid vectors that have special promoters located 5' to their cloning site. The promoters are recognised by DNA-dependent RNA polymerases and so produce RNA rather than DNA. RNA transcripts formed in this way can be used as RNA probes or to identify and quantify mRNA production *in vitro*.

IN VIVO *EXPRESSION*

Fertilised oocytes contain two pronuclei. Into one of these a gene, in the form of cloned DNA, can be microinjected using a finely drawn pipette. The injected (foreign) DNA becomes randomly integrated into the genome and is present in multiple copies in a head-to-head tandem arrangement in oocyte DNA. Expression of foreign DNA will occur if there are sufficient copy numbers and the environment into which DNA has become integrated is suitable. Transgenic animals, usually mice, with the foreign DNA can be identified by screening their DNA with PCR. This is the least sophisticated of the transgenic mouse models since there is no control over where the injected DNA inserts in the genome or how much DNA (genes) is integrated into the host DNA.

The next level of complexity with a transgenic animal is the knockout or knockin mouse. This requires the inactivation or insertion of a specific gene by homologous recombination. The approach to this is described in more detail in Chapter 5 and Figure 5.3. The conditional transgenic provides the most sophisticated of the transgenic animals since the inserted gene can be regulated by specific stimuli. Although the above descriptions refer to mice, considerable work is now being undertaken in other transgenic models, one of which, the zebrafish, is described in Chapter 5.

Transgenic mice provide very useful *in vivo* expression systems to test the function or significance of genes, gene sequences, promoters or the phenotypic effects of mutations in genes. Their main disadvantage reflects the time and effort required to produce and maintain these animals.

MICROARRAYS

Microarrays are an important new development since the second edition of *Molecular Medicine*. They are described in detail in Chapter 5 and Figures 5.1 and 5.2. This technology provides a snapshot of the transcriptome (i.e., all mRNA species in a cell), allowing the expression of hundreds to thousands of genes to be assessed at any point in time. Usually, microarrays are studied in different cells or tissues, and comparisons in terms of gene expression are made. For example, an accepted cut-off for gene expression in microarrays is greater than twofold

(this means an up-regulated gene) or less than 0.5-fold (that is, a down-regulated gene). Microarrays therefore generate large data sets. Two obstacles that must be overcome before this technology becomes more mainstream are (1) high cost and (2) better software to allow analysis and interpretation of the data generated. It should be noted that microarrays are only screens. They identify likely changes in the transcriptome that will then need to be confirmed with more specific measures, for example, real-time Q-PCR for each individual gene detected as changing its expression on the microarray.

The terms "microarrays" and "chips" are used interchangeably. Historically, **microarrays** refer to the spotting of DNA onto glass slides in the form of a grid. These spots are hybridisation targets for cDNA obtained from cells or tissues. In effect, a microarray is an ASO (see above) that analyses hundreds to thousands of genes simultaneously. On the other hand, **chips** historically referred to the spotting of oligonucleotide arrays on a solid medium including glass. They are usually commercially produced. This approach ensures better reproducibility and quality. Chips can be designed according to need, for example, a chip to detect gene changes in cells from prostate cancer.

Microarrays are known as **closed systems** because the range of genes that will be tested is limited (in terms of

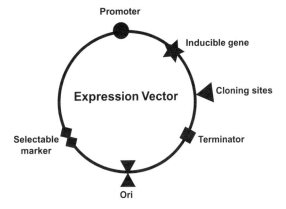

Fig. A.17   DNA expression vector.
The vector depicted has an origin of replication (ori) that allows it to replicate autonomously. There is a selectable marker (usually an antibiotic resistance gene) that gives a growth advantage to the host (e.g., *E. coli*) containing the vector. The bacterial promoter ensures that transcription will occur. An inducible bacterial-specific gene is useful since it allows expression to be controlled. The DNA sequence (or gene) to be expressed is inserted into the multiple cloning site. Translation is stopped by placing a termination signal downstream of the cloning site. A fusion product (protein from the inducible gene plus protein from the cloned gene) is generated, and it protects the non-bacterial protein from degradation. These two components need to be separated and the expressed protein purified from bacterial products.

277

what the microarray or chip has spotted onto it). There are also **open systems** to allow the assessment of all mRNA species. One such assay is called SAGE (Serial Analysis of Gene Expression). Unlike the closed microarrays, the SAGE approach does not require prior knowledge of what genes are being targeted. This technique takes all the mRNA species in a cell (i.e., the transcriptome) and converts them to cDNA by RT-PCR. Each cDNA is characterised by sequencing a small 10 base pair segment called a SAGE tag. The sequence (and how many times it is detected on analysis) will provide the identity of the underlying gene (as well as a semi-quantitative estimate of its expression level). Thus, SAGE has one potential advantage over the closed microarray because it allows new genes to be discovered.

FURTHER READING

Preamble

Crocker J. Demystified . . . molecular pathology in oncology. Molecular Pathology 2002; 55:337–347 (*although the title suggests oncology as the subject, this article provides an easy-to-read summary of most techniques described in the appendix*).

Bryant-Greenwood P. Molecular diagnostics in obstetrics and gynecology. Clinical Obstetrics and Gynecology 2002; 45:605–621 (*provides a basic and brief overview of many of the relevant techniques in molecular medicine*).

www.acmg.net Internet address for the American College of Medical Genetics (*provides useful material under "Standards and Guidelines for Clinical Genetics Laboratories," 2003 edition*).

DNA and RNA

Ginzinger DG. Gene quantification using real time quantitative PCR: an emerging technology hits the mainstream. Experimental Hematology 2002; 30:503–512 (*provides useful description of Q-PCR and the way real-time PCR works for this*).

McGovern MM, Benach M, Wallenstein S, Boone J, Lubin IM. Personnel standards and quality assurance practices of biochemical genetic testing laboratories in the United States. Archives Pathology Laboratory Medicine 2003; 127:71–76 (*provides some interesting observations about skills and quality practices in genetic testing laboratories*).

Mullis KB. The unusual origin of the polymerase chain reaction. Scientific American 1990; 262:56–65 (*easy-to-read account describing the history as well as the basis for PCR*).

Gene Discovery

Bird T, Jarvik G, Wood NW. Genetic association studies: genes in search of diseases. Neurology 2001; 57:1153–1154 (*brief summary of association studies including their strengths, weaknesses and ways in which to enhance them*).

Collins FS. Positional cloning moves from perditional to traditional. Nature Genetics 1995; 9:347–350 (*although this is now a historical article, it provides some insight into how positional cloning developed*).

Ewens WJ, Spielman RS. Locating genes by linkage and association. Theoretical Population Biology 2001; 60:135–139 (*provides a more mathematically based discussion of linkage and association studies*).

Morton NE. Genetic epidemiology, genetic maps and positional cloning. Philosophical Transactions of the Royal Society of London 2003; 358:1701–1708 (*provides an interesting perspective of genetic analysis through to positional cloning and where this is now leading, with a number of ethically related questions about the technology*).

DNA Characterisation

Bartlett JMS, Stirling D (eds). PCR protocols. Methods in Molecular Biology, 2nd edition. 2003. Human Press, New Jersey (*reference for those who want to understand fully some of the DNA mutation methods*).

Cotton RGH, Edkins E, Forrest S (eds). Mutation detection. A Practical Approach. 1998. IRL Press, Oxford (*comprehensive but technical overview of the various methods involved in DNA mutation testing*).

Gite S, Lim M, Carlson R, Olejnik J, Zehnbauer B, Rothschild K. A high throughput nonisotopic protein truncation test. Nature Biotechnology 2003; 21:194–197 (*describes the PTT as well as ways in which to improve this test*).

O'Donovan MC et al. Blind analysis of denaturing high performance liquid chromatography as a tool for mutation detection. Genomics 1998; 52:44–49 (*brief introduction to dHPLC*).

Romppanen E-L. Oligonucleotide ligation assay: applications to molecular diagnosis of inherited disorders. Scandinavian Journal of Clinical Laboratory Investigation 2001; 61:123–130 (*nice summary of the OLA assay*).

Functional Genomics

Cox RD, Brown SDM. Rodent models of genetic disease. Current Opinion in Genetics & Development 2003; 13:278–283 (*brief overview of the various rat and mouse models used in modern genomics research*).

Patino WD, Mian OY, Shizukuda Y, Hwang PM. Current and future applications of SAGE to cardiovascular medicine. Trends in Cardiovascular Medicine 2003; 13:163–168 (*gives an overview of the SAGE technique and compares it to microarrays*).

Wildsmith SE, Elcock FJ. Microarrays under the microscope. Molecular Pathology 2001; 54:8–16 (*nice review of the practical aspects of microarrays*).

GLOSSARY AND ABBREVIATIONS

GLOSSARY

Adult stem cells   Cells that are capable of maintaining, generating and replacing terminally differentiated cells within their own specific tissue following normal cell turnover or in response to injury. Adult stem cells are now considered to have the capacity to differentiate beyond their own tissue boundaries; this capacity is called *plasticity* (see also *embryonic stem cells*; Table 6.13).

Allele specific oligonucleotides (ASOs)   Oligonucleotides that are constructed with DNA sequences homologous to specific alleles. Two ASOs can be made which differ in sequence at only one nucleotide base, thereby distinguishing a mutant allele with a point mutation from its corresponding wild-type allele (see Figure A.10).

Alleles   (Abbreviation for *allelomorph*.) Alternative forms of the same gene.

Allogeneic   From one person to another who is genetically dissimilar but of the same species (see also *autologous, heterologous*).

Alu repeat   The most common interspersed repeat (~300 bp in size) in human DNA, accounting for about 5% of total DNA. So named because it is cleaved by the restriction enzyme *Alu*I.

Amino acids   The building blocks of proteins. Each amino acid is encoded by a nucleotide triplet (see also *codon*; Table 2.1).

Amniocentesis   Aspiration of amniotic fluid during pregnancy.

Amplification   Multiple copies of a DNA sequence.

Aneuploid   Any chromosome number that is not an exact multiple of the haploid number (23 in humans). Examples of aneuploidy include the presence of an extra copy of a single chromosome—e.g., trisomy 21, (Down's syndrome)—or the absence of a single chromosome—

e.g., monosomy, as found in Turner syndrome 45,X (see also *ploidy*).

Anneal   To form double-stranded nucleic acid from single stranded forms.

Annotation of DNA sequence/gene   The conversion of raw sequence data to knowledge of gene location on a chromosome and the function of that gene, i.e., characterisation of a gene (see Figure 4.16).

Antibody   A protein produced by higher vertebrates following exposure to a foreign substance (called an *antigen*). The Y-shaped antibodies bind to antigens and neutralise them. Antibodies can be polyclonal or monoclonal in origin (the latter is derived from a single cell, and so each antibody is identical). More correct term is *immunoglobulin*.

Anticipation   Increasing severity or earlier age at onset of a genetic disease in successive generations (see Figure 3.4).

Antigenic drift   A more subtle form of variation than antigenic shift (see next entry). Involves accumulation of mutations in the haemagglutinin and neuraminidase genes in the influenza A and B viruses. This reduces the effectiveness of previously acquired antibodies to influenza and so facilitates the spread of the virus.

Antigenic shift   A mechanism seen only with influenzae A viruses to escape the human immune response. It results from the replacement of the haemagglutinin or the neuraminidase with novel subtypes that have not been around in humans for some time. These subtypes come from waterfowl, a large reservoir of influenzae viruses. The consequence is a pandemic, i.e., worldwide epidemic (see also *antigenic drift*).

Antisense   Antisense DNA is the non-coding strand of DNA. The latter functions as the template for mRNA production, which then contains the sequence present on the sense strand. Antisense RNA or antisense oligonucleotides have sequences that are complementary to mRNA and so interfere with the latter's function (see Figure 6.6).

Apoptosis   A process involving programmed cell death. It has also been called *genetically determined cell suicide*. Leads to death of cells during development or cell turnover in the adult. Mutations in genes involved in apoptosis are associated with prolonged cell survival including cancer.

Association   (Ch. 4)

Assortive mating   Sexual reproduction in which the pairing of mates is not random, i.e., members of a particular group that are more (less) likely to mate with other members of that group produce positive (negative) assortive mating.

Attenuated virus   A virus that has become less pathogenic following passage outside its natural host.

Autologous   From the same person (see also *allogeneic, heterologous*).

Autosomal disease   Disease that is the result of an abnormality affecting the 22 pairs of autosomes (non-sex chromosomes).

Bacteriophage "phage"   A virus that infects bacteria (see also *clone, vector*).

Base pair   A measurement of length for DNA. Includes a nucleotide base with its complementary base; i.e., adenine (A) would bind to thymine (T) or cytosine (C) to guanine (G) (see *complementary*, Figure 2.1).

Beneficence   (Ch. 10)

Bioinformatics   The application of tools involving computation and analysis to capture, store and interpret biological data (see Chapters 1, 5).

Candidate gene   A gene that would be a good starter to initiate a search for the genetic basis of an inherited disorder of unknown origin, e.g., the myosin genes in muscle disorders.

Cap   Post-transcriptional change to the 5′ end of the growing mRNA molecule in which a modified nucleotide (4 methylguanosine) is added. Has a functional role since it is recognised by ribosomes as the initiation signal for protein synthesis.

Carcinogen   Physical or chemical agent that induces cancer.

Carrier   An individual who is heterozygous for a mutant allele that causes a genetic disorder in the homozygous or hemizygous states.

Cell cycle   The timed sequence of events occurring in a eukaryotic cell between mitotic divisions. Divided into M (mitotic), S (DNA synthetic), G_1 and G_2 (gap or pause phases) and G_o (resting phase). The times for each component differ between cell lines (see Figure 4.13).

Centimorgan (cM)   Distance between DNA loci as determined on a genetic map. A distance of 1 cM indicates two markers are inherited separately 1% of the time. In terms of the physical map, 1 cM is very approximately equal to 1 Mb (Mb; see *megabase*). Name is derived from TH Morgan.

Centromere   The heterochromatic constricted portion of a chromosome where the chromatids are joined (see also *heterochromatin, telomeres*; Figure 2.10).

Chimaera   An individual composed of a mixture of genetically different cells. A chimaera is distinguished from a mosaic on the basis that the cells in the former are derived from different zygotes, e.g., transgenic mouse formed by the embryonic stem cell approach (see also *mosaicism, transgenic*; Figure 5.3).

Chips   (App)

Chorionic villus sampling or **sample (CVS)**   Biopsy of the chorion frondosum during pregnancy to obtain a source of fetal tissue for prenatal diagnosis.

Chromatin   The complex of DNA and protein in which the genetic material is packaged inside the cells of organisms with nuclei (eukaryotes).

Cis-acting locus   A region of a gene or nearby locus that affects its function. Could include regions such as the promoter or an enhancer (see also *trans-acting locus*).

Clinical haemochromatosis   (Ch. 3)

Clone   To clone DNA means to take a gene or part of a gene and isolate it from the remainder of genomic DNA and then produce genetically identical material. The cloned DNA can be produced in unlimited amounts (see also *functional cloning* and *positional cloning*; Figures A.4, A.5, A.6). The term *cloning* has wider ramifications since it also applies to the genetic duplication of cells or even whole organisms, including potentially a human. One method that has been used to clone various animal species is called *somatic cell nuclear transfer*.

Codon Three adjacent nucleotide bases in DNA/RNA that encode for an amino acid (see Table 2.1).

Comparative genomic hybridisation (CGH) Microarrays used to look for genomic gains or losses particularly in cancer (see Table 5.2).

Complementary The specific binding between the purine-pyrimidine base pairs of double-stranded nucleic acid. Thus, adenine (purine) will covalently bind to thymine (pyrimidine) and guanine (purine) to cytosine (pyrimidine) in a 1-to-1 ratio (see also *base pair*; Figure 2.1).

Complementary DNA (cDNA) DNA that is synthesised from an mRNA template. The enzyme required for this is reverse transcriptase (see Figure 2.2).

Compound (heterozygote) An individual with two different mutant alleles at a locus.

Concordance Both members of a twin pair demonstrating the same phenotype or trait (see also *discordance*).

Congenic Inbred strains that differ from one another in a small chromosomal segment, cf., syngeneic inbred strains that are identical except for sexual differences.

Congenital Present at birth.

Conservation (DNA) The finding that a DNA sequence is present in a wide range of phylogenetically distant organisms suggests functional significance since it is unlikely that during evolution a region of DNA would have remained unaltered unless it had a specific and important function; e.g., it is a gene. The *ras* proto-oncogene illustrates this since it is conserved in organisms as divergent as humans and yeast.

Constitutional (cells) Cells that would be representative of an organism; e.g., in DNA testing for loss of heterozygosity, examples of constitutional cells which would provide a baseline for the DNA polymorphisms would be lymphocytes (if the cancer is non-haematological) or fibroblasts which could be obtained from a skin biopsy.

Constitutive (genes) Genes that are expressed following interaction between a promoter and RNA polymerase without additional regulation. Also called *housekeeping genes* since often expressed in all cells at low levels. In contrast are inducible genes, e.g., metallothionein, which expresses following exposure to some heavy metals.

Consultand The person seeking or referred for genetic counselling (see also *proband*).

Contigs Overlapping clone sets that represent a continuous region of DNA.

Contiguous gene syndromes A group of disorders associated with malformation patterns, often with mental retardation and growth abnormalities. The clinical heterogeneity found in these disorders usually reflects the involvement of a number of physically related but otherwise distinct genes.

CpG islands Regions of 1–2 kb containing a high density of hypomethylated cytosine residuals associated with guanine. CpG islands are frequently found at the 5′ end of genes (see also *methylation*).

Cyclins Families of interacting proteins involved in the regulation of the cell cycle. So named because their levels are cell-cycle dependent.

Cytokines Proteins (but not antibodies) released by some cells in response to contact with an antigen, e.g., interleukin-2 (IL2). Cytokines function as intracellular mediators, e.g., generation of immune response seen with IL2 (see Figure 6.2).

Decoding Identifying the function of a gene from its DNA sequence.

Deletion Loss of a segment of DNA or chromosome (see also *interstitial deletion, microdeletion*; Figures 2.12, A.2, A.3).

Diploid The chromosome number found in somatic cells. In humans this will be 46, i.e., twice the number present in the germ cells (see also *haploid*; Figure 2.9).

Discordance Members of a twin pair not demonstrating the same phenotype or trait (see also *concordance*).

Disomy See *uniparental disomy*.

Dizygotic twins Twins (fraternal) produced from two separate ova fertilised by different sperms (see also *monozygotic twins*; Box 4.1).

DNA bank (Ch. 3)

DNA chip See *microarray*.

DNA thermal cycler (App)

Dominant A genetic disorder is said to have dominant inheritance if the mutant phenotype is produced when only one of the two normal (wild-type) alleles at a particular locus is mutated (see also *recessive*; Figure 3.2).

Dominant negative effect   Inactivation of one of the two tumour suppressor gene loci can produce what appears to be a dominant effect if the mutant protein inhibits the normal product from the remaining normal allele.

e antigen   Hepatitis B virus e antigen (HBeAg)—a part of the core antigen of the hepatitis B virus (HBcAg) that is secreted into the serum through cellular secretion pathways. HBeAg correlates strongly with infectivity.

Electroporation   The use of a pulsed electric field to introduce DNA into cells in culture.

Embryonic stem cells (ES cells)   In the embryo's blastocyst stage before implantation, the inner cell mass contains all the cells that will make up the fetus. Some of these cells are pluripotential because they will give rise to all types of somatic cells as well as the germ cells. When these pluripotential stem cells are grown *in vitro*, they are called *embryonic stem cells* (see also *adult stem cells*; Table 6.13).

Enhancers   DNA sequences with the following properties: (1) They increase transcriptional activity. (2) They are effective even if inverted in position. (3) They operate over long distances.

***Env* gene**   A gene that encodes envelope protein of a retrovirus (see Figure 8.2).

Epigenetic   Heritable change in the pattern of gene expression mediated by mechanisms other than changes in the primary DNA sequence of the gene. The changes can be inherited, for example, imprinting, but they do not involve an alteration in genetic information. When considering epigenetics in the post Human Genome Project era, a more suitable term might be *epigenomics*.

Episomal   In gene therapy, refers to vectors that remain free in the target cell without being taken up into the host's genome.

Euchromatin   Non-condensed, light-appearing bands following staining to produce G (Giemsa) banding of chromosomes. More likely to contain transcriptionally active DNA (see *heterochromatin*; Figures 2.9, 2.10).

Eukaryotes   Organisms ranging from yeast to humans that have nucleated cells.

Eutherian   Placental mammals.

Exclusion   (Ch. 9)

Exon   That segment in a gene which codes for a polypeptide and is represented in the mRNA.

Expressed sequence tag (EST)   A small DNA segment that can be amplified by PCR. This segment functions as a unique identifier for a region of the genome.

Expressivity   The severity of a phenotype. Variable expressivity is a feature of autosomal dominant disorders.

Familial   A condition that is more common in relatives of an affected individual than in the general population (see Breast Cancer in Chapter 4).

Fetal medicine units   (Ch. 7)

Fetal therapy   (Ch. 7)

Fingerprints   Dermatoglyphic fingerprints: Derived from the ridged skin patterns of the fingers. DNA fingerprints: Obtained from multiple microsatellite DNA polymorphisms (see also *minisatellites, microsatellites*; Chapter 9).

Five prime (5′)   The 5′ position of one pentose ring in DNA is connected to the 3′ position of the next pentose via a phosphate group. The phosphodiester-sugar backbone of DNA consists of 5′–3′ linkages, and this is the direction that the nucleotide bases are transcribed (see Figure 2.1).

Flanking (markers, DNA)   DNA markers on either side of a locus; DNA sequences on either side of a gene.

Fluorescence *in situ* hybridisation (FISH)   Nonisotopic method to label DNA probes for *in situ* hybridisation. The ability to utilise multiple fluorochromes in the same reaction increases the utility of this procedure. The resolving power of FISH is further enhanced if interphase chromosomes are studied.

Footprinting   Technique that identifies sites where there is protein bound to DNA. This complex then becomes resistant to degradation by nucleases.

Forward genetics   Mutagenesis of the genome produces different phenotypes. From them, it becomes possible to detect the underlying genes; i.e., no prior knowledge of genes is needed, and so the model is phenotype driven (see the discussion of ENU mouse in Chapter 5 and contrast forward genetics with reverse genetics as exemplified by the transgenic mouse).

Frameshift mutation   A mutation in DNA such as a deletion or insertion which interferes with the normal codon (triplet base) reading frame. All codons 3′ to the mutation will have no meaning. For example, the triplets GGT-TCT-GTT code for amino acids glycine, serine and valine, respectively. A deletion of one nucleotide (e.g., a G of the GGT) would disrupt the reading frame to give GTT-CTG-TT, etc. The protein product will terminate when a new stop codon is reached.

Functional cloning   Cloning strategy in which knowledge of a gene's product (function) is used to clone the gene. Now rarely used and replaced by positional cloning (see also *clone, positional cloning*; Box 1.1).

Functional genomics   See *genome*.

G418   Neomycin analogue that kills cells unless they are neomycin resistant or carry the gene for neomycin resistance (see Figure 6.10).

Gag gene   Group-specific antigen that encodes core protein for a retrovirus (see Figure 8.2).

Gain of function   (Ch. 3)

Ganciclovir   Prodrug that can be phosphorylated to its active metabolite by thymidine kinase from the herpes simplex 1 virus (HSV-tk). The active metabolite causes cell death by inhibiting DNA synthesis (see Figure 6.5).

G-banding   G (for Giemsa) banding used to identify chromosomal bands in a karyotype. Spreads of cells in metaphase are treated with trypsin and then stained with Giemsa (see Figure 2.9).

Gene   A sequence of DNA nucleotide bases coding for a polypeptide.

Gene therapy   The transfer of genetic material (DNA/RNA) into the cells of an organism to treat disease or for research purposes.

Gene transfer   (Ch. 6)

Genetic engineering   Colloquial term for recombinant DNA technology: The experimental or industrial applications of technologies that can alter the genome of a living cell.

Genetic map   An indirect measure of distance, constructed by determining how frequently two markers (DNA polymorphisms, physical traits or syndromes) are inherited together. Distances in genetic maps are measured in terms of centimorgans (see also *physical map*).

Genome   The complete (haploid) genetic material (DNA) of an organism. Hence, **genomics**—the study of the structure of the genome including its DNA sequence. **Functional genomics** is an additional variation in which function is included (in contrast is **genetics**, which is the study of single genes and their functions).

Genocopy (or **genetic mimic**)   This term is more difficult to define and overlaps with phenocopy. It can refer to a disorder with a similar phenotype due to abnormalities at different genetic loci or genetic mechanisms. For example, hereditary elliptocytosis (a disorder of the red blood cell leading to haemolytic anaemia of variable severity) is caused by mutations in different genes at four loci, but each gene encodes a protein involving the red blood cell membrane skeleton.

Genotype (genotypic)   The genetic constitution of an organism. In terms of DNA markers, it refers to the genetic constitution of alleles at a specific locus, e.g., the two haplotypes (see also *haplotype*; Figure 2.15).

Germ cells   Cells that differentiate early in embryogenesis to form ova and sperm.

G proteins   (Abbreviation for *guanine-binding proteins*.) These proteins play an important role in relaying messages from the cell surface to the nucleus. They act by binding GTP (guanosine triphosphate), which leads to activation of a second messenger system such as adenyl cyclase. There are many G proteins including the product of the *ras* proto-oncogenes. G proteins are self-regulating since the GTP-G protein complex is hydrolysed to inactive GDP-G protein by GTPase activity of the G protein. More than 100 receptors convey messages through G proteins (see also *ras, signal transduction*).

Guthrie spot   Term used (incorrectly) to describe the blood spot taken from newborns by heal prick. The blood spot is needed for newborn screening of genetic and metabolic disorders. The name is derived from the newborn screen for phenylketonuria, which utilises a test called the *Guthrie bacterial inhibition assay*.

Haematopoietic   Related to the blood; blood forming.

Haemoglobinopathies   Inherited disorders involving globin, the protein component of haemoglobin. Divided into the thalassaemia syndromes (e.g., α or β thalassaemia) and the variant haemoglobins (e.g., sickle cell anaemia [HbS]).

Haploid   The chromosome number found in gametes. In humans, this will be 23, i.e., one member of each chromosome pair (see also *diploid*).

Haplotype   A set of closely linked DNA markers at one locus inherited as a unit (see Figure 2.15).

Helper retrovirus (Ch. 6)

Hemizygous Having only one copy of a given genetic locus; e.g., a male is hemizygous for DNA markers on the X chromosome (see Figure 3.19).

Hereditary haemochromatosis (Ch. 3)

Heterochromatin Condensed, dark-appearing bands following G (Giemsa) banding of chromosomes. Contains predominantly repetitive DNA (see also *euchromatin, centromere*; Figures 2.9, 2.10).

Heteroduplex Hybrid DNA involving two strands that are different; e.g., there may be a base-mismatch (see also *homoduplex*; Figure A.14).

Heterologous Belonging to another species, e.g., the use of salmon sperm DNA to block non-specific hybridisation by human DNA.

Heteroplasmy The presence of more than one type of mitochondrial DNA in a cell. There are thousands of molecules of mitochondria DNA per cell. If there is mutant mitochondrial DNA, it can be present in varying amounts. Some cells might have predominantly wild-type DNA; others, predominantly mutant DNA (called *homoplasmy*); and others are said to be heteroplasmic because there is a mixture of both. Thus, phenotypic variation between cells is possible.

Heterozygote (heterozygous) An individual with two different alleles (e.g., gene, polymorphic marker) at a single locus (see also *homozygote*; Figure 3.9).

Homeobox (Also called *homeodomain*; *homeo*— Greek for "alike") A sequence of about 180 bp near the 3' end of some homeotic genes. The 60 amino acid peptide encoded by the homeobox is a DNA-binding protein (see also *HOX genes*).

Homoduplex Hybrid DNA involving two strands that are identical (see also *heteroduplex*, Figure A.14).

Homologous recombination A form of gene targeting on the basis of recombination between DNA sequences in the chromosome and newly introduced identical DNA sequences (see also *homology*; Figure 6.10).

Homology Fundamental similarity, matched; e.g., homologous (the same) chromosomes pair at meiosis; homology between DNA sequences means close similarity.

Homozygote (homozygous) An individual with two identical alleles (e.g., gene, polymorphism) at

a single locus (see also *heterozygote*; Figure 3.9).

Hot spots Regions in genes or DNA where mutations occur with unusually high frequency.

Housekeeping (genes) Genes that are expressed in virtually all cells since they are fundamental to the cell's functions.

HOX genes Family of genes involved in transcriptional regulation of embryonic development. *HOX* genes contain homeoboxes and determine the shape of the body along the antero-posterior axis of the embryo. Mutations in *HOX* genes cause a part of the body to be replaced by a structure normally found elsewhere. Conserved DNA sequences within these genes are called *homeoboxes*. All vertebrates including humans have *HOX* gene complexes located on different chromosomes. Another gene family involved in development is the *PAX* genes, the conserved sequence for which is called the *paired box* (see *PAX genes, homeobox*; Figure 7.1).

Human cloning See *clone*.

Human Genome Project Multicentred, multinational, multibillion dollar project. It officially started in 1990, and its completion was announced in 2000. The project's major aims were to provide a complete sequence of the human genome and the genomes of a number of model organisms.

Human leukocyte antigen (HLA) HLA is encoded for by a multigene complex occupying approximately 3500 kb of DNA on the short arm of chromosome 6. Antigens belonging to the HLA system are found on the surface of all cells except the red blood cells. HLA is concerned with normal immunological responses and plays a vital role in graft rejection or acceptance following transplantation. Also known as *major histocompatibility complex* or *MHC* (see Figure 4.2).

Hybridisation The pairing, through complementary nucleotide bases (A with T and G with C), of RNA/DNA strands to produce an RNA/RNA or RNA/DNA or DNA/DNA hybrid (see Figures 2.1, A.10).

Illegitimate transcription Low transcription of a tissue-specific transcribing gene in non-specific cells, e.g., the detection of mRNA for the β myosin heavy chain gene (a muscle-specific gene) in peripheral blood lymphocytes. Also called *ectopic* or *leaky* RNA.

Immunophenotyping (Also called *cell marker analysis*.) Typing of cells with immunological markers such as monoclonal antibodies.

Imprinting   Reversible modification of DNA that leads to differential expression of maternally and paternally inherited DNA or homologous chromosomes (see *uniparental disomy*; Figures 4.4, 4.5).

Inactivated (killed) microorganisms   (Ch. 6)

Informative (polymorphism)   Means a polymorphism is heterozygous and so able to distinguish two alleles. In a parental mating, at least one parent must be heterozygous for a polymorphism to be potentially informative. If both parents are heterozygous, the polymorphism is fully informative—if there is a key individual to help assign which marker co-segregates with disease, etc. (see Figures 2.15, A.16).

***In silico* cloning**   The modern version of positional cloning. Here, the discovery of a new gene is made by using knowledge of the human DNA sequence in the various databases and computer software.

***In situ* hybridisation**   Hybridisation of a DNA probe (labelled with ³H, fluorescein or a chemical such as biotin) to a metaphase chromosome spread or a tissue section on a slide.

***In situ* PCR**   (Ch. 2)

Interleukins   Proteins secreted by mononuclear leukocytes that induce the growth and differentiation of other haematopoietic cells.

Interspersed repeats   (Ch. 9)

Interstitial deletion   Loss of DNA or part of a chromosome that does not occupy a terminal position.

Intron   Segment of DNA that is transcribed but does not contain coding information for a polypeptide (also called *intervening sequence* or *IVS*). It is spliced out of the transcript before mature mRNA is formed.

Isoforms   Functionally related proteins that differ slightly in their amino acid sequence.

Isozymes (isoenzymes)   Different forms of an enzyme.

Junk (DNA)   (Ch. 9)

Karyotype   An individual's or a cell's chromosomal constitution (number, size and morphology). Determined by examination of chromosomes with light microscopy and the use of stains (see Figure 2.9).

Kilobase (kb)   One thousand base pairs in a sequence of DNA.

Kilodalton (kDa)   One thousand daltons. A unit that measures the molecular weight of proteins. One dalton approximates to the molecular weight of a hydrogen atom. The molecular weight of a protein will be based on the sum of the atomic weights of the elements that comprise it.

Library   A large number of recombinant DNA clones that have been inserted into a vector for the purpose of cloning a segment of DNA (see Figure A.5).

Ligand   A molecule that binds to a complementary site on a cell or other molecule.

Linkage   The tendency to inherit together two or more non-allelic genes or DNA markers than are to be expected by independent assortment. Genes/DNA markers are linked because they are sufficiently close to each other on the same chromosome (see Figures 2.14, 2.15, A.16).

Linkage disequilibrium   Preferential association of linked genes/DNA markers in a population, i.e., the tendency for some alleles at a locus to be found with certain alleles at another locus on the same chromosome with frequencies greater than would be expected by chance alone (e.g., HLA-DQ and HLA-DR alleles).

Lipofection   An *in vivo* or *in vitro* way to transfer DNA into a cell's nucleus. The gene of interest is mixed with a cationic lipid suspension and then mixed with the cell of interest.

Liposomes   Synthetic spherical vesicles with a lipid bilayer. Function as artificial membrane systems to deliver DNA, etc., into cells.

LOD score   Statistical test to determine whether a set of linkage data are linked or unlinked. LOD is an abbreviation of the log_{10} *of the odds* favouring linkage. For genetic disorders that are not X-linked, A LOD score of +3 (1000:1 odds of linkage) indicates linkage, whereas a score of −2 is odds of 100:1 against linkage.

Long PCR   (Ch. 2)

Lymphoproliferative disorders   Lymphomas and leukaemias of lymphocyte origin.

Lyonisation   (Ch. 3)

Medical informatics   (Ch. 5)

Megabase (Mb)   One million base pairs in a sequence of DNA.

Meiosis Process in which diploid germ cells undergo division to form the haploid chromosome number (see also *mitosis*).

Messenger RNA (mRNA) Transfers the genetic information from DNA to the ribosomes. Contains the template for polypeptide production.

Metabolomics The global metabolic profile in any cell, tissue or organism. Some prefer the term *metabonomics*.

Metastasis A secondary tumour arising from cells carried from the primary tumour to a distant locus.

Methylation (of DNA) Vertebrate DNA contains a small proportion of 5-methylcytosine, which arises from methylation of cytosine bases where they occur in the sequence CpG. The methylation status of DNA correlates with its functional activity: Inactive genes are more heavily methylated and vice versa (see also *CpG islands*).

MHC See *human leukocyte antigen*.

Microarray (Also called *DNA chip*.) Ordered, high-density arrangements of nucleic acid spots. Each spot represents a different DNA probe attached to an immobile surface. Probes can be cDNA or oligonucleotides. This technology allows the simultaneous measurement of transcriptional activity for hundreds to tens of thousands of genes (i.e., transcriptome) in any cell under any condition (see Figures 5.1, 5.2).

Microdeletion DNA or chromosomal deletion that is not detectable by conventional techniques such as microscopy (cytogenetics) or Southern blotting (DNA mapping).

MicroRNA See *RNA interference (RNAi)*.

Microsatellites As for minisatellites except that the polymorphism allele size is smaller (e.g., <1 kb) and the basic core repeat unit involves a two to four nucleotide base pair repeat motif. Also known as simple sequence repeats (SSRs) or short tandem repeats (STRs). One example is repeats of the motif AC—ACACACACAC, etc. (see Table 9.2, Figure 2.13).

Microsatellite instability (Ch. 3)

Minisatellites Repeat DNA segments that comprise short head-to-tail tandem repeats giving the variable number of tandem repeat (VNTR) type polymorphisms with approximate size between 1–30 kb. VNTRs can be of two types: single locus or multilocus. The latter were used previously to construct a DNA fingerprint of an indi-

vidual (see also *microsatellites, satellite DNA*; Table 9.2, Figure 2.13).

Missense mutation A single DNA base change that leads to a codon specifying a different amino acid, e.g., the base change of GGT (glycine) to GTT (valine).

Mitosis Somatic cell division; process in which chromosomes duplicate and segregate during cell division (see also *meiosis*).

Modifying genes (Ch. 3)

Molecular biology (Ch. 1)

Molecular medicine (Ch. 1)

Monoclonal Derived from a single clone, i.e., monoclonal antibody, monoclonal lymphocyte population (see also *polyclonal*).

Monozygotic twins Genetically identical twins formed by the division into two at an early stage in development of an embryo derived from a single fertilised egg (see *dizygotic twins*; Box 4.1).

Mosaicism A condition in which an individual or tissue has two or more cell lines of different genetic or chromosomal constitution. In contrast to a chimaera, both cell lines in a mosaic are derived from the same zygote (contrast with chimaera).

Multifactorial disorders Diseases that result from an interaction of environmental factors with multiple genes at different loci (see also *polygenic inheritance*, which is sometimes used in the same sense as *multifactorial*; see Figure 4.1).

Multidrug resistance (MDR) Development of simultaneous resistance to multiple structurally unrelated chemotherapeutic agents (see also *P-glycoprotein*).

Multiplex PCR (Ch. 2)

Murine Of the mouse (Latin, *mus*).

Mutation An alteration in genetic material. This could be a single base change (point mutation) to more extensive losses of DNA (deletions) (see also *missense mutation, nonsense mutation*; Tables 2.3, 3.15).

Nanoscience The study of the fundamental principles of molecules and structures with at least one dimension approximately between 1 and 100 nanometres— 1 nanometre (nm) is 1×10^{-9} of a metre (see Chapter 5).

Necrosis (Ch. 4)

Negative predictive value (NPV) (Ch. 8)

Nested PCR (Ch. 2)

Newborn screening (Ch. 7)

Nonsense mutation A single DNA base change resulting in a premature stop codon (TAA, TGA, TAG), e.g., TCG (serine) to TAG (stop).

Northern blotting Procedure to transfer RNA from an agarose gel to a nylon membrane (see also *Southern blotting, western blotting*).

Nosocomial Hospital acquired.

Nucleases Enzymes that break down nucleic acid. There are DNAase (DNase) and RNAase (RNase) enzymes. RNA, in particular, is susceptible to RNAases so that preparation of RNA requires a lot more care compared to the more robust DNA.

Nucleic acid amplification technique (NAT) Term used predominantly in microbiology to refer to the amplification of nucleic acids with various techniques, including PCR, ligase chain reaction, nucleic acid sequence-based amplification and other methods.

Nucleotide The monomeric component of DNA or RNA comprising a base (A—adenine; T—thymine; U—uracil; G—guanine or C—cytosine), a pentose sugar (deoxyribose or ribose) and a phosphate group (see Figure 2.1).

Oligonucleotides Small single-stranded segments of DNA typically 20–30 nucleotide bases in size that are synthesised *in vitro*. Uses include DNA sequencing, DNA amplification and DNA probes (see also *primer, allele specific oligonucleotides*).

Oncogenes Genes associated with neoplastic proliferation following a mutation or perturbation in their expression (see *proto-oncogenes, ras*).

***Online Mendelian Inheritance in Man* (OMIM)** An encyclopaedia of phenotypes for genetic traits, disorders and gene loci established by Victor McKusick. Originally available on hard copy (MIM—*Mendelian Inheritance in Man*), but now the Internet version (OMIM) is a must because of the frequent changes that occur.

Orthologous Genes or proteins found in different species that are so similar in their nucleotide or amino acid sequences that they are likely to have originated from a single ancestral gene. Such genes play a core function. For example, the β globin genes in many species are nearly identical (see also *paralogous*).

P53 A tumour suppressor gene, mutations of which are frequently found in human cancers. The correct name for this gene is *TP53* although *P53* is popularly used (see also *tumour suppressor genes*).

Packaging cells (Ch. 6)

Palindrome A DNA sequence that is identical in either direction. The example given is the recognition sequence for the restriction enzyme *Sal*I. Further examples are found in Table A.1.

*Sal*I   5′-GTCGAC-3′
         3′-CAGCTG-5′

Paralogous Genes in the *same* species that are so similar in their nucleotide sequences that they are assumed to have originated from a single ancestral gene; for example, the β and δ globin genes are paralogs. These genes have overlapping function and arose during evolution through duplication (see also *orthologous, HOX genes*; Figure 7.1).

Parthenogenesis The development of an egg that has been activated in the absence of sperm.

Pathogenesis The steps involved in development of a disease.

***PAX* genes** (Abbreviation for *paired box*.) Genes that play a role in the development of many tissues. These genes encode transcription factors involved in early embryological development. The paired box is a conserved DNA binding domain that resembles the paired genes of *Drosophila* (see also *HOX gene*; Figure 7.1, Table 7.3).

Penetrance All or nothing phenomenon relating to the expression of a gene. Calculated by the proportion of affected individuals among the carriers of a particular genotype. For example, if 20 out of a 100 individuals who have a known DNA mutation show the corresponding clinical phenotype, then the penetrance for this disorder is 20% (at a certain age group).

P-glycoprotein A glycoprotein associated with multidrug resistance. A member of the ATP-binding cassette transporter proteins. P-glycoprotein allows the active extrusion of a variety of compounds out of cells. The gene for P-glycoprotein is *MDR1* (see also *multidrug resistance*).

Pharmacogenetics The differential effects of a drug *in vivo* in patients, depending on the presence of inherited gene variants (see *pharmacogenomics*, Table 5.4).

Pharmacogenomics The differential effects of compounds *in vivo* or *in vitro* on gene expression, among the entirety of expressed genes. However, as indicated in Table 5.4, a consistent definition of pharmacogenetics and pharmacogenomics is difficult to obtain. With time, these two terms are likely to become synonymous.

Phase The combination in which polymorphic markers have been inherited within the context of a family study.

Phenocopy An environmentally produced phenotype that mimics one caused by a genetic mutation; for example, small red blood cells (microcytosis) are similar findings in thalassaemia (genetic) and iron deficiency (environmental). Another example quoted is Huntington's disease–like 1 (HDL1), which is said to be a phenocopy of Huntington's disease but is caused by familial prion disease and not an expansion of a CAG triplet repeat. The second example might also be called a *genocopy* (see also *genocopy*).

Phenome Overall phenotypic characteristics of an organism based on the interaction of the complete genome with the environment (see Figure 1.4). For consistency with the other . . . omics (genomics, proteomics, transcriptomics, metabolomics, epigenomics), the study of the phenome would be phenomics.

Phenotype (phenotypic) The observed appearance of a gene or an organism determined by the genotype and its interaction with the environment.

Physical transfer (Ch. 6)

Physical map A map that can be constructed in different ways, but in contrast to genetic maps, it represents measurements of physical length (bp, kb, Mb). Types of physical maps include cytogenetic, pulsed field gel electrophoresis, fluorescence *in situ* hybridisation, contigs—e.g., cosmid or YAC (see also *genetic map*).

Plasmid Cytoplasmic, autonomously replicating extra-chromosomal circular DNA molecule. Used as vectors for cloning. *In vivo*, plasmids are found in bacteria where they can code for antibiotic resistance factors (see also *episomal, vector*).

Plasticity Used in the context of stem cell plasticity. The potential for tissue-specific adult stem cells to differentiate into cells of a different tissue. For example, bone marrow differentiates into haematopoietic cells, but there

is some evidence that it can also differentiate into brain, liver and various other tissues. *Transdifferentiation* is another term to describe how stem cells from one tissue can change to adopt the developmental fate of another tissue.

Pleiotropy Different effects of a gene on apparently unrelated characteristics such as the phenotype, organ systems or functions.

Ploidy The number of chromosomes in a cell. Euploid, the correct number; aneuploid, an abnormally high or low number; polyploid, a multiple of the euploid number.

***Pol* gene** Gene that encodes reverse transcriptase enzyme of a retrovirus (see Figure 8.2).

Polyclonal Derived from more than one cell (see also *monoclonal*).

Polygenic inheritance Trait that results from an interaction of multiple genes at different loci (see also *multifactorial disorders*).

Polymerase RNA polymerases are enzymes catalysing the formation of RNA using DNA as a template. DNA polymerases are enzymes that can synthesise DNA from four nucleotide precursors (dATP, dTTP, dCTP and dGTP) provided a template or primer is available to start off the process. Functions of the DNA polymerases include DNA repair and DNA replication. Reverse transcriptase is also a DNA polymerase (see Figure 2.2).

Polymerase chain reaction (PCR) DNA method allowing amplification of a targeted DNA sequence (see Figure 2.5).

Polymorphisms (DNA) A segment of DNA that can occur in two or more forms. This variation may be detected through size differences, or the alleles can be distinguished by changes in the nucleotide sequence. Polymorphic variations result from point mutations (see RFLP) or insertions of repetitive DNA sequences (see VNTR). In terms of human genetics, polymorphisms are inherited along mendelian lines in a family and by definition should occur at a 1% or more frequency within a population (see *restriction fragment length polymorphism [RFLP], variable number of tandem repeats [VNTR], simple sequence repeat [SSR], single nucleotide polymorphism [SNP]*, Table 9.2).

Positional cloning Cloning of a gene on the basis of its chromosomal position rather than its functional properties. Previously called *reverse genetics* (see *clone, functional cloning*; Box 3.3).

Positive predictive value (PPV) (Ch. 8)

Predictive DNA test Enables a mutation in DNA to be detected in a clinically normal individual, and from this "predict" the development of a genetic disease at some future time. Another name for this type of test is presymptomatic DNA test. Some prefer to use both terms, e.g., *predictive* DNA test is reserved for identifying mutations that increase a person's risk but provide no certainty of developing a disorder with a genetic basis. An example of this would be a breast cancer DNA test. In contrast, *presymptomatic* DNA test can determine whether a person will develop a genetic disorder before any signs or symptoms appear. Examples given for the latter test include Huntington's disease and haemochromatosis. However, the distinction seems unnecessarily complex since the Huntington's disease DNA test could be either depending on the number of repeats (see Table 3.4). If anything, a haemochromatosis DNA test is more predictive than presymptomatic because of the many other factors that influence development of this disease. Therefore, for simplicity, the term predictive is used to describe both types of tests.

Predisease (Ch. 2)

Preimplantation genetic diagnosis (PGD) A form of prenatal diagnosis that involves sampling cells from the developing embryo. Once a genetic diagnosis is made, the embryos shown to be unaffected are transferred to the patient by IVF. One form of PGD involves the sampling of 1–2 cells from the 8–12 cell stage blastomere.

Premutations (Ch. 3)

Prenatal diagnoses (Ch. 7)

Presymptomatic DNA test See predictive DNA test.

Primer A short oligonucleotide segment that pairs with a complementary single-stranded DNA sequence. The double-stranded segment formed has a free 3′ terminus, which provides the template for extension into a second strand (see *oligonucleotides*; Figures 2.8, A.7; contrast it with probe).

Prions Transmissible proteins that have two shapes—the normal form (benign, showing few β sheets) and the pathologic form that has a significant component of β sheets. The pathologic form can spread from one organism to another and interfere with the shape of the normal protein, leading to neurodegenerative diseases such as Creutzfeldt-Jakob disease (CJD) in humans and bovine spongiform encephalopathy (BSE) in cattle.

Proband (or **propositus** or **index case**) The affected individual from whom a pedigree is constructed (see also *consultand*; Figure 3.13).

Probe A single-stranded segment of DNA or RNA labelled with a radioactive or chemical substance. The probe will bind to its complementary single-stranded target sequence. Hybrids formed are detectable by autoradiography or by chemical changes. There are a number of different probes: genomic, cDNA, RNA, oligonucleotide. The naming of probes has led to confusion. Therefore, an attempt to induce uniformity has been made by naming loci to which probes will hybridise; e.g., D15S10 indicates human chromosome 15 locus 10. A number of DNA probes could hybridise to this locus (see Figure A.10).

Prokaryotes Bacteria and certain algae with cells that are not nucleated.

Promoter DNA sequence located immediately 5′ to a gene, which indicates the site for transcription initiation. May influence the amount of mRNA produced and the tissue specificity. Examples of promoters are the TATA, CCAAT boxes (see also *Cap*).

Proteome All the proteins in a cell, tissue or organism at any given time. Hence, the field of proteomics, which unlike the constancy found in DNA (all nucleated cells have the same DNA content), is made considerably more complex because the proteome is different for each cell, and even the same cell can change its protein profile depending on environmental effects, e.g., infected and non-infected cell.

Proto-oncogenes Normal genes comprising a number of functionally different classes that are involved in cellular growth control. Altered forms of the proto-oncogenes are called *oncogenes*.

Provirus Virus that is integrated into the chromosome of its host cell and can be transmitted from one generation to another without causing lysis of the host (see *retrovirus, reverse transcriptase*).

Pulsed field gel electrophoresis (PFGE) A type of gel electrophoresis allowing large fragments of DNA to be separated by altering the angle at which the electric current is applied. Now used mostly for microbiologic DNA fingerprinting.

Quantitative PCR (Q-PCR) The use of PCR to quantitate DNA (see *real-time PCR*, Figure 2.7).

Quantitative trait loci (QTL) Genes that control the expression of traits demonstrate quantitative inheritance

289

and are mapped to QTLs. Complex characteristics—for example, height in humans and hypertension—are QTLs that result from the interaction of a number of genes on separate chromosomes (the environment also plays a role).

Ras A family of proto-oncogenes (H-*ras*-1, K-*ras*-2 and N-*ras*) that encode for a protein called p21. p21 binds to GTP/GDP and has GTPase activity. *Ras*-derived proteins play a physiological role in regulation of cellular proliferation. Mutations in *ras* are found in a number of cancers (see also *G proteins*).

Real-time PCR A variant of PCR that monitors the amplified product as it is produced after each cycle in real time. Has proven to be an invaluable technique for quantitative PCR (see Figure 2.7).

Recessive The products of both normal (wild-type) alleles at a particular locus are non-functional in a recessive disorder (see also *dominant*; Figure 3.9).

Recombinant DNA (rDNA) technology (Ch. 1)

Recombination Crossing over (breakage and rejoining) between two loci, which results in new combinations of genetic markers/traits at those loci; e.g., one locus has four genetic markers linearly arranged as a-b-c-d, and the second locus is b-b-c-a. Recombination involving these two regions between the b-c markers would give new genetic combinations, i.e., a-b-c-a and b-b-c-d (see Figure 2.15).

Repair genes A group of genes that monitor and repair DNA errors. There are three major repair pathways: (1) mismatch repair, (2) nucleotide excision repair and (3) base excision repair (see Table 3.6).

Reproductive cloning (Also called *Cloning-to-Produce-Children*.) The use of technology such as somatic cell nuclear transfer to produce an animal (child) with a virtually identical genetic makeup to another animal (human) existing or previously existing (see also *clone*, *therapeutic cloning*; Figure 6.9, Table 6.14).

Restriction endonucleases (enzymes) Enzymes that recognise specific short DNA sequences and cleave DNA at these sites (see Table A.1).

Restriction fragment length polymorphism (RFLP) Biallelic DNA polymorphism that results from the presence or absence of a restriction endonuclease site (see *polymorphisms*; Table 2.6, Figure 2.13).

Restriction map A series of restriction endonuclease recognition sites associated with a DNA locus or gene (see Figure A.2).

Retrovirus RNA virus that utilises reverse transcriptase during its life cycle. After infecting the host cell, the retroviral (RNA) genome is transcribed into DNA, which is then integrated into host DNA. In this way the retrovirus can replicate (see also *provirus*, *reverse transcriptase*; Figure 4.8).

Reverse genetics A term once used to describe positional cloning (see also *clone, functional cloning, in silico cloning*; Figure A.4). Reverse genetics is now used to describe animal models that are genotype based, i.e., the transgenic mouse in which a gene is manipulated to provide information on the resulting phenotype (see Chapter 5).

Reverse transcriptase Enzyme that enables synthesis of single-stranded DNA (called *cDNA*) from an RNA template (see also *polymerase*; Figure 2.2).

Ribosomal RNA (rRNA) The RNA content of ribosomes. The latter are small cellular particles that are the site of protein synthesis in the cytoplasm.

Ribotyping The use of rRNA-specific DNA probes for hybridising to restriction fragments to distinguish bacteria on the basis of their rRNA patterns. Polymorphic bands so produced allow typing of bacteria at the strain level.

RNA interference (RNAi) (Also called *RNA mediated gene silencing* and *post-transcriptional gene silencing*.) The process by which small RNA species can regulate gene function. These species include small double-stranded RNA (dsRNA) species known as small interfering RNAs (siRNA) that target mRNA and micro RNAs (miRNA) that work at the level of translation (see Figure 2.3).

RT-PCR (Ch. 2)

Satellite DNA Short head-to-tail tandem repeats that incorporate specific DNA motifs (see *microsatellites, minisatellites*; Table 2.6, Figure 2.13, Table 9.2).

Screening (genetic) Testing individuals on a population basis to identify those who would be at risk for disease or transmission of a genetic disorder (see Tables 2.8, 2.9).

Sensitivity The proportion of those with disease who test positive (see also *specificity*; Tables 2.10, 8.3).

Sequence tagged site (STS) A way to provide unambiguous identification of DNA markers generated by the Human Genome Project. STSs comprise short, single-copy DNA sequences that characterise mapping landmarks on the genome.

Sequencing (DNA) Establishing the identity and order of nucleotides in a segment of DNA. The gold standard in characterising a mutation (see Figures A.7, A.8).

Sibship A group comprising the brothers and sisters (siblings) in a family.

Signal transduction Transfer of signals from extracellular factors and their surface receptors by cytoplasmic messengers to modulate events in the nucleus (see also *G proteins*).

Simple sequence repeat (SSR) See *microsatellites*.

Single nucleotide polymorphism (SNP—*pronounced snip*) Variations of a single nucleotide at a given position in the genome in a population that, by definition, occurs at a frequency greater than 1% (see also *polymorphism*). SNPs occur approximately 1 in each 600–1000 bp, so in the human genome the number of SNPs will be significantly greater than 3.3×10^6 (see Figure 2.13).

siRNA (small interfering RNA) see *RNA interference (RNAi)*.

Somatic cells Any cells in an organism which are not germ cells, i.e., sperm or eggs.

Somatic cell gene therapy (Ch. 6)

Somatic cell genetic disorders One of the five groups of genetic disorders. Defects in DNA are found in specific somatic cells. An example of this type of disorder is sporadic cancer. By comparison, the four other categories (single gene, polygenetic, multifactorial and chromosomal disorders) have the genetic abnormality present in all cells including the germ cells.

Somatic cell hybrid A hybrid formed from the fusion together of different cells. They usually come from different species; e.g., human and rodent hybrids are frequently used for human gene mapping.

Somatic cell nuclear transfer (SCNT) The introduction of nuclear material from a somatic cell (the donor DNA) into an oocyte (recipient) that has had its nucleus removed. This yields a product (clone) that has the genetic (nuclear) constitution which is virtually identical

to the donor of the somatic cell (see also *clone, human cloning*; Figure 6.9).

Somatic mutation A mutation occurring in any cell that will not become a germ cell.

Southern blotting Named after E Southern. Describes the procedure for transferring denatured (i.e., single-stranded) DNA from an agarose gel to a solid support membrane such as nylon (see *northern blotting, western blotting*).

SOX Sry-type HMG box. Developmental and differentiation genes with transcription factor activity and characterised by having a HMG (high mobility group) DNA binding domain.

Specificity The proportion of those without disease who test negative (see also *sensitivity*; Tables 2.10, 8.3).

Splicing Removing introns to produce mature mRNA.

Sporadic No obvious genetic cause.

SRY Sex determining region Y. The gene on the Y chromosome that specifies maleness in humans.

SSR (Ch. 9)

Start codon Nucleotide codon (ATG) that is positioned at the beginning of a gene sequence in eukaryotes. Prokaryotes do not have such a start codon, and so ATG is translated into the amino acid methionine.

Stem cell Cell with the capacity for (1) self-renewal, i.e., giving rise to more stem cells, and (2) generating different progeny (see *embryonic stem cells, adult stem cells*).

Sticky ends Fragments of double-stranded DNA with a few bases not paired; i.e., they anneal with greater efficiency than blunt-end fragments (see Table A.1).

Stop codons Nucleotide codons (TAA, TGA and TAG) positioned at the 3′ end of a gene sequence that indicate the termination of a polypeptide.

STR (Ch. 9)

Subunit vaccines (Ch. 6)

Synonymous Because the genetic code is degenerate (a change in the third nucleotide may still produce the same amino acid, Tables 2.1, 2.5), single base changes may not alter the amino acid sequence, and they are called synonymous (i.e., same sense) codons. In contrast are non-synonymous changes, which produce a different amino acid.

Syntenic genes Genetic loci or genes that lie on the same chromosome or same DNA strand.

Tandem repeats Small sections of repetitive DNA in the genome, arranged in head-to-tail formation.

Targeting (Ch. 6)

Telomeres The two ends of a chromosome (see *centromere*; Figure 2.10)

Thalassaemia syndromes (Ch. 3)

Therapeutic cloning (Also called *Cloning-for-Biomedical-Research*.) Production of a human embryo by a process such as somatic cell nuclear transfer for the purpose of research or for the extraction of stem cells. The ultimate goal in this type of cloning is to gain scientific knowledge for medical research (see also *clone, reproductive cloning*; Figure 6.9, Table 6.14).

Therapeutic transfer (Ch. 6)

Trans-acting locus In contrast to the cis-acting locus, the trans-acting locus involves a second but distinct gene that can influence another gene's function through production of a regulatory-type protein.

Transcription Synthesis of a single-stranded RNA molecule from a double-stranded DNA template in the nucleus (see *polymerase, translation*).

Transcriptome All mRNA species in a cell, tissue or organism at a given time. Like the proteome, this is very variable compared to the fixed genome (see also *microarray*).

Transdifferentiation See *plasticity*.

Transduction (gene) Transmission of genetic material from one cell to another by viral infection.

Transduction (signal) See *signal transduction*.

Transfection Acquisition of new genetic markers by incorporation of added DNA into eukaryotic cells by physical or viral-dependent means (see Figure 6.1).

Transfer RNA (tRNA) Provides the link between mRNA and rRNA. Each tRNA can combine with a specific amino acid and also bind to the relevant mRNA codon

(see also *codon, messenger RNA [mRNA], ribosomal RNA [rRNA], translation*).

Transformation (of bacteria) Acquisition of new genetic markers by incorporation of added DNA into bacteria.

Transformation (of cells) Sudden change in a cell's normal growth properties into those found in a tumour cell.

Transgenic The presence of foreign DNA in the germline. Transgenic animals are produced by experimental insertion of cloned genetic material into the animal's genome. This can be done by microinjection of DNA into the pronucleus of a fertilised egg or through utilisation of embryonic stem cells. A proportion of transgenic animals will express the foreign gene and transmit it to their progeny (see *embryonic stem cells*).

Transition Change of a purine (i.e., adenine or guanine) to a purine or a pyrimidine (i.e., cytosine or thymine) to a pyrimidine (see also *transversion*).

Translation Cytoplasmic production of a polypeptide from the triplet codon information on mRNA (see *transcription*).

Translocation The presence of a segment of a chromosome on another chromosome (see Figure 4.10).

Transposable elements DNA sequences that can move from one chromosomal site to another.

Transposon A type of transposable element that is flanked by repeat sequences. Transposons usually possess genes, for example, with resistance to antibiotics. Transposons allow for genetic recombination, thereby enhancing genetic diversity. An interesting observation from the Human Genome Project was the finding that hundreds of human genes appear to have derived from transposable elements.

Transversion Change of a purine to a pyrimidine or vice versa (see also *transition*).

Tumour suppressor genes (Also called *recessive oncogenes, anti-oncogenes, growth suppressor genes*.) Normal genes with one component of their function being the suppression of tumourigenesis (see *P53*; Figures 4.11, 4.12).

Two-hit model of tumourigenesis (Ch. 4)

Uniparental disomy The inheritance of two copies of a chromosome from the one parent. This can be isodisomy (both chromosomes from the one parent are identical copies) or heterodisomy (the two chromosomes are different copies of the same chromosome). Described with a number of human chromosomes, e.g., 7, 11, 15, 16 (see also *imprinting*; Figure 4.6).

Variable number of tandem repeats (VNTR)   A multiallelic DNA polymorphism that results from insertions or deletions of DNA between two restriction sites (see *polymorphisms*; Table 2.6, Figure 2.13, Table 9.2).

Variant haemoglobins   (Ch. 3)

Vector   Cloning vehicle (i.e., plasmid, phage, cosmid or YAC) into which DNA to be cloned can be inserted (see Figure A.5).

Viral transfer   (Ch. 6)

Western blotting (immunoblotting)   A technique used to separate and identify proteins (see *northern blotting*, *Southern blotting*).

Wild-type (gene)   The form of the gene normally present in nature.

X-chromosome inactivation   Random inactivation of one of the two female X chromosomes during early embryonic development. Thus, cells in a female are mosaic in respect to which of the X chromosomes is functional.

X-linked   (Ch. 3)

Yeast artificial chromosome (YAC)   A cloning vector that allows large segments of DNA (e.g., 300 kb in size) to be cloned.

Zebrafish   A new model organism allowing the study of genetic disease particularly in relation to development. One unusual property of this model is that organs are visible during development.

Zoonoses   Infections transmitted from animals to humans.

Zygote   The diploid cell resulting from union of the haploid male and haploid female gametes, i.e., fertilised ovum.

ABBREVIATIONS

A	Adenine nucleotide base	**CDK**	Cyclin dependent kinase
AIDS	Acquired immunodeficiency syndrome	*C. elegans*	*Caenorhabditis elegans*
ADA	Adenosine deaminase	*CFTR*	Cystic fibrosis transmembrane conductance regulator, i.e., cystic fibrosis gene
APC	Adenomatous polyposis coli		
ALL	Acute lymphoblastic leukaemia	**CGH**	Comparative genomic hybridisation
ART	Assisted reproductive technologies	**CHO**	Chinese hamster ovary (cell line)
ASO	Allele specific oligonucleotide	**CJD**	Creutzfeldt-Jakob disease
ATM	Ataxia telangiectasia mutated	**vCJD**	variant Creutzfeldt-Jakob disease
bp	Base pair	**cM**	Centimorgan
BLAST	Basic Local Alignment Search Tool	**CMV**	Cytomegalovirus
BRCA1	Breast cancer 1	**CSF**	Colony stimulating factor
BSE	Bovine spongiform encephalopathy	**Ct**	Cycle threshold (for Q-PCR)
C	Cytosine nucleotide base	**CVS**	Chorionic villus sample or sampling

cDNA	Complementary or copy DNA		**H-2**	mouse equivalent of MHC/HLA
DHFR	Dihydrofolate reductase		**HA**	Haemagglutinin (influenza A)
DGGE	Denaturing gradient gel electrophoresis		**HAART**	Highly active antiretroviral therapy
DHPLC	(also written dHPLC). Denaturing high performance liquid chromatography		**HbF**	Haemoglobin F (fetal haemoglobin)
DNA	Deoxyribonucleic acid		**HbS**	Haemoglobin S (sickle haemoglobin)
dsDNA	Double-stranded DNA		**HBV**	Hepatitis B virus
dsRNA	Double-stranded RNA		**HCV**	Hepatitis C virus
DZ	Dizygotic (twin)		**HGP**	Human Genome Project
EBV	Epstein-Barr virus		**HIV**	Human immunodeficiency virus
ELISA	Enzyme linked immunosorbent assays		**HLA**	Human leukocyte antigen
ELSI	Ethical, legal and social implications		*HOX*	Homeobox
ENU	N-ethyl-N-nitrosourea		**HPV**	Human papilloma virus
EPO	Erythropoietin		**HPFH**	Hereditary persistence of fetal haemoglobin
ES cells	Embryonic stem cells		**HSV**	Herpes simplex virus
EST	Expressed sequence tag		**HTML**	HyperText Markup Language
FAP	Familial adenomatous polyposis		**HTTP**	HyperText Transfer Protocol
FASTA	Fast all		**ICSI**	Intracytoplasmic sperm injection
FDA	Food and Drug Administration		**IEC**	Institutional Ethics Committee
FISH	Fluorescence *in situ* hybridisation		**IgG**	Immunoglobulin G
FTP	File transfer protocol		**IgM**	Immunoglobulin M
5′ → 3′	(five prime to three prime) Direction of transcription		**IP**	Internet Protocol or Intellectual Property
G	Guanine nucleotide base		**IVF**	*In vitro* fertilisation
GAS	Group A streptococci		**IVS**	Intervening sequence (= intron)
G-CSF	Granulocyte colony stimulating factor		**kb**	Kilobase
GTP	Guanosine triphosphate		**kDa**	Kilodalton
GDP	Guanosine diphosphate		**LDL**	Low density lipoprotein
GM	Genetically modified (food)		**LOH**	Loss of heterozygosity

LTR	Long terminal repeat (of a retrovirus)		**PTT**	Protein truncation test
MALDI-TOF	Matrix assisted laser desorption ionisation—time of flight		**Q-PCR**	Quantitative PCR
Mb	Megabase		**QTL**	Quantitative Trait Loci
MDR	Multidrug resistance		**RAC**	Recombinant DNA Advisory Committee
MDR1	Gene for P-glycoprotein		**RB1**	Retinoblastoma gene
MHC	Major histocompatibility complex		**rDNA**	Recombinant DNA
miRNA	Micro RNA		**RFLP(s)**	Restriction fragment length polymorphism(s)
MoAb	Monoclonal antibody		**rh**	Recombinant human
mRNA	Messenger ribonucleic acid (RNA)		**RNAi**	RNA interference
mtDNA	Mitochondrial DNA		**rRNA**	Ribosomal RNA
MRSA	Methicillin resistant *Staphylococcus aureus*		**RT-PCR**	Reverse transcriptase PCR
MZ	Monozygotic (twin)		**SAGE**	Serial analysis of gene expression
NA	Neuraminidase (influenza A)		**SARS**	Severe acute respiratory syndrome
NAT	Nucleic acid amplification technique		**SARS-CoV**	SARS coronavirus
neo	Neomycin		**SCID**	Severe combined immunodeficiency
NIH	National Institutes of Health		**SCNT**	Somatic cell nuclear transfer
nt	Nucleotide		**siRNA**	Small interfering RNA
NPV	Negative predictive value		**SIV**	Simian immunodeficiency virus
OMIM	*Online Mendelian Inheritance in Man*		**SNP**	single nucleotide polymorphism
^{32}P	Radioactive phosphorus		**SOX**	Sry-type HMG box
PAX	Paired box		**ssDNA**	Single-stranded DNA
PCR	Polymerase chain reaction		**SSCP**	Single-stranded conformation polymorphism
PERV	Pig endogenous retrovirus		**SSR**	Simple sequence repeat (microsatellite)
PFGE	Pulsed field gel electrophoresis		**SRY**	Sex determining region Y
PGD	Preimplantation genetic diagnosis		**STR**	Short tandem repeat (microsatellite)
PON-1	Paraoxonase-1		**T**	Thymine nucleotide base
PPV	Positive predictive value		**s**	Thymidine kinase

TP53	P53 tumour suppressor gene
TSG	Tumour suppressor gene
URL	Uniform Resource Locator
VNTR(s)	Variable number of tandem repeat(s)
WT1	Wilms' tumour gene
WWW	World Wide Web
YAC(s)	Yeast artificial chromosome(s)

FURTHER READING

King RC, Stansfield WD. A Dictionary of Genetics. Oxford University Press, 6th edition. 2002. Oxford, UK (*excellent source of information in genetics*).

http://www.genome.gov/glossary.cfm (*Internet reference to the National Human Genome Research Institute's Talking Glossary of Genetic Terms—fairly basic and limited, but well presented and some terms are illustrated*).

INDEX